Fisheries Management and Conservation

Fisheries Management and Conservation

Edited by **Simon Oakenfold**

R CALLISTO
REFERENCE

New York

Published by Callisto Reference,
106 Park Avenue, Suite 200,
New York, NY 10016, USA
www.callistoreference.com

Fisheries Management and Conservation
Edited by Simon Oakenfold

© 2016 Callisto Reference

International Standard Book Number: 978-1-63239-752-2 (Hardback)

Printed in the United States of America.

Contents

Preface

The main aim of this book is to educate learners and enhance their research focus by presenting diverse topics covering this vast field. This is an advanced book which compiles significant studies by distinguished experts. This book addresses successive solutions to the challenges arising in the area of application, along with it; the book provides scope for future developments.

Fisheries management is mainly concerned with protecting and managing fishery resources to maintain a balance between exploitation and replenishment. This subject heavily draws upon fisheries science which is a multidisciplinary field that combines oceanography, freshwater biology, conservation, population dynamics, management and economics to present a coherent status of fisheries. This book attempts to assist those with a goal of delving into the fields of fisheries management and conservation. It attempts to further enlighten the readers about the new concepts in this field. Some of the diverse topics covered in this book address the varied branches that fall under this discipline. It will prove to be an asset to agriculturists, oceanographers, marine biologists, professionals, researchers and students associated with the field of fisheries at various levels.

It was a great honour to edit this book, though there were challenges, as it involved a lot of communication and networking between me and the editorial team. However, the end result was this all-inclusive book covering diverse themes in the field.

Finally, it is important to acknowledge the efforts of the contributors for their excellent chapters, through which a wide variety of issues have been addressed. I would also like to thank my colleagues for their valuable feedback during the making of this book.

Editor

Coastal Migration and Homing of Roanoke River Striped Bass

Jody L. Callihan,[*][1] **Julianne E. Harris,**[2] **and Joseph E. Hightower**
North Carolina Cooperative Fish and Wildlife Research Unit, Department of Applied Ecology,
North Carolina State University, 127 David Clark Labs, Campus Box 7617, Raleigh,
North Carolina 27695, USA

Abstract

Anadromy in Roanoke River Striped Bass *Morone saxatilis* has been documented; however, the specifics of the ocean migration and the degree of homing in this population remain unstudied and would greatly benefit the management of this economically important species. To this end, we telemetered and released 19 large Roanoke River Striped Bass (750–1,146 mm TL) on their spawning grounds during the springs of 2011 and 2012. Data from a large-scale acoustic telemetry array along the U.S. Atlantic coast (480 total receivers, including the Roanoke River) were used to evaluate the seasonal migration and distribution of telemetered fish, their degree of homing and skipped spawning, their migration speeds, and the environmental drivers of migration timing. We found that large Roanoke River Striped Bass (>900 mm TL) rapidly emigrated (~59 km/d) after spawning to distant (>1,000 km) northern ocean waters (New Jersey to Massachusetts), where they spent their summers. They then migrated southward in the fall to overwintering habitats off Virginia and North Carolina and completed their migration circuit the following spring by returning to the Roanoke River to spawn. Our results showed no evidence of straying or skipped spawning, as all migrants successfully returned (homed) to the Roanoke River the next spring to spawn. Cooler ocean water temperatures in 2013 delayed the spring spawning run by nearly 3 weeks relative to a year of average spring temperatures (2012). Our study provides novel information that aids the management of Striped Bass at both small (e.g., setting of fishing seasons in the Roanoke River) and large spatial scales (e.g., stock identification of Roanoke River fish in the mixed-stock ocean fishery) and more broadly highlights the utility of large-scale cooperative telemetry arrays in studying fish migration.

Many fish undertake migrations during some stage of their life. Migration has been defined as "those movements, often nonrandom or directed, that result in an alternation between two or more separate habitats, occur with a regular periodicity and involve a large fraction of the population" (Northcote 1978, 1984). Migrations occur at various spatial and temporal scales and are related to activities such as feeding, seeking environmental refugia or shelter, and reproduction (Wootton

1998). From an evolutionary perspective, migration is favored when the benefits of moving to a different habitat outweigh the costs of this behavior and therefore positively impact fitness (Gross 1987; Hendry et al. 2004).

Anadromy is a specific type of migration in which fish are born in freshwater and subsequently emigrate to ocean habitats, where they spend most of their lives (feeding and overwintering), but return to freshwater environments to reproduce

Subject editor: Michelle Heupel, James Cook University, Queensland, Australia

*Corresponding author: jody.callihan@ferc.gov
[1]Present address: Federal Energy Regulatory Commission, 888 First Street Northeast, Washington, D.C. 20426, USA.
[2]Present address: U.S. Fish and Wildlife Service, Columbia River Fisheries Program Office, 1211 Southeast Cardinal Court, Suite 100, Vancouver, Washington 98683, USA.

(McDowall 1987; McDowall 2001). Most anadromous species are believed to exhibit a strong "homing" behavior, or the ability to return from distant ocean waters to the same freshwater system (river) in which they previously spawned or were born. Direct evidence of homing in anadromous fishes is largely restricted to the well-studied salmonids of the Pacific coast (Hartman and Raleigh 1964; Quinn 1993; Candy and Beacham 2000; Quinn et al. 2006) and also American Shad *Alosa sapidissima* (Melvin et al. 1986; Hendricks et al. 2002). However, even in those studies the extent of ocean migration is unknown, mainly due to the difficulty of tracking the movements of individual organisms over vast expanses of the ocean or along its coastline. The knowledge of ocean migration distances is important because "homing" generally implies that fish move considerable distances away from their spawning habitats and therefore must invoke some type of guidance mechanisms (e.g., orientation to celestial bodies or geomagnetic fields, olfaction) to return "home" (Leggett 1977; Dittman and Quinn 1996). To qualify as homing, the extent (distance) of ocean migration should be such that it allows for potential straying into other known spawning systems and should not be locally restricted to nearshore ocean waters just outside the mouth of the natal estuary (sensu Huntsman 1937).

Striped Bass *Morone saxatilis* is a common species along the Atlantic coast of the USA that exhibits variation in its migration behavior (degree of anadromy) both among and within populations. Striped Bass populations to the south of Cape Hatteras, North Carolina, are believed to be nonanadromous riverine residents (Raney 1952; Dudley et al. 1977), whereas populations to the north of Cape Hatteras exhibit anadromy and originate from four principal spawning systems including the Hudson River, Delaware River, Chesapeake Bay, and Roanoke River (Boreman and Lewis 1987; Waldman et al. 1997; Welsh et al. 2007; Able et al. 2012; Callihan et al. 2014; Kneebone et al. 2014). Within these populations, the degree of anadromy has been shown to vary as a function of fish size (Waldman et al. 1990; Dorazio et al. 1994; Callihan et al. 2014) and year-class strength (i.e., density-dependent movement; Merriman 1941; Dunning et al. 2006). Additionally, intrapopulation variability in lifetime migration behaviors (e.g., resident versus anadromous) irrespective of fish size or sex has been found in Striped Bass, particularly the Hudson River population (Secor and Piccoli 1996; Secor et al. 2001; Zlokovitz et al. 2003); this strategy is thought to promote population resiliency (the contingent hypothesis: Clark 1968; Secor 1999). Relative to other spawning populations, the occurrence of anadromy in Roanoke River Striped Bass has only been recently documented in the primary literature (Callihan et al. 2014), and the details of this migration are poorly understood and warrant further investigation.

Roanoke River Striped Bass exhibit a strong, size-dependent ocean emigration pattern after spawning. Spawning occurs in the upper Roanoke River from river kilometer (rkm) 195 (measuring from its confluence with the Albemarle Sound) to just below the fall line (rkm 209) once water temperatures reach 18°C in the spring (Hassler et al. 1981; Rulifson 1990; Carmichael et al. 1998). The Roanoke River spawning run consists of mature adults; that is, females > age 4 (>450 mm total length [TL]) and males > age 3 (>350 mm TL) (Trent and Hassler 1968; Olsen and Rulifson 1992; Boyd 2011). Based on tag returns of adult Striped Bass captured and released on the Roanoke River spawning grounds across an 18-year period (1991–2008), Callihan et al. (2014) found it was predominantly the large adults (>900 mm TL) in the population that emigrated to distant ocean waters (>1,000 km north of the release site). The smallest adults (<600 mm TL) were mainly recaptured in the Albemarle Sound estuary, and an intermediate size-group of 700–850 mm TL appeared to emigrate from freshwater and utilize nearby North Carolina ocean waters during the summer (Callihan et al. 2014).

While Callihan et al. (2014) provided convincing evidence of size-based ocean emigration, their study lacked the resolution needed to evaluate more specific details of the coastal migration of large Roanoke River Striped Bass. Multiple relocations of fish throughout the year are necessary to address questions such as the degree of homing, variability among the migration trajectories of individual fish, and environmental drivers of migration timing. However, these questions cannot be addressed with conventional tagging alone because multiple recaptures are rare and therefore only two data points (tagging and recapture) are available for most fish (Pine et al. 2003). Passive acoustic telemetry is an evolving technology that provides unique data on fish migration that can complement and expand upon the more coarse information obtained with conventional tagging studies. In particular, the scalability of this technology and relative ease of sharing detection data among researchers has promoted the development of large-scale cooperative telemetry arrays (Grothues et al. 2009; Welch et al. 2009; Pautzke et al. 2010; Welch et al. 2011; Wood et al. 2012; Kneebone et al. 2014) that constitute an unprecedented means to study fish migration.

In this 3-year study (May 2011–June 2014), we used data from the Atlantic Cooperative Telemetry (ACT) Network, which includes our local receiver array in the Roanoke River, to investigate the migration and homing behaviors of the migratory component of the Roanoke River Striped Bass population (fish > 900 mm TL) identified by Callihan et al. (2014). Specifically, we examined the seasonal migration and distribution, degree of homing, effect of temperature on the timing of the spring spawning run, and postspawning migration speeds of large Roanoke River Striped Bass. We also estimated the degree of "skipped spawning" in large Striped Bass, which is defined as the extent of nonannual spawning by mature fish (Rideout et al. 2005). These new data inform the management of this economically important species at both smaller spatial scales, when fish are concentrated on the spawning grounds and highly vulnerable to exploitation (state-level jurisdiction), as well as at larger spatial scales, when Roanoke River Striped

Bass contribute to mixed-stock ocean fisheries during the non-spawning period (multistate and federal-level jurisdictions).

METHODS

Fish Tagging

A total of 19 Striped Bass (750–1,146 mm TL; mean = 1,032 mm TL) were captured, telemetered, and released on the Roanoke River spawning grounds (Figure 1) during the spring spawning seasons of 2011 (May 2; $n = 6$ fish) and 2012 (April 19; $n = 13$ fish). Striped Bass were captured on the spawning grounds with a boat-mounted electrofisher (Smith-Root 7.5 GPP; 1,000 V of direct current, 4–5 A) operating at a pulse rate of 60 pulses/s. Netted fish were transported

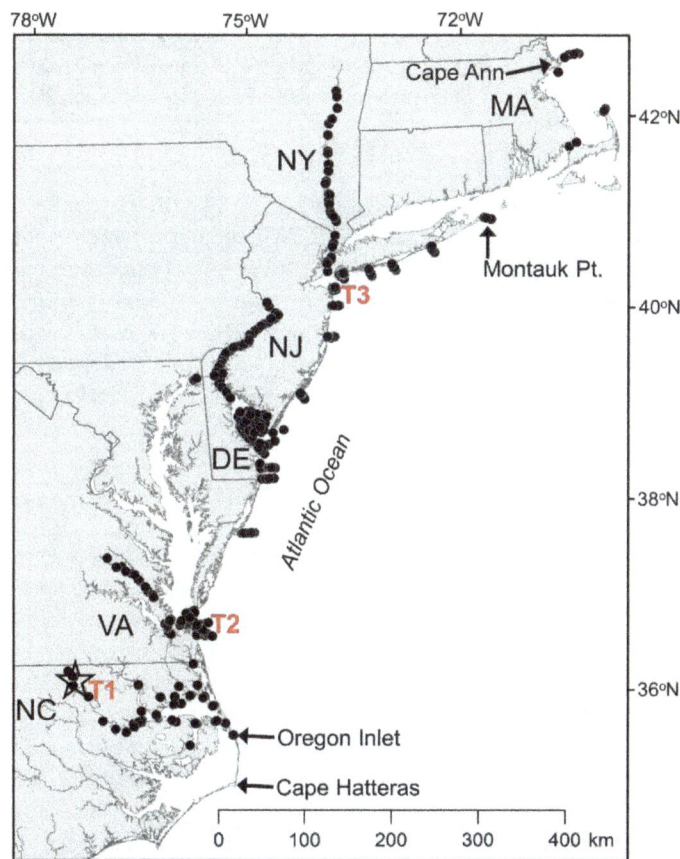

FIGURE 1. Map of the study area, showing the U.S. Atlantic coast from the Oregon Inlet, North Carolina (NC), to Cape Ann, Massachusetts (MA). Adult Striped Bass were captured, telemetered, and released on the upper Roanoke River spawning grounds (star on map) in May 2011 and April 2012. Black circles indicate the locations of acoustic receivers ($n = 480$) from which detection data were available; these receivers are part of the Atlantic Cooperative Telemetry Network. The T1, T2, and T3 (red text) denote the locations of water temperature stations in the upper Roanoke River (U.S. Geological Survey gauge 0208062765), along coastal Virginia (VA) (National Oceanic and Atmospheric Administration Buoy 44099), and along coastal New York (NY) (National Oceanic and Atmospheric Administration Buoy 44065), respectively. Additional abbreviations are as follows: NJ = New Jersey and DE = Delaware.

(<2 km) in a live well to the tagging vessel, sexed via expression of gonadal products, surgically implanted with a Vemco V13-1 L acoustic transmitter, and immediately released at the site of tagging. All fish > 900 mm TL ($n = 17$; 939–1,146 mm TL) were females, and the two smallest fish (750 and 873 mm TL) were males. The transmitters we used in 2011 had an average delay (the time between successive transmissions) of 60 s and a manufacturer-estimated battery life of 632 d. The transmitters we used in 2012 had an average delay of 90 s and an estimated battery life of 890 d, with the exception of two transmitters that were used from the previous year and implanted into fish F6 and F7 (the estimated battery lives for these transmitters were adjusted for shelf time between the 2011 and 2012 tagging events). Striped Bass were also externally tagged with an internal anchor tag (Floy Model FM-84) that indicated a US$100 reward would be given to fishers who reported information on recaptured fish (e.g., date, time, location of capture).

The lack of smaller adults (350–750 mm TL) in our study was due to the targeted sampling of large Striped Bass (>900 mm TL) on the Roanoke River spawning grounds. A major advantage of tagging fish on the spawning grounds is that the spawning population (stock) being studied is known (Waldman et al. 1988; Callihan et al. 2014). The fish we telemetered on the spawning grounds were part of a larger study on mortality and reporting rates of Striped Bass in the Roanoke River and Albemarle Sound (Harris and Hightower 2014). In that study, electrofishing on the spawning grounds was found to be the most effective method to capture large Striped Bass, presumably because this is where fish were concentrated and exhibited more restricted movements. All smaller Striped Bass ($n = 142$ fish ranging in size from 445 to 695 mm TL; mean = 517 mm TL) telemetered by Harris and Hightower (2014) were captured and released in western Albemarle Sound prior to the spawning season. Although the spawning population of those individuals could not be confirmed (because they were not captured on the spawning grounds), it is interesting to note that none of these smaller fish were detected outside of Albemarle Sound (mean time at liberty = 7.5 months). One larger fish (a 905-mm female) that was released in western Albemarle Sound did emigrate to the ocean 3 weeks postrelease but did not appear to be part of the Roanoke River spawning population as it was intermittently detected off the coast of Long Island, New York, for a period of 1.5 years, from May 2011 through October 2012.

Receiver Arrays

Our receiver array in the Roanoke River ($n = 19$ receivers) is part of the collaborative ACT Network along the U.S. Atlantic coast (http://www.theactnetwork.com/). This program involves data-sharing of detections from acoustically tagged organisms released by researchers from Georgia to Maine. For the purposes of our study, detection data (from the Striped

TABLE 1. Summary of receiver deployment and operation history during the study period: May 2, 2011, to July 8, 2014. The number of receivers that were operational in each detection area (see Figure 2 for area definitions) at the start of the study is provided, and as some receivers were added to the coastal arrays during the second year of this study, the deployment dates for these "partial" receivers are also provided. If receivers in a given array were not operational year-round, their seasonal dates of operation are listed, as is the period for which detection data were available from each array. Abbreviations are as follows: NA = not available, NC = North Carolina, NJ = New Jersey, and NY = New York.

Array	Number of receivers operational at start of study	Number of receivers added to array in year 2	Deployment dates for partial receivers	Operational year-round?	Detection data availability
Roanoke River, NC	16	3	Mar 15, 2012	Yes	May 2, 2011→Jul 8, 2014
Albemarle Sound, NC	61	NA	NA	Yes	May 2, 2011→May 1, 2014
Chesapeake Bay	NA	68	Dec 1, 2012	Yes	Dec 1, 2012→Feb 28, 2014
Delaware coast	12	27	Jan 1, 2012	Yes	May 2, 2011→Dec 31, 2013
Delaware Bay and River	94	18	Jan 1, 2012	Yes	May 2, 2011→Dec 31, 2013
NJ–NY coast	21	101	Jan 7, 2012	Yes	May 2, 2011→Jun 30, 2014
Hudson River	12	21	Apr 1, 2012	No (Apr–Oct)	May 2, 2011→Dec 31, 2013
Massachusetts coast	26	NA	NA	No (Apr–Oct)	May 2, 2011→Dec 31, 2013

Bass we released on the Roanoke River) and receiver operation data were available from other researchers' arrays (461 receivers; Vemco VR2 and VR2W receivers) deployed from Albemarle Sound, North Carolina, northward to Cape Ann, Massachusetts (Figure 1; Table 1). Most of these arrays were deployed and active before our first tagging event in spring 2011, with the exception of the Chesapeake Bay array, which was deployed in December 2012 (Table 1). Receivers were added to some arrays during the second year (2012) of our study (Figure 2; Table 1). Most notably, 101 receivers were added to the New Jersey–New York coastal array, which was expanded in January 2012 both northward (to Montauk Point) and southward (along the coast of New Jersey) from the western end of Long Island, where 21 receivers were initially deployed in 2011 (Figure 2; Table 1). Most arrays were operational year-round, except the two northernmost arrays in Massachusetts and the Hudson River, which were seasonally operational from April to October (Table 1). Although the occasional loss of individual receivers occurred in all arrays, rarely were entire receiver lines lost or compromised. The one exception was during winter 2013 (December 2013–February 2014), when only 4 of 21 receivers offshore of the mouth of Chesapeake Bay were operational due to receiver loss or failure (water damage) during the previous summer and fall. Detection data were available from all arrays north of North Carolina from May 2011 through at least December 2013; data were available from Chesapeake Bay and the New Jersey/New York coast through February 2014 and June 2014, respectively (Table 1).

Due to the wide geographic (multiple habitats) and temporal (years) scope of our study, it was difficult to define an "average" detection range for acoustic receivers. The theoretical maximum range, in calm ocean waters, for the transmitters we used (power output = 147 dB re 1 μPa at 1 m) is 539 m for VR2 and VR2W receivers, which operate at a frequency of 69 kHz (http://vemco.com/range-calculator/). However, the detection range is generally higher in freshwater than in marine habitats (Pincock et al. 2010; Pincock and Johnston 2012) and also strongly dependent on sea state (Lembo et al. 2002; Mathies et al. 2014) and turbidity (Callihan 2011).

Processing of Telemetry Data

We screened raw detection data for false detections prior to data analyses. Typically, detections that are isolated in space and time (e.g., a single detection at a given receiver in a 24-h period or less) are flagged as potential false positives in fish telemetry studies (Heupel et al. 2006; Dagorn et al. 2007; Pincock 2012). However, due to the rapid movements of Striped Bass in our study (see Results), it is entirely possible that a telemetered fish could pass by a receiver or receiver line and emit only one transmission before the fish is out of detection range. Therefore, instead of using a rigid criterion to identify false detections, we viewed animations of the successive detection locations of each fish in ArcMap (version 10.1; using time-enabled shapefiles) to ensure there was a logical sequence of detections. In a few instances, a fish was detected at two geographically disparate locations (>200 km) at essentially the same time (<1 h apart). In these situations, the false detection was easily identified as that which did not agree spatially with prior and subsequent detections of the fish being examined (see Video S.1 in the supplemental file for an example animation). Using this rationale, we deemed 0.14% of the 55,762 total raw detections as false and excluded them from analyses.

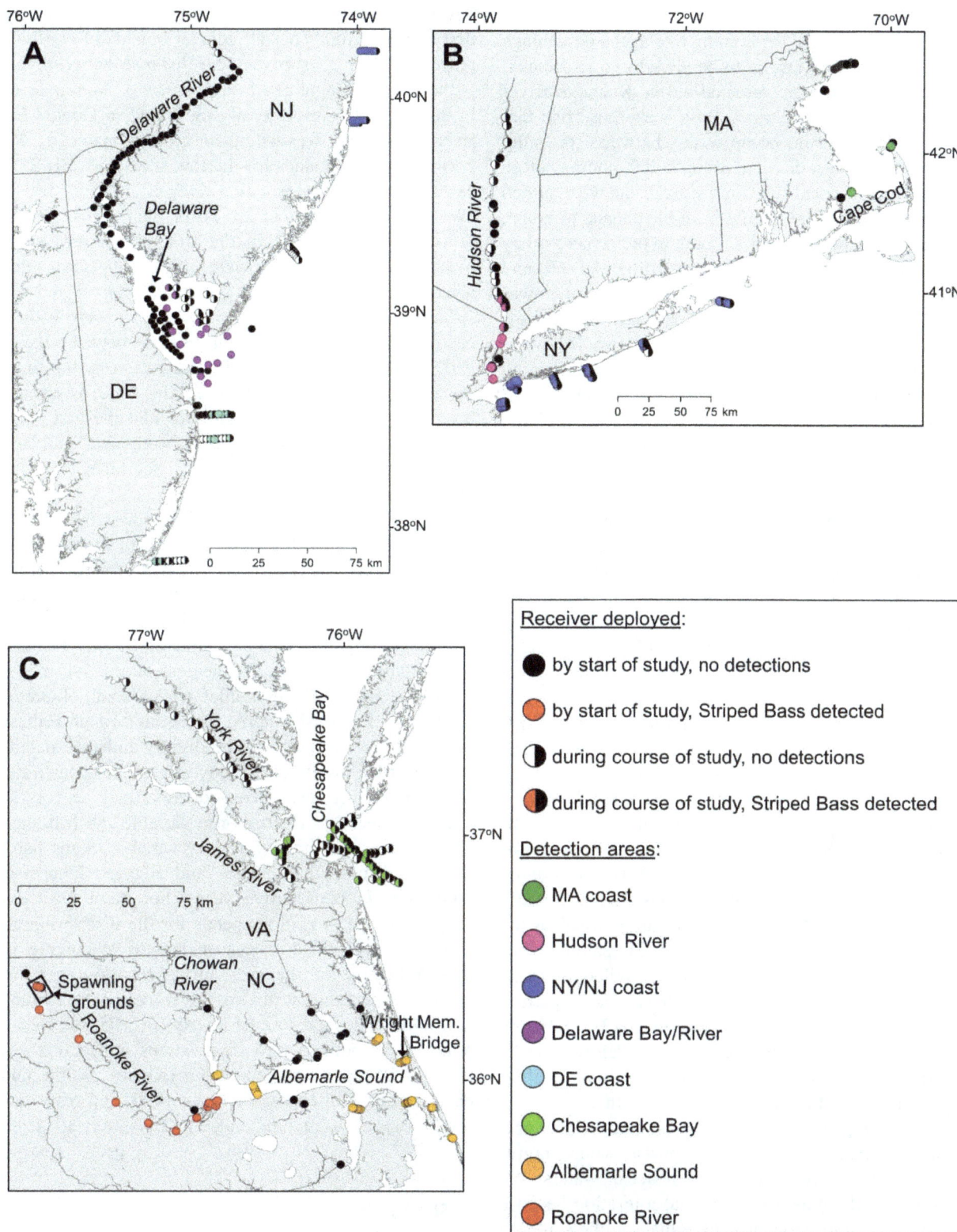

FIGURE 2.   Receiver deployment and Striped Bass detection locations in (A) the Delaware (DE)–New Jersey (NJ) coastal region and Delaware Bay and River, (B) coastal New York (NY)–Massachusetts (MA) and the Hudson River, and (C) North Carolina (NC) to lower Chesapeake Bay. As indicated in the legend, fill patterns of circles (i.e., completely filled with any color versus half black) are used to denote whether receivers were deployed and operational by the start of the study (May 2011) or deployed during the course of the study (after May 2011); the different fill colors represent the detection areas used to illustrate movement patterns in Figure 3. Table 1 provides additional information on the dates of receiver deployments in each area and their operation seasonality. The black rectangle in the upper Roanoke River in panel (C) encompasses the spawning grounds of Roanoke River Striped Bass.

Data Analyses

Homing.—We calculated the homing rate as the percentage of "migrant" Striped Bass known to be alive through the next spring (April–May) that were detected on the Roanoke River spawning grounds. Migrant Striped Bass were those fish that moved to (were detected in) ocean waters between spawning events. Four fish were last detected more than 9 months before the start (April 1) of the next spawning season, and we assumed these fish died and therefore did not have the chance to home. Three of these fish appeared to be in route to the ocean as they were last detected at the Wright Memorial Bridge (F1, F9) and mouth of the Roanoke River (F8) in May after migrating downriver postspawning; the other fish (F10) was last detected off the coast of New York in June. In addition, two fish were reported as being harvested by fishers prior to the following spring and were therefore excluded from homing analyses. All other Striped Bass were detected at least 13 months postrelease and were therefore eligible for the homing analysis (i.e., were available for detection through the following spring).

In addition to providing information on homing, the acoustic monitoring of all major spawning systems of migratory Striped Bass permitted an evaluation of skipped spawning. We assumed Striped Bass skipped spawning in a given year if they were not detected in any spawning system (Roanoke River, Chesapeake Bay, Delaware River, or Hudson River) during the spring (April–May) after their release. We only included in this analysis fish known to be alive (detected) through the end of the next spawning season (late May).

Timing of the spawning run.—To examine interannual differences in the timing of the spawning run, we used detection patterns to quantify and compare (among years) the dates Striped Bass arrived and departed from the Roanoke River. For these analyses, we only included fish that made the complete spawning run (i.e., detected in the Albemarle Sound or ocean waters both before and after spawning). We considered the "arrival date" as the day Striped Bass were initially detected in the Roanoke River and the "departure date" as the date of last detection in the river. Due to low sample sizes ($n = 5$ fish per spawning year), a two-sample Wilcoxon exact test was used to test for differences in arrival and departure dates between spawning years (2012 versus 2013). Only one male (M1) was available for these analyses. Therefore, we performed statistical tests with and without this individual to account for its potential bias (sex effects) on results.

Postspawning migration speeds.—We estimated the migration speeds of Striped Bass during their postspawning migration from the Roanoke River to northern ocean waters. We focused on this northern leg of the coastal migration because fish were detected more frequently there than on the southern leg (see Results), thus providing more accurate estimates of migration speeds. The starting and ending points of the "postspawning migration" for each fish were as follows: (1) the date and location of the last detection on the Roanoke

River spawning grounds and (2) the date and location of the first detection on a coastal receiver outside of North Carolina, respectively. We estimated the distance between these endpoints using Google Earth and ArcMap. Nonlinear distances in the river were more easily measured in Google Earth; the river kilometers we estimated closely matched, within 1–3 km, published values for the few sites ($n = 3$ U.S. Geological Survey gauges) for which river kilometers were available (Wehmeyer and Wagner 2011) for the exact locations where we deployed receivers in the Roanoke River. Straight-line measurements in ArcMap were sufficient to estimate distances across the open waters of Albemarle Sound and along the Atlantic coast. Migration speeds (km/d) were estimated for each fish by dividing the total distance between the start and end points of the postspawning migration by the time it took to complete this migration (i.e., the time between the last detection on the spawning grounds and the first detection in ocean waters). Migration speeds were also standardized to body lengths per second (BL/s) to facilitate comparisons of our results to other studies.

In calculating migration speeds, we assumed all fish exited the Albemarle Sound through the Oregon Inlet. While only 1 of 14 migrants was detected at the pair of receivers at Oregon Inlet, this was likely due to the low detection efficiency and occasional receiver loss in this high-energy environment (M. Loeffler, North Carolina Division of Marine Fisheries, personal communication). The nearest alternative exit point to ocean waters was Hatteras Inlet, 67 km south of Oregon Inlet. Therefore, if any fish entered the ocean through Hatteras Inlet, our migration speeds would be slightly underestimated.

To test for an effect of body size on postspawning migration speeds, we used least-squares linear regression with migration speed as the response variable and fish length (mm TL) at release as the explanatory variable. Some fish were at liberty for more than 1 year and engaged in postspawning migrations in consecutive years. For these individuals, we only estimated migration speeds for the first spring and summer after their release because lengths at tagging were the most reflective of the size during the migration. We also excluded one fish that did not undergo a postspawning migration until its second year at liberty (fish F5, see Results). We removed as outliers any observations with Cook's distance (Cook's D) values exceeding $4/n$ (where n = the number of observations; Bollen and Jackman 1990) and with studentized residuals $> |2|$, as recommended by Belsley et al. (1980).

RESULTS

Detection Summary

A total of 55,762 valid detections were logged from the 19 Striped Bass released in the Roanoke River in the springs of 2011 and 2012. Detection locations ranged from the upper

Roanoke River, North Carolina, to Cape Cod, Massachusetts, a distance of ~1,200 km. Most Striped Bass (13 of 19, or 68%) were detected more than 1 year (≥390 d) after being released; three of these fish (M2, F16, and F17) were detected more than 2 years postrelease (Figure 3). Of the other six fish, three were last detected ~2 months (59–66 d) postrelease, and three were last detected just 11–21 d after release. Two of the fish detected for ~2 months (F11 and F12) were reported as being harvested by recreational and commercial fishers in coastal Massachusetts and Albemarle Sound, respectively (Figure 3). The three fish detected for only a short period (≤21 d) appeared to survive the tagging process as they all moved downriver and were detected at the mouth of the Roanoke River. Two of these individuals (F1 and F9) moved across Albemarle Sound and were last detected at the Wright Memorial Bridge (Figure 2C) and possibly moved into the ocean.

Coastal Migration

Large Striped Bass (>900 mm TL) exhibited a strong seasonal migration pattern along the U.S. Atlantic coast. After emigrating from the Roanoke River spawning grounds in May, fish migrated northward to ocean waters off northern New Jersey and New York, where they resided during the summer (June–September) (Figure 4). Only two fish were detected north of Montauk Point, New York, in the vicinity of Cape Cod; there were no detections to the north of Cape Cod on the Cape Ann receiver line (Figure 2B). A southward fall migration began in October as indicated by the progression of detections, New York–New Jersey to Delaware to Virginia, from October to December (Figure 4). Although detections during winter were sparse, especially in 2013, fish appeared to overwinter off the coasts of Virginia and North Carolina as there were no detections north (Figure 4) or south of this area

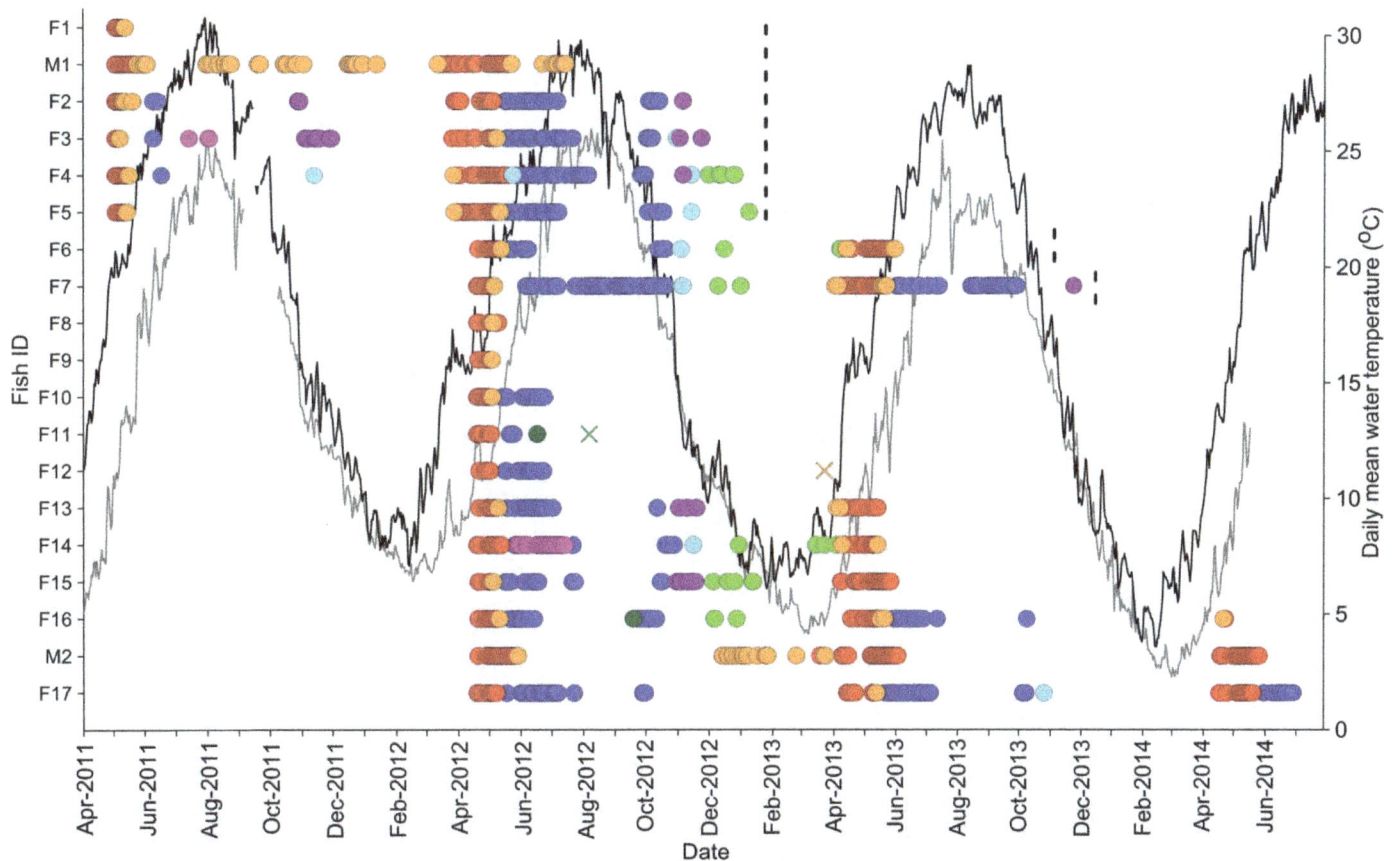

FIGURE 3. Detection histories of individual Striped Bass telemetered and released on the upper Roanoke River in the springs of 2011 (May 2; n = 6 fish; 873–1,104 mm TL) and 2012 (April 19; n = 13 fish; 750–1,146 mm TL). The fish IDs preceded by an "F" are females; "M" prefixes denote males. Detection locations are color coded by the geographic areas in Figure 2; if a given fish was detected at any receivers within a specific area, a color-filled circle for that area (e.g., orange for the Albemarle Sound) is shown for that day. The vertical dashed line to the right of the detection history for each fish represents the estimated expiration date of the transmitter battery provided by the manufacturer; lack of a dashed line indicates that the transmitter for that individual was projected to be active beyond the end of the study (July 8, 2014). An "X" denotes Striped Bass that were reported as being harvested by fishermen and are color coded to the geographic area (Figure 2) in which harvest occurred. The black and gray lines represent the mean daily water temperatures for the upper Roanoke River, North Carolina, and coastal waters off New York, respectively (see Figure 1 for locations of temperature stations; also note that temperature data was unavailable off New York after May 17, 2014).

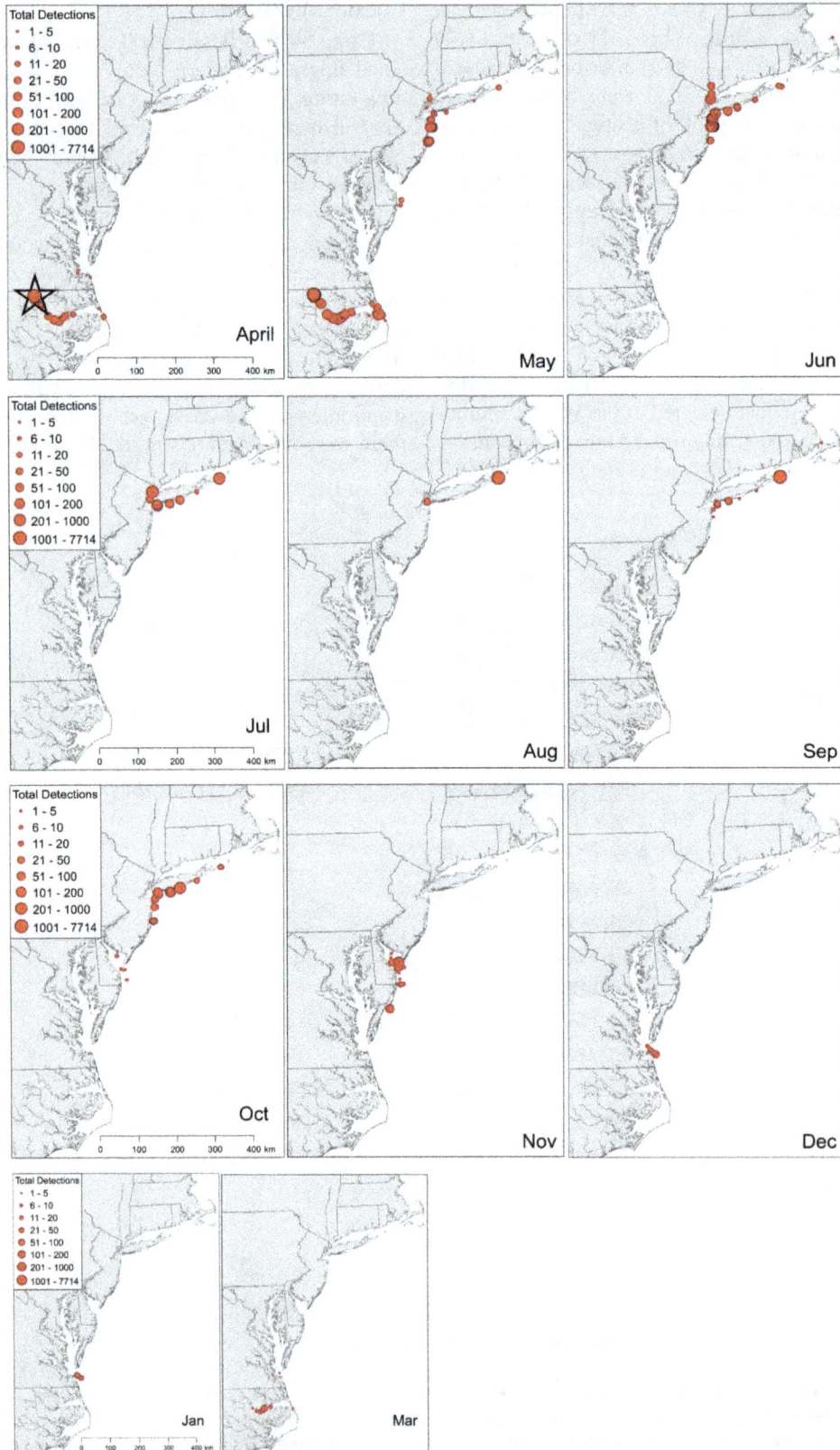

FIGURE 4.   Monthly detections of Striped Bass > 900 mm TL ($n = 17$ females) telemetered and released in the upper Roanoke River (star on map) on May 2, 2011, and April 19, 2012. The detection data were pooled across fish and years (May 2011 to July 2014); bubble sizes are positively scaled to the total number of monthly detections at each receiver station as indicated in the legend. Note there were no detections in February.

from December to March. Receiver arrays were deployed year-round (2011–present) in coastal rivers and nearshore areas of South Carolina ($n = 47$ receivers; Santee–Cooper River) and Georgia ($n = 120$ receivers; Altamaha River), but the Striped Bass released in our study were not detected by these arrays (B. Post, South Carolina Department of Natural Resources, personal communication; D. Peterson, University of Georgia, personal communication). During late March and April, Striped Bass completed their migration circuit and returned to the Roanoke River from ocean waters to spawn (Figures 3, 4).

The seasonal migration pattern revealed in our study was remarkably consistent across fish. All Striped Bass > 900 mm TL that were known to be at liberty for at least 1 year ($n = 11$) engaged in a similar seasonal, coastwide migration pattern (Figure 3), regardless of their release year (2011 or 2012). Moreover, the two females at liberty for more than 2 years (F16 and F17) repeated the same seasonal migration pattern in consecutive years (Figure 3).

Although the two smallest fish in our study, males of 750 and 873 mm TL, did not appear to migrate to northern ocean waters, their detection pattern suggests that they left the Albemarle Sound after spawning. Both males (M1 and M2) moved down the Roanoke River in late May and crossed the Albemarle Sound in 2–3 d as they were detected at the Oregon Inlet (M1) and Wright Memorial Bridge (M2) and not detected thereafter for 59 and 198 d, respectively (Figure 3). Given the lack of detections on coastal arrays to the north and south of North Carolina, these males likely utilized nearshore ocean waters off North Carolina during the summer (which lacked receivers, Figure 1). Male M1 was sporadically detected at Oregon Inlet from August to January. Both males (M1 and M2) were detected in the Albemarle Sound during winter and made the spawning run in the Roanoke River the following spring (Figure 3).

Homing and Skipped Spawning

Roanoke River Striped Bass exhibited a high degree of homing. The estimated homing rate was 100% as all fish ($n = 11$) that migrated to distant northern ocean waters and were alive through the following spring returned to the Roanoke River spawning grounds that spring (2012 or 2013). There were no detections in the Delaware River, and the two fish (F3 and F14) detected in the Hudson River were detected at locations (lower river, higher salinity; Figure 2B) and times (summer, June–August; Figure 3) at which spawning does not occur. Although two fish (F6 and F14) were detected at the mouth of the Chesapeake Bay during early April, these fish were detected 4–5 d later in Albemarle Sound (Figure 3), then made the > 200-km spawning run up the Roanoke River.

We found no evidence of skipped spawning. All Striped Bass known to be alive through the next spawning season ($n = 13$) participated in the spring spawning run up the Roanoke River. Interestingly, the three fish (F16, F17, and M2) available for detection during two successive spawning seasons after their release made the spawning run in consecutive years (2013 and 2014; Figure 3).

Timing of the Spawning Run

The spring spawning run occurred later in 2013 than in 2012. On average, Striped Bass arrived in the Roanoke River 19 d later in 2013 (April 13) than in 2012 (March 25). This difference in arrival dates was significant regardless of whether male M1 was included in analyses (exact Wilcoxon tests: $P = 0.008$ with male, $P = 0.016$ without male). Striped Bass did not enter the river until the ocean waters off of Virginia warmed to and remained above 9–10°C, which occurred 1 month later in 2013 (April 8) than in 2012 (March 8) (Figure 5). Upon reaching the spawning grounds, Striped Bass did not leave until river temperatures reached at least 18°C, which occurred on May 1 in 2012 and May 11 in 2013 (Figure 5). Most fish (70%, or 7 of 10) left the spawning grounds < 6 d after this temperature threshold was reached (Figure 5). However, a few fish ($n = 3$, including male M1) remained on the spawning grounds for longer periods and did not emigrate downriver until 15–17 d after temperatures reached 18°C. Striped Bass exited the Roanoke River 1 week earlier in 2012 (mean departure date = May 12) than in 2013 (mean departure date = May 19), but the difference in departure times was only significant, albeit marginally so, when male M1 was excluded from the analyses (exact Wilcoxon tests: $P = 0.08$ without male, $P = 0.24$ with male).

Postspawning Migration Speeds

Northward migration speeds ranged from 23.8 to 79.6 km/d, or 0.26 to 0.80 BL/s. Estimates of migration speeds during 2011 (range = 23.8–26.9 km/d; mean = 25.5 km/d; $n = 3$ fish) were more than two-fold lower than in 2012 (range = 31.9–79.6 km/d; mean = 59.3 km/d; $n = 9$ fish). We considered the 2011 data biased low given the similarity in fish size and water temperature between years and therefore excluded the 2011 data from the regression analysis. In addition, one observation from 2012 (24.8 km/d) had a Cook's D value of 0.60 (greater than the cutoff of 0.40) and a studentized residual of −2.5 and was therefore considered an outlier and removed from the analysis. Despite the relatively small size range of fish examined (939–1,146 mm TL), migration speeds showed a strong ($r^2 = 0.78$) positive relationship ($P = 0.002$) with fish length (Figure 6). The fastest migration speed of nearly 80.0 km/d was achieved by the largest fish in the study, an 1,146-mm female (F15) that migrated from the Roanoke River spawning grounds to coastal New Jersey (off Shark River Inlet), a distance of 837 km, in just over 10 d.

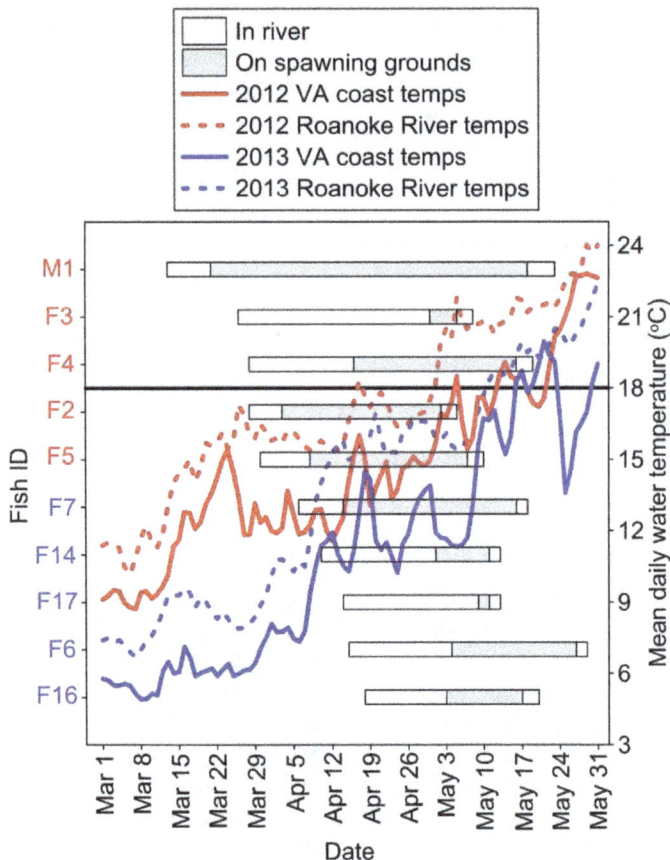

FIGURE 5. Timing of the Roanoke River spawning run of Striped Bass (>873 mm TL) in relation to spring water temperatures in the upper Roanoke River and along coastal Virginia (VA) (see Figure 1 for locations of temperature stations). Detection patterns in the Roanoke River were used to determine the periods of upriver and downriver migration (white portions of the horizontal bars) and when fish resided on the spawning grounds (the gray-filled portions of the bars). The thick black line indicates the minimum spawning temperature of Roanoke River Striped Bass (18°C) determined by Rulifson (1990). The fish IDs in red represent fish released in 2011 that made the 2012 spawning run, and the fish IDs in blue denote fish released in 2012 that made the 2013 spawning run. The fish IDs preceded by "F" indicate females and "M" indicates males. Data are only included for Striped Bass known to enter and subsequently exit the Roanoke River (i.e., those fish detected in coastal waters or the Albemarle Sound prior to entering the river and also detected in either of those regions following their exit from the river).

DISCUSSION

Our study provides novel information on the life history of Roanoke River Striped Bass. Most notably, our study is the first to document homing in this population. We demonstrated that large Striped Bass (>900 mm TL) emigrate rapidly from the Roanoke River spawning grounds to northern ocean waters from New Jersey to Massachusetts, where they spend their summers. Then they migrate southward in the fall to overwintering habitats offshore of North Carolina and Virginia and subsequently return to the Roanoke River to spawn the next spring (there was no evidence of skipped spawning). Furthermore, we found that temperature had a strong effect on the

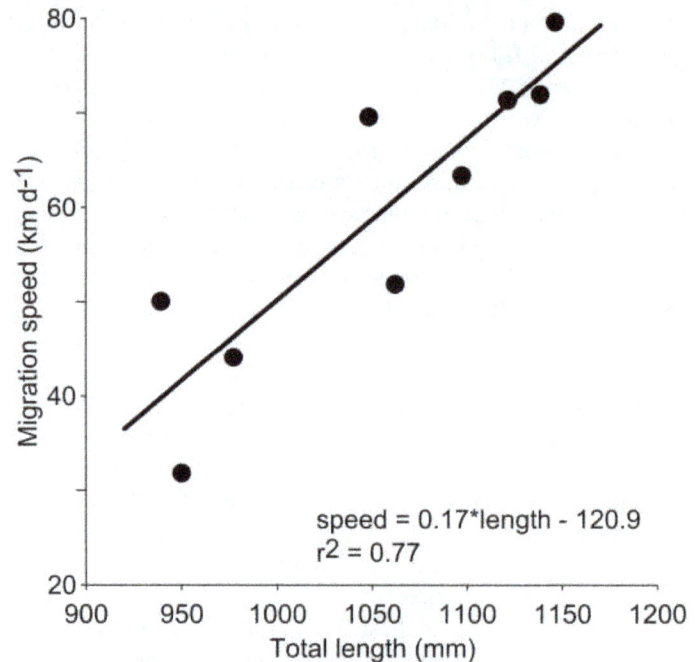

FIGURE 6. Relationship between migration speed and fish length during the 2012 postspawning migration of Striped Bass from the Roanoke River, North Carolina, to northern ocean waters off New Jersey and New York. Migration speeds were estimated for each fish by dividing the distance between their last detection on the spawning grounds and first detection in ocean waters off New Jersey or New York by the time between these detections.

timing of the Roanoke River spawning run, as entry to the Roanoke River from ocean habitats occurred almost 3 weeks later in a cold year (2013) relative to a year of average spring temperatures. Our results inform the management of this economically important species at both small (state-level) and large (U.S. Atlantic coast) spatial scales and also provide impetus for future research avenues.

Coastal Migration and Homing

The movement patterns of Roanoke River Striped Bass revealed by our study typify migration. First, movements were rapid and directed (i.e., nonrandom). After release, telemetered Striped Bass migrated rapidly (mean = 59.3 km/d; maximum = 79.6 km/d) to northern ocean waters (there were no detections to the south of Oregon Inlet), and fish successfully returned (homed) to the Roanoke River to spawn the next spring. The strong directionality of postspawning movements was also observed by Callihan et al. (2014), who noted that tag returns of large Roanoke River Striped Bass (>900 mm TL) occurred exclusively to the north of Oregon Inlet. The migration speeds estimated in this study (0.39–0.80 BL/s) fall within the range of sustained swimming speeds (0.4–1.5 BL/s) reported for the active migration phases of other anadromous fishes including salmonids, shads (family Clupeidae), and sturgeons (family Acipenseridae) (Beamish 1978; Bernatchez and

Dodson 1987; Quinn 1988), further confirming the migratory behavior of Roanoke River Striped Bass. Dingle and Drake (2007) noted that another characteristic of migration is its preemptive nature, namely that "habitats are abandoned before their quality has declined too seriously." Inshore water temperatures in North Carolina routinely approximate 30°C in midsummer (Figure 3); these temperatures are especially unsuitable for Striped Bass > 900 mm TL (Coutant 1985) due to the increased metabolic demand posed by a large body size (Hartman and Brandt 1995). Therefore, the rapid postspawning migration of large Striped Bass to cooler ocean waters to the north, which is preemptive and occurs well in advance of midsummer (by nearly 2 months), likely provides a metabolic reprieve and places fish in a more ideal environment for growth (Callihan et al. 2014). Northcote (1978, 1984) also stated that migration involves a large fraction of the population. Although the coastal migration of Roanoke River Striped Bass mainly involves fish > 900 mm TL, this size-group has comprised nearly 20% of the mature female population in recent years (Callihan et al. 2014) and likely contributes substantially to reproductive output in the population. Finally, the migration of Roanoke River Striped Bass clearly involves an alteration between habitats (ocean for feeding and freshwater habitats for spawning) that occurs with a regular periodicity (on a seasonal basis).

An effect of water temperature on the timing of the spawning run (cooler temperatures delayed the 2013 spawning run of Roanoke River Striped Bass) has been previously demonstrated in Striped Bass and other anadromous species. Peer and Miller (2014) analyzed 25 years of gill-net catch data for the Chesapeake Bay spawning grounds (upper Chesapeake Bay and Potomac River) and found that females occurred on the spawning grounds later in cooler years. Furthermore, Douglas et al. (2009) determined via acoustic telemetry that adult Striped Bass arrived on the spawning grounds in the Miramichi River, Canada, about 1 week later in the spring of 2005 (cooler year) versus 2004. A negative relationship between the timing of the spawning run (day of year) and water temperature (i.e., warm water causes earlier spawning) has also been documented in American Shad (Leggett and Whitney 1972; Quinn and Adams 1996). Interannual variability in water temperature has a greater effect on run timing in anadromous species such as American Shad and Striped Bass, whose progeny experience environmental conditions similar to those of adults during the spawning run due to their much shorter hatching times than salmonids, whose run timings are less plastic and primarily under genetic control (Quinn and Adams 1996). Based on these findings, it appears reasonable to conclude that cooler water temperatures were the primary cause for the delayed spawning run of Roanoke River Striped Bass in 2013 versus 2012, especially given the similar size of telemetered Striped Bass participating in the run those years (2012: 873–1,104 mm TL; 2013: 939–1,146 mm TL). Warming ocean temperatures likely act as a cue for Roanoke River Striped

Bass to enter the Albemarle Sound estuary from their offshore wintering grounds and subsequently move upriver to spawn. It should also be noted that no fish left the spawning grounds until river temperatures reached 18°C, which is the minimum reported spawning temperature for this population (Rulifson 1990).

Our study is the first to document homing in Roanoke River Striped Bass, and in general, our results agree with genetic studies of this species along the U.S. Atlantic coast. The subsequent recapture of tagged fish in the same spawning system they were previously released in cannot be taken as true evidence of homing because the location(s) of tagged fish during the interim (i.e., between spawning events) is unknown. The Striped Bass telemetered in our study not only returned to the same river they previously spawned in (were released in) but in the interim underwent an extensive coastal migration (>1,000 km) past other major spawning systems (Chesapeake Bay, Delaware River, Hudson River), which they did not enter for the purposes of spawning (i.e., no spawning runs were made in those systems). The strong tendency of Roanoke River Striped Bass to return to their river of previous spawning should result in reproductive isolation of this population, which has been confirmed by genetic studies. Gauthier et al. (2013) genetically analyzed young-of-the-year Striped Bass from each major spawning system along the U.S. Atlantic coast, including the Roanoke River as well as the Chesapeake Bay, Delaware River, and Hudson River. They found significant genetic divergence among samples from these major spawning systems (Gauthier et al. 2013) indicating a high degree of, but not "perfect," homing of adults to their natal system (Bielawski and Pumo 1997). Gauthier et al. (2013) concluded that most of the limited, contemporary gene flow that does occur is from the large spawning population of the Chesapeake Bay into other systems and that migratory Striped Bass originating in other systems, including the Roanoke River, return exclusively to their natal rivers to spawn, as we found.

Waldman et al. (2012) hypothesized that the latitudinal limits of the coastal migration of Striped Bass were related to population origin in a manner that minimized migration costs but still placed fish in a favorable environment for growth (i.e., migratory Striped Bass from more southerly populations do not migrate as far north in summer and those from more northerly populations do not migrate as far south in winter). While their results (genetic-based stock compositions) were equivocal and did not conclusively support this prediction, they recommended further testing of their hypothesis (Waldman et al. 2012). In our study, Roanoke River Striped Bass, which are believed to be the southernmost migratory (anadromous) population along the U.S. Atlantic coast (Boreman and Lewis 1987), were not detected north of Cape Cod, which provides some support for the hypothesis of Waldman et al. (2012). However, conventional tagging data indicate that Roanoke River Striped Bass occasionally use coastal waters north of Cape Cod based on tag returns from this area, the farthest being from Sheepscot Bay, Maine (1,350 km from the Roanoke River release site) (J. L. Callihan, unpublished data). Future data collected by the ACT Network should provide further insight into this research question.

Receiver Operation Histories

While large-scale acoustic telemetry arrays can provide novel information on fish migration, two of our findings highlight the utility of evaluating receiver operation histories when analyzing such data. First, postspawning migration speeds were more than two-fold lower in 2011 than 2012, despite the fact that telemetered Striped Bass were of similar size between years (2011: mean size = 1,054 mm TL; 2012: mean size = 1,053 mm TL). The disparity in migration speed estimates between years was likely due to the reduced spatial coverage of the New Jersey–New York array in 2011 ($n = 21$ active receivers) relative to 2012 ($n = 122$ active receivers). After reaching their summer foraging grounds (which appear to be in the vicinity of Long Island for Roanoke River Striped Bass), fish likely make reduced movements. Therefore, by the time Striped Bass were in the vicinity of the 2011 receivers (a small cluster on the western end of Long Island), they were probably already in foraging mode, whereas in 2012 fish were first detected farther south (in New Jersey) while they were still in transit to the foraging grounds, thus providing more accurate estimates of sustained swimming speeds during active migration. This finding clearly highlights the implications of receiver array design and location on the accuracy of migration speed estimates. When the 2011 data were removed from the regression analysis, a strong positive relationship was observed between fish size and migration speed, as has been shown in many fish species; this is due primarily to the fact that larger fish have a greater stride length, or the distance moved with one tail beat (Bainbridge 1958; Sambilay 1990). Secondly, there were no detections off the mouth of Chesapeake Bay in winter 2013, possibly implying that Striped Bass overwintered farther offshore that year, which seems plausible given that ocean temperatures were cooler in 2013. However, the lack of detections could also be attributed to the reduced receiver coverage ($n = 4$ operational receivers) in that region during winter 2013. Therefore, an interannual difference in overwintering distribution cannot be inferred from these detection data because of the potential confounding effect of reduced receiver coverage.

Management Implications

The high degree of homing demonstrated by Roanoke River Striped Bass could aid fishery managers in determining the stock composition of the mixed-stock ocean fishery along the U.S. Atlantic coast during the nonspawning period (summer to winter). By releasing telemetered Striped Bass on feeding or overwintering grounds and determining what river they returned to for spawning, managers could identify the composition of the migratory mixed stock. Additional data from fish tagged as part of the United States Fish and Wildlife Service Cooperative Tagging Program (ASMFC 2013) would also help identify the composition of the migratory stock. This "river of return" method could also complement genetic-based stock identification tools to further investigate if and how the stock composition of migratory Striped Bass varies across space (e.g., coastal sampling location) and time (e.g., sampling year) (sensu Waldman et al. 2012).

Our results on the effect of water temperature on the run timing of Roanoke River Striped Bass inform local-scale management within North Carolina. Since 2008, the North Carolina Wildlife Resources Commission has used a fixed open season, March 1 to April 30, in the Roanoke River to control fishing effort and limit the number of females that are harvested before they have a chance to spawn. Females have been shown to arrive on the spawning grounds about 10–14 d later (early May) than males (mid to late April) (Carmichael et al. 1998), hence the seasonal closure of the fishery on April 30. Long-term water temperature data were available for the upper Roanoke River, in the vicinity of the spawning grounds (Figure 1), for 14 of the past 15 years (1999–2013, except 2007). Based upon these data, the mean day on which river temperatures exceeded the 18°C minimum spawning temperature was May 2. During most years (79%, or 11 of 14 years) this threshold was reached by the first week of May and in some years (50%) as early as the latter half of April. Therefore, in most years, females likely arrived in the Roanoke River well before the season was closed. For example, in 2012, a year of average spring temperatures (18°C by May 1), the four telemetered females that participated in the spawning run arrived in the lower Roanoke River during March 26 to March 30. There are regulations in place to protect large, prime-spawning females (e.g., only 1 of 2 fish allowed to be kept each day during the open season can exceed 686 mm TL); however, temperature could be used as an adaptive cue to further manage female harvest if deemed necessary in the future.

Large Roanoke River Striped Bass do not appear to exhibit skipped spawning. Accordingly, estimates of spawning stock biomass made from adult collections on the Roanoke River spawning grounds should not need to be adjusted upwards to account for skipped spawning (Jørgensen et al. 2006; Rideout and Tomkiewicz 2011; Skjæraasen et al. 2012), at least for this segment of the population (fish > 900 mm TL). Secor and Piccoli (2007) found that although skipped spawning in Chesapeake Bay Striped Bass was minimal overall, younger adults were more likely to skip spawning than older (larger) fish based on lifetime otolith microchemistry profiles. Therefore, skipped spawning warrants further investigation in smaller adult Roanoke River Striped Bass before it can be discounted for the entire population.

In closing, our study demonstrated the utility of large-scale and long-term acoustic telemetry arrays for examining fish migration. When accompanied with receiver operation data, such studies can provide robust and novel data on the migration dynamics of fishes. For instance, our study advanced the current knowledge of Roanoke River Striped Bass life history by providing the first direct evidence of homing in this population. Furthermore, our telemetry-based results have important and immediate implications for management, including stock identification, the setting of fishing seasons, and the effect that skipped spawning has (or does not have in our case) on population biomass estimates. As acoustic telemetry technology and the network of researchers using this approach continue to evolve, we are likely to learn much more about the migration and movements of fish and other aquatic organisms that can enhance their management and conservation.

ACKNOWLEDGMENTS

We are grateful to the following researchers for providing detections on their receiver arrays of the Striped Bass we released on the Roanoke River, in addition to providing operation data for their arrays: Lori Brown, Keith Dunton, Dewayne Fox, Michael Frisk, Ben Gahagan, William Hoffman, Michael Loeffler, and Carter Watterson. We also thank the numerous individuals from the North Carolina Wildlife Resources Commission, the North Carolina Division of Marine Fisheries, and North Carolina State University for their assistance collecting and tagging Striped Bass as well as downloading stationary receivers in the Roanoke River, especially Ladd Bayliss, Kevin Dockendorf, Charlton Godwin, Michael Fisk, Jared Flowers, Kyle Hussey, Jeremy McCargo, Tyler Moore, Ben Ricks, and Joseph Smith. Ben Gahagan provided helpful comments on a previous draft of this paper. Research funding was provided by North Carolina's Marine Resources Fund, through sales of the Coastal Recreational Fishing License. The Cooperative Fish and Wildlife Research Unit is jointly supported by North Carolina State University, North Carolina Wildlife Resources Commission, U.S. Geological Survey, U.S. Fish and Wildlife Service, and Wildlife Management Institute. Any use of trade, firm, or product names is for descriptive purposes only and does not imply endorsement by the U.S. Government. Sampling was conducted under the Institutional Animal Care and Use Committee protocol 10-145-O.

REFERENCES

Able, K. W., T. M. Grothues, J. T. Turnure, D. M. Byrne, and P. Clerkin. 2012. Distribution, movements, and habitat use of small Striped Bass (*Morone saxatilis*) across multiple spatial scales. U.S. National Marine Fisheries Service Fishery Bulletin 110:176–192.

ASMFC (Atlantic States Marine Fisheries Commission). 2013. 2013 Atlantic Striped Bass benchmark stock assessment. ASMFC, 57th SAW Assessment Report, Arlington, Virginia.

Bainbridge, R. 1958. The speed of swimming of fish as related to size and to the frequency and amplitude of the tail beat. Journal of Experimental Biology 35:109–133.

Beamish, F. W. H. 1978. Swimming capacity. Pages 101–187 *in* W. S. Hoar and D. J. Randall, editors. Fish physiology VII locomotion. Academic Press, New York.

Belsley, D. A., E. Kuh, and R. E. Welsch. 1980. Regression diagnostics identifying influential data and sources of collinearity. Wiley, New York.

Bernatchez, L., and J. J. Dodson. 1987. Relationship between bioenergetics and behavior in anadromous fish migrations. Canadian Journal of Fisheries and Aquatic Sciences 44:399–407.

Bielawski, J. P., and D. E. Pumo. 1997. Randomly amplified polymorphic DNA (RAPD) analysis of Atlantic coast Striped Bass. Heredity 78:32–40.

Bollen, K. A., and R. Jackman. 1990. Regression diagnostics: an expository treatment of outliers and influential cases. Pages 257–291 *in* J. Fox and J. S. Long, editors. Modern methods of data analysis. Sage, Newbury Park, California.

Boreman, J., and R. R. Lewis. 1987. Atlantic coastal migration of Striped Bass. Pages 331–339 *in* M. J. Dadswell, R. J. Klauda, C. M. Moffitt, R. L. Saunders, R. A. Rulifson, and J. E. Cooper, editors. Common strategies of anadromous and catadromous fishes. American Fisheries Society, Symposium 1, Bethesda, Maryland.

Boyd, J. B. 2011. Maturation, fecundity, and spawning frequency of the Albemarle/Roanoke Striped Bass stock. Master's thesis. East Carolina University, Greenville, North Carolina.

Callihan, J. L. 2011. Spatial ecology of adult Spotted Seatrout, *Cynoscion nebulosus*, in Louisiana coastal waters. Doctoral dissertation. Louisiana State University, Baton Rouge.

Callihan, J. L., C. H. Godwin, and J. A. Buckel. 2014. Effect of demography on spatial distribution: movement patterns of the Albemarle Sound–Roanoke River stock of Striped Bass (*Morone saxatilis*) in relation to their recovery. U.S. National Marine Fisheries Service Fishery Bulletin 112:131–143.

Candy, J. R., and T. D. Beacham. 2000. Patterns of homing and straying in southern British Columbia coded-wire tagged Chinook Salmon (*Oncorhynchus tshawytscha*) populations. Fisheries Research 47:41–56.

Carmichael, J. T., S. L. Haeseker, and J. E. Hightower. 1998. Spawning migration of telemetered Striped Bass in the Roanoke River, North Carolina. Transactions of the American Fisheries Society 127:286–297.

Clark, J. 1968. Seasonal movement of Striped Bass contingents of Long Island Sound and the New York Bight. Transactions of the American Fisheries Society 123:950–963.

Coutant, C. C. 1985. Striped Bass, temperature, and dissolved oxygen: a speculative hypothesis for environmental risk. Transactions of the American Fisheries Society 114:31–61.

Dagorn, L., K. N. Holland, and D. G. Itano. 2007. Behavior of Yellowfin (*Thunnus albacares*) and Bigeye (*T. obesus*) tuna in a network of fish aggregating devices (FADs). Marine Biology 151:595–606.

Dingle, H., and V. A. Drake. 2007. What is migration? Bioscience 57:113–121.

Dittman, A. H., and T. P. Quinn. 1996. Homing in Pacific salmon: mechanisms and ecological basis. Journal of Experimental Biology 199:83–91.

Dorazio, R. M., K. A. Hattala, C. B. McCollough, and J. E. Skjeveland. 1994. Tag recovery estimates of migration of Striped Bass from spawning areas of the Chesapeake Bay. Transactions of the American Fisheries Society 123:950–963.

Douglas, S. G., G. Chaput, J. Hayward, and J. Sheasgreen. 2009. Prespawning, spawning, and postspwning behavior of Striped Bass in the Miramichi River. Transactions of the American Fisheries Society 138:121–134.

Dudley, R. G., A. W. Mullis, and J. W. Terrell. 1977. Movements of adult Striped Bass (*Morone saxatilis*) in the Savannah River, Georgia. Transactions of the American Fisheries Society 106:314–322.

Dunning, D. J., J. R. Waldman, Q. E. Ross, and M. T. Mattson. 2006. Dispersal of age-2+ Striped Bass out of the Hudson River. Pages 287–294 *in* J.

Waldman, K. Limburg, and D. Strayer, editors. Hudson River fishes and their environment. American Fisheries Society, Symposium 51, Bethesda, Maryland.

Gauthier, D. T., C. A. Audemard, J. E. L. Carlsson, T. Y. Darden, M. R. Denson, K. S. Reece, and J. Carlsson. 2013. Genetic population structure of US Atlantic coast Striped Bass (*Morone saxatilis*). Journal of Heredity 104:510–520.

Gross, M. R. 1987. Evolution of diadromy in fishes. Pages 14–25 *in* M. J. Dadswell, R. J. Klauda, C. M. Moffitt, R. L. Saunders, R. A. Rulifson, and J. E. Cooper, editors. Common strategies of anadromous and catadromous fishes. American Fisheries Society, Symposium 1, Bethesda, Maryland.

Grothues, T. M., K. W. Able, J. Carter, and T. W. Arienti. 2009. Migration patterns of Striped Bass through nonnatal estuaries of the U.S. Atlantic coast. Pages 135–150 *in* A. J. Haro, K. L. Smith, R. A. Rulifson, C. M. Moffitt, R. J. Klauda, M. J. Dadswell, R. A. Cunjak, J. E. Cooper, K. L. Beal, and T. S. Avery, editors. Challenges for diadromous fishes in a dynamic global environment. American Fisheries Society, Symposium 69, Bethesda, Maryland.

Harris, J. E., and J. E. Hightower. 2014. Estimating mortality rates for Albemarle Sound–Roanoke River Striped Bass using an integrated modeling approach. North Carolina Division of Marine Fisheries, Final Report, Morehead City.

Hartman, K. J., and S. B. Brandt. 1995. Comparative energetics and the development of bioenergetics models for sympatric estuarine piscivores. Canadian Journal of Fisheries and Aquatic Sciences 52:1647–1666.

Hartman, W. L., and R. F. Raleigh. 1964. Tributary homing of Sockeye Salmon at Brooks and Karlus lakes, Alaska. Journal of the Fisheries Research Board of Canada 21:485–503.

Hassler, W. W., N. L. Hill, and J. T. Brown. 1981. The status and abundance of Striped Bass, *Morone saxatilis*, in the Roanoke River and Albemarle Sound, North Carolina, 1956–1980. North Carolina Department of Natural Resources and Community Development, Division of Marine Fisheries, Special Scientific Report 38, Morehead City.

Hendricks, M. L., R. L. Hoopes, D. A. Arnold, and M. L. Kaufmann. 2002. Homing of hatchery-reared American Shad to the Lehigh River, a tributary to the Delaware River. North American Journal of Fisheries Management 22:243–248.

Hendry, A. P., T. Bohlin, B. Jonsson, and O. K. Berg. 2004. To sea or not to sea? Anadromy versus non-anadromy in salmonids. Pages 92–125 *in* A. P. Hendry and S. C. Stearns, editors. Evolution illuminated: salmon and their relatives. Oxford University Press, Oxford, UK.

Heupel, M. R., J. M. Semmens, and A. J. Hobday. 2006. Automated acoustic tracking of aquatic animals: scales, design and deployment of listening station arrays. Marine and Freshwater Research 57:1–13.

Huntsman, A. G. 1937. "Migration" and "homing" of salmon. Science 85:313–314.

Jørgensen, C., B. Ernande, Ø. Fiksen, and U. Dieckmann. 2006. The logic of skipped spawning in fish. Canadian Journal of Fisheries and Aquatic Sciences 63:200–211.

Kneebone, J., W. S. Hoffman, M. J. Dean, D. A. Fox, and M. P. Armstrong. 2014. Movement patterns and stock composition of adult Striped Bass tagged in Massachusetts coastal waters. Transactions of the American Fisheries Society 143:1115–1129.

Leggett, W. C. 1977. The ecology of fish migrations. Annual Review of Ecological Systems 8:285–308.

Leggett, W. C., and R. R. Whitney. 1972. Water temperature and the migrations of American Shad. U.S. National Marine Fisheries Service Fishery Bulletin 70:659–670.

Lembo, G., M. T. Spedicato, F. Økland, P. Carbonara, I. A. Fleming, R. S. McKinley, E. B. Thorstad, M. Sisak, and S. Ragonese. 2002. A wireless communication system for determining site fidelity of juvenile Dusky Groupers *Epinephelus marginatus* (Lowe, 1834) using coded acoustic transmitters. Hydrobiologia 483:249–257.

Mathies, N. H., M. B. Ogburn, G. McFall, and S. Fangman. 2014. Environmental interference factors affecting detection range in acoustic telemetry studies using fixed receiver arrays. Marine Ecology Progress Series 495:27–38.

McDowall, R. M. 1987. The occurrence and distribution of diadromy among fishes. Pages 1–13 *in* M. J. Dadswell, R. J. Klauda, C. M. Moffitt, R. L. Saunders, R. A. Rulifson, and J. E. Cooper, editors. Common strategies of anadromous and catadromous fishes. American Fisheries Society, Symposium 1, Bethesda, Maryland.

McDowall, R. M. 2001. Anadromy and homing: two life-history traits with adaptive synergies in salmonid fishes? Fish and Fisheries 2:78–85.

Melvin, G. D., M. J. Dadswell, and J. D. Martin. 1986. Fidelity of American Shad, *Alosa sapidissima* (Clupeidae), to its river of previous spawning. Canadian Journal of Fisheries and Aquatic Sciences 43:640–646.

Merriman, D. 1941. Studies on the Striped Bass (*Roccus saxatilis*) of the Atlantic coast. U.S. National Marine Fisheries Service Fishery Bulletin 50.

Northcote, T. G. 1978. Migratory strategies and production in freshwater fishes. Pages 326–359 *in* S. D. Gerking, editor. Ecology of freshwater fish production. Blackwell Scientific Publications, Oxford, UK.

Northcote, T. G. 1984. Mechanisms of fish migration in rivers. Pages 317–355 *in* J. D. McCleave, G. P. Arnold, J. J. Dodson, and W. H. Neill, editors. Mechanisms of migration in fishes, Plenum, New York.

Olsen, E. J., and R. A. Rulifson. 1992. Maturation and fecundity of Roanoke River–Albemarle Sound Striped Bass. Transactions of the American Fisheries Society 121:524–537.

Pautzke, S. M., M. E. Mather, J. T. Finn, L. A. Deegan, and R. M. Muth. 2010. Seasonal use of a New England estuary by foraging contingents of migratory Striped Bass. Transactions of the American Fisheries Society 139:257–269.

Peer, A. C., and T. J. Miller. 2014. Climate change, migration phenology, and fisheries management interact with unanticipated consequences. North American Journal of Fisheries Management 34:94–110.

Pincock, D. G. 2012. False detections: what they are and how to remove them from detection data. Vemco. Available: http://vemco.com/wp-content/uploads/2012/11/false_detections.pdf. (August 2014)

Pincock, D. G., and S. V. Johnston. 2012. Acoustic telemetry overview. Pages 1–33 *in* N. S. Adams, J. W. Beeman, and J. H. Eiler, editors. Telemetry techniques: a user guide for fisheries research. American Fisheries Society, Bethesda, Maryland.

Pincock, D., D. Welch, S. McKinley, and G. Jackson. 2010. Acoustic telemetry for studying migration movements of small fish in rivers and the ocean-current capabilities and future possibilities. Pages 105–117 *in* K. Wolf and J. O'Neal, editors. Tagging, telemetry, and marking measures for monitoring fish populations [online publication]. Pacific Northwest Aquatic Monitoring Partnership, Special Publication. Available: http://www.pnamp.org/document/3637. (August 2014).

Pine, W. E., K. H. Pollock, J. E. Hightower, T. J. Kwak, and J. A. Rice. 2003. A review of tagging methods for estimating fish population size and components of mortality. Fisheries 28(10):10–23.

Quinn, T. P. 1988. Estimated swimming speeds of migrating adult Sockeye Salmon. Canadian Journal of Zoology 66:2160–2163.

Quinn, T. P. 1993. A review of homing and straying of wild and hatchery-produced salmon. Fisheries Research 18:29–44.

Quinn, T. P., and D. J. Adams. 1996. Environmental changes affecting the migratory timing of American Shad and Sockeye Salmon. Ecology 77:1151–1162.

Quinn, T. P., I. J. Stewart, and C. P. Boatright. 2006. Experimental evidence of homing to site of incubation by mature Sockeye Salmon, *Oncorhynchus nerka*. Animal Behaviour 72:941–949.

Raney, E. C. 1952. The life history of the Striped Bass, *Roccus saxatilis* (Walbaum). Bulletin of the Bingham Oceanographic Collection 14:5–97.

Rideout, R. M., G. A. Rose, and M. P. M. Burton. 2005. Skipped spawning in female iteroparous fishes. Fish and Fisheries 2005:50–72.

Rideout, R. M., and J. Tomkiewicz. 2011. Skipped spawning in fishes: more common than you might think. Marine and Coastal Fisheries: Dynamics, Management, and Ecosystem Science [online serial] 3:176–189.

Rulifson, R. A. 1990. Abundance and viability of Striped Bass eggs spawned in Roanoke River, North Carolina, in 1989. Report to the U.S. Environmental Protection Agency and North Carolina Department of Environment, Health, and Natural Resources, Project APES 90-11. Available: http://thescholarship.ecu.edu/handle/10342/2845. (August 2014).

Sambilay, V. C. Jr. 1990. Interrelationships between swimming speed, caudal fin aspect ratio and body length of fishes. ICLARM (International Center for Living Aquatic Resources Management) Fishbyte 8:16–20.

Secor, D. H. 1999. Specifying divergent migrations in the concept of stock: the contingent hypothesis. Fisheries Research 43:13–34.

Secor, D. H., and P. M. Piccoli. 1996. Age- and sex-dependent migrations of Striped Bass in the Hudson River as determined by chemical microanalysis of otoliths. Estuaries 19:778–793.

Secor, D. H., and P. M. Piccoli. 2007. Oceanic migration rates of upper Chesapeake Bay Striped Bass (*Morone saxatilis*), determined by otolith microchemical analysis. U.S. National Marine Fisheries Service Fishery Bulletin 105:62–73.

Secor, D. H., J. R. Rooker, E. Zlokovitz, and V. S. Zdanowicz. 2001. Identification of riverine, estuarine, and coastal contingents of Hudson River Striped Bass based upon otolith elemental fingerprints. Marine Ecology Progress Series 211:245–253.

Skjæraasen, J. E., R. D. M. Nash, K. Korsbrekke, M. Fonn, T. Nilsen, J. Kennedy, K. H. Nedreaas, A. Thorsen, P. R. Witthames, A. J. Geffen, H. Høie, and O. S. Kjesbu. 2012. Frequent skipped spawning in the world's largest cod population. Proceedings of the National Academy of Sciences of the USA 109:8995–8999.

Trent, L., and W. W. Hassler. 1968. Gill net selection, migration, size and age composition, sex ratio, harvest efficiency, and management of Striped Bass in the Roanoke River, North Carolina. Chesapeake Science 9:217–232.

Waldman, J., L. Maceda, and I. Wirgin. 2012. Mixed-stock analysis of wintertime aggregations of Striped Bass along the mid-Atlantic coast. Journal of Applied Ichthyology 28:1–6.

Waldman, J. R., D. J. Dunning, Q. E. Ross, and M. T. Mattson. 1990. Range dynamics of Hudson River Striped Bass along the Atlantic coast. Transactions of the American Fisheries Society 119:910–919.

Waldman, J. R., J. Grossfield, and I. Wirgin. 1988. Review of stock discrimination techniques for Striped Bass. North American Journal of Fisheries Management 8:410–425.

Waldman, J. R., R. A. Richards, W. B. Schill, I. Wirgin, and M. C. Fabrizio. 1997. An empirical comparison of stock identification techniques applied to Striped Bass. Transactions of the American Fisheries Society 126:369–385.

Wehmeyer, L. L., and C. R. Wagner. 2011. Relation between flows and dissolved oxygen in the Roanoke River between Roanoke Rapids dam and Jamesville, North Carolina, 2005–2009. U.S. Geological Survey, Scientific Investigations Report 2011-5040, Reston, Virginia.

Welch, D. W., M. C. Melnychuk, J. C. Payne, E. L. Rechisky, A. D. Porter, G. D. Jackson, B. R. Ward, S. P. Vincent, C. C. Wood, and J. Semmens. 2011. In situ measurement of coastal ocean movements and survival of juvenile Pacific salmon. Proceedings of the National Academy of Sciences of the USA 108:8708–8713.

Welch, D. W., M. C. Melnychuk, E. R. Rechisky, A. D. Porter, M. C. Jacobs, A. Ladouceur, R. S. McKinley, and G. D. Jackson. 2009. Freshwater and marine migration and survival of endangered Cultus Lake Sockeye Salmon (*Oncorhynchus nerka*) smolts using POST, a large-scale acoustic telemetry array. Canadian Journal of Fisheries and Aquatic Sciences 66:736–750.

Welsh, S. A., D. R. Smith, R. W. Laney, and R. C. Tipton. 2007. Tag-based estimates of annual fishing mortality of a mixed Atlantic coastal stock of Striped Bass. Transactions of the American Fisheries Society 136:34–42.

Wood, C. C., D. W. Welch, L. Godbout, and J. Cameron. 2012. Marine migratory behavior of hatchery-reared anadromous and wild non-anadromous Sockeye Salmon revealed by acoustic tags. Pages 289–311 in J. R. McKenzie, B. Parsons, A. C. Seitz, R. K. Kopf, M. G. Mesa, and Q. Phelps, editors. Advances in fish tagging and marking technology. American Fisheries Society, Bethesda, Maryland.

Wootton, R. J. 1998. Ecology of teleost fishes. Kluwer Academic Publishers, London.

Zlokovitz, E. R., D. H. Secor, and P. M. Piccoli. 2003. Patterns of migration in Hudson River Striped Bass as determined by otolith microchemistry. Fisheries Research 63:245–259.

2

Oogenesis and Fecundity Type of Gray Triggerfish in the Gulf of Mexico

Erik T. Lang*[1]

Riverside Technology (Contracting for the National Marine Fisheries Service),
Panama City Laboratory, 3500 Delwood Beach Road, Panama City, Florida 32408, USA

Gary R. Fitzhugh

National Marine Fisheries Service, Panama City Laboratory, 3500 Delwood Beach Road, Panama City,
Florida 32408, USA

Abstract

The fecundity of Gray Triggerfish *Balistes capriscus* has been difficult to estimate, as few imminently spawning or recently spawned females have been detected. Our study focused on verifying the pattern of oogenesis and fecundity type in Gray Triggerfish. During 1999–2012, females ($n = 1,092$) were collected from the eastern Gulf of Mexico, and subsets of these fish were used to calculate condition indices and assess ovarian histology. The gonadosomatic index, hepatosomatic index, and Fulton's condition factor indicated that liver and somatic energy stores increased prior to spawning and were depleted throughout the spawning period, characteristic of a capital pattern of energy storage and allocation to reproduction. Typical of a capital breeding pattern, we also observed (1) a hiatus in oocyte size distribution and (2) group-synchronous oogenesis, which are both traits of a determinate fecundity type. However, evidence that fecundity was not set prior to spawning included the observation of "de novo" vitellogenesis during the spawning season; secondary oocytes increased in number and failed to increase in mean size over time. Thus, Gray Triggerfish exhibit an indeterminate fecundity type with mixed reproductive traits that may characterize species exhibiting female parental care in warmwater environments. Further, we estimated the secondary oocyte growth rate (37 µm/d) based upon the time lag of postovulatory follicle (POF) degeneration. Using oocyte growth rate and the proportion of females bearing POFs, the interspawning interval was estimated to range from 8 to 11 d, indicating that 8–11 batches/female could be produced during the estimated 86-d reproductive period. The hiatus in oocyte size distribution was used to define a minimum size (250 µm) from which to distinguish an advancing batch of secondary growth oocytes. Batch fecundity (BF) ranged from 0.34 to 1.99 million eggs and was significantly related to FL (mm): $BF = 8,703.69 \cdot FL - 1,776,483$ ($r^2 = 0.56$).

The Gray Triggerfish *Balistes capriscus* is an economically important species in recreational and commercial fisheries of the Gulf of Mexico (hereafter, Gulf). Like other Gulf species, measures of reproductive potential for Gray Triggerfish are needed to support assessments of stock status, and reproductive potential estimated from fecundity metrics is most accurate (Paulik 1973; Tomkiewicz et al. 2003; Lambert 2008; McBride et al. 2015). Unfortunately, particular fecundity

Subject editor: Anthony Overton, East Carolina University, Greenville, North Carolina

*Corresponding author: elang@wlf.la.gov
[1]Present address: Louisiana Department of Wildlife and Fisheries, 2000 Quail Drive, Baton Rouge, Louisiana 70808, USA.

methodologies have been applied to some species without validation, thus necessitating corrections (e.g., Arocha 2002; Gordo et al. 2008; Fitzhugh et al. 2012). An understanding of oogenesis allows for classification of a species' fecundity type and provides insight on spawning frequency (Murua and Saborido-Rey 2003; Witthames et al. 2009).

Although information on Gray Triggerfish reproduction exists in research reports, the pattern of oogenesis and fecundity type in this species have not been validated. In a 2005 Gulf stock assessment, the Gray Triggerfish was assumed to be an asynchronous indeterminate spawner that produces numerous batches (Ingram 2001; SEDAR 09 2006). Previous unpublished studies had difficulties in determining Gray Triggerfish fecundity and/or spawning frequency (Wilson et al. 1995; Hood and Johnson 1997; Ingram 2001; Moore 2001), primarily because females with advanced-stage oocytes (oocyte maturation [OM]) were very rare among the collected samples and because postovulatory follicles (POFs) were detected in few females (Ofori-Danson 1990; Wilson et al. 1995; Hood and Johnson 1997; Ingram 2001; Moore 2001). This has generated uncertainty and a broad range in estimates of spawning frequency.

Part of the difficulty in classifying reproductive traits and fecundity type may lie in the rather unique biology of the Gray Triggerfish. It is a nesting species with a haremic mating system wherein the males and females share short-term parental care (Simmons and Szedlmayer 2012). Considering this mating system, a better understanding of fecundity type for Gray Triggerfish may depend on determining the female reproductive energy strategy utilized from the income–capital breeding typology (Stearns 1992; Jönsson 1997; Jager et al. 2008; McBride et al. 2015). Capital breeders are commonly cold-water species that experience dynamic seasonality. Before the reproductive season begins, energy for reproduction is acquired and stored during periods when food may be abundant (Jager et al. 2008). In income breeders, a rapid transfer of energy to reproduction may occur when ecosystem productivity is high yet the timing of productivity is less predictable in time and space (Santos et al. 2010; McBride et al. 2015). It is also possible for species to exhibit mixed capital and income traits depending on the degree of available surplus energy and demands from activities such as migration and parental care (Jager et al. 2008; McBride et al. 2015). In several fish species, the accessory behaviors of parental care have been associated with high energy demands, which can affect other life history traits (Kuwamura 1997; Donelson et al. 2008; Jager et al. 2008). The objectives of the present study were to (1) verify fecundity type for Gray Triggerfish in the Gulf, (2) classify energy storage in light of the mating system, (3) calculate spawning frequency, and (4) postulate a method to estimate annual fecundity for this species.

METHODS

Sample collection and condition indices.—Hook-and-line sampling of Gray Triggerfish was conducted to observe a time

TABLE 1. Number of samples obtained for each body index measure (Fulton's condition factor [K], gonadosomatic index [GSI], and hepatosomatic index [HSI]) calculated for female Gray Triggerfish from various fisheries-independent and fisheries-dependent sources in the eastern Gulf of Mexico, 1999–2012. Gear types were hook and line (HL; includes bandit reel), trap (TR), and spear (SP).

Source	Sampling gear	K	GSI	HSI
Fisheries independent	HL + TR	420	552	204
Fisheries dependent	HL + SP	672	77	34
Total		1,092	629	238

series of reproductive development during May–July 2012 in Gulf waters offshore of Panama City, Florida. All captured specimens were retained, measured for FL to the nearest millimeter, and weighed to the nearest gram. Dorsal spines were extracted for age estimation; livers were excised and weighed; gonads were removed, macroscopically assessed for sex and reproductive state, and weighed; and ovaries were fixed in 10% neutral buffered formalin (<24 h on ice) for further processing in the laboratory (see below). Length, weight, macroscopic sex and reproductive state, and (in some cases) gonad weight, liver weight, and ovary tissue samples were also measured or obtained from specimens collected by trapping, spearing, or hook and line in various fishery-independent and fishery-dependent surveys within the northern Gulf (1999–2012; Table 1). The gonadosomatic index (GSI), hepatosomatic index (HSI), and Fulton's condition factor (K; Ricker 1975) based upon weight (g) and FL (cm) were used to examine the relationship between energy storage and reproduction:

$$\text{GSI} = \left(\frac{\text{Gonad weight}}{\text{Total weight} - \text{Gonad weight}} \right) \times 100,$$

$$\text{HSI} = \left(\frac{\text{Liver weight}}{\text{Total weight} - \text{Liver weight}} \right) \times 100,$$

and

$$K = 100 \times \left(\frac{\text{Total weight}}{\text{FL}^3} \right).$$

Oocyte staging and measurements.—Oocyte stage and oocyte diameter in Gray Triggerfish were used to categorize fecundity type as determinate or indeterminate. The key criteria used to test for determinate fecundity followed Hunter et al. (1992) and Murua and Saborido-Rey (2003). The criteria include (1) a hiatus in the size distribution of developing oocytes, (2) an increase in secondary oocyte diameter through the spawning season, (3) a decrease in the number of secondary growth oocytes through the spawning season, and (4) a secondary oocyte growth rate (G) that does not allow for "de novo" vitellogenesis within one spawning period.

Oocyte stage and size data were obtained by examining an approximately 150-mg subsample removed from the posterior region of all available ovaries following the method of Harris et al. (2002). Each subsample was weighed, and the oocytes were disassociated with forceps and plated as a whole mount in a petri dish (150-mm diameter) to be imaged with an EPSON V750 scanner. The top or cover of the scanner rested on the petri dish, parallel with the bottom of the scanner. The scanner's software (EPSON Scan) was set to scan "positive film" at 2,400 dpi in "professional" mode, which allowed for a transmitted light image with high resolution.

The image was then uploaded into ObjectJ, a plug-in for ImageJ (version 1.47 s), to conduct image analysis. Primary growth oocytes were difficult to count and measure in the scans because of their small size (<100 μm) and low degree of clarity (Figure 1). Secondary growth oocytes (including cortical alveolar oocytes and vitellogenic oocytes) from 2012

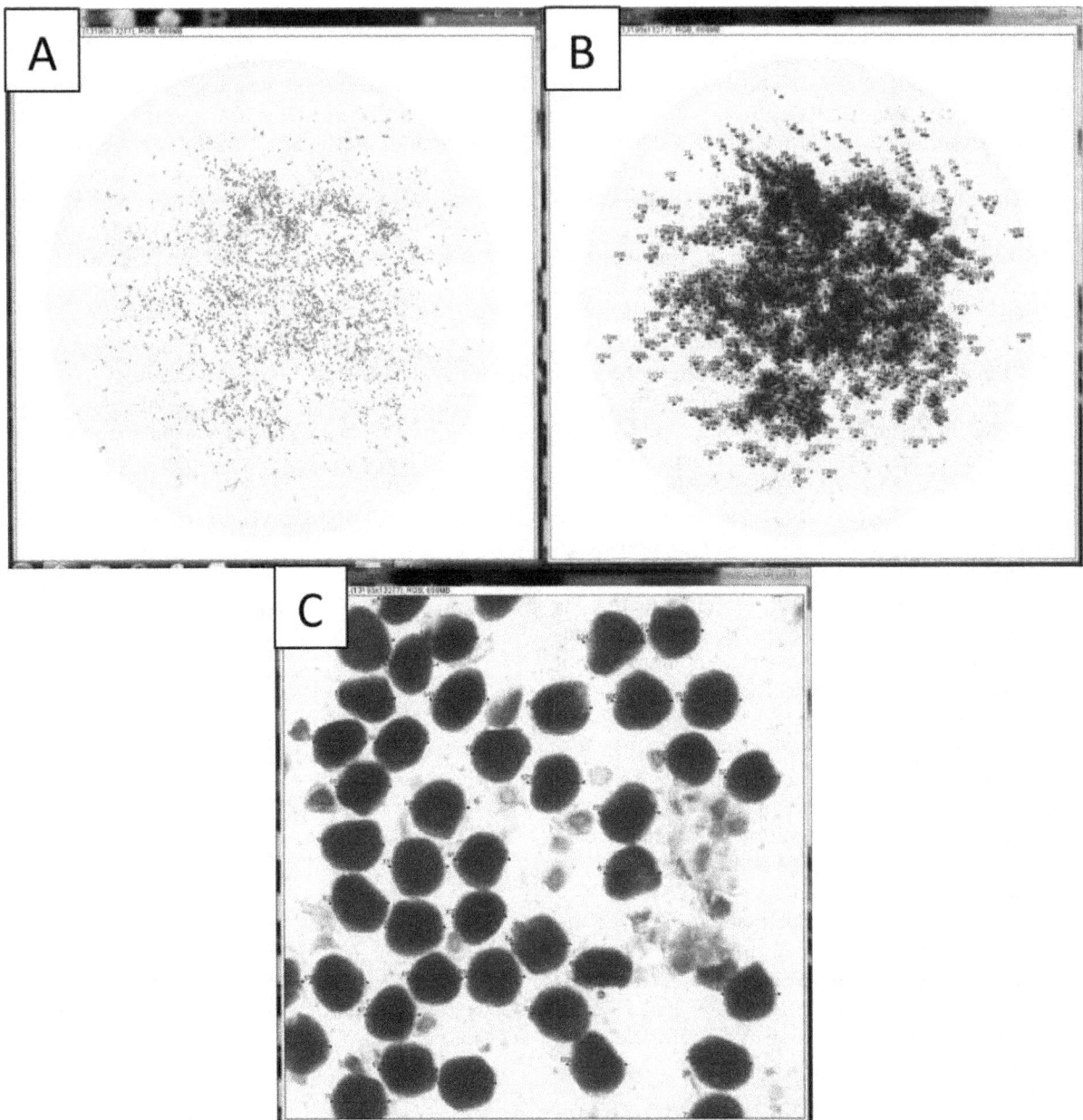

FIGURE 1.   Tissue subsamples ($\sim$150 mg) were extracted from Gray Triggerfish ovaries, and disassociated oocytes were scanned in a 150-mm Petri dish. Screen captures are shown for (A) the lowest scan magnification, (B) ObjectJ delineation of secondary growth oocytes, and (C) the highest scan magnification at 2,400-dpi resolution. Secondary growth oocytes appear dark against the background using transmitted light.

samples were automatically counted and measured by using a macro within ObjectJ (Figure 1; //simon.bio.uva.nl/objectj/7a-Examples.html). Counts of secondary oocytes were used to estimate batch fecundity (BF) based upon an oocyte size criterion (see Results) and were expressed as somatic relative fecundity (eggs/g of ovary-free body weight; Kjesbu et al. 1998) and oocyte density (secondary oocyte count/g of ovary weight) for comparison with other studies. Oocyte size frequency histograms were used to assess a hiatus in oocyte development, whereas oocyte enumeration and diameter were regressed against time within the spawning season to resolve determinate fecundity criteria 2 and 3. Three randomly selected females were chosen from each month (May–July; i.e., a total of 9 females) to represent the oocyte size distribution. All regressions were performed using R version 2.13.1.

Postovulatory follicle measurements.—Ovary samples from all years were processed histologically with hematoxylin and eosin stain. Using an ocular microscope camera (MiniVid; 5 MP), photos were taken of each histology slide in which POFs were present. Five random images were chosen from all possible nonrepetitive images of each histology slide with oocytes filling the frame. Each oocyte that represented the leading stage of oogenesis (cortical alveolar or vitellogenic oocytes) and that was sectioned through the germinal vesicle was measured for diameter (Foucher and Beamish 1980). The area of every POF present within the five random images was measured and plotted as a predictor of leading oocyte stage diameter.

Oocyte growth and interspawning interval.—Oocyte growth was determined using methods similar to those of Ganias et al. (2011). The POF area was converted to a percentage of POF duration by using known POF durations from oceanic species with a temperature regime similar to that of Gray Triggerfish (Jackson et al. 2006). Simple linear regression was used to investigate changes in oocyte size, number, and growth over time.

The G of secondary oocytes was calculated using the difference in oocyte size (O_i) after some period of growth (time lag; t_i) and the average size of oocytes from the spawning batch at the beginning of spawning batch development for the whole population (β_o; Ganias et al. 2011). The average oocyte sizes at the beginning of the spawning cycle were observed in fish with the most recent POFs. Because POFs are resorbed relatively quickly in Gulf warmwater fishes, the t_i could be estimated from the percent change in POF area over an expected 24-h duration ($\leq$24 h in 24–28.5°C water; Jackson et al. 2006):

$$G = \frac{O_i - \beta_o}{t_i}.$$

Spawning frequency was calculated using both the interspawning interval (ISI; Ganias et al. 2011) and POF (Parker 1980) methods. The ISI method assumes that the growth of secondary oocytes is group synchronous and that G is constant; thus, the ISI is the difference between the minimum (O_b) and maximum (O_e) observed sizes of secondary oocytes divided by G:

$$\text{ISI} = \frac{O_e - O_b}{G}.$$

The POF method is based upon the proportion of females bearing POFs relative to the total number of mature females sampled during the spawning season. Maturity was designated based upon the presence of vitellogenic oocytes. Because we estimate that POF duration is 24 h, the inverse of the proportion of females bearing POFs yields the expected spawning interval in days (Parker 1980). Due to the often-opportunistic collection of samples during surveys, the sample sizes varied from year to year and were ad hoc with respect to the timing of reproductive development and spawning. Samples were grouped across years to increase the sample size for the POF method. This grouping necessarily assumes that there is no difference in spawning intensity—and, hence, prevalence of POFs—among years.

RESULTS

Field Sampling and Condition Indices

In total, 1,092 female Gray Triggerfish were collected from various fishery-independent and fishery-dependent sources in the northern Gulf during 1999–2012, and 85% of these females were captured from depths of 10–50 m (Table 1). Of the collected females, 629 had gonad weight data for use in calculating GSI; 238 had liver weight data for use in HSI estimation; and 531 were histologically examined for POFs. The HSI peaked in May, 1 month before peak spawning (peak GSI); decreased sharply from May to September; and then began to increase again (Figure 2). Seasonally, K from all 1,092 females decreased after June, consistent with the HSI and GSI evidence that energy stores increased during the spring, were highest at the onset of spawning, and were depleted throughout the spawning period (Figure 3).

Oocyte Size Analysis

Examination of oocyte size frequency distributions suggested a group-synchronous pattern of oocyte development, which is commonly associated with determinate fecundity (Figure 4). The individual oocyte size frequencies for seven of the nine randomly selected females were bimodally distributed, and four of those seven females showed a clear hiatus between the cortical alveolar stage and the vitellogenic oocyte stage. Females that exhibited a unimodal peak in oocyte distribution had smaller oocytes on average, with early vitellogenic oocytes that we expect would have disassociated in size from

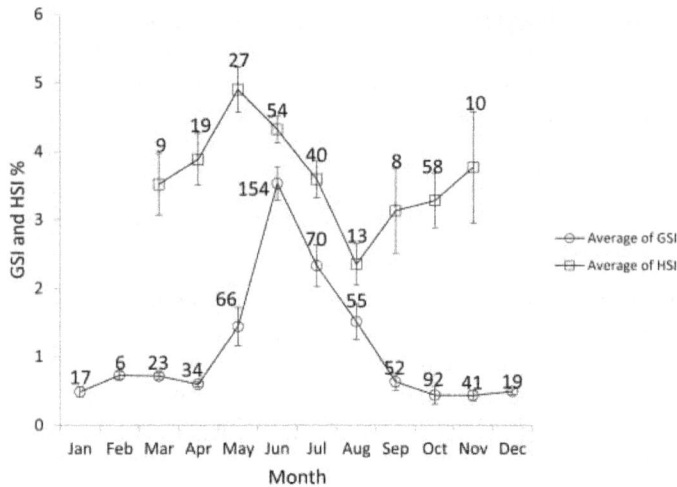

FIGURE 2. Mean (±SE) gonadosomatic index (GSI) and hepatosomatic index (HSI) of female Gray Triggerfish collected from the northern Gulf of Mexico, 1999–2012. Sample sizes are indicated for each month.

the primary growth and cortical alveolar stages over time. However, linear regressions of secondary oocyte diameter ($r^2 = 0.05$, $P = 0.148$; Figure 5) or the number of secondary growth oocytes ($r^2 = 0.01$, $P = 0.491$; Figure 6) versus time during the spawning season were not significant. Secondary growth oocytes did not increase in diameter and did not decrease in number as the spawning season progressed; therefore, contrary to our results on oocyte size frequency distributions, this finding did not support a determinate fecundity type for Gray Triggerfish.

Oocyte Growth and Interspawning Interval

Despite more intensive sampling during the spawning season in some years (e.g., 2012), POFs were rarely found—occurring in only 31 of the 531 females that were sampled for histology. Over 50% of those 31 recently spawned females possessed POFs with a collapsed lumen and thus a smaller POF surface area and a skewed size distribution. However,

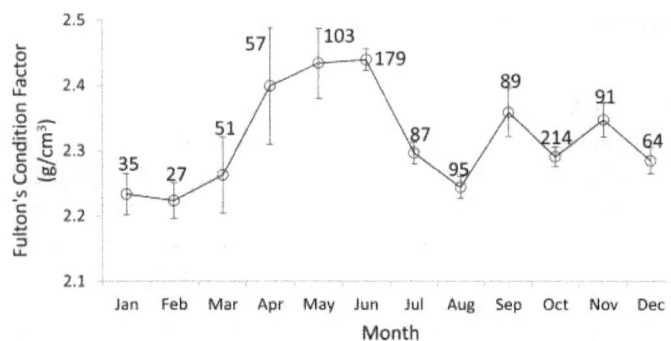

FIGURE 3. Mean (±SE) Fulton's condition factor for female Gray Triggerfish collected from the northern Gulf of Mexico, 1999–2012. Sample sizes are indicated for each month.

data on POF surface area for seven recently spawned females collected in 2001 were tested and fulfilled the linear regression assumptions for normality among means (Shapiro–Wilk normality test: $W = 0.8957$, $P = 0.306$). A Shapiro–Wilk normality test also indicated that the diameter of secondary growth oocytes was normally distributed ($W = 0.9288$, $P = 0.541$). Growth of secondary oocytes exhibited an inverse linear trend with POFs: as secondary oocytes grew larger, the surface area of POFs diminished ($r^2 = 0.63$, $P = 0.034$; Figure 7).

By estimating the G of secondary growth oocytes, we calculated the ISI to be 11 d. This computation used the POF duration from oceanic fish species that reside in similar water temperatures (Jackson et al. 2006) and used the POF "clock" to estimate t_i. We found that the maximum POF surface area for a Gray Triggerfish was 12 μm^2. Given that a POF shrinks as it ages and may last about 24 h (1 d), we observed a 74% decrease in POF surface area during an oocyte size regression series from 2001 (Figure 7). Thus, the proportional decrease in POF surface area over time served as the denominator to standardize to 24 h the corresponding growth in secondary oocytes (estimated $G = 37$ $\mu m/d$). The resulting ISI of 11 d was rapid enough to allow for de novo vitellogenesis to occur multiple times during the spawning season, which would indicate an indeterminate fecundity type. Therefore, three of the four hypothesis tests supported indeterminate fecundity in Gray Triggerfish.

Using the POF method, we calculated an average ISI of 7.6 d. Females were observed on 43 sampling dates between May 26 and August 19 during 1999–2012. The spawning interval was calculated as the inverse of the proportion of females with POFs ($n = 31$) among the total number of mature females ($n = 236$). Thus, the proportion with POFs (31/236) was 0.131, and the inverse was 7.6. For the 86-d spawning season, estimated as the interval during which POFs were detected, the POF method predicted that a female could produce up to 11 batches/season, and the ISI method predicted up to 8 batches/season.

Batch Fecundity Estimation

Gray Triggerfish undergoing OM and/or oocyte hydration were rarely collected during routine sampling, thus adding to the challenges in estimating fecundity. Of the 236 maturing females that were collected during the spawning season, only one was a ripe female undergoing OM. Because of the group-synchronous development, the population of advancing vitellogenic oocytes can be considered equivalent to a batch. However, all fish with POFs did possess cortical alveolar oocytes and newly formed vitellogenic oocytes with diameters less than 250 μm (Figure 7). Therefore, to distinguish the oldest batch, only secondary growth oocytes of at least 250 μm in diameter were counted.

Batch fecundity ranged from 0.34 to 1.99 million eggs, and somatic relative fecundity ranged from 590 to 2,686 eggs/g of

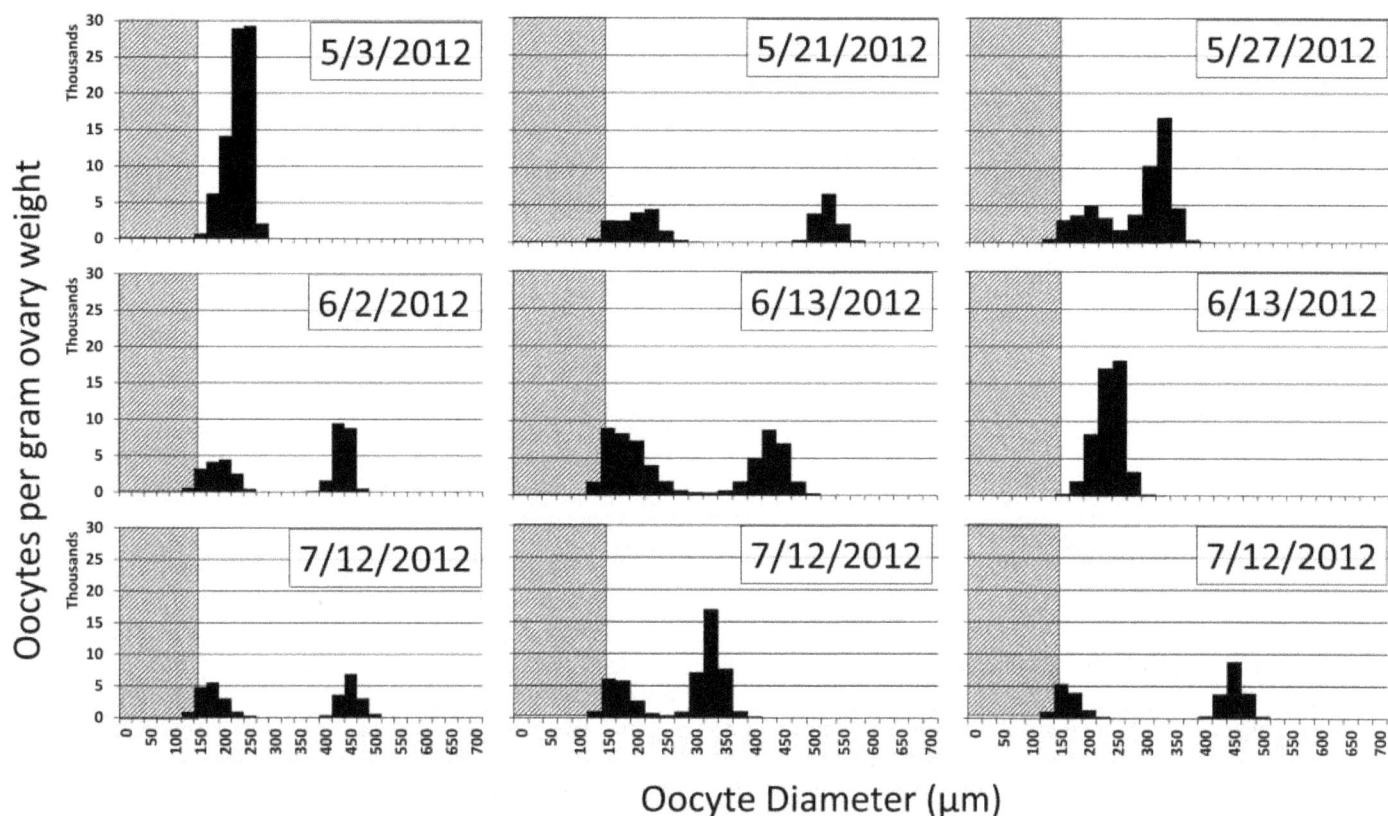

FIGURE 4. Oocyte size frequency (number of oocytes per gram of ovary weight) for nine individual Gray Triggerfish collected from the Gulf of Mexico in May–July 2012 (three randomly selected females per month). The date of capture is displayed for each fish. The hatched bar indicates the diameters of primary growth oocytes.

ovary-free body weight in 266–386-mm FL specimens (Table 2). A portion (56%) of the variance in BF was significantly explained by FL: BF = 8,703.69·FL − 1,776,483 ($r^2 =$ 0.56, $P < 0.0001$; Figure 8).

DISCUSSION

Similar to balistids worldwide, Gray Triggerfish in the Gulf spawn in pairs, are nest builders that establish and defend territories, and guard their eggs (Kuwamura 1997; Kawase 2003; Simmons and Szedlmayer 2012). Perhaps related to this reproductive strategy, Gray Triggerfish exhibit group-synchronous secondary oocyte development, which is unusual for a warmwater species. Wallace and Selman (1981) defined group-synchronous development as the presence of two populations of oocytes that are clearly distinguishable from one another: a larger, more synchronous population; and a smaller, more heterogeneous one. Gray Triggerfish clearly exhibited this pattern, with a distinct size hiatus between cortical alveolar oocytes and vitellogenic oocytes.

A group-synchronous oocyte development pattern is typically associated with determinate fecundity (Murua and Saborido-Rey 2003; McBride et al. 2015), but we found

evidence that Gray Triggerfish display indeterminate fecundity. The number of vitellogenic oocytes did not decrease and their diameter did not increase through the spawning season. Both of these results suggested that de novo vitellogenesis was occurring, thus violating the criteria for a determinate fecundity type. Although rare in the literature, other species have exhibited both group-synchronous oocyte development and the potential for de novo vitellogenesis; these include the Spiny Damselfish *Acanthochromis polyacanthus*, Mediterranean Sardine *Sardina pilchardus sardina*, and Blackmouth Angler *Lophiomus setigerus* (Nakazono 1993; Yoneda et al. 1998; Ganias et al. 2004).

Perhaps most diagnostic, we estimated that the G of secondary growth oocytes (37 μm/d) was rapid enough to yield multiple batches within an estimated 86-d spawning season. Although there are few published values of secondary oocyte G, some comparisons are available from higher-latitude fishes. For two species from the British Isles, reported G-values were approximately 3.2 μm/d for Sole *Solea solea* (based upon an increase in mean diameter over 143 d; Witthames and Greer-Walker 1995) and approximately 1.3 μm/d for Atlantic Mackerel *Scomber scombrus* (Greer-Walker et al. 1994). In both cases, secondary oocyte development proceeded slowly enough that de novo vitellogenesis could not occur within the

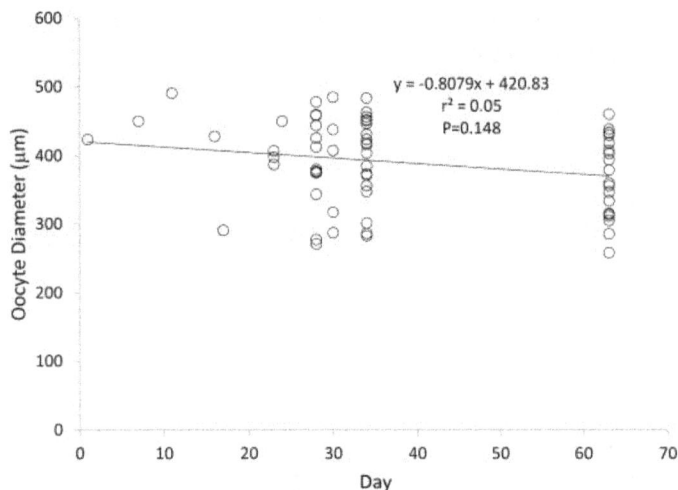

FIGURE 5. Regression of mean secondary oocyte diameter (μm) per individual female Gray Triggerfish against day of collection within the 2012 spawning season (day 1 = May 11).

spawning season; hence, the fecundity type was considered determinate.

Gray Triggerfish exhibited traits that are common in capital-breeding species. Condition indices (HSI and K) peaked just before or close to the onset of spawning, followed by declines throughout the reproductive period, similar to observations in other species with a capital pattern (Alonso-Fernández and Saborido-Rey 2012). The HSI of Gray Triggerfish exceeded the GSI in all months, which emphasizes the importance of the liver for energy storage and mobilization (Htun-Hun 1978; Rinchard and Kestemont 2003). In more extreme capital-breeding species, the seasonal increase in condition indices may occur well outside of the reproductive period (e.g., Htun-Hun 1978; Yoneda et al. 2001; Alonso-Fernández

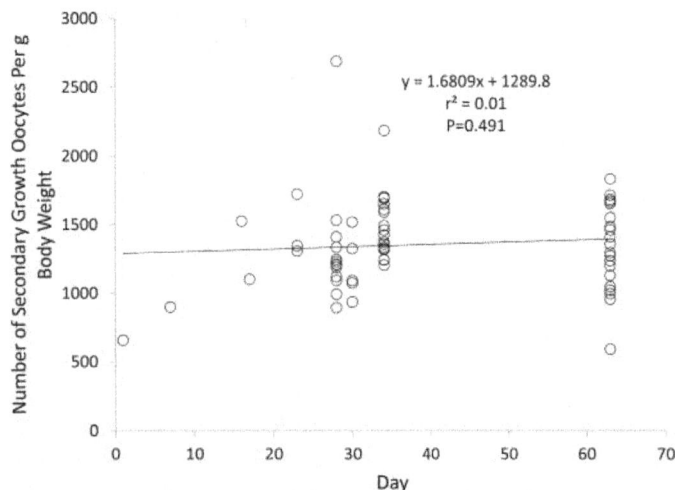

FIGURE 6. Regression of the number of secondary growth oocytes per gram of body weight in female Gray Triggerfish against day of collection within the 2012 spawning season (day 1 = May 11).

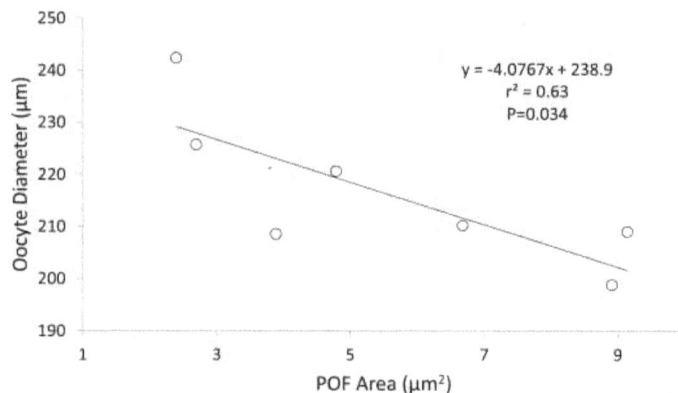

FIGURE 7. Regression of mean secondary oocyte diameter (μm) against mean postovulatory follicle (POF) area (μm^2) in seven female Gray Triggerfish selected from 2001 samples (see text).

and Saborido-Rey 2012). By contrast, income-breeding species lack seasonal patterns in condition indices or exhibit only weak seasonal patterns (e.g., Domínguez-Petit et al. 2010). Although we postulate that Gray Triggerfish are capital breeders, we cannot reject the possibility that food intake during the spawning season may partially provide the energy needed for egg production. However, the liver in Gray Triggerfish seems to be important for energy mobilization and may help to ensure that oocyte development proceeds even when energy intake is low (e.g., Allen and Wootton 1982; Alonso-Fernández and Saborido-Rey 2012) or when there is a hiatus in feeding during parental care.

We found that Gray Triggerfish in the Gulf show attributes of indeterminate fecundity associated with warmwater environments. However, the species also exhibits group-synchronous oocyte development and a capital breeding pattern associated with territoriality and female parental care—traits that are most often linked with determinate fecundity (McBride et al. 2015). Gray Triggerfish possibly show determinate fecundity elsewhere in their range, which extends from Nova Scotia to Argentina in the western Atlantic (Hoese and Moore 1998). Building on the review by McBride et al. (2015), our findings support a conclusion that such mixed reproductive attributes in fish may be more common than previously thought and may be predicted to occur in lower latitudes for species with energetically demanding accessory reproductive activities, such as female parental care.

Batch Fecundity

Our findings on the reproductive strategy of Gray Triggerfish in the Gulf support use of a fecundity methodology based upon batch size and spawning frequency. Application of an indeterminate fecundity methodology is the cautionary approach when there is uncertainty about a species' fecundity type (Gordo et al. 2008; Lowerre-Barbieri et al. 2011;

TABLE 2.   Fecundity metrics and associated FL information for female Gray Triggerfish used in fecundity analysis (OFBW = ovary-free body weight; batch fecundity = number of secondary growth oocytes ≥ 250 μm in diameter within the ovary).

Statistic	FL (mm)	Somatic relative fecundity (oocytes/g of OFBW)	Oocyte density (oocytes/g of ovary weight)	Batch fecundity	Secondary growth oocyte diameter (μm)
Minimum	266	590	12,653	339,605	250 (limit)
Maximum	386	2,686	47,688	1,990,861	590
Mean ± SE	311.3 ± 3.6	1,356.9 ± 41	24,468.82 ± 1,085.64	932,908.7 ± 41,458.6	370 ± 0.1

Fitzhugh et al. 2012). However, female Gray Triggerfish are rarely sampled in OM or the hydrated oocyte stage, which is typically the basis for identifying a spawning batch. The difficulty of detecting OM in female Gray Triggerfish seems to be a common problem (Wilson et al. 1995; Hood and Johnson 1997; Moore 2001). This could be related to (1) the lack of oocyte hydration during proteolysis or (2) reduced feeding prior to spawning, such that females undergoing OM have a low susceptibility to hook-and-line capture (Wilson et al. 1995; Moore 2001; Simmons and Szedlmayer 2012). MacKichan and Szedlmayer (2007) reported an egg diameter of 620 ± 3 (mean ± SE) μm from Gray Triggerfish nest excavations, which is close to the mean diameter of ova (602 ± 0.72 μm) from the one female we detected as undergoing yolk coalescence and considered to be in spawning condition. In contrast, broadcast spawners with positively buoyant eggs, such as the Red Snapper *Lutjanus campechanus*, have an egg diameter of about 820 μm (Rabalais et al. 1980). A lack of oocyte hydration would result in a much shorter period during which to collect imminently spawning females, which in turn would lead to a low incidence of actively spawning

individuals in sample collections. A similar difficulty occurs with the Mediterranean Sardine and is attributed to a relatively long (>10-d) spawning interval (Ganias et al. 2004). Due to the infrequent capture of Gray Triggerfish undergoing OM, we used the spawning batch size separation method (Ganias et al. 2004) to calculate BF. The oocyte size hiatus that develops as the secondary growth oocytes advance provides the basis upon which to define the spawning batch.

Other investigators of Gray Triggerfish fecundity have applied various criteria based on oocyte size or stage to define the spawning batch. Ingram (2001) reported 400 μm as the diameter that delimited the beginning of a spawning batch based upon the assumption that this size corresponded to the onset of OM. Using this criterion, oocyte density was estimated as 8,015 ± 247 (mean ± SE) oocytes/g of ovary (Ingram 2001, cited in SEDAR 09 2006). However, we found no evidence of OM occurring for oocytes with diameters of 590 μm or less. The only imminently spawning female in our collections had a mean oocyte diameter of 602 ± 0.72 μm for oocytes undergoing OM. Given our finding that secondary oocyte development is group synchronous, the 400-μm criterion would likely identify only a partial batch. Hood and Johnson (1997) indicated that the spawning batch could be identified as the standing stock of vitellogenic oocytes within an ovary—essentially the same conclusion we made. Their results for BF ranged from 0.2 to 1.2 million eggs, with a density of 13,809 ± 6,122 (mean ± SE) oocytes/g of ovary. Our estimated oocyte density was 24,468 ± 1,086 oocytes/g of ovary, which was much higher than the estimates from either Hood and Johnson (1997) or Ingram (2001). The difference between our estimate and that of Ingram (2001) can be readily accounted for by the difference in the minimum oocyte size used to define a spawning batch (250 μm versus 400 μm). The difference between our oocyte density estimate and that calculated by Hood and Johnson (1997) is more difficult to explain, and we can only speculate that factors we did not account for (e.g., year or season effects) were important. Although there are other published estimates of Gray Triggerfish fecundity (Manooch and Raver 1984; Bernardes and Dias 2000), the details provided on methods and oocyte size or stage criteria for fecundity counts were not sufficient for a comparison with our results. The range of our BF estimates

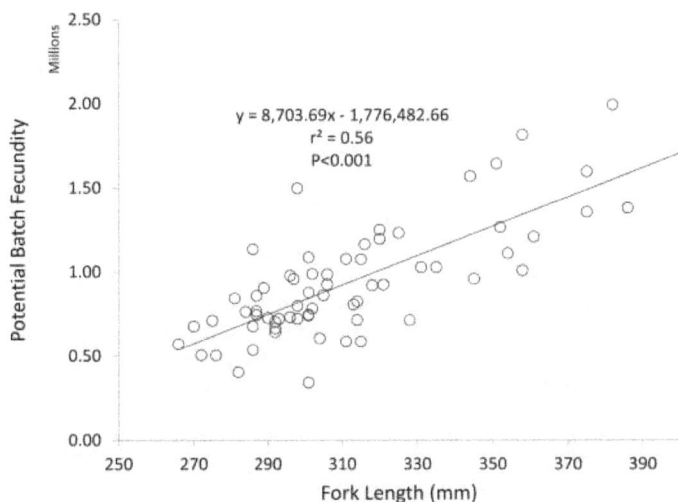

FIGURE 8.   Regression of potential batch fecundity against fork length (mm) of female Gray Triggerfish. Batch fecundity was defined as the number of secondary growth oocytes at least 250 μm in diameter within the ovary.

(0.34–2.0 million eggs) was similar to counts of eggs from nine nests (0.42–1.4 million eggs) examined by MacKichan-Simmons (2008).

Spawning Frequency

As found in many studies, spawning frequency can be difficult to assess due to low sample sizes and uneven sampling through the reproductive season (Stratoudakis et al. 2006; Ganias et al. 2011). Similar to the rarity of female Gray Triggerfish undergoing OM, the occurrence of females exhibiting POFs was comparatively low. This necessitated aggregation of the data across years, which assumes no year effect, such that the proportion of females bearing POFs could be used to calculate the spawning interval (7.6 d via the POF method). However, in one of the study years (2001), samples were evenly distributed over June and July and thus permitted spawning interval estimation via the ISI method (11 d). This estimate depends on two assumptions: (1) that POF degeneration offers a clock by which oocyte growth can be estimated, and the maximum POF surface area we observed (12 μm^2) is representative of a new day-0 POF in Gray Triggerfish; and (2) that the relationship between the rate of POF degeneration and secondary oocyte growth is constant over the spawning season. We think the first assumption is reasonable based on the corresponding observations that cortical alveolar oocytes were the leading gamete stage and that provisioning of yolk for the subsequent batch had not yet begun. Furthermore, the relationship between POF degeneration and secondary oocyte growth was linear for our example from 2001. For the second assumption, there is evidence from the literature that POF degeneration and secondary oocyte growth are both temperature dependent (Kjesbu et al. 1998; Ganias 2012). We also expect that water temperature—and thus rates of POF degeneration and secondary oocyte growth—may change over a several-month-long spawning season. However, a simple assumption may be that any changes in the rate of oocyte growth are proportional to changes in the rate of POF degeneration. Further work is needed to test this assumption. For the purposes of the present paper, the POF and ISI methods were needed to support each other due to the low sample size of individuals that contained POFs. A study of Yellowmargin Triggerfish *Pseudobalistes flavimarginatus* found that spawning grounds were vacated for 10 d between spawns on Australia's Great Barrier Reef (Gladstone 1994). This would be similar to the 11-d spawning interval we calculated for Gray Triggerfish. Earlier estimates of spawning interval for this species vary greatly: from 15 to 37 d in the U.S. South Atlantic (Moore 2001) to 3.7 d in the Gulf (Ingram 2001).

Female Gray Triggerfish in the Gulf exhibited mixed reproductive attributes that were apparently associated with the warmwater environment and with the energetically demanding behaviors required for defense and parental care. Fecundity type in the Gulf should be considered indeterminate, and the

hiatus in the size distribution of advancing secondary oocytes can be used to identify a spawning batch. Our estimated scope for the number of batches produced during a spawning season is lower (8–11 d) than that estimated elsewhere in the Gulf (c.f. Ingram 2001), but additional research is needed. In particular, age or size dependency in batch number will be difficult to assess due to the nesting behavior and parental care exhibited by Gray Triggerfish.

Further field collection efforts could improve estimates of Gray Triggerfish fecundity. Due to the nesting and territorial behavior of this species, scuba and spearfishing are necessary to readily identify and collect females that are in spawning condition. A survey could be designed to improve our knowledge of the timing and duration of spawning markers by targeting prespawn and postspawn females in conjunction with (1) visual observations of color phase and the size of females on nests and (2) measurements of environmental variables (perhaps most importantly, water temperature).

ACKNOWLEDGMENTS

We are grateful to Steve Theberge and Jenny Ragland for taking high-resolution scans of whole-mount oocytes and to Steve Theberge for measuring POF surface area from histology slides. In addition, we thank Doug DeVries, Patrick Raley, and Chris Gardner for aiding in the capture of Gray Triggerfish in the field; and Bill Walling for the collection of Gray Triggerfish biological samples at Panama City fish houses. Funding for this project was partially provided by the Marine Fisheries Initiative Program.

REFERENCES

Allen, J. R. M., and R. J. Wootton. 1982. The effect of ration and temperature on the growth of the Three-spined Stickleback, *Gasterosteus aculeatus* L. Journal of Fish Biology 20:409–422.

Alonso-Fernández, A., and F. Saborido-Rey. 2012. Relationship between energy allocation and reproductive strategy in *Trisopterus luscus*. Journal of Experimental Marine Biology and Ecology 416–417:8–16.

Arocha, F. 2002. Oocyte development and maturity classification of Swordfish from the north-western Atlantic. Journal of Fish Biology 60:13–27.

Bernardes, R. A., and J. F. Dias. 2000. Aspectos da reproducao do peixe-porco, *Balistes capriscus* (Gmelin) (Actinopterygii, Tetraodontiformes, Balistidae) coletado na costa sul do Estado de Sao Paulo, Brasil. [Observations about the reproduction of the Grey Triggerfish, *Balistes capriscus* (Gmelin) (Actinopterygli, Tetraodontiformes, Balistidae) collected from the southern coast of the state of Sao Paulo, Brazil.] Revista Brasileira de Zoologia 17:687–696.

Domínguez-Petit, R., F. Saborido-Rey, and I. Medina. 2010. Changes in proximate composition, energy storage and condition of European Hake (*Merluccius merluccius*, L. 1758) through the spawning season. Fisheries Research 104:73–82.

Donelson, J. M., M. I. McCormick, and P. L. Munday. 2008. Parental condition affects early life-history of a coral reef fish. Journal of Experimental Marine Biology and Ecology 360:109–116.

Fitzhugh, G. R., K. W. Shertzer, G. T. Kellison, and D. M. Wyanski. 2012. Review of size- and age-dependence in batch spawning: implications for stock assessment of fish species exhibiting indeterminate fecundity. U.S. National Marine Fisheries Service Fishery Bulletin 110:413–425.

Foucher, R., and R. Beamish. 1980. Production of nonviable oocytes by Pacific Hake (*Merluccius productus*). Canadian Journal of Fisheries and Aquatic Sciences 37:41–48.

Ganias, K. 2012. Thirty years of using the postovulatory follicles method: overview, problems and alternatives. Fisheries Research 117–118:63–74.

Ganias, K., C. Nunes, T. Vavalidis, M. Rakka, and Y. Stratoudakis. 2011. Estimating oocyte growth rate and its potential relationship to spawning frequency in teleosts with indeterminate fecundity. Marine and Coastal Fisheries: Dynamics, Management, and Ecosystem Science [online serial] 3:119–126.

Ganias, K., S. Somarakis, A. Machias, and A. Theodorou. 2004. Pattern of oocyte development and batch fecundity in the Mediterranean Sardine. Fisheries Research 67:13–24.

Gladstone, W. 1994. Lek-like spawning, parental care and mating periodicity of the Triggerfish *Pseudobalistes flavimarginatus* (Balistidae). Environmental Biology of Fishes 39:249–257.

Gordo, L., A. Costa, P. Abaunza, P. Lucio, A. T. G. W. Eltink, and I. Figueiredo. 2008. Determinate versus indeterminate fecundity in Horse Mackerel. Fisheries Research 89:181–185.

Greer-Walker, M., P. R. Witthames, and I. Bautista de los Santos. 1994. Is the fecundity of the Atlantic Mackerel (*Scomber scombrus*: Scombridae) determinate? Sarsia 79:13–26.

Harris, P. J., D. M. Wyanski, D. B. White, and J. L. Moore. 2002. Age, growth, and reproduction of Scamp, *Mycteroperca phenax*, in the southwestern North Atlantic, 1979–1997. Bulletin of Marine Science 70:113–132.

Hoese, H., and R. H. Moore. 1998. Fishes of the Gulf of Mexico, 2nd edition. Texas A&M University Press, College Station.

Hood, P. B., and A. K. Johnson. 1997. A study of the age structure, growth, maturity schedules and fecundity of Gray Triggerfish (*Balistes capriscus*), Red Porgy (*Pagrus pagrus*), and Vermilion Snapper (*Rhomboplites aurorubens*) from the eastern Gulf of Mexico. Marine Fisheries Initiative, Final Report FO499-95-F, St. Petersburg, Florida.

Htun-Hun, M. 1978. The reproductive biology of the Dab *Limanda limanda* (L.) in the North Sea: gonadosomatic index, hepatosomatic index, and condition factor. Journal of Fish Biology 13:369–378.

Hunter, J. R., B. J. Macewicz, N. C. H. Lo, and C. A. Kimbrell. 1992. Fecundity, spawning, and maturity of female Dover Sole *Microstomus pacificus*, with an evaluation of assumptions and precision. U.S. National Marine Fisheries Service Fishery Bulletin 90:101–128.

Ingram, G. W. 2001. Stock structure of Gray Triggerfish, *Balistes capriscus*, on multiple spatial scales in the Gulf of Mexico. Doctoral dissertation. University of South Alabama, Mobile.

Jackson, M. W., D. L. Nieland, and J. H. Cowan. 2006. Diel spawning periodicity of Red Snapper *Lutjanus campechanus* in the northern Gulf of Mexico. Journal of Fish Biology 68:695–706.

Jager, H. I., K. A. Rose, and A. Vila-Gispert. 2008. Life history correlates and extinction risk of capital-breeding fishes. Hydrobiologia 602:15–25.

Jönsson, K. I. 1997. Capital and income breeding as alternative tactics of resource use in reproduction. Oikos 78:57–66.

Kawase, H. 2003. Spawning behavior and biparental egg care of the Crosshatch Triggerfish, *Xanthichthys mento* (Balistidae). Environmental Biology of Fishes 66:211–219.

Kjesbu, O. S., P. R. Witthames, P. Solemdal, and M. Greer-Walker. 1998. Temporal variations in the fecundity of Arcto-Norwegian cod (*Gadus morhua*) in response to natural changes in food and temperature. Journal of Sea Research 40:303–321.

Kuwamura, T. 1997. Evolution of female egg care in haremic triggerfish, *Rhinecanthus aculeatus*. Ethology 103:1015–1023.

Lambert, Y. 2008. Why should we closely monitor fecundity in marine fish populations? Journal of Northwest Atlantic Fishery Science 41:93–106.

Lowerre-Barbieri, S. K., N. J. Brown-Peterson, H. Murua, J. Tomkiewicz, D. M. Wyanski, and F. Saborido-Rey. 2011. Emerging issues and methodological advances in fisheries reproductive biology. Marine and Coastal Fisheries: Dynamics, Management, and Ecosystem Science [online serial] 3:71–91.

MacKichan, C. A., and S. T. Szedlmayer. 2007. Reproductive behavior of the Gray Triggerfish, *Balistes capriscus*, in the northeastern Gulf of Mexico. Proceedings of the Gulf and Caribbean Fish Institute 59:214–217.

MacKichan-Simmons, C. A. 2008. Gray Triggerfish, *Balistes capriscus*, reproductive behavior, early life history, and competitive interactions between Red Snapper, *Lutjanus campechanus*, in the northern Gulf of Mexico. Doctoral dissertation. Auburn University, Auburn, Alabama.

Manooch, C. S. III, and D. Raver Jr. 1984. Fisherman's guide: fishes of the southeastern United States. North Carolina State Museum of Natural History, Raleigh.

McBride, R. S., S. Somarakis, G. R. Fitzhugh, A. Albert, N. A. Yaragina, M. J. Wuenschel, A. Alonso-Fernández, and G. Bailone. 2015. Energy acquisition and allocation to egg production in relation to fish reproductive strategies. Fish and Fisheries 16:23–57.

Moore, J. L. 2001. Age, growth and reproductive biology of the Gray Triggerfish (*Balistes capriscus*) from the southeastern United States, 1992–1997. Master's thesis. University of Charleston, Charleston, South Carolina.

Murua, H., and F. Saborido-Rey. 2003. Female reproductive strategies of marine fish species of the North Atlantic. Journal of Northwest Atlantic Fisheries Science 33:23–31.

Nakazono, A. 1993. One-parent removal experiment in the brood-caring damselfish, *Acanthochromis polyacanthus*, with preliminary data on reproductive biology. Australian Journal of Marine and Freshwater Research 44:699–707.

Ofori-Danson, P. 1990. Reproductive ecology of the triggerfish, *Balistes capriscus*, from the Ghanaian coastal waters. Tropical Ecology 31:1–11.

Parker, K. 1980. A direct method for estimating Northern Anchovy, *Engraulis mordax*, spawning biomass. U.S. National Marine Fisheries Service Bulletin 78:541–544.

Paulik, G. J. 1973. Studies of the possible form of the stock-recruitment curve. Rapports et Proces-Verbaux des Reunions du Conseil International pour l'Exploration de la Mer 164:302–315.

Rabalais, N. N., S. C. Rabalais, and C. R. Arnold. 1980. Description of eggs and larvae of laboratory reared Red Snapper (*Lutjanus campechanus*). Copeia 1980:704–708.

Ricker, W. E. 1975. Computation and interpretation of biological statistics of fish populations. Fisheries Research Board of Canada Bulletin 191.

Rinchard, J., and P. Kestemont. 2003. Liver changes related to oocyte growth in Roach, a single spawner fish, and in Bleak and White Bream, two multiple spawner fish. International Review of Hydrobiology 88:68–76.

Santos, R. N., S. Amadio, and E. J. G. Ferreira. 2010. Patterns of energy allocation to reproduction in three Amazonian fish species. Neotropical Ichthyology 8:155–162.

SEDAR (Southeast Data Assessment and Review) 09. 2006. Stock assessment report of Gray Triggerfish in the Gulf. SEDAR, Charleston, South Carolina.

Simmons, C. M., and S. T. Szedlmayer. 2012. Territoriality, reproductive behavior, and parental care in Gray Triggerfish, *Balistes capriscus*, from the northern Gulf of Mexico. Bulletin of Marine Science 88:197–209.

Stearns, S. C. 1992. The evolution of life histories. Oxford University Press, Oxford, UK.

Stratoudakis, Y., M. Bernal, K. Ganias, and A. Uriarte. 2006. The daily egg production method (DEPM): recent advances, current applications and future challenges. Fish and Fisheries 7:35–57.

Tomkiewicz, J., M. J. Morgan, J. Burnett, and F. Saborido-Rey. 2003. Available information for estimating reproductive potential of northwest Atlantic groundfish stocks. Journal of Northwest Atlantic Fishery Science 33:1–21.

Wallace, R. A., and K. Selman. 1981. Cellular and dynamic aspects of oocyte growth in teleosts. American Zoologist 21:325–343.

Wilson, C. A., D. L. Nieland, and A. L. Stanley. 1995. Age, growth, and reproductive biology of Gray Triggerfish (*Balistes capriscus*) from the northern Gulf of Mexico commercial harvest. Marine Fisheries Initiative, Final Report 8, St. Petersburg, Florida.

Witthames, P. R., and M. Greer-Walker. 1995. Determinacy of fecundity and oocyte atresia in Sole (*Solea solea*) from the channel, the North Sea and the Irish Sea. Aquatic Living Resources 8:91–109.

Witthames, P. R., A. Thorson, H. Murua, F. Saborido-Rey, L. N. Greenwood, R. Domínguez, M. Korta, and O. S. Kjesbu. 2009. Advances in methods for determining fecundity: application of the new methods to some marine fishes. U.S. National Marine Fisheries Service Fishery Bulletin 107:148–164.

Yoneda, M., M. Tokimura, H. Fujita, N. Takeshita, K. Takeshita, M. Matsuyama, and S. Matsuura. 1998. Ovarian structure and batch fecundity in *Lophiomus setigerus*. Journal of Fish Biology 52:94–106.

Yoneda, M., M. Tokimura, H. Fujita, N. Takeshita, K. Takeshita, M. Matsuyama, and S. Matsuura. 2001. Reproductive cycle, fecundity, and seasonal distribution of the Anglerfish *Lophius litulon* in the East China and Yellow seas. U.S. National Marine Fisheries Service Fishery Bulletin 99:356–370.

3

Quantifying Delayed Mortality from Barotrauma Impairment in Discarded Red Snapper Using Acoustic Telemetry

Judson M. Curtis* and Matthew W. Johnson[1]
Harte Research Institute for Gulf of Mexico Studies, Texas A&M University–Corpus Christi,
6300 Ocean Drive, Corpus Christi, Texas 78412-5869, USA

Sandra L. Diamond
Department of Biology, Texas Tech University, Lubbock, Texas 79409, USA; and
School of Science and Health, Hawkesbury Campus, University of Western Sydney, Locked Bag 1797,
Penrith, NSW 2751, Australia

Gregory W. Stunz
Harte Research Institute for Gulf of Mexico Studies, Texas A&M University–Corpus Christi,
6300 Ocean Drive, Corpus Christi, Texas 78412-5869, USA

Abstract
Red Snapper *Lutjanus campechanus* is the most economically important reef fish in the Gulf of Mexico, and despite being intensively managed, the stock remains overfished. These fish are susceptible to pressure-related injuries (i.e., barotrauma) during fishing that compromise survival after catch and release. Barotrauma-afflicted fish may not only experience immediate mortality but also delayed mortality after returning to depth. This variability and unknown fate leads to uncertainty in stock assessment models and rebuilding plans. To generate better estimates of immediate and delayed mortality and postrelease behavior, Red Snapper were tagged with ultrasonic acoustic transmitters fitted with acceleration and depth sensors. Unique behavior profiles were generated for each fish using these sensor data that allowed the classification of survival and delayed mortality events. Using this information, we compared the survival of Red Snapper released using venting, nonventing, and descending treatments over three seasons and two depths. Red Snapper survival was highest at cooler temperatures and shallower depths. Fish released using venting and descender tools had similar survival, and both these groups of fish had higher survival than nonvented surface-released fish. Overall, Red Snapper had 72% survival, 15% immediate mortality, and 13% delayed mortality, and all fish suffering from delayed mortality perished within a 72-h period after release. Results from these field studies enhance the understanding of the delayed mortality and postrelease fate of Red Snapper regulatory discards. Moreover, these data support the practice of using venting or descender devices to increase the survival of discarded Red Snapper in the recreational fishery and show that acoustic telemetry can be a valuable tool in estimating delayed mortality.

Subject editor: Don Noakes, Vancouver Island University, Nanaimo, British Columbia

*Corresponding author: judd.curtis@tamucc.edu
[1]Present address: Bureau of Ocean Energy Management, 1201 Elmwood Park Boulevard, New Orleans, Louisiana 70123, USA.

The success of catch-and-release fishing as a management tool is predicated upon the assumption that discarded fish will survive (Bartholomew and Bohnsack 2005; Cooke and Suski 2005; Arlinghaus et al. 2007). Many offshore reef fish species in deepwater environments routinely experience barotrauma when brought rapidly to the surface during fishing and, consequently, suffer an increased risk of discard mortality in catch-and-release fisheries (Rummer 2007). Certainly, the development of techniques that avoid or minimize injury or mortality associated with barotrauma has the potential to improve the management and recovery timelines for many reef fish species.

Gulf of Mexico Red Snapper *Lutjanus campechanus* is an ideal model species on which to test methods to reduce barotrauma-related injuries. This species commonly experiences severe barotrauma (Rummer and Bennett 2005), and a large proportion of the total catch may be discarded (Dorf 2003; Campbell et al. 2013). Red Snapper is considered the most economically important reef fish species in the Gulf of Mexico and has been heavily managed since the fishery was first classified as overfished in 1988 (Goodyear 1988; Hood et al. 2007). Management strategies enacted by the Gulf of Mexico Fishery Management Council (GMFMC) for the recreational fishery have included reducing bag limits, shortening fishing seasons, and setting minimum size limits with the goal of reducing fishing pressure and allowing stocks to rebound (see Hood et al. 2007 for comprehensive fishery management history). However, with the stock not yet fully rebuilt and almost two decades remaining in the rebuilding phase, management strategies have become increasingly strict and more controversial (Cowan et al. 2010). An unintended consequence of these tightened regulations has been an increase in the frequency of "regulatory discards" —fish that are required by law to be released because they do not meet size, season, or bag requirements.

Minimizing death after release is a common aim for fishery managers. One management strategy enacted by the GMFMC to increase survival in reef fish was a requirement to vent the swim bladder prior to release (GMFMC 2007). More recently, there has been some skepticism over the efficacy of venting in reducing discard mortality (Wilde 2009; Scyphers et al. 2013; Campbell et al. 2014), and studies specific to Red Snapper have shown positive (Gitschlag and Renaud 1994), neutral (Render and Wilson 1994, 1996), and negative (Burns et al. 2002) effects of venting on survival. An alternative to venting, and potentially a more effective release method, is rapid recompression using descender devices. This technique involves rapidly descending the fish back to depth on a weighted line prior to release to rapidly recompress the swim bladder and alleviate any barotrauma symptoms without having to vent the fish. Additionally, this method also avoids releasing the fish at the surface, where increased risk of predation exists (Burns et al. 2004). The venting regulation has since been rescinded (GMFMC 2013), which allows for the use of descender devices; however, the efficacy of these devices in reducing discard mortality in the Gulf of Mexico Red Snapper fishery warrants further research.

The results of studies quantifying discard mortality in the Gulf of Mexico recreational Red Snapper fishery remain highly variable—the latest estimate of discard mortality from a meta-analysis of studies ranges from 0% to 91% (Campbell et al. 2013). This large variability is influenced by multiple factors, including season, fishery sector, geographical region, and water depth, and is further convoluted by interactions among these factors (Gingerich et al. 2007). Moreover, the majority of these studies have only assessed immediate discard mortality, or mortality that is observed from surface observations within several seconds postrelease, while delayed mortality is unknown. Although Red Snapper that are capable of resubmerging unassisted after catch and release are presumed to survive, this assumption is largely untested, and there is evidence that the ability to swim away is unrelated to survival (Bettoli and Osborne 1998; St John and Syers 2005; Diamond and Campbell 2009). A substantial proportion of fish may undergo delayed mortality hours to several days after a supposed successful release (Rummer and Bennett 2005). Studies attempting to estimate delayed mortality have used field caging experiments (Gitschlag and Renaud 1994; Render and Wilson 1994; Diamond and Campbell 2009; Roach et al. 2011) or in-laboratory hyperbaric chamber simulations (Rummer and Bennett 2005; Burns 2009; Drumhiller et al. 2014). While considerable success has been achieved using these methods, these study designs do not allow for tracking postrelease survival over longer time periods and they have an inherent bias because they exclude predatory effects, prevent foraging, and restrict natural movement (Campbell et al. 2013).

One method to alleviate the artifact biases associated with estimating delayed mortality using passive tagging or cage studies is through the use of ultrasonic acoustic telemetry (Campbell et al. 2014). This technique has already been extremely successful tracking the movements, long-term residency, and site fidelity of Red Snapper (Szedlmayer and Schroepfer 2005; Peabody and Wilson 2006; Westmeyer et al. 2007; Topping and Szedlmayer 2011a) but has not yet been used to quantify discard mortality in the recreational fishery. Transmitters equipped with accelerometer and depth sensors allow researchers to monitor the postrelease survival and behavior of fish. For fish experiencing barotrauma, these tags can provide information on presence or absence, mortality (no acceleration), postrelease depth preference, and activity level compared with fish not experiencing barotrauma. There have been no published tagging studies that used these advanced acoustic tags to examine the physiological responses of Red Snapper, particularly as they relate to regulatory discards and examining delayed mortality. Using this tagging methodology not only allows us to avoid cage artifacts but also to replicate postrelease fishing practices most reflective of the actual fishery and approximate the most natural behavioral characteristics of the fish.

The primary goal of this study was to quantify the extent of immediate and delayed mortality due to barotrauma impairment in the Red Snapper recreational fishery using surface observations and acoustic telemetry. Specifically, we tested (1) whether certain release treatments are more favorable for increasing survival after catch and release and if using descender devices or venting tools are a better alternative to not venting, (2) whether the season of capture associated with differences in water temperatures and the presence of thermoclines influences survival, and (3) if the depth of capture influences survival. This study will help managers better understand how delayed mortality may factor into overall discard mortality estimates and determine which release strategies maximize the chances of survival for Red Snapper discarded by recreational anglers.

METHODS

Release treatments.—Four standing oil and gas platforms approximately 50 km east of Port Aransas, Texas, were selected as study sites for these experiments (Figure 1). Sites MU-762-A and MU-759-A (approximately 27°45′N,

96°35′W) reside at a water depth of 50 m and sites MI-685-B and MI-685-C (approximately 27°55′N, 96°35′W) at a water depth of 30 m. Prior to sampling, fish were randomly assigned to one of four release treatments: (1) vented surface release, (2) nonvented surface release, (3) descended bottom release, and (4) control (no barotrauma). Surface-released fish were released into an open-bottom 1.0-m³ holding cage with mesh walls to protect fish from predation and enable retrieval of fish (and transmitters) that experienced immediate mortality at the surface. The number of immediate surface mortalities after catch and release was recorded for each trial and incorporated into the analyses. Vented surface-released fish were punctured in the abdomen posterior to the pectoral fin using a venting tool (Team Marine USA prevent fish venting tool), tagged, and released at the surface. Descended bottom-released fish were not vented prior to tagging but, instead of being released at the surface, were forced back to depth quickly using a weighted line with an inverted barbless hook (Shelton Fish Descender) attached to the fish's jaw. Once at the seafloor, fish were released with a slight upward pull of the line to release the hook from the jaw.

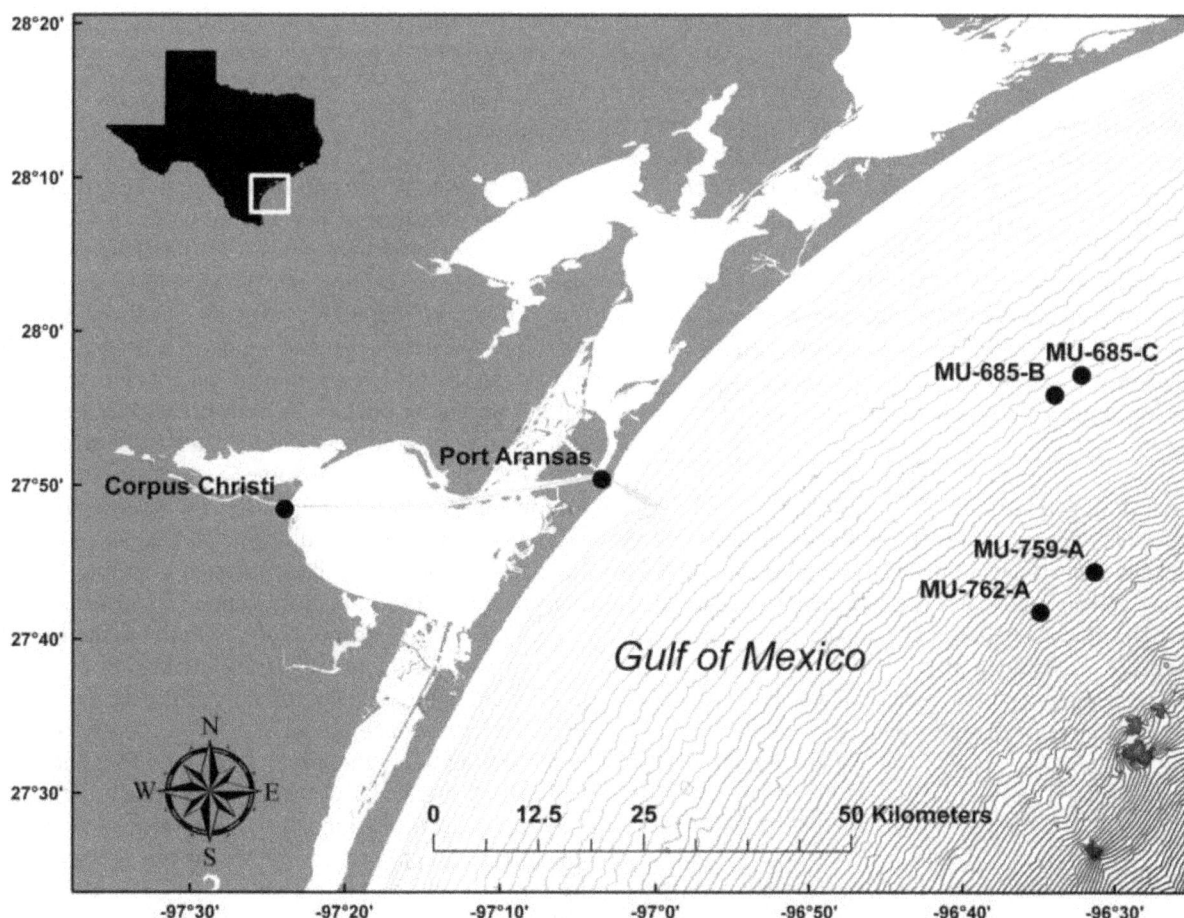

FIGURE 1. Study sites (standing oil and gas platforms) in the Gulf of Mexico off the southern Texas coast, where field tagging experiments occurred. Sites MU-685-B and MU-685-C reside at water depths of 30 m and sites MU-762-A and MU-759-A at 50 m.

Control fish showed no evidence of barotrauma prior to tagging and release. To achieve this, fish were captured using single hook and line at the 30-m platforms prior to experimental trials, transported to the Texas A&M AgriLife Research Mariculture Laboratory in Port Aransas, Texas, and held in 6.4-m^3 tanks. Fish were treated for parasites using copper (II) sulfate and were fed three times weekly to satiation with a diet of squid *Loligo* sp. and sardines *Sardinella* sp. Fish recovered and began feeding quickly (typically within 24 h), and the condition and behavior of these fish were closely monitored. After a 3-week holding period, fish appeared healthy and acclimated to surface pressure, and it was assumed that any effects of barotrauma from capture had healed. Fish were then transported in oxygenated live wells to the study sites, where they were tagged and released along with the fish assigned to the other release treatments in randomized order.

Fish tagging.—Red Snapper were captured from the seafloor at each site by experienced anglers using a rod and reel equipped with 6/0 Lazer Sharp circle hooks baited with squid, scad *Trachurus* sp., or sardines. This gear type and bottom fishing strategy are the standard fishing practices used in the recreational Red Snapper fishery. Fish were measured for maximum total length (mm) and assessed (presence or absence) for six externally visible barotrauma symptoms: everted stomach, swollen and hard abdomen, exophthalmia (eyes forced from orbits), distended intestines, subcutaneous gas bubbles, and bleeding from the gills. A barotrauma impairment score (scale: 0–1) was calculated by summing the number of visible symptoms divided by six—the total number of possible symptoms (Diamond and Campbell 2009). All fish that were tagged had been captured by hooking in the mouth. Fish that appeared obviously moribund or deceased after capture were not tagged in order to eliminate the possibility of hook-induced mortality from our study. The focal point of this study was to examine only barotrauma effects on discard and delayed mortality; thus, we controlled for the effects of mortality from other causes, such as hook mortality, by purposefully selecting fish for which barotrauma was the only evident stressor. Significant differences in total length and barotrauma impairment among release treatments were tested using an analysis of variance (ANOVA; $\alpha = 0.05$).

Red Snapper were externally tagged with Vemco V9AP ultrasonic coded transmitters (V9AP-2H; 46×9 mm; 69 kHz; random delay interval: 30–90 s; estimated battery life: 45 d) containing built-in acceleration and pressure (i.e., depth) sensors. To measure acceleration, the V9AP tags calculate a value (m/s^2) that represents the root mean square acceleration on three axes (*X*, *Y*, and *Z*) averaged over a fixed time interval:

$$\text{m/s}^2 = \sqrt{x^2 + y^2 + z^2} \text{ averaged over time (T).} \quad (1)$$

Depth was calculated by an algorithm that converts pressure sensors to a depth value (maximum depth = 100 m). Because one goal of our study was to explore survival under a variety of release treatments, fish were rapidly (<3 min)

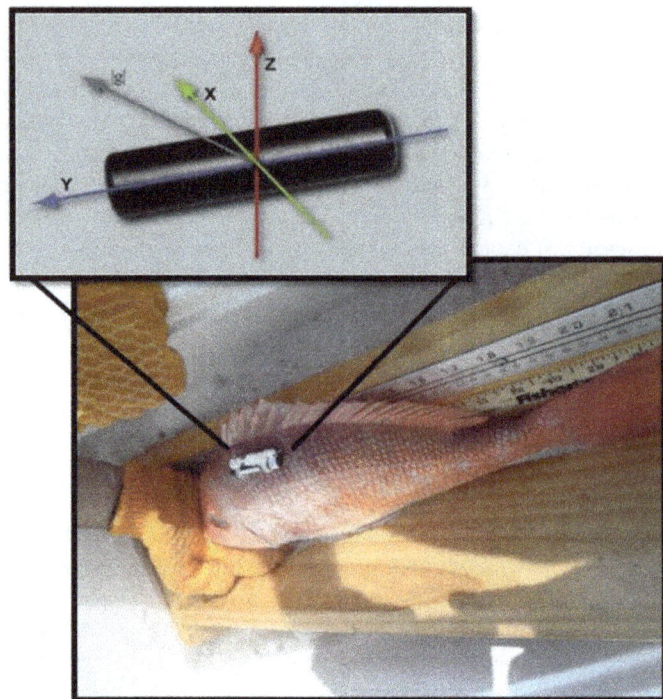

FIGURE 2. Acoustic transmitters were externally attached to prevent the unavoidable venting associated with internal tag implantation methods (lower panel). Vemco V9AP accelerometer tags measure the animal's acceleration signal (|g|) along three axes (*X*, *Y*, and *Z*) averaged over a fixed time interval (upper panel).

tagged externally without anesthesia to best replicate normal catch-and-release practices and minimize artifacts associated with tagging-related surgeries. One challenge was to prevent the unavoidable venting that is associated with the incision and suture procedures of traditional internal tag implantation; therefore, we developed and validated a protocol to attach tags to fish externally. Tags were positioned below the anterior dorsal spines approximately 2–3 cm below the dorsal edge, and fish were punctured between the 2nd and 3rd pterygiophores below the anterior dorsal spines using a sterile stainless steel hollow surgical needle. A plastic cinch-up external Floy tag was passed through one hollow needle, attached to the acoustic transmitter, and passed back through a second hollow needle between the 4th and 5th pterygiophores and secured so that the orientation of the transmitter was parallel to the fish and on the opposite side of the point of attachment (Figure 2). Fish were held in a tagging cradle with their gills submerged in oxygenated water to reduce potential injury or stress from emersion while still allowing the fish to ventilate during the tagging procedure. An externally visible dart tag containing identification and reward information was also inserted into the posterior dorsal spine region in the event that the fish were recaptured by anglers. During preliminary trials ($n = 20$), tag presence did not impair fish behavior and tag retention using our external attachment method was 100% for at least 20 d after the fish were released (Johnson et al. 2015).

Experimental design.—Three tagging trials occurred during winter 2010, summer 2010, and spring 2011. Winter and summer trials were performed at a water depth of 50 m on site MU-762-A. Twenty fish were tagged and released on site during each season using one of three release treatments: control, nonvented surface release, or descended bottom release. However, because of the repeated inability of nonvented fish to resubmerge during the summer trial, we added a vented surface release treatment. We subsequently included this treatment in our spring trial and incorporated a second depth into the experimental design to test for differences between capture depths of 30 and 50 m. Thirty-two fish were tagged at each depth, with all four release treatments included. Two Vemco VR2W-69kHz acoustic monitoring receivers were attached to platform crossbeams by scuba divers at each study site. Receivers were placed at depths of approximately 20 and 30 m for 50-m sites and at 15 and 25 m for 30-m sites. The detection range of VR2W receivers in this environment from previous studies conducted by our research group (authors' unpublished data) and other studies (Topping and Szedlmayer 2011a; Kessel et al. 2014) shows that after 500 m there is a substantial drop-off in detection efficiency. Therefore, we assumed a maximum detection range of 500 m for this study, which, combined with the known high site fidelity of Red Snapper (Szedlmayer and Schroepfer 2005; Westmeyer et al. 2007; Topping and Szedlmayer 2011a), ensured that we were able to detect tagged fish that remained on site for the duration of our experiment. During each sampling event, we measured water temperature, salinity, dissolved oxygen, and conductivity using a Manta2 water quality multiprobe (Eureka Environmental Engineering). Hourly sea surface temperatures for 10 d after tagging were obtained from the National Oceanic and Atmospheric Administration–National Data Buoy Center station 42020 (26°58′N, 96°42′W). Significant differences in sea surface temperatures among seasons were tested using an ANOVA ($\alpha = 0.05$).

Fate classification.—The VR2W receivers were retrieved from the study sites after approximately 60 d, and data were uploaded to Vemco VUE software and exported for analysis to R version 3.0.2 (R Development Core Team 2013). Acceleration and depth profiles for each fish were plotted over time using tag sensor data. Using these unique acoustic profiles along with surface observations, the fate of each individual was classified into one of four categories: survival, surface mortality, delayed mortality, or unknown. Fish experiencing mortality at the surface were retrieved and transmitters reused. These surface mortalities did not yield an acoustic profile but were counted towards estimates of total overall mortality; therefore, mortality equaled the sum of immediate mortality witnessed by surface observations plus delayed mortality as indicated by acoustic returns. Fish that did not register sufficient detections (≤ 5 pings) were classified as unknown because it was not possible to classify these events as either survival or delayed mortality. These fish were omitted from subsequent analyses, which reduced the sample size; however, we wanted to be certain the fate of the fish was accurately assigned. Fish classified as survivors exhibited active acoustic profiles with frequent bursts in acceleration and changes in depth. These included both resident fish that remained on site continuously for the duration of the tag life and fish that were determined to have emigrated from the array (Heupel and Simpfendorfer 2002). Emigrants showed similar active acceleration and depth profiles before sudden cessation of detections. Delayed mortality events were classified by initially active acceleration and depth movements followed by a sudden drop-off to zero acceleration and depth equal to the seafloor within 3 d.

Survival analysis.—Percent survival was calculated using the binomial distribution for two outcomes: survival and mortality. Survival estimates ($\hat{S}$) were calculated following equations in Pollock and Pine (2007):

$$\hat{S} = \frac{x}{n}, \qquad (2)$$

with a standard error of

$$\text{SE}(\hat{S}) = \sqrt{\frac{\hat{S}(1 - \hat{S})}{n}}, \qquad (3)$$

where x is the number of survivors and n is the total number of tagged fish minus the fish classified as unknown (i.e., $n =$ survivors + surface mortalities + delayed mortalities). Percent survival among release treatments at 50 m was compared among the three seasons, and the effect of capture depth at 30 m versus 50 m on the fate of discarded Red Snapper was compared during the spring.

The Cox proportional hazards model (Cox 1972), built into the "survival" package in R (Therneau and Grambsch 2000), was used to examine the relationship between survival and multiple explanatory variables (Sauls 2014). The Cox model is a semiparametric regression method for survival data. It provides an estimate of the treatment effect on survival after adjustment for other covariates in the model and gives an estimation of the hazard ratio (in this case the proportional risk of death) among levels within each of these explanatory variables. For survival analysis, this method is advantageous over logistic regression models because it can account for survival times and censored data, whereas regression models do not. Additionally, hazard ratios between covariates may be estimated without needing to specify the underlying baseline hazard, which may not be known. The Cox proportional hazards model is given by the following:

$$h(t) = h_0(t) \exp \left(\sum_{i=1}^{p} \beta_i X_i \right), \qquad (4)$$

where $h_0(t)$ is an unspecified function representing the baseline hazard, β_i is the regression coefficients, and X_i is the explanatory variables or covariates in the model. A stepwise logistic regression using Akaike information criteria values was performed to determine which covariates to include in the Cox proportional hazards model.

RESULTS

Fish Tagging

A total of 111 Red Snapper ranging from 280 to 651 mm total length (mean $\pm$ SE $= 446 \pm 8$ mm) were captured and tagged over three seasonal trials. No significant differences existed in total length among release treatments (ANOVA: $F_{3, 106} = 2.13$, $P = 0.10$). Fish released under vented, nonvented, and descended release treatments had a mean $\pm$ SE barotrauma impairment score of 0.32 ± 0.02 and were not significantly different (ANOVA: $F_{2, 89} = 0.41$, $P = 0.66$). All control treatment fish had a barotrauma impairment score of 0 at the time of release.

Temperature was plotted against depth using the observed hydrographic water data to determine if thermoclines in the water column were present and at what depths they occurred (Figure 3). Winter 2010 had a thoroughly mixed water column at a constant temperature of 24°C. Water temperatures from 22°C to 31°C occurred in the summer 2010 profile, with a steep thermocline observed beginning at 25 m and continuing to the seafloor. Spring 2011 had a temperature range of 3.5°C (23.5°C at the surface to 20.0°C at the seafloor), with a thermocline beginning at a depth of 20 m. Mean sea surface temperatures during the first 10 trial days for each season were significantly different (ANOVA: $F_{2, 716} = 5,102$, $P < 0.001$; Figure 4). Winter temperature was relatively constant over 10 d and averaged 23.0 ± 0.4°C (mean $\pm$ SD). Summer temperatures also remained constant for 10 d and averaged 30.5 ± 0.4°C. In the spring, temperature had a slight increasing trend over 10 d and averaged 25.0 ± 0.7°C.

Fate Classification

The classification fates from all trials are presented by season, depth, and release method (Table 1). In the spring season when multiple sites were included in the experimental design, there were no site-to-site differences; therefore, the two 30-m and two 50-m sites were each pooled together. Surface mortalities ($n = 13$) were immediate and were caused by the inability to resubmerge unassisted, typically because of overly positive buoyancy from gas expansion in the swim bladder in nonvented fish. Sixty-two fish survived and exhibited active acceleration and depth profiles. Survivors included both residents that remained on site continuously (Figure 5A) and emigrants that left the array (Figure 5B). There were 11 fish that experienced delayed mortality (Figure 5C), and 25 fish were

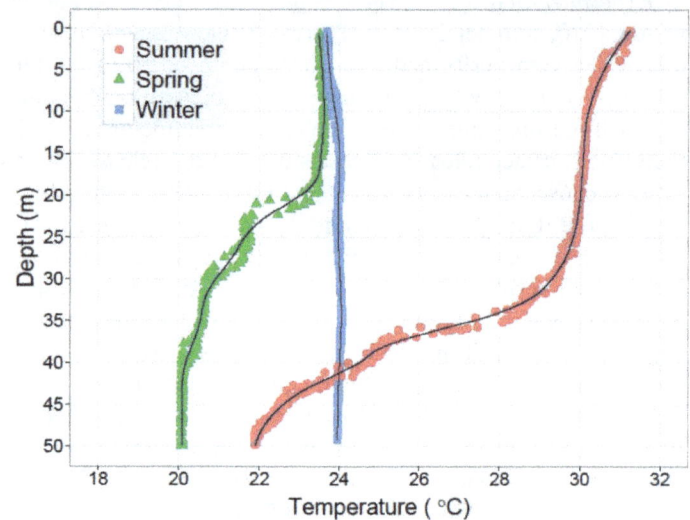

FIGURE 3. Temperature versus depth data collected on the day of tagging for each seasonal trial using a Manta2 water quality multiprobe at oil and gas platform site MU-762-A (water depth of 50 m). The black smoothing lines were fitted to the temperature data using a Loess model.

classified as fate unknown (Figure 5D). To examine the time elapsed to a delayed mortality event, the acceleration and depth acoustic profiles of all fish classified as suffering delayed mortality were plotted over time. By approximately the third day, all fish showed acceleration values of 0 and a depth equivalent to the bottom depth (Figure 6). After this time period

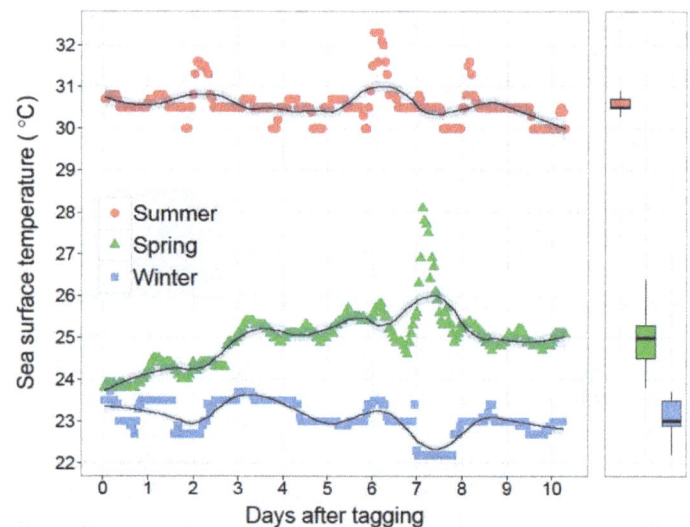

FIGURE 4. Sea surface temperatures during the first 10 d fish were at liberty for three seasonal tagging trials. The data was obtained from the National Oceanic and Atmospheric Administration–National Data Buoy Center buoy 42020 (26°58′N, 96°42′W). The boxplots show the distribution of temperature data for each season; the thick horizontal line in each box indicates the median, the box dimensions represent the 25th–75th percentiles, and the error bars represent the 10th and 90th percentiles. The black smoothing lines were fitted to the temperature data using a Loess model.

TABLE 1. Summary of the results of Red Snapper experimental trials. Tagged indicates the number of fish tagged and released, including those that perished on the surface, while fate unknown indicates fish whose fate was unclassifiable as a survivor or a mortality. Sample size (n) equals the number of fish tagged minus those whose fate was unknown. Surface mortality indicates fish that perished at the surface, and delayed mortality indicates fish that exhibited delayed mortality (perished in < 3 d). Survivor indicates fish that exhibited long-term (> 3-d) survival. The survival estimate ($\hat{S}$) is calculated from equation (2), with the standard error (SE) of the survival estimate calculated from equation (3). Note that "n/a" denotes that the vent treatment was not performed in the winter season.

Trial and total	Tagged	Fate unknown	n	Surface mortality	Delayed mortality	Survivor (x)	$\hat{S}$	SE ($\hat{S}$)
Winter – 50 m	22	4	18	2	2	14	0.78	0.10
Control	4	1	3	0	0	3	1.00	0.00
Descend	8	2	6	0	2	4	0.67	0.19
Nonvent	10	1	9	2	0	7	0.78	0.14
Vent	n/a	n/a	n/a	n/a	n/a	n/a	n/a	n/a
Summer – 50 m	25	5	20	5	4	11	0.55	0.11
Control	3	0	3	0	1	2	0.67	0.27
Descend	9	3	6	0	1	5	0.83	0.15
Nonvent	8	1	7	4	2	1	0.14	0.13
Vent	5	1	4	1	0	3	0.75	0.22
Spring – 50 m	32	9	23	3	5	15	0.65	0.10
Control	6	4	2	0	0	2	1.00	0.00
Descend	8	2	6	0	2	4	0.67	0.19
Nonvent	10	0	10	3	1	6	0.60	0.15
Vent	8	3	5	0	2	3	0.60	0.22
Spring – 30 m	32	7	25	3	0	22	0.88	0.06
Control	6	3	3	0	0	3	1.00	0.00
Descend	7	0	7	0	0	7	1.00	0.00
Nonvent	10	2	8	2	0	6	0.75	0.15
Vent	9	2	7	1	0	6	0.86	0.13
Total	111	25	86	13	11	62	0.72	0.05

elapsed, there were no further delayed mortality events. All delayed mortality events occurred in trials at 50 m; there was no delayed mortality at 30 m.

Survival Analysis

Based on the classifications described from acoustic profiles, survival was compared among release treatments over all seasons and depths ($n = 86$). Survival was highest for control fish, followed in decreasing order by those with a descended bottom release, vented surface release, and nonvented surface release (Figure 7). All release treatments in winter and spring had similar survival; however, fish released nonvented during summer experienced much lower survival than those in vented or descended release treatments (Table 1; Figure 8). Between the two depths in spring, control fish experienced 100% survival, and survival was higher for every experimental release treatment at 30 m (Table 1). Immediate surface mortality was highest in the summer and lowest in the winter (Figure 9). Delayed mortality was higher in the summer and spring at 50 m than in the winter and did not occur in the spring at the 30-m depth. Overall, there was 72% survival, 15% surface

mortality, and 13% delayed mortality for all fish in this study (Table 1).

Stepwise logistic regression using Akaike information criteria values identified release method, season, depth, and total length as significant covariates to be used in the Cox proportional hazards model, and these covariates had a significant effect on survival (Log-rank test: $\chi^2 = 20.98$, df = 7, $P < 0.01$, $n = 86$). Based on the calculated hazards ratio, descended fish were 2.3 times, vented fish 3.7 times, and nonvented fish 6.9 times as likely to perish as control fish; nonvented fish were 3.0 times and 1.9 times more likely to perish than descended and vented fish, respectively; fish released in winter were 1.6 times and in summer 5.0 times as likely to perish as fish released in the spring; and fish released after capture from a 50-m depth were 2.5 times as likely to perish as fish caught at a 30-m depth (Table 2). Decreases in total length resulted in a slightly less risk of mortality; smaller fish had higher chances of survival than larger fish.

DISCUSSION

There has been considerable debate regarding the best release practices for increasing survival in catch-and-release

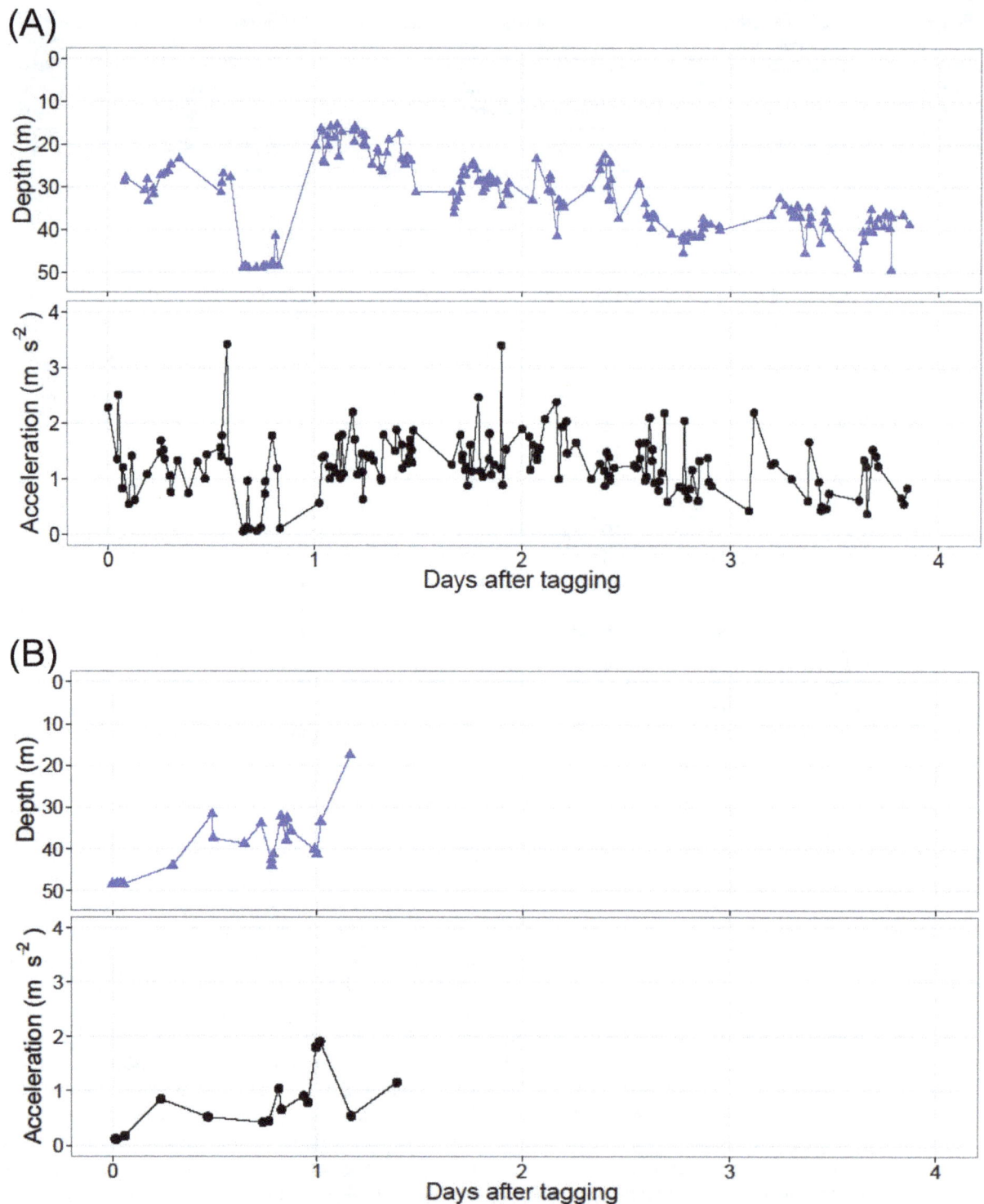

FIGURE 5. Acoustic telemetry depth and acceleration profiles of tagged Red Snapper for 4 d following their catch and release. Each point represents individual acoustic detections, and they are connected by lines for visualization. Triangles represent the depth profile for each fish, and solid dots represent acceleration. (A) Resident survivors showed active acceleration and depth profiles and remained on site continuously, while (B) emigrant survivors exhibited active profiles similar to the resident survivors but left the array. (C) Delayed mortality profiles showed that before 3 d the fish had fallen to the seafloor and perished, showing no further vertical movement or acceleration, and (D) unknown profiles did not contain sufficient data to classify them as either survivor or delayed mortality.

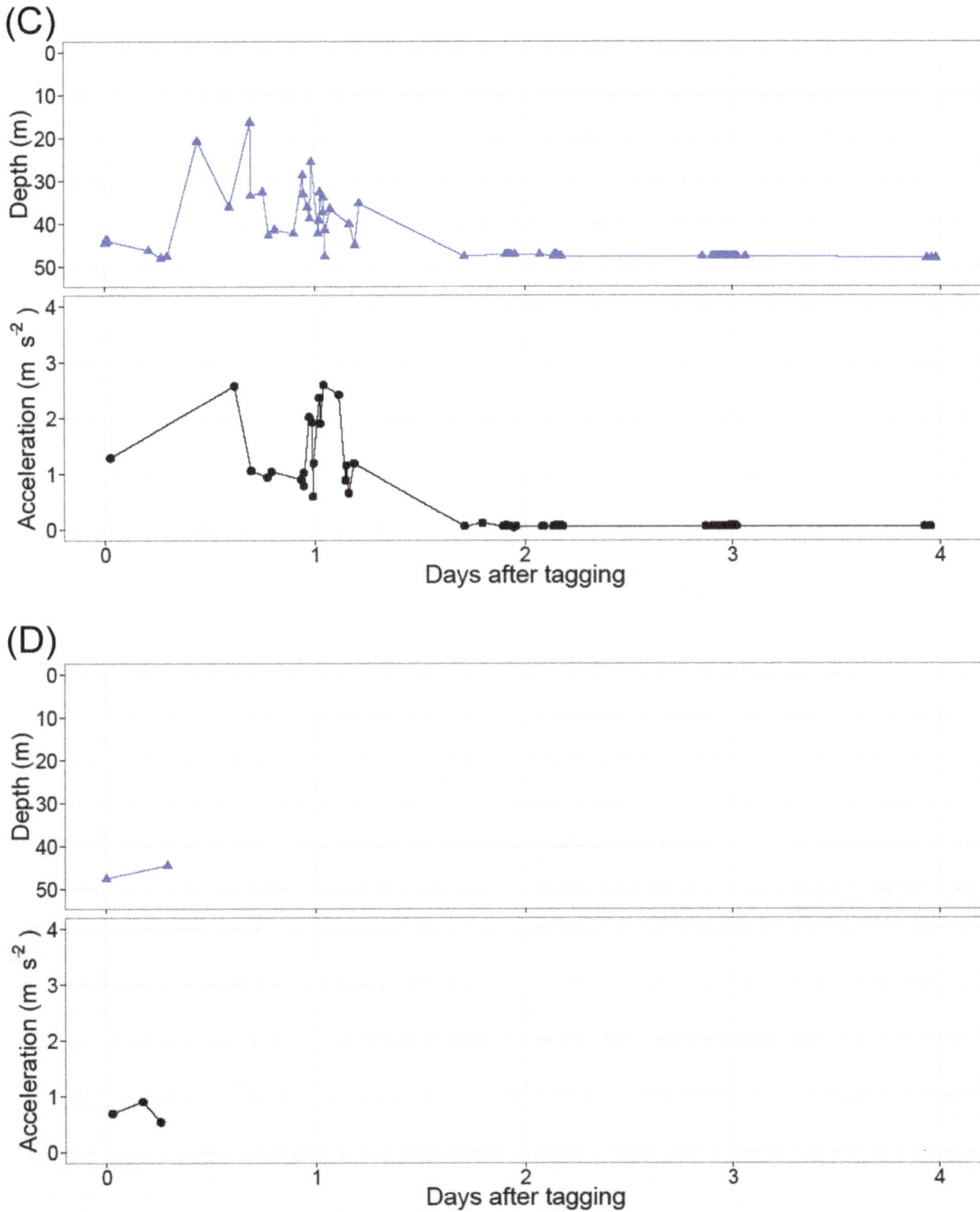

FIGURE 5. Continued.

fisheries, with differing results depending on species, season, depth of capture, angler experience, fish size, and a variety of other factors. For Red Snapper in the Gulf of Mexico, the question of venting or not venting has recently been at the forefront of this debate, with contradictory results among different studies (Wilde 2009) and confusion at the management and regulation level. This uncertainty has subsequently contributed to the GMFMC rescinding the requirement of venting prior to release after establishing this requirement only 5 years prior. Recently, highly

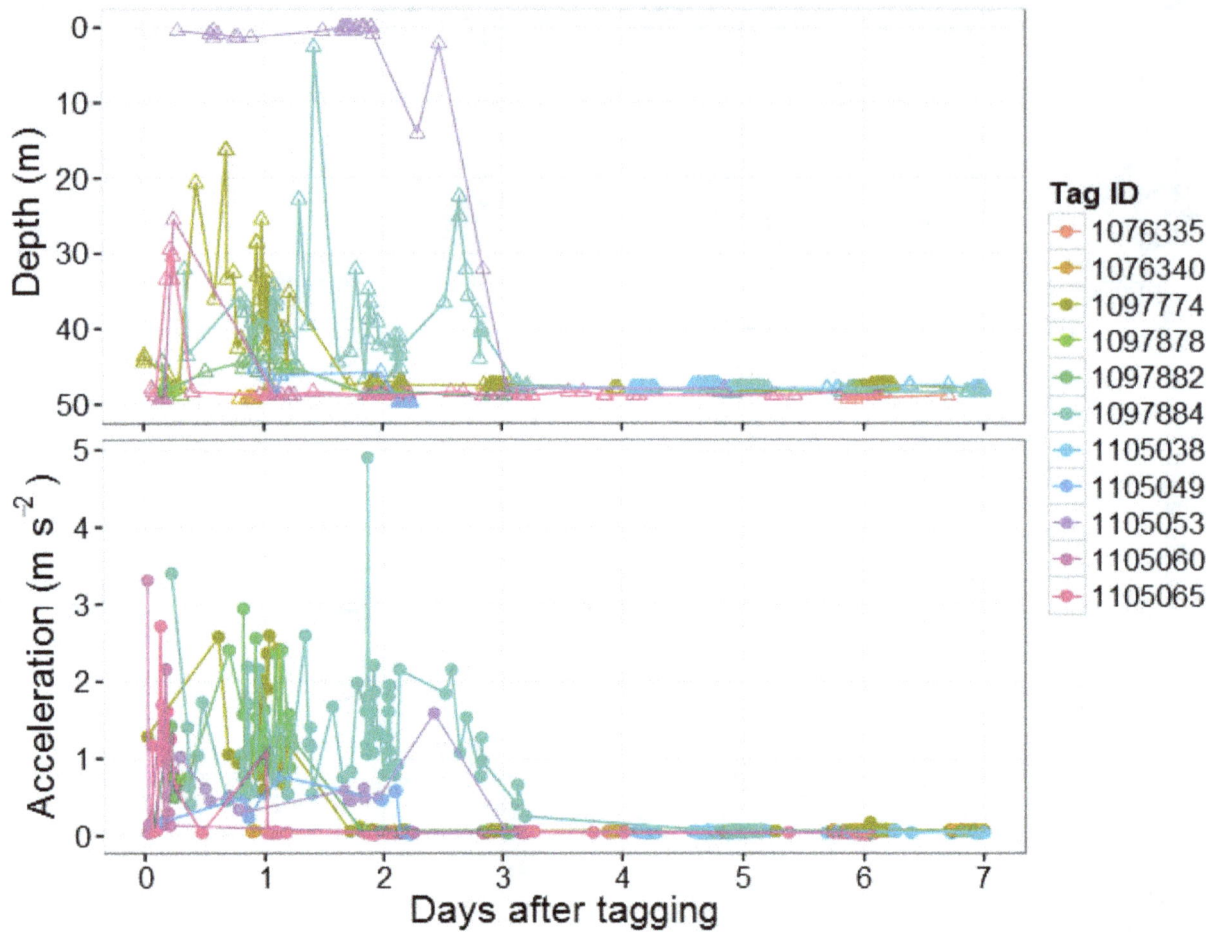

FIGURE 6.   Acoustic telemetry depth and acceleration profiles of all acoustically tagged Red Snapper classified as delayed mortality for 1 week after catch and release ($n = 11$). The points represent individual acoustic detections and are connected by lines for visualization. The triangles in the upper panel represent depth; the solid dots in the lower panel represent acceleration. All delayed mortality events occurred in trials at a water depth of 50 m.

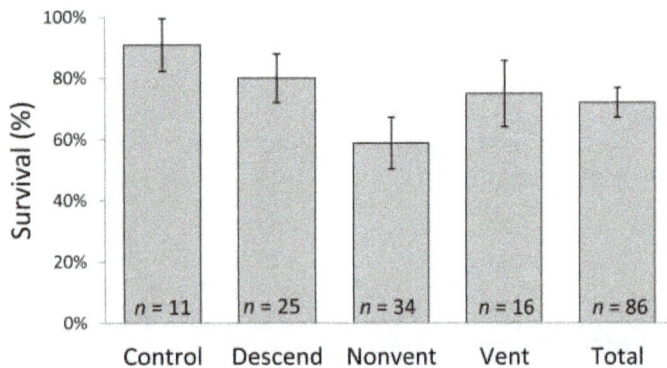

FIGURE 7.   Percent survival (error bars indicate SE) of Red Snapper classified by acoustic profiles over all seasons and depths (summer, winter, spring–50 m, spring–30 m). Fish classified as fate "unknown" from acceleration and depth profiles are omitted in the analysis; therefore, sample size (n) for each group is equal to the number of fish tagged minus the unknowns. The four release treatments included the following: control fish (i.e., no barotrauma), descended release (Shelton fish descender), nonvented surface release, and vented surface release.

controlled laboratory experiments using hyperbaric chambers strongly advocated for venting in reducing discard mortality (Drumhiller et al. 2014); however, this study did not consider the impact of season (i.e., water temperature) on survival. The field observations here clearly showed survival was highly dependent upon season. While the majority of nonvented fish in our study survived catch and release during the winter and spring trials, only one fish survived in summer. Additionally, the largest number of immediate surface mortality events occurred in summer and the bulk of those mortalities were from fish that were unvented and released at the surface. Render and Wilson (1994) observed a similar interaction between season and release treatment. With the recreational fishing season occurring during the summer months (GMFMC 2015), the threat of immediate surface mortality is magnified by the number of anglers fishing for Red Snapper. Thus, using appropriate release methods to reduce the risk of mortality is imperative for increasing postrelease survival.

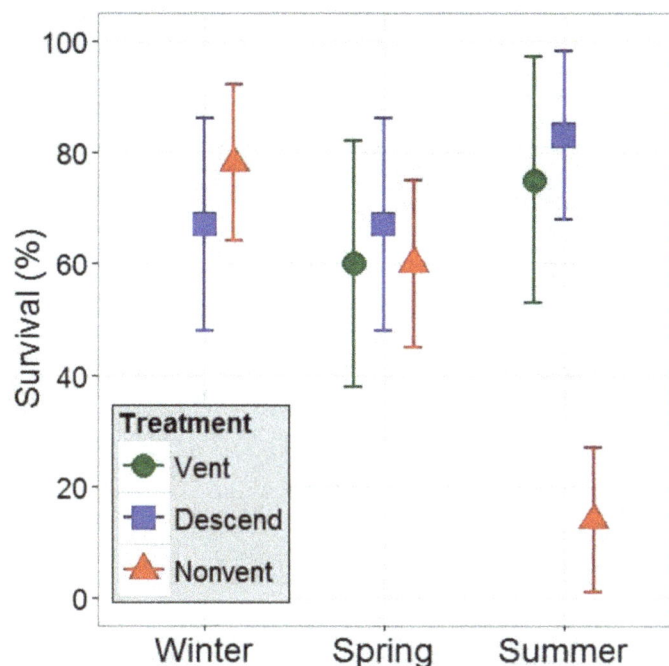

FIGURE 8. Percent survival (error bars indicate SE) of Red Snapper during field trials for three seasons: winter, spring (50-m sites only), and summer. Fish classified as fate "unknown" from acceleration and depth profiles are omitted in the analysis; therefore, the sample size (*n*) for each group is equal to the number of fish tagged minus the unknowns. The release treatments included the following: descended release (Shelton fish descender), nonvented surface release, and vented surface release. Note that the vented treatment was not performed in winter.

Temperature, Depth, and Release Treatments

Thermal stress occurs when captured fish are displaced and released in water temperatures that extend beyond their temperature tolerance range or in temperatures in which they are not acclimated (Cooke and Suski 2005; Gingerich et al. 2007; Diamond and Campbell 2009; Gale et al. 2013). Thermal stress caused by elevated water temperatures causes numerous

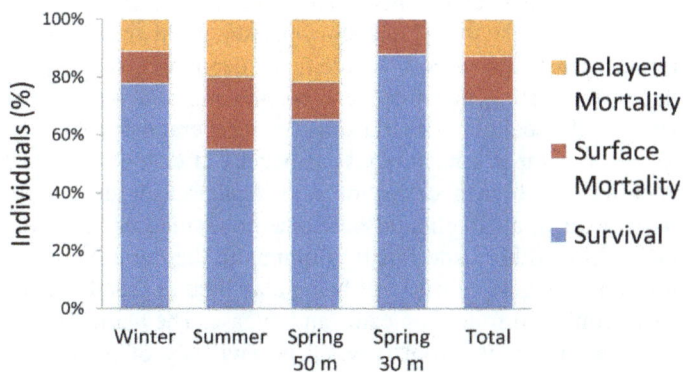

FIGURE 9. Stacked bar graph grouping all release treatments that shows the known fates of individuals by season based on acoustic profiles (survival, surface mortality, or delayed mortality). Each column is reported as a percentage out of 100%. Winter and summer trials were both performed at sites with a water depth of 50 m.

physiological and behavioral changes that can have profound effects on cellular function and metabolic activity (Fry 1971; Prosser 1991; Cooke and Suski 2005). Additionally, levels of dissolved oxygen are depressed at higher water temperatures and this may cause additional physiological problems in catch-and-release fisheries (Arlinghaus et al. 2007). High surface-to-bottom temperature differentials decreased survival in Black Rockfish *Sebastes melanops* (Hannah et al. 2012) and Red Snapper (Diamond and Campbell 2009). In our study, fish tagged and released in the summer season were five times as likely to perish as fish tagged in spring and two and a half times as likely as fish tagged in winter. The increased risk of mortality associated with higher sea surface temperatures during summer is likely exacerbated by large temperature differentials produced by the summer thermocline. Fish caught and released in the summer were brought from bottom temperatures of 22°C up to 31°C at the surface, a differential of 9°C. In contrast, spring fish experienced a much smaller 3.5°C differential and winter fish experienced < 1°C differential, and these fish had considerably higher survival rates. Summer sea surface temperatures approached the 33°C upper tolerance limit of Red Snapper (Moran 1988). Coupled with the additional physiological stress associated with a 9°C water temperature change (i.e., thermocline), these warmer surface waters in summer likely played a significant role in reducing Red Snapper survival after catch and release.

Rapid recompression strategies using descender devices showed positive benefits for Red Snapper in this study. These fish were three times as likely to survive as fish that were similarly nonvented but released at the surface. Descended fish were also over one and a half times more likely to survive than vented fish. Previous studies involving descender devices have proven them beneficial for increasing postrelease survival in several species of Pacific rockfish *Sebastes* sp. (Jarvis and Lowe 2008; Hochhalter and Reed 2011; Rogers et al. 2011; Hannah et al. 2012; Pribyl et al. 2012) and Australian snapper *Lutjanus* sp. (Sumpton et al. 2010; Butcher et al. 2012). The reversal of barotrauma injuries through rapid recompression shows similar benefits for Red Snapper in the Gulf of Mexico. Additionally, the survival of descended fish in our study showed less seasonal variability than with other release treatments. While sea surface temperatures during tagging trials significantly differed by season, water temperatures at the seafloor were stable throughout the year. Returning fish to these cooler water temperatures by using descending devices seems to further enhance postrelease survival and appears to be particularly important when seasonal thermoclines create stratification in the water column.

The severity of barotrauma symptoms increases with capture depth (Alós 2008; Hannah et al. 2008a; Brown et al. 2010; Campbell et al. 2010b; Butcher et al. 2012), and the majority of deepwater catch-and-release studies have identified this variable as the greatest predictor of release mortality.

TABLE 2. Cox proportional hazards model using treatment, season, depth, and total length (TL) as covariates. The hazard ratio shows the proportional risk of each level of a particular treatment against the baseline risk of mortality. For the continuous covariate of total length, the proportional risk is based on a difference of one unit (i.e., 1 mm).

Covariate	Coefficient (b)	SE	Hazard ratio (e^b)	95% CI for e^b	P
Control	baseline				
Descend	0.818	1.010	2.267	0.263–19.574	0.457
Vent	1.311	1.138	3.708	0.399–34.847	0.249
Nonvent	1.927	1.043	6.868	0.889–53.056	0.065
Spring	baseline				
Winter	0.487	0.735	1.627	0.385–6.871	0.508
Summer	1.607	0.683	4.986	1.308–19.013	0.019
30 m	baseline				
50 m	0.925	0.684	2.522	0.660–9.641	0.176
TL	−0.009	0.004	0.991	0.983–0.999	0.039

Results from our study concur with previous findings as fish captured in the shallower (30-m) depth were more likely to survive than those captured at the deeper (50-m) depth. Survival during the spring trials when two depths were compared was 88% at 30 m and 65% at 50 m. Both of these estimates fall within the range of the SEDAR31-DW22 meta-analysis estimates (Campbell et al. 2012) but are nearer the lower boundary. A similar depth influence was documented in Pacific rockfish *Sebastes* spp. (Hannah et al. 2008b), West Australian Dhufish *Glaucosoma hebraicum* (St John and Syers 2005), Painted Comber *Serranus scriba* (Alós 2008), Gag *Mycteroperca microlepis* (Burns et al. 2002; Rudershausen et al. 2007; Sauls 2014), and most pertinently Red Snapper, for which depth was the most important factor in determining release mortality (Campbell et al. 2013).

The apparent correlation between mortality and depth is most likely due to the link between depth and the extent of barotrauma injuries caused by catastrophic decompression (Rummer 2007; Campbell et al. 2010a; Pribyl et al. 2011). The severity of barotrauma symptoms typically increases with depth, as increased pressure causes higher volumetric expansion of internal gases. However, in some studies visible barotrauma symptoms from fish caught in deeper waters appeared reduced or absent (Brown et al. 2010; Campbell et al. 2013). Further examination revealed that this absence of visible barotrauma injuries can occur when the swim bladder ruptures from overexpansion of gases (Rummer 2007; Rogers et al. 2008; Roach et al. 2011; Campbell et al. 2013; Kerwath et al. 2013). This allows internal organs (i.e., stomach or intestines) that would otherwise be displaced to remain inside the body cavity so that the fish may appear healthy and unafflicted by barotrauma injuries upon surfacing when in fact their survival chances are severely depressed. Furthermore, fish with ruptured swim bladders may have neutral or negative buoyancy allowing them to easily resubmerge and presumably

survive, when in fact they simply sink to the bottom and perish. Thus, we recommend cautious use of fish condition indices as proxies for predicting postrelease survival in fish suffering from barotrauma as these indices may have a tendency to underreport overall discard mortality because the visible extent of barotrauma symptoms present may not be indicative of the ultimate fate of the fish.

Estimates of Delayed Mortality

A unique aspect in integrating accelerometer and depth sensors into acoustic transmitters was the ability to detect exactly when delayed mortality was occurring. The total mortality estimate of 28% (surface + delayed) is similar to previous estimates of discard mortality found at these depths in SEDAR33 (Campbell et al. 2013), though typically studies from this meta-analysis did not include estimates of delayed mortality. The 11 fish that experienced delayed mortality in our trials persisted for less than 3 d before perishing. At that point in time, acceleration values became 0 and depth reflected the site depth, illustrating that fish were not moving and were likely lying on the seafloor. The transmitters of several fish continued to transmit these data for days to weeks after mortality had occurred. Without acceleration and depth sensor data these fish would in all likelihood have been classified as survivors that exhibited high site fidelity throughout the duration of the transmitter tag life, instead of being classified as fish that perished within 3 d following catch and release. The ability to differentiate mortality from survival is obviously of paramount importance in tagging studies that assess postrelease mortality but also for those estimating site fidelity, residency time, and migration patterns. Acoustic tags that lack sensor data and only relay presence or absence information may be insufficient to answer questions addressing these topics. Based on the finding of delayed mortality occurring at 3 d, we recommend that any studies assessing postrelease mortality of Red Snapper should monitor fish for a minimum time period of 3 d to ensure that lingering effects of the catch-and-release process that may cause mortality are accurately documented.

This study was able to account for barotrauma-induced delayed mortality in addition to surface mortality through the use of ultrasonic acoustic telemetry. Previous researchers estimating delayed mortality of Red Snapper in the field have relied on caging experiments, which have an inherent bias because they exclude predatory effects, prevent foraging, and restrict natural movement (Campbell et al. 2013). In such studies, separating the influence of caging effects from barotrauma affliction in estimating mortality is difficult. Delayed mortality estimates in caging studies ranged from 20% to 71% at depths from 20 to 50 m (Gitschlag and Renaud 1994; Render and Wilson 1994; Diamond and Campbell 2009). Acoustic telemetry allowed us to estimate delayed mortality in fish that were unrestricted in movement and behavior. Comparatively, we found delayed mortality estimates ranging from 0% to 22%. The estimates of survival in this study were higher than those reported from caging studies, suggesting that the effect of caging itself may be an influential factor that contributes to postrelease mortality. The exclusion of predators should enhance survival, but this is seemingly less important than the need to move unrestricted, presumably to forage. Predator abundance is typically low and highly variable, so the benefits of caging are minimal when compared with the energetic requirements needed to survive. Using acoustic telemetry eliminates one bias associated with caging practices and allows fish to behave unhindered, thus representing a more natural postrelease scenario.

Limitations of Acoustic Telemetry

A primary challenge in using acoustic telemetry for estimating delayed mortality when compared with passive mark–recapture methods is a limitation of sample size. The inherent cost of these transmitters restricts the use of large sample sizes and complex study designs, which are possible to attain using traditional passive tags. Additionally, the risk of tag collisions using this acoustic technology restricted our sample size on an individual site. The use of other acoustic technologies on the market that do not incur tag collision issues may provide a solution to this limitation, and future studies by our research group seek to explore these options. Nonetheless, the tradeoff of low sample size is far outweighed by the fact that investigators can remotely determine the fate of the fish in most cases and do not have to rely on recaptures or the unknowns associated with unrecovered fish. Certainly some disadvantages are associated with the detection limits of acoustic receivers and the variability in detection efficiency because of environmental fluctuation, and these should be accounted for through

rigorous range testing (How and de Lestang 2012; Kessel et al. 2014). However, using reef fish that exhibit high site fidelity, such as Red Snapper (Szedlmayer and Schroepfer 2005; Westmeyer et al. 2007; Topping and Szedlmayer 2011b), increases the likelihood of detection as they typically remain within the range of receivers positioned on the structure.

Many acoustic telemetry studies have noted that a substantial portion of tagged fish have an immediate postrelease emigration event, likely in response to capture and handling stress (Schroepfer and Szedlmayer 2006; Lowe et al. 2009; Topping and Szedlmayer 2011a). This rapid emigration quickly moves fish outside the detection range, with few to zero acoustic transmissions being detected. Without this acoustic information, and if fish are never recaptured, the fate of these emigrants remains unknown. In the present study, 25 of 111 (22.5%) individuals recorded too few acoustic detections to classify fate with any confidence. These fish classified as unknown were omitted from inclusion in the survival analysis, which reduced the experimental sample size. However, the number of unknown fish was fairly consistent across seasons and release treatments, and because of this trend, the omission of the unknowns would not bias one group unfairly with a disproportionate sample size compared with the other groups. Despite the low sample size, several patterns still emerged, and future replication using acoustic tagging would help further support these discard mortality estimates.

Conclusions and Implications

Of central importance to effective fisheries management is the ability to accurately estimate population demographic parameters for stock assessments. For Red Snapper in the Gulf of Mexico, a high level of uncertainty has surrounded estimates of discard mortality, which represents an important parameter due to the high volume of discards that occur in this fishery. Historically, managers have focused on immediate mortality but have not incorporated delayed mortality into population models. If delayed loss is not accounted for in stock assessment models, it is likely that total mortality will be underestimated. Until recently, researchers faced inherent limitations with the methods involved in making these mortality estimates. We have shown that acoustic telemetry possesses the ability to overcome some of these challenges, but results must endure further replication to overcome the inherent low sample sizes before implementation into the stock assessment process.

ACKNOWLEDGMENTS

We thank the many people that contributed to this study: members of the Fisheries and Ocean Health Lab at the Harte Research Institute for Gulf of Mexico Studies; M. Robillard, J. Williams, L. Payne, K. Drumhiller, P. Jose, W. McGlaun, A. Lund, R. Pizano, and R. Palacios for logistical support during field trials; B. Grumbles, Captain R. Sanguehel, and deckhands of the Scatcat at Fisherman's Wharf, Port Aransas, Texas; Captain M. Miglini and the scuba team for acoustic hydrophone deployment and retrieval; Captain P. Young and deckhands of MoAzul Sportfishing; A. Lawrence, T. Ussery, and technicians at the Texas AgriLife Mariculture Laboratory, Port Aransas, Texas; and M. Holland and other support staff at VEMCO for sharing expertise during acoustic transmitter design and for technical support through the duration of the study. We thank the two anonymous reviewers for providing editorial comments and feedback that greatly enhanced the quality of this manuscript. This research was supported through the National Oceanic and Atmospheric Administration Marine Fisheries Initiative Grant Number NA10NMF4330126.

REFERENCES

Alós, J. 2008. Influence of anatomical hooking depth, capture depth, and venting on mortality of Painted Comber (*Serranus scriba*) released by recreational anglers. ICES Journal of Marine Science 65:1620–1625.

Arlinghaus, R., S. J. Cooke, J. Lyman, D. Policansky, A. Schwab, C. Suski, S. G. Sutton, and E. B. Thorstad. 2007. Understanding the complexity of catch-and-release in recreational fishing: an integrative synthesis of global knowledge from historical, ethical, social, and biological perspectives. Reviews in Fisheries Science 15:75–167.

Bartholomew, A., and J. A. Bohnsack. 2005. A review of catch-and-release angling mortality with implications for no-take reserves. Reviews in Fish Biology and Fisheries 15:129–154.

Bettoli, P. W., and R. S. Osborne. 1998. Hooking mortality and behavior of Striped Bass following catch and release angling. North American Journal of Fisheries Management 18:609–615.

Brown, I. W., W. D. Sumpton, M. McLennan, D. Mayer, M. J. Campbell, J. M. Kirkwood, A. R. Butcher, I. Halliday, A. Mapleston, D. Welch, G. A. Begg, and B. Sawynok. 2010. An improved technique for estimating short-term survival of released line-caught fish, and an application comparing barotrauma-relief methods in Red Emperor (*Lutjanus sebae* Cuvier 1816). Journal of Experimental Marine Biology and Ecology 385:1–7.

Burns, K. M. 2009. J and circle hook mortality and barotrauma and the consequence for Red Snapper survival. Southeast Data Assessment and Review, SEDAR24-RD47, North Charleston, South Carolina.

Burns, K. M., C. C. Koenig, and F. C. Coleman. 2002. Evaluation of multiple factors involved in release mortality of undersized Red Grouper, Gag, Red Snapper, and Vermilion Snapper. Mote Marine Laboratory, Technical Report 790, Sarasota, Florida.

Burns, K. M., R. R. Wilson, and N. F. Parnell. 2004. Partitioning release mortality in the undersized Red Snapper bycatch: comparison of depth vs. hooking effects. Mote Marine Laboratory, Technical Report 932, Sarasota, Florida.

Butcher, P. A., M. K. Broadhurst, K. C. Hall, B. R. Cullis, and S. R. Raidal. 2012. Assessing barotrauma among angled snapper (*Pagrus auratus*) and the utility of release methods. Fisheries Research 127:49–55.

Campbell, M. D., W. B. Driggers, and B. Sauls. 2012. Release mortality in the Red Snapper fishery: a synopsis of three decades of research. Southeast Data, Assessment, and Review, SEDAR31-DW22, North Charleston, South Carolina.

Campbell, M. D., W. B. Driggers, B. Sauls, and J. F. Walter. 2013. Release mortality in the Red Snapper fishery: a meta-analysis of three decades of research. Southeast Data, Assessment, and Review, SEDAR33-RD21, North Charleston, South Carolina.

Campbell, M. D., W. B. Driggers, B. Sauls, and J. F. Walter. 2014. Release mortality in the Red Snapper fishery: a synopsis of three decades of research. Southeast Data, Assessment, and Review, SEDAR31-DW22, North Charleston, South Carolina.

Campbell, M. D., R. Patino, J. Tolan, R. Strauss, and S. L. Diamond. 2010a. Sublethal effects of catch-and-release fishing: measuring capture stress, fish impairment, and predation risk using a condition index. ICES Journal of Marine Science 67:513–521.

Campbell, M. D., J. Tolan, R. Strauss, and S. L. Diamond. 2010b. Relating angling-dependent fish impairment to immediate release mortality of Red Snapper (*Lutjanus campechanus*). Fisheries Research 106:64–70.

Cooke, S. J., and C. D. Suski. 2005. Do we need species-specific guidelines for catch-and-release recreational angling to effectively conserve diverse fishery resources? Biodiversity and Conservation 14:1195–1209.

Cowan, J. H., C. B. Grimes, W. F. Patterson, C. J. Walters, A. C. Jones, W. J. Lindberg, D. J. Sheehy, W. E. Pine III, J. E. Powers, M. D. Campbell, K. C. Lindeman, S. L. Diamond, R. Hilborn, H. T. Gibson, and K. A. Rose. 2010. Red Snapper management in the Gulf of Mexico: science- or faith-based? Reviews in Fish Biology and Fisheries 21:187–204.

Cox, D. R. 1972. Regression models and life-tables. Journal of the Royal Statistical Society 34:187–220.

Diamond, S. L., and M. D. Campbell. 2009. Linking "sink or swim" indicators to delayed mortality in Red Snapper by using a condition index. Marine and Coastal Fisheries: Dynamics, Management, and Ecosystem Science [online serial] 1:107–120.

Dorf, B. A. 2003. Red Snapper discards in Texas coastal waters—a fishery dependent onboard survey of recreational headboat discards and landings. Fisheries, Reefs, and Offshore Development 36:155–166.

Drumhiller, K. L., G. W. Stunz, M. W. Johnson, M. R. Robillard, and S. L. Diamond. 2014. Venting or rapid recompression increase survival and improve recovery of Red Snapper with barotrauma. Marine and Coastal Fisheries: Dynamics, Management, and Ecosystem Science [online serial] 6:190–199.

Fry, F. E. J. 1971. The effect of environmental factors on the physiology of fish. Pages 1–98 *in* W. S. Hoar and D. J. Randall, editors. Fish physiology, environmental relations and behavior, volume VI. Academic Press, New York.

Gale, M. K., S. G. Hinch, and M. R. Donaldson. 2013. The role of temperature in the capture and release of fish. Fish and Fisheries 14:1–33.

Gingerich, A. J., S. J. Cooke, K. C. Hanson, M. R. Donaldson, C. T. Hasler, C. D. Suski, and R. Arlinghaus. 2007. Evaluation of the interactive effects of air exposure duration and water temperature on the condition and survival of angled and released fish. Fisheries Research 86:169–178.

Gitschlag, G. R., and M. L. Renaud. 1994. Field experiments on survival rates of caged and released Red Snapper. North American Journal of Fisheries Management 14:131–136.

GMFMC (Gulf of Mexico Fishery Management Council). 2007. Amendment 27 to the Gulf of Mexico Reef Fish Fishery Management Plan. GMFMC, Tampa, Florida.

GMFMC (Gulf of Mexico Fishery Management Council). 2013. Framework action to set the annual catch limit and bag limit for Vermilion Snapper, set annual catch limit for Yellowtail Snapper, and modify the venting tool requirement. GMFMC, Tampa, Florida.

GMFMC (Gulf of Mexico Fishery Management Council). 2015. Regional management of recreational Red Snapper. Updated draft for Amendment 39 to the Fishery Management Plan for the reef fish resources of the Gulf of Mexico. GMFMC, Tampa, Florida.

Goodyear, C. P. 1988. Recent trends in the Red Snapper fishery of the Gulf of Mexico. National Oceanic and Atmospheric Administration, National Marine Fisheries Service, Southeast Fisheries Science Center, CRD 87/88-16, Miami Laboratory, Miami.

Hannah, R. W., S. J. Parker, and K. M. Matteson. 2008a. Escaping the surface: the effect of capture depth on submergence success of surface-released

Pacific rockfish. North American Journal of Fisheries Management 28:694–700.

Hannah, R. W., P. S. Rankin, and M. T. O. Blume. 2012. Use of a novel cage system to measure postrecompression survival of northeast Pacific rockfish. Marine and Coastal Fisheries: Dynamics, Management, and Ecosystem Science [online serial] 4:46–56.

Hannah, R. W., P. S. Rankin, A. N. Penny, and S. J. Parker. 2008b. Physical model of the development of external signs of barotrauma in Pacific rockfish. Aquatic Biology 3:291–296.

Heupel, M. R., and C. A. Simpfendorfer. 2002. Estimation of mortality of juvenile Blacktip Sharks, *Carcharhinus limbatus*, within a nursery area using telemetry data. Canadian Journal of Fisheries and Aquatic Sciences 59:624–632.

Hochhalter, S. J., and D. J. Reed. 2011. The effectiveness of deepwater release at improving the survival of discarded Yelloweye Rockfish. North American Journal of Fisheries Management 31:852–860.

Hood, P. B., A. J. Strelcheck, and P. Steele. 2007. A history of Red Snapper management in the Gulf of Mexico. Pages 267–284 *in* W. F. Patterson III, J. H. Cowan Jr., G. R. Fitzhugh, and D. L. Nieland, editors. Red Snapper ecology and fisheries in the U.S. Gulf of Mexico. American Fisheries Society, Symposium 60, Bethesda, Maryland.

How, J. R., and S. de Lestang. 2012. Acoustic tracking: issues affecting design, analysis and interpretation of data from movement studies. Marine and Freshwater Research 63:312–324.

Jarvis, E. T., and C. G. Lowe. 2008. The effects of barotrauma on the catch-and-release survival of southern California nearshore and shelf rockfish (Scorpaenidae, *Sebastes* spp.). Canadian Journal of Fisheries and Aquatic Sciences 65:1286–1296.

Johnson, M. W., S. L. Diamond, and G. W. Stunz. 2015. External attachment of acoustic tags to deepwater reef fishes: an alternate approach when internal implantation affects experimental design. Transactions of the American Fisheries Society 144:851–859.

Kerwath, S. E., C. G. Wilke, and A. Gotz. 2013. The effects of barotrauma on five species of South African line-caught fish. African Journal of Marine Science 35:243–252.

Kessel, S. T., S. J. Cooke, M. R. Heupel, N. E. Hussey, C. A. Simpfendorfer, S. Vagle, and A. T. Fisk. 2014. A review of detection range testing in aquatic passive acoustic telemetry studies. Reviews in Fish Biology and Fisheries 24:199–218.

Lowe, C. G., K. M. Anthony, E. T. Jarvis, L. F. Bellquist, and M. S. Love. 2009. Site fidelity and movement patterns of groundfish associated with offshore petroleum platforms in the Santa Barbara Channel. Marine and Coastal Fisheries: Dynamics, Management, and Ecosystem Science [online serial] 1:71–89.

Moran, D. 1988. Species profiles: life histories and environmental requirements of coastal fishes and invertebrates (Gulf of Mexico)–Red Snapper. U.S. Fish Wildlife Service Biological Report 82(11.83).

Peabody, M. B., and C. A. Wilson. 2006. Fidelity of Red Snapper (*Lutjanus campechanus*) to petroleum platforms and artificial reefs in the northern Gulf of Mexico. U.S. Department of the Interior, Minerals Management Service, Gulf of Mexico OCS Region, OCS Study MMS 2006-005, New Orleans, Louisiana.

Pollock, K. H., and W. E. Pine. 2007. The design and analysis of field studies to estimate catch-and-release mortality. Fisheries Management and Ecology 14:123–130.

Pribyl, A. L., M. L. Kent, S. J. Parker, and C. B. Schreck. 2011. The response to forced decompression in six species of Pacific rockfish. Transactions of the American Fisheries Society 140:374–383.

Pribyl, A. L., C. B. Schreck, M. L. Kent, K. M. Kelley, and S. J. Parker. 2012. Recovery potential of Black Rockfish, *Sebastes melanops* Girard, recompressed following barotrauma. Journal of Fish Diseases 35:275–286.

Prosser, C. L. 1991. Environmental and metabolic animal physiology. Wiley-Liss, New York.

R Development Core Team. 2013. R: a language and environment for statistical computing. R Foundation for Statistical Computing, Vienna. Available: http://www.R-project.org/. (September 2015).

Render, J. H., and C. A. Wilson. 1994. Hook-and-line mortality of caught and released Red Snapper around oil and gas platform structural habitat. Bulletin of Marine Science 55:1106–1111.

Render, J. H., and C. A. Wilson. 1996. Effect of gas bladder deflation on mortality of hook-and-line caught and released Red Snappers: implications for management. Pages 244–253 *in* F. Arreguin-Sanchez, J. L. Munro, M. C. Balgos, and D. Pauly, editors. Biology, fisheries, and culture of tropical groupers and snappers. International Center of Living Aquatic Resources Management, Conference Proceedings 48, Manila.

Roach, J. P., K. C. Hall, and M. K. Broadhurst. 2011. Effects of barotrauma and mitigation methods on released Australian Bass *Macquaria novemaculeata*. Journal of Fish Biology 79:1130–1145.

Rogers, B. L., C. G. Lowe, and E. Fernández-Juricic. 2011. Recovery of visual performance in Rosy Rockfish (*Sebastes rosaceus*) following exophthalmia resulting from barotrauma. Fisheries Research 112:1–7.

Rogers, B. L., C. G. Lowe, E. Fernández-Juricic, and L. R. Frank. 2008. Utilizing magnetic resonance imaging (MRI) to assess the effects of angling-induced barotrauma on rockfish (*Sebastes*). Canadian Journal of Fisheries and Aquatic Sciences 65:1245–1249.

Rudershausen, P. J., J. A. Buckel, and E. H. Williams. 2007. Discard composition and release fate in the snapper and grouper commercial hook-and-line fishery in North Carolina, USA. Fisheries Management and Ecology 14:103–113.

Rummer, J. L. 2007. Factors affecting catch and release (CAR) mortality in fish: insight into CAR mortality in Red Snapper and the influence of catastrophic decompression. Pages 123–144 *in* W. F. Patterson III, J. H. Cowan Jr., G. R. Fitzhugh, and D. L. Nieland, editors. Red Snapper ecology and fisheries in the U.S. Gulf of Mexico. American Fisheries Society, Symposium 60, Bethesda, Maryland.

Rummer, J. L., and W. A. Bennett. 2005. Physiological effects of swim bladder overexpansion and catastrophic decompression on Red Snapper. Transactions of the American Fisheries Society 134:1457–1470.

Sauls, B. 2014. Relative survival of Gags *Mycteroperca microlepis* released within a recreational hook-and-line fishery: application of the Cox regression model to control for heterogeneity in a large-scale mark–recapture study. Fisheries Research 150:18–27.

Schroepfer, R. L., and S. T. Szedlmayer. 2006. Estimates of residence and site fidelity for Red Snapper *Lutjanus campechanus* on artificial reefs in the northeastern Gulf of Mexico. Bulletin of Marine Science 78:93–101.

Scyphers, S. B., F. J. Fodrie, F. J. Hernandez, S. P. Powers, and R. L. Shipp. 2013. Venting and reef fish survival: perceptions and participation rates among recreational anglers in the northern Gulf of Mexico. North American Journal of Fisheries Management 33:1071–1078.

St John, J., and C. J. Syers. 2005. Mortality of the demersal West Australian Dhufish, *Glaucosoma hebraicum* (Richardson 1845) following catch and release: the influence of capture depth, venting and hook type. Fisheries Research 76:106–116.

Sumpton, W. D., I. W. Brown, D. G. Mayer, M. F. McLennan, A. Mapleston, A. R. Butcher, D. J. Welch, J. M. Kirkwood, B. Sawynok, and G. A. Begg. 2010. Assessing the effects of line capture and barotrauma relief procedures on post-release survival of key tropical reef fish species in Australia using recreational tagging clubs. Fisheries Management and Ecology 17:77–88.

Szedlmayer, S. T., and R. L. Schroepfer. 2005. Long-term residence of Red Snapper on artificial reefs in the northeastern Gulf of Mexico. Transactions of the American Fisheries Society 134:315–325.

Therneau, T. M., and P. M. Grambsch. 2000. Modeling survival data: extending the Cox model. Springer, New York.

Topping, D. T., and S. T. Szedlmayer. 2011a. Site fidelity, residence time and movements of Red Snapper *Lutjanus campechanus* estimated with long-term acoustic monitoring. Marine Ecology Progress Series 437:183–200.

Topping, D. T., and S. T. Szedlmayer. 2011b. Home range and movement patterns of Red Snapper (*Lutjanus campechanus*) on artificial reefs. Fisheries Research 112:77–84.

Westmeyer, M. P., C. A. Wilson, and D. L. Nieland. 2007. Fidelity of Red Snapper to petroleum platforms in the northern Gulf of Mexico. Pages 105–121 *in* W. F. Patterson III, J. H. Cowan Jr., G. R. Fitzhugh, and D. L. Nieland, editors. Red Snapper ecology and fisheries in the U.S. Gulf of Mexico. American Fisheries Society, Symposium 60, Bethesda, Maryland.

Wilde, G. R. 2009. Does venting promote survival of released fish? Fisheries 34:20–28.

4

Ontogenetic and Sex-Specific Shifts in the Feeding Habits of the Barndoor Skate

Joseph D. Schmitt*

Department of Fish and Wildlife Conservation, Virginia Polytechnic Institute and State University, 100 Cheatham Hall, Blacksburg, Virginia 24060, USA

Todd Gedamke, William D. DuPaul, and John A. Musick

Virginia Institute of Marine Science, College of William and Mary, Post Office Box 134, Gloucester Point, Virginia 23062, USA

Abstract

Diet analysis is critical in understanding the flow of energy within marine food webs and is necessary for trophic ecosystem modeling and subsequent ecosystem-based management recommendations. This study represents the first comprehensive diet description for the Barndoor Skate *Dipturus laevis*, the largest rajid species found on the continental shelf in the northwestern Atlantic Ocean. Stomach contents were extracted from 273 individual skate caught as bycatch in the commercial scallop fishery on Georges Bank and a total of 31 prey species were identified. The Barndoor Skate feeds primarily upon sand shrimp *Crangon septemspinosa*, the rock crab *Cancer irroratus*, the Acadian hermit crab *Pagurus acadianus*, and teleost fish. Length-specific analysis revealed four significant feeding groups (ANOVA: $P < 0.01$). Skate < 35 cm TL were specialized feeders foraging solely on caridean shrimp, and as size increased (35–75 cm TL), they began to feed upon rock crab and then the Acadian hermit crab. At lengths ranging from 85 to 105 cm TL, no caridean shrimp were found in the skate's diet and the prevalence of crustaceans decreased. Large skate (>105 cm TL) began to prey heavily upon teleost fish, yet also continued to consume larger crustaceans. Significant sex-specific differences in food habits were also observed in the biggest skate (>105 cm TL): males fed primarily on teleost fish (~80%); however, females maintained a diet of approximately equal amounts of fish and crustaceans. These sex-specific feeding patterns and differential food niche utilization may be mitigated by sexually dimorphic dentition.

The Barndoor Skate *Dipturus laevis* is the largest member of the family Rajidae found on the continental shelf in the northwestern Atlantic Ocean, reaching a maximum length of 152 cm and a weight of 20 kg (Bigelow and Schroeder 1953). The species has been reported to range from Cape Hatteras, North Carolina, to the Grand Banks of Newfoundland, Gulf of St. Lawrence, and Nova Scotia (Leim and Scott 1966; McEachran and Musick 1975). It ranges from shallow coastal waters to depths greater than 450 m and tolerates water temperatures ranging from 1.2°C to 20°C (Bigelow and Schroeder 1953; McEachran and Musick 1975; Kulka et al. 2002). While the Barndoor Skate has no commercial value, it is often taken

Subject editor: Donald Noakes, Vancouver Island University, Nanaimo, British Columbia

*Corresponding author: jds2012@vt.edu

as a bycatch species on Georges Bank and overfishing has been reported to be threatening the survival of the species as a whole (Casey and Myers 1998). As an elasmobranch, the Barndoor Skate was believed to be particularly vulnerable to fishing pressure due to its large size and presumed late maturation. Over the last 10 years, however, a dramatic recovery in the population has been observed (Gedamke 2006). Although a reduction in fishing pressure is clearly a critical component in the recovery of the species, it may only be a single factor in a more complex picture. To accurately evaluate the population dynamics of the Barndoor Skate, more information about its life history and trophic interactions (i.e., predator–prey relationships) must be incorporated into future analyses.

Skate represent the most diverse group of chondrichthyan fishes with nearly 250 described species (Ebert and Compagno 2007), yet skate are poorly studied compared with other marine elasmobranchs (Orlov 1998). Skate are believed to play important roles in structuring demersal marine communities due to their wide distribution and high abundance (Ebert and Bizzarro 2007), yet studies pertaining to the diet composition and trophic roles of skate are few (Garrison and Link 2000; Bulman et al. 2001; Davenport and Bax 2002; Morato et al. 2003; Braccini and Perez 2005). The Barndoor Skate is no exception. In 2007, Ebert and Bizzarro (2007) reported the standardized trophic level for 60 different species of skate, including the Barndoor Skate, and their trophic level calculations were based on stomach contents from three specimens. There is clearly a paucity of information pertaining to the food habits and trophic role of the Barndoor Skate.

Over the last 40 years, Georges Bank and the Gulf of Maine have undergone major changes (Fogarty and Murawski 1998; Collie and Delong 1999). Commercially important species such as Atlantic Cod *Gadus morhua*, Haddock *Melanogrammus aeglefinus*, and hake *Urophycis* spp. were replaced with elasmobranchs of lower market value including dogfishes and skate; meanwhile Atlantic Mackerel *Scomber scombrus* and Atlantic Herring *Clupea harengus* populations grew to historic levels (Collie and Delong 1999). While the decline in the groundfish community was largely attributed to overfishing, predation can be a large contributor to prerecruit mortality on Georges Bank (Sissenwine et al. 1984), further emphasizing the need to expand our knowledge of trophic interactions on Georges Bank.

The overall life history of the Barndoor Skate has only recently been investigated (Gedamke et al. 2005), yet very little is known about their food habits or their ecological role in the Gulf of Maine. Understanding the feeding habits of the Barndoor Skate can bring valuable insight into predator–prey relationships and can contribute to future studies of trophic interactions (Caddy and Sharp 1986), with the hope that future studies will not be weakened by limited data as in the Ebert and Bizzarro (2007) study. The present study will provide vital information concerning the trophic role of the Barndoor Skate and will assist in the designation of essential fish habitat,

which is defined legally as "those waters and substrate necessary to fish spawning, breeding, feeding, or growth to maturity (16 U.S.C. 1802[10])." Our work addresses a research need identified in both the 2001 stock assessment report (NEFMC 2001) and the essential fish habitat source documents (NOAA 2003) to "investigate trophic interactions between skate species in the complex, and between skates and other groundfish."

This study represents the first comprehensive analysis of the food habits of the Barndoor Skate in the Gulf of Maine. The goals of this study are threefold: (1) quantify the diet composition of the barndoor skate, (2) evaluate the possible ontogenetic shifts in prey items, and (3) explore the possibility that sexually dimorphic mature dentition influences prey selection.

METHODS

Field collections.—All specimens included in this study were collected onboard commercial scallop vessels on Georges Bank. A majority of the specimens were collected in the southern portion (south of 41°30′N) of closed area II while a limited number were collected from the Nantucket Lightship closed area (Figure 1). Both of these areas were closed to the

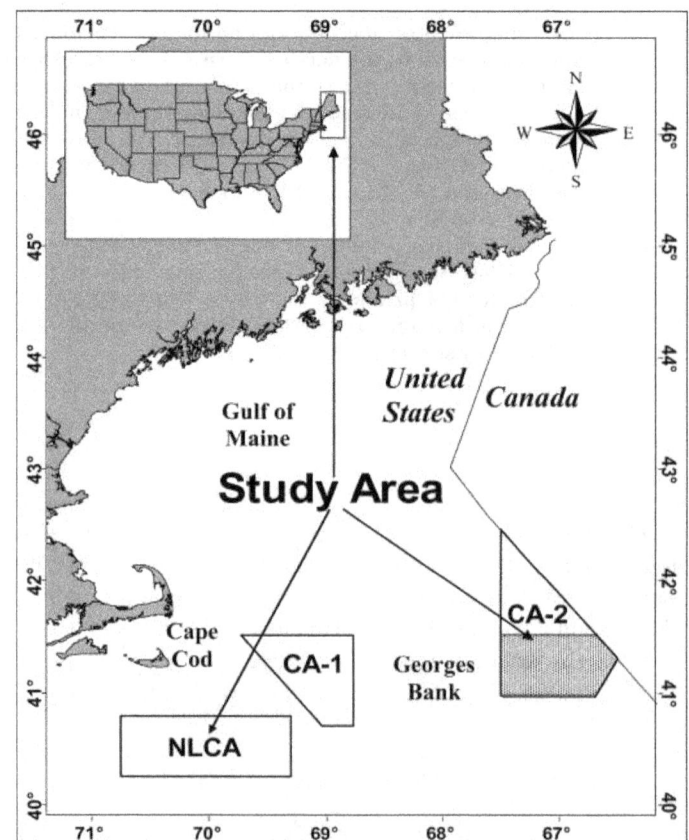

FIGURE 1. Sample locations for Barndoor Skate. CA-2: Georges Bank closed area II; CA-1: Georges Bank closed area I; NLCA: Nantucket Lightship closed area. The 91.4-m isobath is indicated by the dotted line. The *x*- and *y*-axes give longitude (W) and latitude (N), respectively.

use of mobile fishing gear in December 1994 in an effort to rebuild groundfish stocks. In June 1999, access to the closed areas began on a limited basis as the development of a rotational management strategy was being explored in the commercial scallop fishery. Data were collected on eight trips between June and November in 1999, 2000, and 2003. Vessels fished with two 4.6-m New Bedford style dredges for sea scallops *Placopecten magellanicus* (Posgay 1957) constructed with a 25.4-cm twine top and either 8.9- or 10.2-cm ring bags. Gear was towed in 55–73 m of water at an average speed of 9.2 km/h. This gear captured a broad range of skate sizes ranging from 20 to 133.5 cm TL. Total length, sex, and maturity stage (mature versus immature) were recorded for all Barndoor Skate specimens upon capture. Sexual maturity was determined by the development of the testes or the presence of eggs in the ovaries along with other criteria (Gedamke et al. 2005). Furthermore, since tooth morphology in the males develops from molariform (i.e., plate-like) to cuspidate dentition (i.e., pointed teeth) at maturity, the presence or development of this secondary sexual character was also noted (Figure 2).

Laboratory processing.—Entire skate stomachs were preserved in a solution of 10% phosphate-buffered formalin and then transferred to 70% ethanol prior to sorting. Stomachs were rinsed to remove all contents and then each prey item was identified to the lowest taxonomic level possible. To assist in the identification of teleost fishes, the museum collection at the Virginia Institute of Marine Science was used for comparative morphology and verification of initial identification. In addition, many of the invertebrates and fish species caught as bycatch were preserved to act as a reference library for the identification of stomach contents. The number of individual prey items was also recorded. In cases where prey were unrecognizable by gross morphology, remaining body parts (i.e., eyes, vertebrae, shell fragments) were used to make individual prey counts (Chipps and Garvey 2007). Following sorting, samples were weighed to obtain wet weights, placed in a drying oven for at least 24 h, and then reweighed to obtain dry weights. Plots and regressions of wet weights versus dry weights were evaluated to determine any differential patterns in the use of one metric versus the other.

Diet analysis.—Percent by weight (%W) was used to indicate which prey items were energetically important to the Barndoor Skate and was calculated as the weight of a given prey item divided by the weight of all prey items. Percent frequency of occurrence (%O) was used to indicate which prey items were routinely utilized by the Barndoor Skate and was calculated as the number of stomachs that contained a particular prey item divided by the total number of stomachs (Chipps and Garvey 2007; Graham et al. 2007).

A total of 31 prey items were recorded (Table 1) and prey items were grouped into three logical ecological groupings: "crabs," "shrimp," and "fish." These three prey groups accounted for 99.77% of the prey items by weight and were

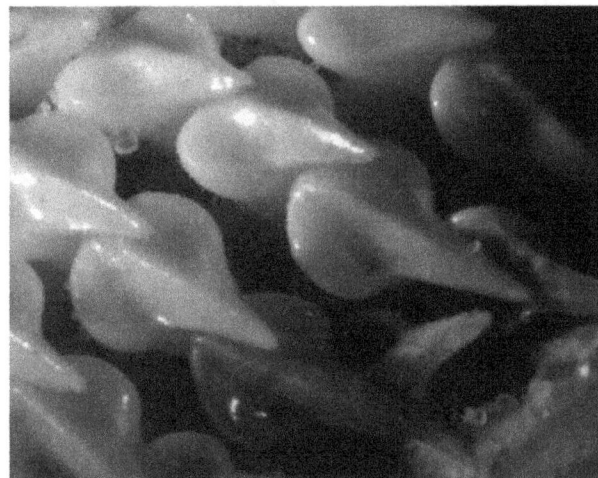

FIGURE 2. Photographs of sexually dimorphic dentition in (**a**) a mature female and (**b**) a mature male Barndoor Skate. The mature female measured 121 cm TL and the mature male measured 116 cm TL.

dominated by five key species: rock crab *Cancer irrotatus*, Acadian hermit crab *Pagurus acadianus*, sand shrimp *Crangon septemspinosa*, Ocean Pout *Macrozoarces americanus*, and Atlantic Herring (Figure 3). Barndoor Skate were pooled into 10-cm length groups and a single-factor ANOVA was used to test the effect of skate TL on the mean proportion by weight for each of the three major prey categories. Prior to testing, proportional weight data were logit-transformed, as this transformation has been demonstrated to be more appropriate than the arcsine transformation for proportion data (Warton and Hui 2011). The logit transformation cannot transform proportions equal to zero or one, so the smallest nonzero value was added to the numerator and denominator of the logit function for zero values. Similarly, the smallest nonzero value was subtracted from the numerator and denominator of the logit function for values equal to one (Warton and Hui 2011). Levene's test for homogeneity of variance indicated equal variances when the smallest length classes (<35 cm TL) were

TABLE 1. Dietary composition of all Barndoor Skate sampled ($n = 273$) on Georges Bank, displayed as percent by number (%N), weight (%W), and frequency of occurrence (%O).

Prey items	%N	%W	%O
Teleost fishes			
Scomber scombrus	0.02	0.33	0.37
Clupea harengus	0.59	10.90	6.59
Myoxocephalus sp.	0.43	5.15	5.13
Urophycis sp.	0.21	6.40	3.30
Urophycis chuss	0.12	5.91	1.10
Macrozoarces americanus	0.14	14.82	2.20
Limanda ferruginea	0.15	2.51	0.73
Paralichthys dentatus	0.02	0.98	0.37
Peprilus triacanthus	0.02	0.34	0.37
Hemitripterus americanus	0.05	1.38	0.73
Paralichthys oblongus	0.07	1.72	0.73
Unidentified fish	0.05	3.32	7.33
Unidentified flatfish	0.17	2.94	2.20
Crustacea			
Caridean shrimp			
Crangon septemspinosa	34.61	1.64	38.83
Pandalus propinquus	0.66	0.01	2.20
Dichelopandalus leptocerus	2.44	0.12	4.76
Pagurid crabs			
Pagurus acadianus	20.79	14.43	38.46
Pagurus pubescens	0.05	0.01	0.73
Cancer crabs			
Cancer irroratus	29.60	17.83	54.95
Cancer borealis	2.13	4.14	6.59
Cancer sp.	0.12	0.06	0.37
Other			
Unidentified crab	0.43	0.31	4.40
Unidentified decapod	0.28	0.19	4.40
Unidentified amphipod	1.11	0.08	1.83
Unidentified isopod	0.66	0.04	1.83
Unidentified barnacle	0.02		0.37
Mollusca			
Unidentified snail	1.07	0.08	8.06
Unidentified bivalve	0.05		0.73
Nematoda			
Unidentified nematode	2.47	0.02	13.19
Trematoda			
Unidentified trematode	0.07		0.37
Unidentified organic matter	0.92	3.62	14.29

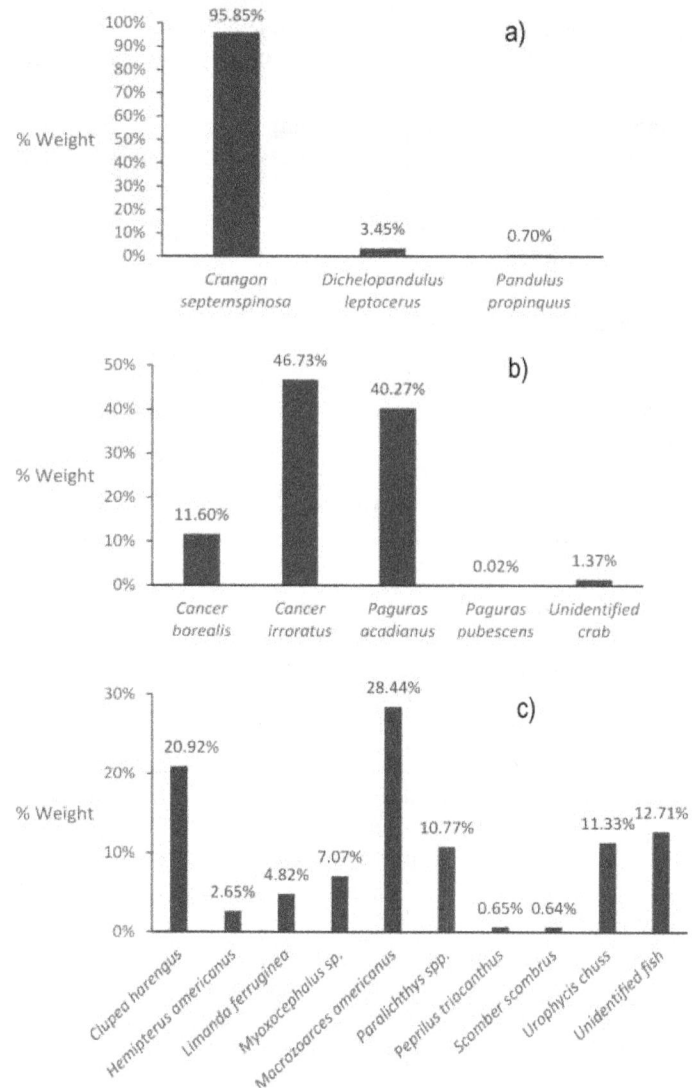

FIGURE 3. Percent by weight contribution of individual species of Barndoor Skate prey to our broader ecological groupings: (**a**) shrimp, (**b**) crabs, and (**c**) fish.

removed, as these length groupings fed solely on shrimp (100% by weight); thus, there was no variance. After transformation, Shapiro–Wilk tests indicated nonnormality for some of the diet data; however, ANOVA is robust to the normality assumption and the validity of the analysis is only slightly affected by a nonnormal distribution (Zar 1999). Proportions were compared using post hoc Tukey's multiple comparison tests to determine at which lengths significant diet shifts occurred. An alpha value of 0.05 ($\alpha = 0.05$) was used for all significance testing. Once significant length-groups were identified, single-factor ANOVA were used to compare the effect of sex on the proportion contribution by weight for each of the three major prey categories within each significant length grouping.

Compound indices such as the index of relative importance (IRI) incorporate multiple single indices and are believed to provide a more balanced understanding of the dietary importance of different prey types (Pinkas et al. 1971; Bigg and Perez 1985; Cortes 1997); however, they are not without controversy. Some authors claim that compound indices are redundant and add little to single indices (MacDonald and Green 1983), and others note that the arbitrary nature of IRI metrics complicates comparisons among species and different food types (Cortes 1997). More recently, IRI has been demonstrated to be intrinsically flawed as it combines mathematically dependent measures; this causes frequently occurring prey items to be overemphasized while rare prey items are underemphasized (Ortaz et al. 2006; Brown et al. 2012). Making matters worse, IRI is nonadditive across taxonomic levels; therefore, IRI values are arbitrarily controlled by the taxonomic resolution chosen by the researcher, which further complicates comparisons among studies (Brown et al. 2012). The intrinsic flaws associated with the IRI have caused researchers to develop a more appropriate compound index known as the prey-specific index of relative importance (PSIRI; Brown et al. 2012). This new index corrects the mathematical flaws associated with IRI and enables researchers to draw comparisons among studies regardless of the taxonomic level chosen by the researchers. The values for %PSIRI were calculated for all the three major prey categories and significant length-classes; %PSIRI is defined as

$$\%\mathrm{PSIRI} = [\%O_i \times (\%PN_i + \%PW_i)]/2,$$

where $\%O_i$ is the frequency of occurrence for prey type i, $\%PN_i$ is the prey-specific percent by number, or the percent by number of prey type i in all stomachs containing prey type i, and $\%PW_i$ is the prey-specific percent by weight, or the percent by weight of prey type i in all stomachs containing prey type i.

RESULTS

Stomach samples were taken from a total of 273 Barndoor Skate, of which 267 (97.8%) contained prey items. A majority ($n = 256$) of the specimens were collected in the southern portion (south of 41°30'N) of closed area II, whereas a limited number ($n = 17$) were collected from the Nantucket Lightship closed area. Samples were taken from 137 females and 136 males ranging from 20 cm to 133.5 cm TL. Linear regressions of wet versus dry weights showed no pattern to the residuals, so only wet weights were used in subsequent analyses.

Ontogenetic diet shifts were immediately obvious upon plotting the 10-mm length-groups versus mean percent weight for the three major prey categories (Figure 4), and statistical analyses indicated significant ontogenetic diet shifts for all three prey categories (ANOVA: $P < 0.001$). Tukey's multiple comparison tests allowed us to stratify skate lengths into four functional feeding groups based on significant dietary shifts in the mean proportion by weight of the major prey categories: <35 cm TL, 35–75 cm TL, 85–105 cm TL, and >105 cm TL. Until skate reached a size of approximately 35 cm, individual stomach samples contained only one species of caridean shrimp. Samples taken from Georges Bank closed area II contained only *Crangon septemspinosa* while samples from the Nantucket Lightship closed area contained only *Dichelopandalus leptocerus*. At lengths ≥ 35 cm TL, skate exhibited a significant shift from the sole utilization of small shrimp to a more diverse diet including rock crab and the Acadian hermit crab (Figure 5). At lengths > 75 cm TL, skate no longer fed upon

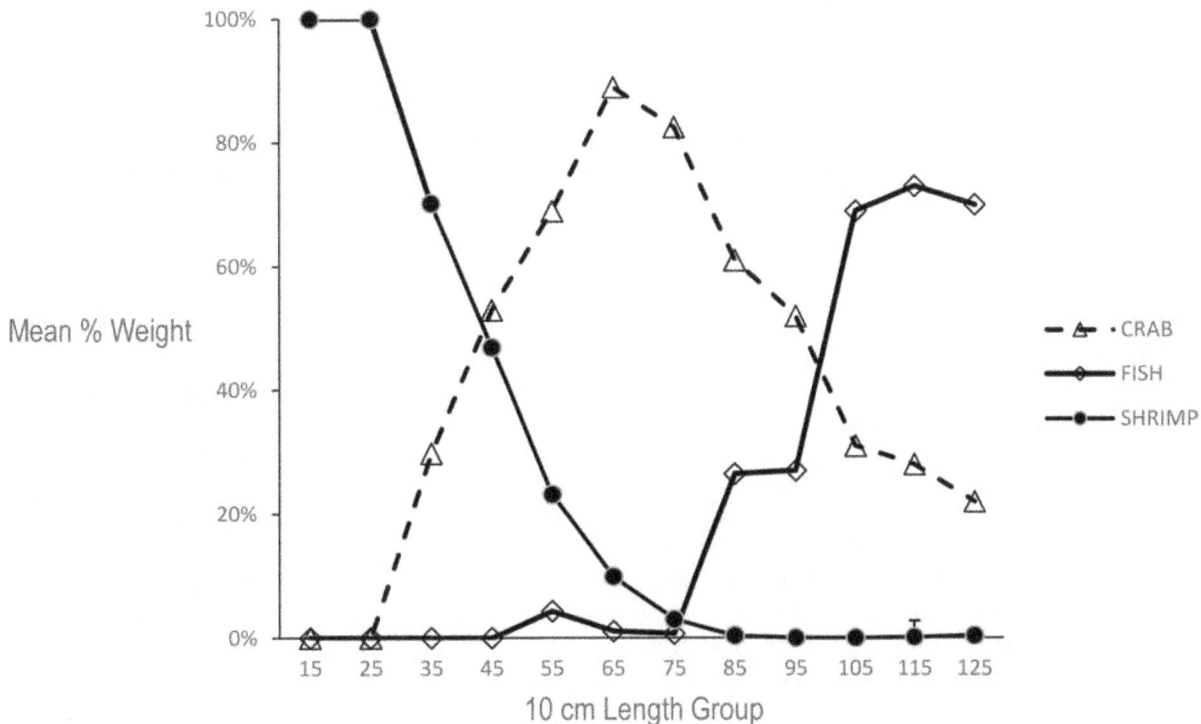

FIGURE 4. Percent by weight contribution of the major prey categories when Barndoor Skate are pooled into 10-cm length-groups.

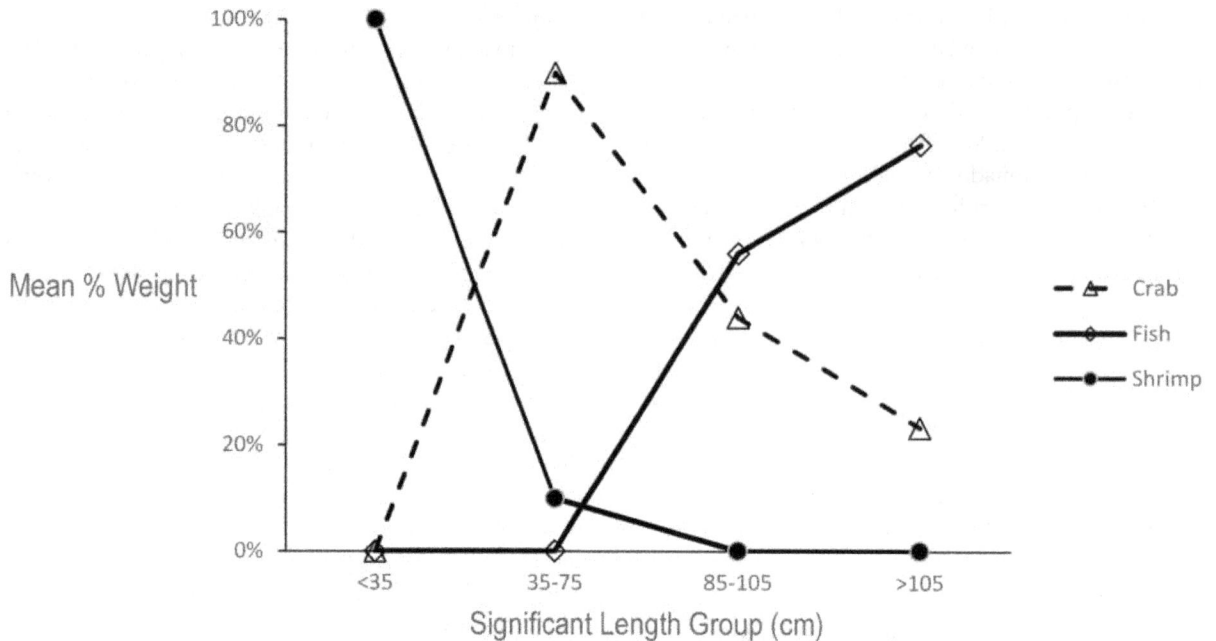

FIGURE 5.   Mean percent weight of each major prey category for all statistically significant length groupings of Barndoor Skate.

caridean shrimp and the prevalence of the rock crab in the diet began to decline in both males and females. At lengths > 105 cm TL, skate incorporated more fish into their diet, though crustaceans were still regularly consumed. A visual comparison of the diet composition for each 10-mm length-group (Figure 4) with the diet composition of the four significant length-groups (Figure 5) demonstrates that this simplification adequately captured the major dietary shifts.

Sex-specific analyses revealed significant feeding patterns for the largest male and largest female skate (ANOVA: $P <$ 0.01). Both sexes fed primarily on shrimp when small, and as they grew, they began to include other larger crustaceans in their diet (Figure 6). At lengths greater than 75 cm TL, however, males began to prey more heavily on teleost fish than did females and this sexually dimorphic feeding pattern became significant for skate > 85 cm TL (ANOVA: $P < 0.01$).

Prey-specific relative importance values were calculated for the three major prey categories when all stomach samples were pooled, and %PSIRI values were also calculated for the four significant length groupings (Figure 7). When stomach samples were pooled, crabs represented the most important prey item for the Barndoor Skate (%PSIRI = 43.43%), followed by shrimp and fish (Table 2). The relative importance of prey items for each length-group largely reflected the results of the gravimetric length-specific analysis, with some exceptions. The same general pattern remained for the smaller skate: shrimp were the most important prey for small skate and crabs were the most important prey type for intermediate-sized skate; however, crabs remained the most important prey item for all but the largest skate (>105 cm TL), which differed from the percent by weight indices (Figure 7).

DISCUSSION

The diet of Barndoor Skate was dominated by a limited number of prey items with clear ontogenetic shifts in food habits. Smaller skate relied entirely on benthic invertebrates while larger skate began including more fish in their diet. Previous studies on skate have elucidated similar patterns, but the behavior does not appear to be consistent, even for species studied in similar geographic regions. In a study of six skate species off the South African coast, Small and Cowley (1992) described three species as crustacean feeders (*Raja miraletus, R. clavata,* and *Cruriraja parcomaculata*), one as a specialist piscivore throughout its size range (*R. alba*) and two having ontogenetic changes in feeding habits. These two species, *R. wallacei* and *R. pullopunctata,* exhibited a pattern consistent with our results and fed primarily on crustaceans when they were small and then became mainly piscivorous when they became large. Ontogenetic changes in diet have also been described for a number of other batoid fishes (McEachran et al. 1976; Ajayi 1982; Platell et al. 1988; Orlov 1998) and have been attributed to morphological constraints (e.g., limited gape, tooth morphology) or better mobility, strength, and overall foraging ability of larger fish.

Our results for smaller skate would appear to support the hypothesis of morphological constraints as distinct shifts in diet were observed. The smallest individuals were specialized feeders foraging entirely on small carid shrimp, although other small prey such as crabs (*Cancer* spp.) would have been available. Individuals began to include other larger crustaceans in their diet once they exceeded 35 cm TL. A similar shift to include Acadian hermit crabs was observed at approximately 45 cm. Considering that all three of these prey are relatively

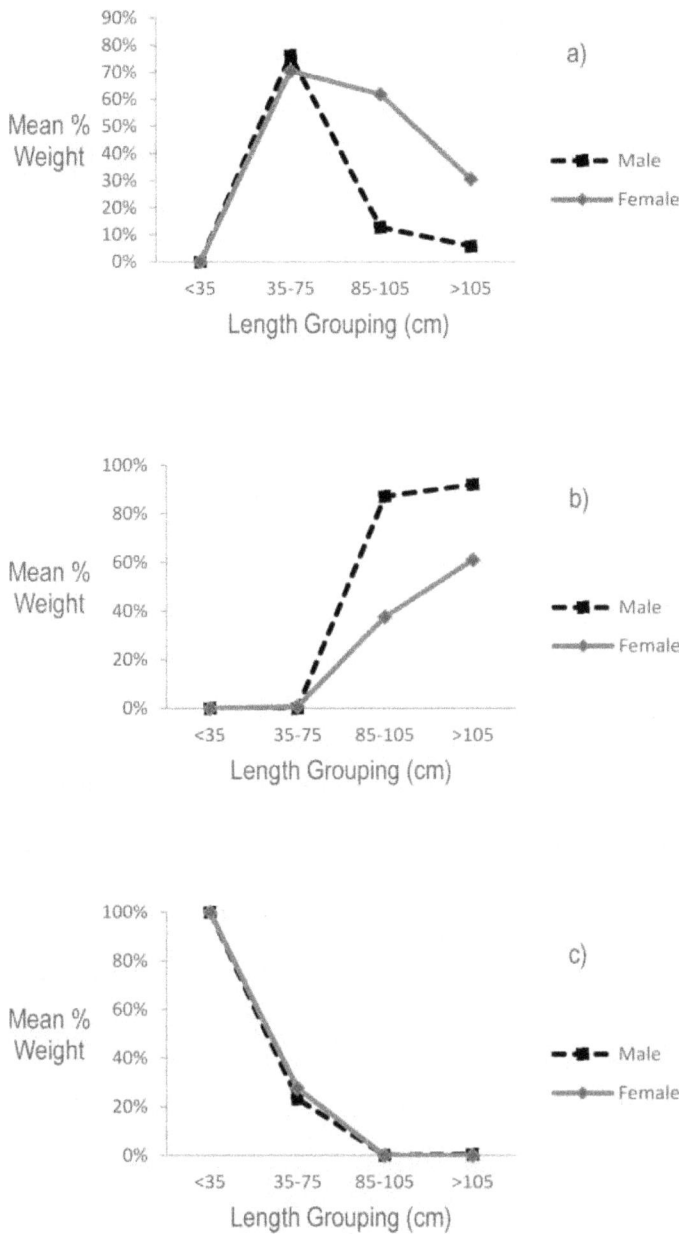

FIGURE 6. Sex-specific feeding patterns of the major prey categories used by Barndoor Skate. Mean percent weight of (**a**) crabs, (**b**) fish, and (**c**) shrimp for the four significant length groupings (<35 cm TL, 35–75 cm TL, 85–105 cm TL, and >105 cm TL).

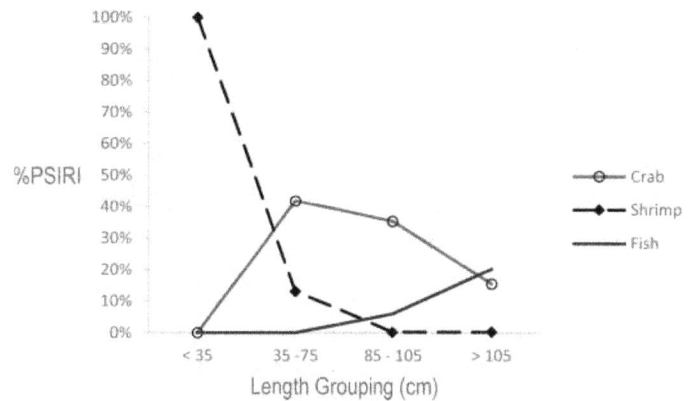

FIGURE 7. Percent prey-specific index of relative importance (%PSIRI) for the major prey categories and the four significant length-groups of the Barndoor Skate.

that the shells of Acadian hermit crabs were crushed prior to being ingested as prey.

The most interesting aspect of the ontogenetic changes in feeding habits we observed is the sex-specific habits of the largest skate. At lengths greater than 105 cm, male skate appeared to preferentially utilize teleost fish as prey (>80% by weight), while large females still preyed heavily upon larger crustaceans.

This sex-specific change in feeding habits became most pronounced at approximately the same size as when the skate reached maturity. Although there are a number of potential causes, the correlation between the size at maturity (males at 108 cm, females at 116 cm: Gedamke et al. 2005) and divergent sex-specific feeding habits is striking. One factor that is likely to play a role is sexual dimorphism in tooth structure. At maturity, female skate retain their molariform (i.e., plate like) teeth while males develop cuspidate dentition (i.e., pointed teeth). The dimorphism in tooth structure in the Barndoor Skate was apparent by even a cursory examination of sampled jaws. We noted that the development of cuspidate dentition in the males coincided with the development of other secondary sexual characteristics (i.e., allometric growth of claspers and development of alar thorns).

TABLE 2. Percent prey-specific index of relative importance (%PSIRI) for the major prey categories and the significant length-groups of the Barndoor Skate.

Length-group	%PSIRI			
	Crab	Shrimp	Fish	Other
All	43.43	16.09	11.68	1.15
<35 cm	0.00	100.00	0.00	0.00
35–75 cm	41.86	13.11	0.06	0.67
85–105 cm	35.38	0.26	6.03	1.40
>105 cm	15.52	0.30	20.24	3.57

slow moving, relatively common in our study area, and should not require great predatory swimming speeds to capture, there must be other factors limiting the Barndoor Skate from utilizing these food sources. Although the size of the mouth may play a role initially, many of the smaller *Cancer* spp. crabs and hermit crabs should be available as prey items. The simplest (but not the only) explanation may be found in the relationship between growth and the increased crushing power of the jaw required to crush thicker-shelled prey. However, very few shell fragments were observed in our samples indicating

In the elasmobranchs, the role of sexual dimorphic dentition is generally attributed to the reproductive behavior of the group and the ability of males to grasp and hold females during mating. Males will bite prospective mates in courtship behavior and during mating to facilitate insertion of the clasper and to maintain intromission (Springer 1960; Tricas and LeFeuvre 1985; Carrier et al. 1994). This behavior has been documented in a number of the batoids including the Atlantic Stingray *Dasyatis sabina* (Kajiura et al. 2000), the Eagle Ray *Aetobatis narinari* (Tricas 1980), the Roughtail Stingray *D. centroura* (Reed and Gilmore 1981), and the Round Stingray *Urolophus halleri* (Nordell 1994). Evolutionarily, the development of sexually dimorphic tooth morphology was likely to have evolved from not only the selective pressures of maximizing reproductive success but also from the selective pressures on both sexes for feeding efficiency.

Feduccia and Slaughter (1974) suggested that the strongly dimorphic tooth morphology in the rajids represents differential niche utilization between the sexes. This phenomenon has been demonstrated in bird, anole, and freshwater fish populations and results in reduced intraspecific competition for food, benefiting the population as a whole (Feduccia and Slaughter 1974). A number of authors studying the food habits of skate have suggested or shown that dentition plays a role in feeding habits (McEachran et al. 1976; Ebert et al. 1991; Smale and Cowley 1992), but as far as we are aware, this has not been confirmed for any elasmobranch.

If prey categories are of importance in determining trophic level, our study, along with that of Ebert and Bizzarro (2007), suggests that the Barndoor Skate falls into the upper trophic-level predator category within their ecosystem and therefore may influence the relative abundance and diversity of co-occurring demersal species (Beddington 1984; Rogers et al. 1999). Due to the ontogenetic shifts observed in our study, we suggest that future trophic-level calculations incorporate different size-classes. This would be beneficial because it would take into account the shift in prey selection with increasing body size. Because prey selection appears to become sexually dichotomous once skate reach maturity, future trophic level calculations should also consider the differences in diet composition by sex. Accurate calculations and comparisons of diet composition for different size-classes, sexes, and life histories are important and would be beneficial in determining the ecological role of the Barndoor Skate within the community food web. This is especially important in the George's Bank region because there is evidence that trophic levels vary between, and within, different ecosystems (Morato et al. 2003). It is possible that the higher proportion of teleosts found in the diets of mature males could place them in a significantly higher trophic level (Holden and Tucker 1974; Quiniou and Andriamirado 1979; Ajayi 1982; Ellis et al. 1996). This is particularly important in the region of George's Bank where skate predation may negatively influence recruitment of commercially important groundfish species (Murawski 1991; Mayo

et al. 1992; Fogarty and Murawski 1998). While there seems to be little evidence that Barndoor Skate are impacting important groundfish species, placement of the Barndoor Skate into an accurate trophic role would help determine their function in structuring the demersal marine communities in which they occur.

Although the correlation between divergent sex-specific feeding habits and sexually dimorphic dentition of Barndoor Skate at maturity was evident, we have not proven the case. Although sexually dimorphic tooth structure is the simplest and most likely explanation, this difference could have been due to other factors including depth segregation, nutritional needs, and/or size differences. For example, mature females may simply have different dietary needs than males, or the benthic feeding strategy of females may conserve energy that can be used for reproduction (Hanchet 1991; Stillwell and Kohler 1993; Simpfendorfer et al. 2001). Caution must be used before generalizing from our results because (1) all samples were collected from a small geographic area, (2) all samples were collected between June and November, and (3) samples were collected on commercial vessels actively fishing in the region. The small geographic area makes the interpretation of our results easier due to a limited number of prey species. Greater variability would have been observed if samples had been taken from a larger geographic area. For example, our samples from the Nantucket Light Ship closed area contained a different species of caridean shrimp than those from Georges Bank closed area II. Both of the prey species, *Crangon septemspinosa* and *Dichelopandalus leptocerus*, are morphologically similar and only reach maximum carapace lengths of 12 and 20 mm, respectively (Squires 1990). In each area, the smallest Barndoor Skate fed on only one species exhibiting clear specialization. If samples had been taken from a larger number of areas, a larger number of prey species would have been recorded for each size-class and interpreting the results may have been more difficult. As such, food habit studies should carefully consider the spatial aspects of sampling and resulting differences in prey availability.

Similarly, samples were only collected between June and November. Feeding patterns may be different at other times of the year. In fact, even our hypothesis pertaining to tooth morphology may have been more difficult to address if samples had been pooled over the entire year. In the Atlantic Stingray, Kajiura and Tricas (1996) showed that the molariform morphology of the teeth in females is stable while male dentition shows a periodic shift from a female-like molariform to a recurved cuspidate form during the reproductive season.

Finally, while the opportunity to sample onboard commercial vessels allowed us to obtain a large number of samples, a significant amount of bycatch was also introduced into the environment. Although this may have facilitated the capture of teleost prey used by our sampled Barndoor Skate population, the differential utilization by males and females would have persisted. Males and females were captured simultaneously in very similar abundances, and prey availability would have been constant for both sexes.

The analytical methods applied in this study were carefully chosen not only to address the problems with pooling data over large spatial scales, but also to deal with the significant limitations of pooling data over a wide range of size-classes. No one method can provide an accurate picture of the feeding habits of a species (Hyslop 1980); thus our application of the common metrics (%N, %W, and %O) to the entire sample set allowed us to identify common prey items, but the details of size-specific prey selection were obscured. Only after a careful analysis of length-specific feeding patterns and reanalysis over distinct size-classes did the ontogenetic shifts in feeding habits become apparent. The combined use of a length-specific graphic analysis and pooled metrics not only allowed for the primary population-wide food sources to be identified but also extracted a compelling picture of the specificity of food preferences at the different life stages of the Barndoor Skate.

ACKNOWLEDGMENTS

We thank D. Rudders, J. D. Lange Jr., B. Carroll, and R. Harshbarger for their collection efforts at sea and the captains and crews of the FV *Celtic*, FV *Alpha Omega*, FV *Barbara Ann*, FV *Heritage*, FV *Tradition*, and the FV *Mary Anne*. We also thank the reviewers whose input greatly improved previous drafts of this manuscript. This research was made possible by funds generated from the 1% set-aside of the scallop total allowable catch (TAC) under Framework Adjustment 13 of the Sea Scallop Fishery Management Plan (FMP) and Framework Adjustment 34 to the Northeast Multispecies FMP.

REFERENCES

Ajayi, T. O. 1982. Food and feeding habits of *Raja* species (Batoidei) in Carmarthen Bay, Bristol Channel. Journal of the Marine Biology Association of the UK 62:215–223.

Beddington, J. R. 1984. The response of multispecies systems to perturbations. Pages 209–225 *in* R. M. May, editor. Exploitation of marine communities. Springer, Berlin.

Bigelow, H. B., and W. C. Schroeder. 1953. Fishes of the western North Atlantic. Sawfishes, guitarfishes, skates and rays. Yale University, Sears Foundation for Marine Research Memoir I Part 2, New Haven, Connecticut.

Bigg, M. A., and M. A. Perez. 1985. Modified volume: a frequency-volume method to assess marine mammal food habits. Marine Mammals and Fisheries 1985:277–283.

Braccini, J. M., and J. E.Perez. 2005. Feeding habits of the sandskate *Psammobatis extenta* (Garman, 1913): sources of variation in dietary composition. Marine and Freshwater Research 56:395–403.

Brown, S. C., J. J. Bizzarro, G. M. Cailliet, and D. A. Ebert. 2012. Breaking with tradition: redefining measures for diet description with a case study of the Aleutian Skate *Bathyraja aleutica* (Gilbert 1896). Environmental Biology of Fishes 95:3–20.

Bulman, C., F. Althaus, X. He, N. J. Bax, and A. Williams. 2001. Diets and trophic guilds of demersal fishes of the southeastern Australian shelf. Marine and Freshwater Research 52:537–548.

Caddy, J. F., and G. D. Sharp. 1986. An ecological framework for marine fishery investigations. FAO (Food and Agriculture Organization of the United Nations) Fisheries Technical Paper 283.

Carrier, J. C, H. L. Pratt, and L. K. Martin. 1994. Group reproductive behaviors in free-living Nurse Sharks, *Ginglymostoma cirratum*. Copeia 1994:646–656.

Casey, J. M., and R. A. Myers. 1998. Near extinction of a large, widely distributed fish. Science 281:690–692.

Chipps, S. R., and J. E. Garvey, 2007. Quantitative assessment of food habits and feeding patterns. Pages 473–513 *in* C. Guy and M. Brown, editors. Analysis and interpretation of freshwater fisheries data. American Fisheries Society, Bethesda, Maryland.

Collie, J. S., and A. K. DeLong. 1999. Multispecies interactions in the Georges Bank fish community. Pages 187–210 *in* B. Baxter, S. Keller, and C. Kaynor, editors. Ecosystem approaches for fisheries management. Alaska Sea Grant College Program, AK-SG-99-01, Anchorage.

Cortes, E. 1997. A critical review of methods of studying fish feeding based on analysis of stomach contents: application to elasmobranch fishes. Canadian Journal of Fisheries and Aquatic Sciences 54:726–738.

Davenport, S. R., and N. J. Bax. 2002. A trophic study of a marine ecosystem off southeastern Australia using stable isotopes of carbon and nitrogen. Canadian Journal of Fisheries and Aquatic Sciences 59:514–530.

Ebert, D. A., and J. J. Bizzarro. 2007. Standardized diet compositions and trophic levels of skates (Chondrichthyes: Rajiformes: Rajoidei). Environmental Biology of Fishes 80:221–37.

Ebert, D. A., and L. J. V. Compagno. 2007. Biodiversity and systematics of skates (Chondrichthyes: Rajiformes: Rajoidei). Environmental Biology of Fishes 80:111–124.

Ebert, D. A., P. D. Cowley, and L. J. Compagno. 1991. A preliminary investigation of the feeding ecology of skates (Batoidea: Rajidae) off the west coast of South Africa. South African Journal of Marine Science 10:71–81.

Ellis, J. R., M. G. Pawson, and S. E. Shackley. 1996. The comparative feeding ecology of six species of shark and four species of ray (Elasmobranchii) in the north-east Atlantic. Journal of the Marine Biology Association of the UK 76:89–106.

Feduccia, A., and B. H. Slaughter. 1974. Sexual dimorphism in skates (Rajidae) and its possible role in differential niche utilization. Evolution 28:164–168.

Fogarty, M. J., and S. A. Murawski. 1998. Large-scale disturbance and structure of marine systems: fishery impacts on Georges Bank. Ecological Applications 8:S6–S22.

Garrison, L. P., and J. S. Link. 2000. Fishing effects on spatial distribution and trophic guild structure of the fish community in the Georges Bank region. ICES Journal of Marine Science 57:723–730.

Gedamke, T. 2006. Development of a stock assessment for the Barndoor Skate (*Dipturus laevis*) on Georges Bank. Doctoral dissertation. College of William and Mary, Virginia Institute of Marine Science, Gloucester Point, Virginia.

Gedamke, T., W. D. DuPaul, and J. A. Musick. 2005. Observations on the life history of the Barndoor Skate, *Dipturus laevis*, on Georges Bank (western North Atlantic). Journal of Northwest Atlantic Fishery Science 35:67–78.

Graham, B. S., D. Grubbs, K. Holland, and B. N. Popp. 2007. A rapid ontogenetic shift in the diet of juvenile Yellowfin Tuna from Hawaii. Marine Biology 150:647–658.

Hanchet, S. 1991. Diet of Spiny Dogfish, *Squalas acanthias* Linnaeus, on the east coast, South Island, New Zealand. Journal of Fish Biology 39:313–323.

Holden, M. J., and R. N. Tucker. 1974. The food of *Raja clavata* Linnaeus 1758, *Raja montagui* Fowler 1910, *Raja naevus* Mueller and Henle 1841, and *Raja brachyura* LaFont 1873 in British waters. Journal du Conseil Permanent International pour l'Exploration de la Mer 35:189–193.

Hyslop, E. J. 1980. Stomach content analysis—a review of methods and their application. Journal of Fish Biology 17:411–429.

Kajiura, S. M., A. P. Sebastian, and T. C. Tricas. 2000. Dermal bite wounds as indicators of reproductive seasonality and behavior in the Atlantic Stingray, *Dasyatis sabina*. Environmental Biology of Fishes 58:23–31.

Kajiura, S. M., and T. C. Tricas. 1996. Seasonal dynamics of dental sexual dimorphism in the Atlantic Stingray, *Dasyatis sabina*. Journal of Experimental Biology 199:2297–2306.

Kulka, D. W., K. T. Frank, and J. E. Simon. 2002. Barndoor Skate in the Northwest Atlantic off Canada: distribution in relation to temperature and depth based on commercial fisheries data. Department of Fisheries and Oceans, Atlantic Fisheries Research Document 2002/073, Ottawa.

Leim, A. H., and W. B. Scott. 1966. Fishes of the Atlantic coast of Canada. Fishery Research Board of Canada Bulletin 155.

Macdonald, J. S., and R. H. Green. 1983. Redundancy of variables used to describe importance of prey species in fish diets. Canadian Journal of Fisheries and Aquatic Sciences 40:635–637.

Mayo, R. K., M. J. Fogarty, and F. M. Serchuk. 1992. Aggregate fish biomass and yield on Georges Bank, 1960–87. Journal of Northwest Atlantic Fisheries Science 14:59–78.

McEachran, J. D., D. F. Boesch, and J. A. Musick. 1976. Food division within two sympatric species - pairs of skates (Pisces: Rajidae). Marine Biology 35:301–317.

McEachran, J. D., and J. A. Musick. 1975. Distribution and relative abundance of seven species of skates (Pisces: Rajidae) which occur between Nova Scotia and Cape Hatteras. U.S. National Marine Fisheries Service Fishery Bulletin 73:110–136.

Morato, T., E. Sola, and M. P. Gros. 2003. Diets of Thornback Ray (*Raja clavata*) and Topeshark (*Galeorhinus galeus*) in the bottom longline fishery of the Azores, northeastern Atlantic. U.S. National Marine Fisheries Service Fishery Bulletin 101:590–602.

Murawski, S. A. 1991. Can we manage our multispecies fisheries? Fisheries 16(5):5–13.

NEFMC (New England Fisheries Management Council). 2001. Stock assessment and fishery evaluation (SAFE) report for the northeast skate complex. NEFMC, Newburyport, Massachusetts.

NOAA (National Oceanic and Atmospheric Administration). 2003. Essential fish habitat source document: Barndoor Skate, *Dipturus laevis*, life history and habitat characteristics. NOAA Technical Memorandum NMFS-NE-173.

Nordell, S. E. 1994. Observations of the mating behavior and dentition of the round stingray, *Urolophus halleri*. Environmental Biology of Fishes 39:219–229.

Orlov, A. M. 1998. The diets and feeding habits of some deep-water benthic skates (Rajidae) in the Pacific waters off the northern Kuril Islands and southeastern Kamchatka. Alaska Fisheries Research Bulletin 5:1–17.

Ortaz, M., P. B. Von Bach, and R. Candia. 2006. The diet of the neotropical insectivorous fish *Creagrutus bolivari* (Pisces: Characidae) according to the "graphic" and "relative importance" methods. Revista de Biologia Tropical 54:1227–1239.

Pinkas, L., M. S. Oliphant, and L. K. Iverson. 1971. Food habits of Albacore, Bluefin Tuna, and Bonito in California waters. California Department of Fish and Game Fish Bulletin 152.

Platell, M. E., I. C. Potter, and K. R. Clarke. 1988. Resource partitioning by four species of elasmobranchs (Batoidea: Urolophidae) in coastal waters of temperate Australia. Marine Biology 131:719–734.

Posgay, J. A. 1957. Sea scallop boats and gear. U.S. Fish and Wildlife Service Fishery Leaflet 442.

Quiniou, L., and G. R. Andriamirado. 1979. Variations of the diet of three species of rays from Douarnenez Bay (*Raja montagui* Fowler, 1910; *Raja brachyura* Lafont, 1873; *Raja clavata* L.,1758). Cybium 3:27–39.

Reed, J. K., and R. G. Gilmore. 1981. Inshore occurrence and nuptial behavior of the Roughtail Stingray, *Dasyatis centroura* (Dasyatidae), on the continental shelf, east central Florida. Northeast Gulf Science 5:59–62.

Rogers, S. I., K. R. Clarke, and J. D. Reynolds. 1999. The taxonomic distinctness of coastal bottom-dwelling fish communities of the Northeast Atlantic. Journal of Animal Ecology 68:769–782.

Simpfendorfer, C. A., A. B. Goodreid, and R. B. McAuley. 2001. Size, sex and geographic variation in the diet of the Tiger Shark, *Galeocerdo cuvier*, from western Australian waters. Environmental Biology of Fishes 61:37–46.

Sissenwine, M. P., E. B. Cohen, and M. D. Grosslein. 1984. Structure of the Georges Bank ecosystem. Rapports et Proces-Verbaux des Reunions du Conseil International pour l'Exploration de la Mer 183:243–254.

Smale, M. J., and P. D. Cowley. 1992. The feeding ecology of skates (Batoidea: Rajidae) off the Cape South Coast, South Africa. South African Journal of Marine Science 12:823–834.

Small, M. J., and P. D. Cowley. 1992. The feeding ecology of skates (Batoidea: Rajidae) off the cape south coast, South Africa. South African Journal of Marine Science 12:823–834.

Springer, S. 1960. Natural history of the Sandbar Shark, *Eulamia milberti*. U.S. National Marine Fisheries Service Fishery Bulletin 61:1–38.

Squires, H. J. 1990. Decapod crustacea of the Atlantic coast of Canada. Canadian Bulletin of Fisheries and Aquatic Sciences 221.

Stillwell, C. E., and N. E. Kohler. 1993. Food habits of the Sandbar Shark *Carcharhinus plumbeus* off the U.S. northeast coast, with estimates of daily ration. U.S. National Marine Fisheries Service Fishery Bulletin 91:138–150.

Tricas, T. C. 1980. Courtship and mating-related behaviors in myliobatid rays. Copeia 1980:553–556.

Tricas, T. C., and E. M. LeFeuvre. 1985. Mating in the Reef White-tip Shark *Triaenodon obesus*. Marine Biology 84:233–237.

Warton, D. I., and F. K. Hui. 2011. The arcsine is asinine: the analysis of proportions in ecology. Ecology 92:3–10.

Zar, J. H. 1999. Biostatistical analysis, 4th edition. Pearson Education, Delhi, India.

5

Temporal and Spatial Dynamics of the Lionfish Invasion in the Eastern Gulf of Mexico: Perspectives from a Broadscale Trawl Survey

Theodore S. Switzer*
Florida Fish and Wildlife Conservation Commission, Fish and Wildlife Research Institute, 100 8th Avenue Southeast, St. Petersburg, Florida 33701, USA

Derek M. Tremain
Florida Fish and Wildlife Conservation Commission, Fish and Wildlife Research Institute, Indian River Field Laboratory, 1220 Prospect Avenue, Suite 285, Melbourne, Florida 32901, USA

Sean F. Keenan, Christopher J. Stafford, Sheri L. Parks, and Robert H. McMichael Jr.
Florida Fish and Wildlife Conservation Commission, Fish and Wildlife Research Institute, 100 8th Avenue Southeast, St. Petersburg, Florida 33701, USA

Abstract

 The recent introduction of invasive Indo-Pacific lionfish species (Red Lionfish *Pterois volitans* and Devil Firefish *P. miles*, hereafter collectively referred to as lionfish) into the western Atlantic Ocean has been extensively documented in both the scientific literature and the media. Nevertheless, much of the information synthesized has been obtained via diver-based surveys and there is likely a depth-related bias to the understanding of the temporal and spatial dynamics of the lionfish invasion. Accordingly, we examined data from a broadscale fisheries-independent trawl survey of bare substrates and low-relief habitats that was initiated in 2008 in the eastern Gulf of Mexico. Lionfish were first observed in the survey in 2010, when two individuals were collected off southwestern Florida. The distribution of lionfish continued to expand northward through the Florida panhandle in 2011 and 2012, when 40 and 29 lionfish were collected, respectively. A dramatic increase in the abundance (391 individuals) and distribution of lionfish occurred in 2013. Evidence from this survey suggests that lionfish first colonized deeper (>30 m) low-relief habitats before populations expanded into shallower waters. The prevalence of lionfish on primarily nonreef habitats at depths beyond those frequented by recreational divers will likely have important implications for efforts to control or eradicate lionfish populations in the region. Moving forward, information from long-term, multispecies surveys such as this will continue to provide valuable insight into the spatial and temporal dynamics of the lionfish invasion and allow us to assess long-term ecological consequences of increasing lionfish abundances.

Range expansion into the western Atlantic Ocean by two invasive Indo-Pacific lionfish species, the Red Lionfish *Pterois volitans* and the Devil Firefish *P. miles* (hereafter collectively referred to as lionfish), has progressed at an unprecedented rate. Lionfish were first reported off southeastern Florida in the mid-1980s; the distribution of reported lionfish sightings remained localized through 1999, after which they rapidly expanded their range (Schofield 2009, 2010; Johnston and Purkis 2011). From 2000 to 2006, lionfish expanded northward along the eastern U.S. coastline, to Bermuda, and subsequently to the Bahamas.

Subject editor: Kenneth Rose, Louisiana State University, Baton Rouge

*Corresponding author: ted.switzer@myfwc.com

Since 2007, lionfish have spread throughout the Caribbean, reaching the Florida Keys in 2009 (Ruttenberg et al. 2012) and the Gulf of Mexico in 2010 (Schofield 2010; Fogg et al. 2013). Often strongly associated with reef habitats (Schultz 1986; Biggs and Olden 2011; Claydon et al. 2012), lionfish in the western Atlantic Ocean have been found to occupy mangrove (Barbour et al. 2010; Claydon et al. 2012; Pimiento et al. 2015), seagrass (Biggs and Olden 2011; Claydon et al. 2012), and lower riverine habitats (Jud and Layman 2012) as well.

Although the invasion and subsequent expansion of lionfish populations throughout the western Atlantic Ocean have been generally well documented (Côté et al. 2013), most studies have relied heavily on data collected by recreational or scientific divers (Schofield 2009, 2010; Ruttenberg et al. 2012; Scyphers et al., in press). As a result, the understanding of the dynamics of the lionfish invasion in waters deeper than those routinely sampled by diver-based surveys (~35 m) is somewhat restricted. Several studies have documented lionfish at depths as great as 100 m (Meister et al. 2005; Whitfield et al. 2007; Lesser and Slattery 2011; Nuttall et al. 2014), where water temperatures are well within the thermal tolerances of the species (Kimball et al. 2004). The effectiveness of diver-based control efforts directed at shallow-water lionfish populations may be undermined by rapidly increasing lionfish populations at greater depths. A more quantitative examination of depth-related lionfish population dynamics is essential for informing population control strategies and quantifying ecological impacts. Accordingly, we analyzed data from a broad-scale, fisheries-independent trawl survey to (1) characterize the range expansion of lionfish in the eastern Gulf of Mexico from 2010 through 2013 and (2) assess depth-associated patterns in lionfish abundance, frequency of occurrence, and size.

METHODS

Study area.—Our analyses focused on data collected in the eastern Gulf of Mexico, from the Dry Tortugas north to the Florida–Alabama border, at depths from 4 to 104 m. Sediment composition in the eastern Gulf of Mexico is dominated by quartz-rich sand on the inner shelf, mollusk-rich sand over a broad area of the middle shelf, and sand rich in coralline algae on the outer shelf (Randazzo and Jones 1997). Although trawlable, nonreef bottom habitat is abundant, most of the natural hard-bottom habitat in the Gulf of Mexico is found off of Florida and the Yucatan Peninsula, with patches of coral and sponge habitat occurring extensively along the West Florida Shelf (WFS) (Briggs 1958; McEachran and Fechhelm 1998). Much of the multibillion-dollar fishing industry in the eastern Gulf of Mexico is derived from species associated with these hard-bottom habitats.

Field methods.—Data were collected as part of the recent Florida expansion of the Southeast Area Monitoring and Assessment Program's (SEAMAP) annual summer groundfish trawl survey. This survey (Eldridge 1988) employs a stratified-random sampling design in which annual sampling effort is proportionally allocated among depth and geographic strata. Initiated in the early 1980s, the SEAMAP groundfish trawl survey originally extended from the Mississippi–Alabama border westward to the Mexico border, encompassing National Marine Fisheries Service (NMFS) statistical reporting zones 11–21. In 2008 and 2009, exploratory summer surveys were conducted from Tampa Bay to Alabama (NMFS zones 5–10) to investigate the feasibility of expanding the SEAMAP groundfish survey into the eastern Gulf of Mexico; this survey was expanded to encompass NMFS zones 2–10 in 2010 (Figure 1). All samples were collected during June and July using a standard 12.8-m SEAMAP shrimp trawl towed at a speed of 3 knots, and tow duration was generally 30 min (bottom sampling area = approximately 1.03 ha/tow). Trawls were typically towed over bare substrates or low-relief habitats to minimize damage to sensitive bottom communities. All lionfish collected were enumerated and measured to the nearest millimeter standard length (SL), and pertinent site information was recorded, including location and water depth. Additional survey details can be found in Rester (2011).

Analytical methods.—To visually explore the patterns of lionfish expansion in the eastern Gulf of Mexico, the locations where trawl sampling was conducted and those where lionfish were collected were plotted annually in a GIS, with symbol size being proportional to the number of lionfish collected. No lionfish were collected in the 2008 or 2009 trawl surveys, so those data were not included in subsequent analyses. For data collected between 2010 and 2013, summary statistics were calculated of the annual trawl sampling effort, frequency of lionfish occurrence (percentage of annual trawl samples that contained at least one lionfish), total number of lionfish collected, and mean number of lionfish collected per trawl. Annual length frequency distributions were also constructed, and a series of Kolmogorov–Smirnov two-sample tests, using the Bonferroni correction for multiple pairwise comparisons ($\alpha = 0.05/3$ or 0.017), were used to compare length frequency distributions between all years excluding 2010 (SAS Institute 2006; Sokal and Rohlf 2012). For 2013 data only, the mean number of lionfish per haul and mean size of lionfish collected (mm SL) were compared among depth bins using one-way analysis of variance (ANOVA) and the Tukey–Kramer adjustment for pairwise comparisons (SAS Institute 2006); depth intervals were chosen to divide the data into six depth quantiles. The frequency of occurrence data, in terms of overall sampling effort as well as samples containing lionfish, were summarized for 10-m depth bins and analyzed via habitat suitability analysis (Baltz 1990) to explore nonlinear patterns of habitat selection along the gradient of depths sampled.

RESULTS

Lionfish were first collected in survey trawls in 2010 (Figure 2; Table 1), when two individuals were captured off

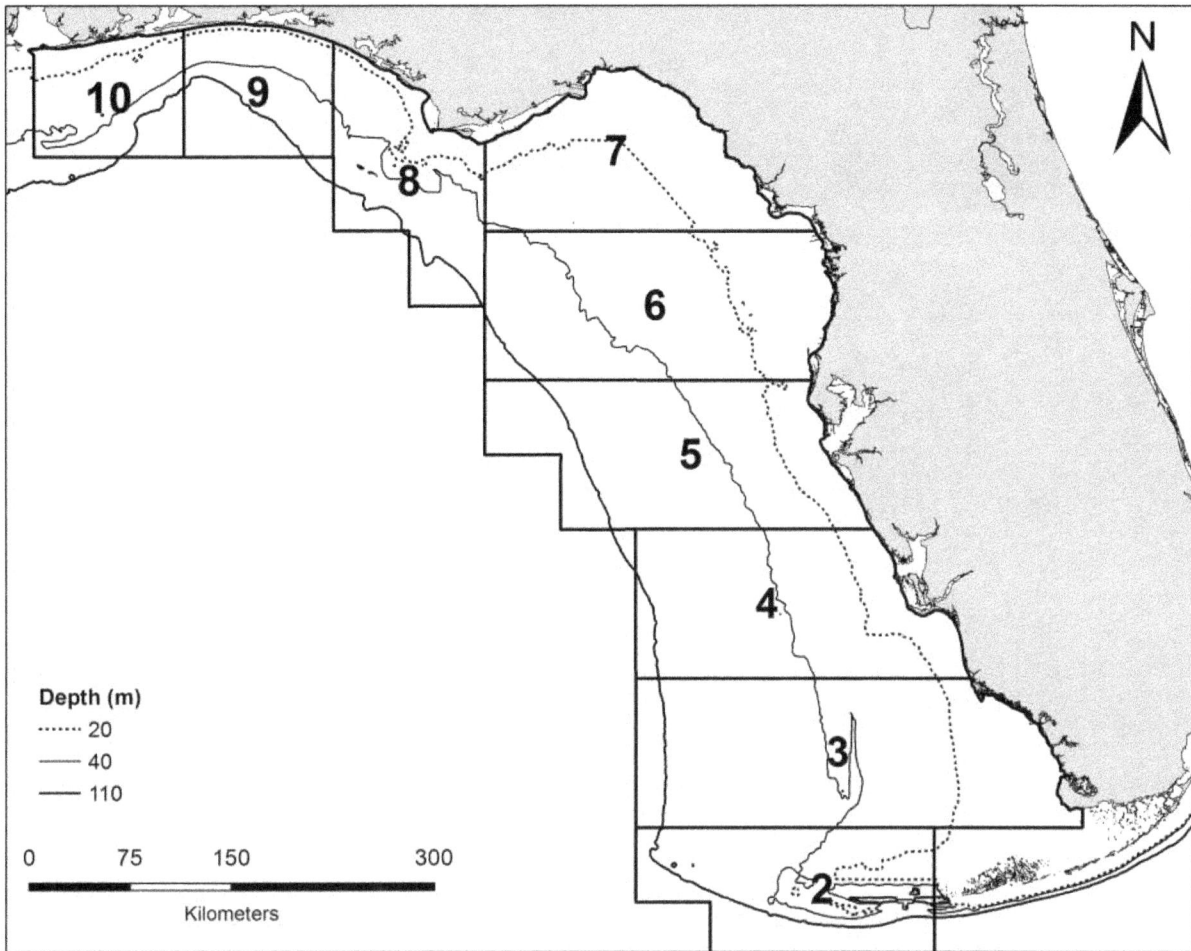

FIGURE 1.   Spatial extent of the Southeast Area Monitoring and Assessment Program (SEAMAP) trawl survey in the eastern Gulf of Mexico. Annual sampling effort is allocated proportionally among statistical reporting zones (2–10) based on the total area of the seafloor from 4 to 110 m.

Florida's southwest coast in depths of approximately 45 m. In 2011 and 2012, lionfish were more abundant ($N = 40$ and $N = 29$, respectively) and catches expanded northward into waters off the Florida panhandle. This expansion first occurred primarily in deeper waters but extended inshore after 2012. A dramatic increase in abundance occurred in 2013 ($N = 391$ lionfish). That year, lionfish were collected in 40% of all trawl samples and at a mean rate of 2.57 individuals/haul (SE, 0.44).

The size of lionfish captured in trawls has increased since the initial invasion into the eastern Gulf of Mexico (Table 1; Figure 3); at $\alpha = 0.017$ for each pairwise test, length frequency distributions did not differ between 2011 and 2012 ($P_{KS} = 0.03$) but did differ between 2011 and 2013 ($P_{KS} < 0.01$), as well as between 2012 and 2013 ($P_{KS} < 0.01$). In 2010, both individuals collected were less than 100 mm SL, but maximum size had exceeded 300 mm by 2012 and 400 mm by 2013. In 2013, both mean lionfish abundance ($F = 9.18$; $P < 0.01$) and size ($F = 3.60$; $P < 0.01$) differed significantly among the depths sampled (Figure 4). The mean

abundance of lionfish was significantly greater in depths from 49 to 67 m than it was in depths less than 35 m, whereas the mean size of lionfish was significantly greater in depths from 49 to 67 m than it was in depths from 27 to 35 m. Overall, lionfish exhibited the highest suitability at depths from 30 to 80 m (Figure 5); no lionfish were collected in waters shallower than 20 m in any year of the survey.

DISCUSSION

This study in the eastern Gulf of Mexico provides the first quantitative description of the lionfish expansion that considers predominantly nonreef habitats and depths beyond those examined in most prior studies. Although the lionfish invasion has been well documented in general, most available literature has emphasized the colonization of shallow habitats (e.g., reefs or various estuarine habitats) at depths accessible to divers. However, lionfish have been observed in waters as deep as 100 m in the western Atlantic Ocean (Meister et al. 2005; Whitfield et al. 2007; Lesser and Slattery 2011) and

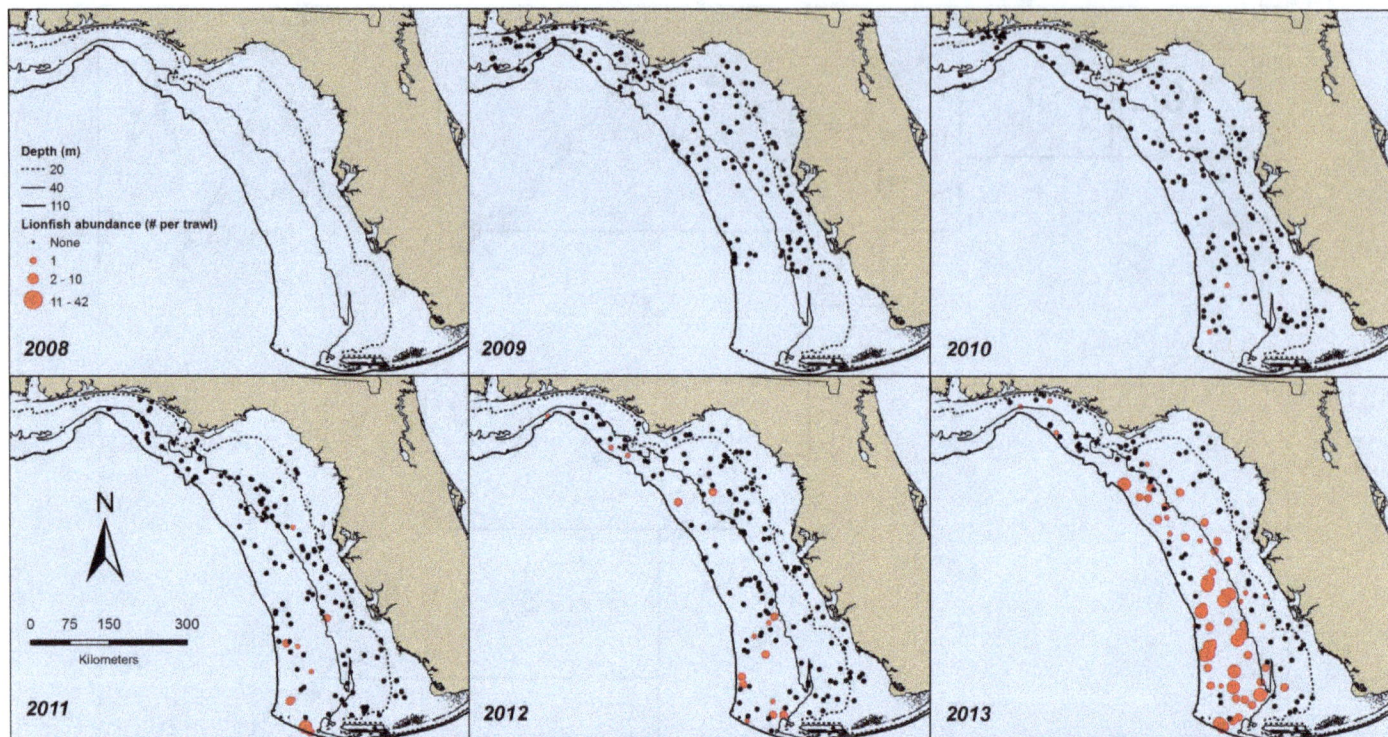

FIGURE 2.   Spatial distribution of lionfish collected during the annual (2008–2013) summer SEAMAP trawl surveys in the eastern Gulf of Mexico. The black circles represent the locations where trawl samples were collected and no lionfish were captured. The red circles represent the locations where lionfish were collected, and the size of each red circle represents the relative number of lionfish collected within each set.

northwestern Gulf of Mexico (Nuttall et al. 2014). Using a stratified-random sampling design across depths to 104 m, we detected the initial expansion into the southeastern Gulf in 2010, which coincides well with the first reports from the lower Florida Keys (Schofield 2010). However, diver-based sightings of lionfish during late 2010 in shallower waters off the Florida panhandle and central western coast (USGS 2014) suggest there were likely multiple and simultaneous expansion pathways into the eastern Gulf of Mexico. Because the SEA-MAP trawl survey did not include the extreme southeastern portion of the gulf before 2010, we cannot confirm or refute the notion that introduction into the gulf was possible as early as 2008, an idea based on the projected age at length (e.g., Barbour et al. 2011) of a single large specimen collected in 2012 (Fogg et al. 2013). Nevertheless, the first reported lionfish sighting from the Dry Tortugas did not occur until late 2009 (Schofield 2010), so it is doubtful that lionfish would have been collected even if trawl effort had been allocated to the region during that period.

The results from this study indicate that in the eastern Gulf of Mexico lionfish likely first settled in deeper habitats along the WFS. Central America is identified as the probable source of lionfish in the gulf (Johnston and Purkis 2011), and deeper WFS habitats would have been the first ones encountered by larvae transported by prevailing currents (e.g., the Yucatan and Loop currents). The incursion of the Loop Current into the

Gulf of Mexico varies during the year, typically attaining its most northerly intrusion during the summer (Sturges and Evans 1983). With Loop Current surface velocities exceeding 60 cm/s in early summer 2010 (Hamilton et al. 2011) and the mean settlement age for planktonic Red Lionfish larvae estimated at 26.5 d (Ahrenholz and Morris 2010), lionfish larvae from the Yucatan could have been transported more than 1,300 km before settlement, placing them along the eastern wall of the Loop Current and in proximity to the WFS. Drift-buoy trajectories recorded during that period also identified cyclonic eddy flows along the eastern Loop Current wall that forced a northward counterflow along the west Florida slope (Hamilton et al. 2011). The overall increase in size from 2010 to 2012, combined with distribution records, provides evidence of a general northward expansion within deeper waters prior to the population expansion inshore. Lionfish become reproductively active within their first year (Morris and Whitfield 2009), and a growing pool of larvae originating from both early colonizers to the WFS and from exponentially growing upstream populations likely facilitated a secondary and more radial population expansion across WFS habitats. Four years after the initial invasion onto the WFS, lionfish densities remain highest in depths of 40–80 m, similar to recent observations in mesophotic depths of the northwestern Gulf of Mexico (80–90 m; Nuttall et al. 2014). In their native range, lionfish are collected most often in shallower waters

TABLE 1. Annual sampling effort and overall catch data for lionfish collected during the summer SEAMAP trawl survey in the eastern Gulf of Mexico.

Year	Total number of samples	Mean (range) sampling depth (m)	Number (and percent) of samples containing lionfish	Total number of lionfish collected	Mean ± SE lionfish per haul	Mean (range) standard length (mm)
2010	161	39 (7–100)	2 (1.2%)	2	0.01 ± 0.01	91 (85–97)
2011	143	38 (4–97)	9 (6.3%)	40	0.28 ± 0.14	174 (129–251)
2012	162	37 (9–101)	16 (9.9%)	29	0.18 ± 0.05	172 (70–337)
2013	152	41 (5–104)	61 (40.1%)	391	2.57 ± 0.44	208 (62–404)

(Kulbicki et al. 2012); as populations continue to increase in abundance, we expect lionfish density in shallower waters to increase.

Although gonads were not analyzed, a significant proportion of lionfish collected in this study were large enough to be reproductively active (Morris and Whitfield 2009), so nonreef habitats in deeper waters may be an important source of lionfish larvae. In general, the SEAMAP trawl survey is restricted to bare substrates or low-relief habitats, but because very little high-resolution habitat information is available for much of the study area, some samples were collected over or near live bottom (sponges, gorgonians, etc.) and higher-relief reef structure. Lionfish use a variety of marine substrates in both their native and nonnative ranges, but they are most commonly associated with structured habitats such as reefs, mangrove swamps, and artificial structure (Barbour et al. 2010; Kulbicki et al. 2012; Schofield et al. 2014). Several studies have documented, to some extent, ontogenetic shifts in habitat affinity.

In studies from Roatán, Honduras (Biggs and Olden 2011), and the Turks and Caicos islands (Claydon et al. 2012), smaller lionfish tended to occupy seagrass habitats, whereas mature individuals were associated with structured reef environments. Consequently, the lionfish abundance and length data collected from low-relief habitats in the eastern Gulf of Mexico may not represent the portion of the population associated with more structured habitats. Furthermore, the size distribution of our catch is strongly influenced by the trawl sampling gear and small postsettlement individuals were not collected in our surveys. Nevertheless, these data provide a conservative and quantitative estimate of the rapid population growth in the eastern Gulf of Mexico. To fully describe the dynamics of lionfish populations in these deeper environments will require data from sampling methods complementary to this trawl survey, such as traps or underwater cameras, that can effectively quantify biota in reef or live-bottom habitats (Bacheler et al. 2013; Dahl and Patterson 2014; Nuttall et al. 2014).

The widespread establishment of lionfish populations in the eastern Gulf of Mexico beyond depths accessible to divers

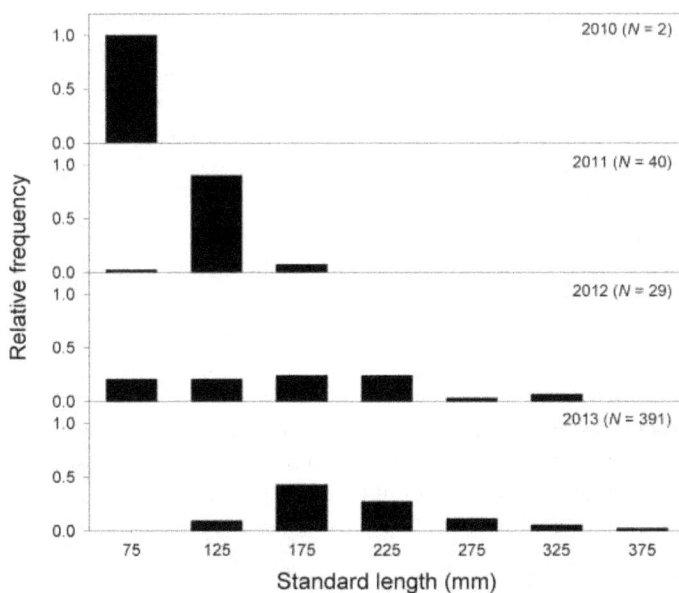

FIGURE 3. Annual length frequency distribution of lionfish collected during the summer SEAMAP trawl surveys in the eastern Gulf of Mexico (2008–2013; no lionfish were collected in 2008 or 2009). Values along the x-axis represent midpoints of 50-mm size-class bins.

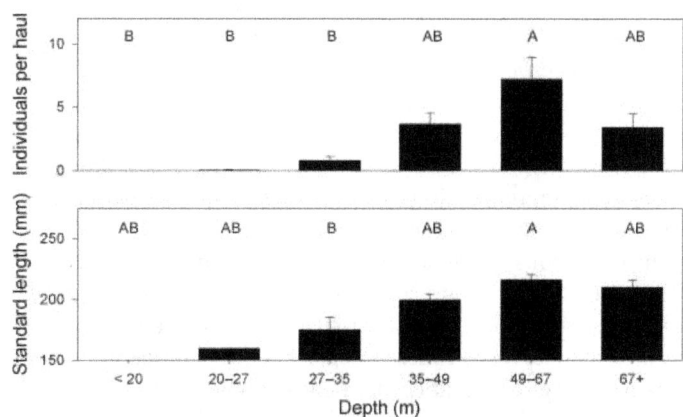

FIGURE 4. By depth, the average number of lionfish per trawl (upper panel) and the average standard length of lionfish (lower panel) collected during the summer 2013 SEAMAP trawl survey in the eastern Gulf of Mexico (error bars indicate SE). Mean values were compared by ANOVA, and the letters above each bar represent groupings as determined via pairwise tests between depth quantiles (means with at least one letter in common are not significantly different).

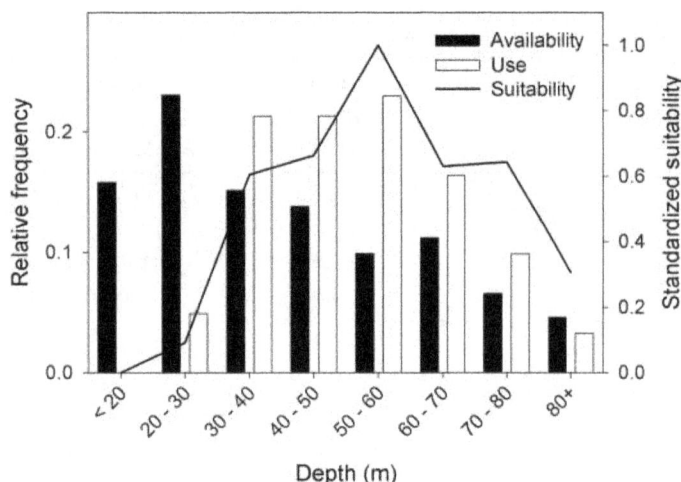

FIGURE 5. Depth-associated habitat suitability for lionfish collected during the summer 2013 SEAMAP trawl survey in the eastern Gulf of Mexico. The black bars represent the relative frequency of depth sampled, the white bars represent the relative frequency of depths within which lionfish were captured, and the black line represents the calculated habitat suitability.

likely has important implications for control strategies and fisheries management. There is little evidence that lionfish populations are vulnerable to biological controls such as predation (Hackerott et al. 2013), and there is very limited information on their susceptibility to parasitism (Ruiz-Carus et al. 2006; Bullard et al. 2011) or disease in nature. A single lionfish, believed to have been released from an aquarium, was collected in 2006 in association with a red tide bloom near Pinellas County, Florida (Schofield 2010), and frequent episodic red tide events in the eastern gulf may provide some level of control in shallow coastal habitats as populations expand into nearshore regions. At present, removal of lionfish by divers is probably the most common method of control, but this method is generally applicable to waters shallower than ~35 m. Lionfish are rarely caught in hook-and-line fisheries but have been reported as incidental catches in some deepwater fisheries (Akins 2012) and are frequent bycatch in commercial trap fisheries (National Marine Fisheries Service, Southeast Fisheries Science Center, Trip Interview Program, personal communication). Recently developed models predict that containment of lionfish populations will prove very difficult if portions of the adult populations remain unexploitable (Arias-González et al. 2011). Accordingly, the development of directed trap fisheries for lionfish may offer alternatives to removal by divers in these deeper habitats. In deeper waters, many ecologically and economically important reef fishes utilize habitats that overlap with those of lionfish, and species such as Red Grouper *Epinephelus morio*, Vermilion Snapper *Rhomboplites aurorubens*, Gray Snapper *Lutjanus griseus*, and Lane Snapper *L. synagris* were often caught in conjunction with lionfish in our survey trawls. The ecological effect of proliferating lionfish populations on these economically important native species and their prey base is unknown, but

recent investigations conducted in shallower waters of the Bahamas document the potential for adverse impacts (Albins and Hixon 2013) on native reef fish recruitment (Albins and Hixon 2008) and prey species' biomass (Côté and Maljkovic 2010; Green et al. 2012). At mesophotic depths, declines in coral reef herbivores caused by lionfish predation or avoidance of lionfish resulted in a phase shift to algae-dominated communities (Lesser and Slattery 2011).

To date, our data suggest that the lionfish expansion in the eastern Gulf of Mexico is still in progress, yet it is unclear how long it will continue. Results from ongoing trawl surveys in the eastern Gulf of Mexico should allow us to document when lionfish abundances eventually level off. Further, this survey, which began before the invasion, should allow us to monitor and assess long-term ecological consequences of increasing lionfish abundances in the eastern Gulf of Mexico.

ACKNOWLEDGMENTS

First and foremost, we acknowledge the countless personnel of the Florida Fish and Wildlife Conservation Commission and volunteers who have assisted with the collection and processing of data since the inception of this survey. We also thank the captain and crew of the RV *Tommy Munro* for their hard work and dedication in assuring both the quality of the data and the safety of our staff. Thanks to B. Crowder, A. Acosta, R. Matheson, and two anonymous reviewers for providing feedback and useful editorial comments that greatly improved the quality of this manuscript. Staff from the Gulf States Marine Fisheries Commission provided critical logistical support for these surveys, as well as for the processing of collected data. Funding for the Florida SEAMAP groundfish survey was provided by the U.S. Department of Commerce, National Oceanic and Atmospheric Administration, National Marine Fisheries Service (grants NA06NMF4350009, NA07NMF4350182, and NA11NMF4350047). Additional support was provided by the U.S. Department of the Interior, U.S. Fish and Wildlife Service, Federal Aid for Sportfish Restoration Project Number F13AF01200 and by proceeds from State of Florida saltwater recreational fishing licenses. The statements, findings, views, conclusions, and recommendations contained in this document are those of the authors, do not necessarily reflect the views of the U.S. Departments of the Interior or of Commerce, and should not be interpreted as representing the opinions or policies of the U.S. Government. The mention of trade names or commercial products does not constitute their endorsement by the U.S. Government.

REFERENCES

Ahrenholz, D. W., and J. A. Morris Jr. 2010. Larval duration of the lionfish, *Pterois volitans,* along the Bahamian archipelago. Environmental Biology of Fishes 88:305–309.

Akins, J. L. 2012. Control strategies: tools and techniques for local control. Pages 24–50 *in* J. A. Morris Jr., editor. Invasive lionfish: a guide to control and management. Gulf and Caribbean Fisheries Institute, Special Publication Series 1, Marathon, Florida.

Albins, M. A., and M. Ass. Hixon. 2008. Invasive Indo-Pacific lionfish *Pterois volitans* reduce recruitment of Atlantic coral-reef fishes. Marine Ecology Progress Series 367:233–238.

Albins, M. A., and M. A. Hixon. 2013. Worst case scenario: potential long-term effects of invasive predatory lionfish (*Pterois volitans*) on Atlantic and Caribbean coral-reef communities. Environmental Biology of Fishes 96:1151–1157.

Arias-González, J. E., C. González-Gándara, J. L. Cabrera, and V. Christensen. 2011. Predicted impact of the invasive lionfish *Pterois volitans* on the food web of a Caribbean coral reef. Environmental Research 111: 917–925.

Bacheler, N. M., C. M. Schobernd, Z. H. Schobernd, W. A. Mitchell, D. J. Berrane, G. T. Kellison, and M. J. M. Reichert. 2013. Comparison of trap and underwater video gears for indexing reef fish presence and abundance in the southeast United States. Fisheries Research 143:81–88.

Baltz, D. M. 1990. Autecology. Pages 585–607 *in* C. B. Schreck and P. B. Moyle, editors. Methods for fish biology. American Fisheries Society, Bethesda, Maryland.

Barbour, A. B., M. S. Allen, T. K. Frazer, and K. D. Sherman. 2011. Evaluating the potential efficacy of invasive lionfish (*Pterois volitans*) removals. PLoS (Public Library of Science) One [online serial] 6(5):e19666.

Barbour, A. B., M. L. Montgomery, A. A. Adamson, E. Díaz-Ferguson, and B. R. Silliman. 2010. Mangrove use by the invasive lionfish *Pterois volitans*. Marine Ecology Progress Series 401:291–294.

Biggs, C. R., and J. D. Olden. 2011. Multi-scale habitat occupancy of invasive lionfish (*Pterois volitans*) in coral reef environments of Roatan, Honduras. Aquatic Invasions 6:347–353.

Briggs, J. C. 1958. A list of Florida fishes and their distribution. Bulletin of the Florida State Museum Biological Sciences 2:223–318.

Bullard, S. A., A. M. Barse, S. S. Curran, and J. A. Morris Jr. 2011. First record of a digenean from invasive lionfish, *Pterois* cf. *volitans*, (Scorpaeniformes, Scopaeniaidae) in the northwestern Atlantic. Journal of Parasitology 97:833–837.

Claydon, J. A. B., M. C. Calosso, and S. B. Traiger. 2012. Progression of invasive lionfish in seagrass, mangrove and reef habitats. Marine Ecology Progress Series 448:119–129.

Côté, I. M., S. J. Green, and M. A. Hixon. 2013. Predatory fish invaders: insights from Indo-Pacific lionfish in the western Atlantic and Caribbean. Biological Conservation 164:50–61.

Côté, I. M., and A. Maljkovic. 2010. Predation rates of Indo-Pacific lionfish on Bahamian coral reefs. Marine Ecology Progress Series 404:219–225.

Dahl, K. A., and W. F. Patterson III. 2014. Habitat-specific density and diet of rapidly expanding invasive Red Lionfish, *Pterois volitans*, populations in the northern Gulf of Mexico. PLoS (Public Library of Science) One [online serial] 9(8):e105852.

Eldridge, P. J. 1988. The southeast area monitoring and assessment program (SEAMAP): a state-federal-university program for collection, management, and dissemination of fishery-independent data and information in the southeastern United States. Marine Fisheries Review 50:29–39.

Fogg, A. Q., E. R. Hoffmayer, W. B. Driggers III, M. D. Campbell, G. J. Pellegrin, and W. Stein. 2013. Distribution and length frequency of invasive lionfish (*Pterois* sp.) in the northern Gulf of Mexico. Gulf and Caribbean Research 25:111–115.

Green, S. J., J. L. Akins, A. Maljkovic, and I. M. Côté. 2012. Invasive lionfish drive Atlantic coral reef fish declines. PLoS (Public Library of Science) One [online serial] 7(3):e32956.

Hackerott, S., A. Valdivia, S. J. Green, I. M. Côté, C. E. Cox, L. Akins, C. A. Layman, W. F. Precht, and J. F. Bruno. 2013. Native predators do not influence invasion success of Pacific lionfish on Caribbean reefs. PLoS (Public Library of Science) One [online serial] 8(7):e68259.

Hamilton, P., K. A. Donohoe, R. R. Leben, A. Lugo Fernández, and R. E. Green. 2011. Loop Current observations during spring and summer 2010: description and historical perspective. Geophysical Monograph 195:117–130.

Johnston, M. W., and S. J. Purkis. 2011. Spatial analysis of the invasion of lionfish in the western Atlantic and Caribbean. Marine Pollution Bulletin 62:1218–1226.

Jud, Z. R., and C. A. Layman. 2012. Site fidelity and movement patterns of invasive lionfish, *Pterois* spp., in a Florida estuary. Journal of Experimental Marine Biology and Ecology 414/415:69–74.

Kimball, M. E., J. M. Miller, P. E. Whitfield, and J. A. Hare. 2004. Thermal tolerance and potential distribution of invasive lionfish (*Pterois volitans/ miles* complex) on the East Coast of the United States. Marine Ecology Progress Series 283:269–278.

Kulbicki, M., J. Beets, P. Chabanet, K. Cure, E. Darling, S. R. Floeter, R. Galzin, A. Green, M. Harmelin-Vivien, M. Hixon, Y. Letourneur, T. Lison de Loma, T. McClanahan, J. McIlwain, G. MouTham, R. Myers, J. K. O'Leary, S. Planes, L. Viglionla, and L. Wantiez. 2012. Distributions of Indo-Pacific lionfishes *Pterois* spp. in their native ranges: implications for the Atlantic invasion. Marine Ecology Progress Series 446:189–205.

Lesser, M. P., and M. Slattery. 2011. Phase shift to algal dominated communities at mesophotic depths associated with lionfish (*Pterois volitans*) invasion on a Bahamian coral reef. Biological Invasions 13:1855–1868.

McEachran, J. D., and J. D. Fechhelm. 1998. Fishes of the Gulf of Mexico. Volume 1: Myxiniformes to Gasterosteiformes. University of Texas Press, Austin.

Meister, H. S., D. M. Wyanski, J. K. Loefer, S. W. Ross, A. M. Quattrini, and K. J. Sulak. 2005. Further evidence for the invasion and establishment of *Pterois volitans* (Teleostei: Scorpaenidae) along the Atlantic Coast of the United States. Southeastern Naturalist 4:193–206.

Morris, J. A. Jr., and P. E. Whitfield. 2009. Biology, ecology, control and management of the invasive Indo-Pacific lionfish: an updated integrated assessment. NOAA Technical Memorandum NOS NCCOS 99.

Nuttall, M. F., M. A. Johnston, R. J. Eckert, J. A. Embesi, E. L. Hickerson, and G. P. Schmahl. 2014. Lionfish (*Pterois volitans* [Linnaeus, 1758] and *P. miles* [Bennett, 1828]) records within mesophotic depth ranges on natural banks in the northwestern Gulf of Mexico. BioInvasions Records 3:111–115.

Pimiento, C., J. C. Nifong, M. E. Hunter, E. Monaco, and B. R. Silliman. 2015. Habitat use patterns of the invasive Red Lionfish *Pterois volitans*: a comparison between mangrove and reef systems in San Salvador, Bahamas. Marine Ecology 36:28–37.

Randazzo, A. F., and D. S. Jones, editors. 1997. The geology of Florida. University of Florida Press, Gainesville.

Rester, J. K. 2011. SEAMAP environmental and biological atlas of the Gulf of Mexico 2009. Gulf States Marine Fisheries Commission, Report 198, Ocean Springs, Mississippi.

Ruiz-Carus, R., R. E. Matheson Jr., D. E. Roberts, and P. E. Whitfield. 2006. The Western Pacific lionfish, *Pterois volitans* (Scorpaenidae), in Florida: evidence for reproduction and parasitism in the first exotic marine fish established in state waters. Biological Conservation 128:384–390.

Ruttenberg, B. I., P. J. Schofield, J. L. Adkins, A. Acosta, M. W. Feeley, J. Blondeau, S. G. Smith, and J. S. Ault. 2012. Rapid invasion of Indo-Pacific lionfishes (*Pterois volitans* and *Pterois miles*) in the Florida Keys, USA: evidence from multiple pre- and post-invasion data sets. Bulletin of Marine Science 88:1051–1059.

SAS Institute. 2006. SAS version 9.00 (TS M0) online documentation. SAS Institute, Cary, North Carolina.

Schofield, P. J. 2009. Geographic extent and chronology of the invasion of non-native lionfish (*Pterois volitans* [Linnaeus 1758] and *P. miles* [Bennett 1828]) in the western North Atlantic and Caribbean Sea. Aquatic Invasions 4:473–479.

Schofield, P. J. 2010. Update on geographic spread of invasive lionfishes (*Pterois volitans* [Linnaeus 1758] and *P. miles* [Bennett 1828]) in the western North Atlantic Ocean, Caribbean Sea, and Gulf of Mexico. Aquatic Invasions 5(Supplement 1):S117–S122.

Schofield, P. J., J. A. Morris Jr, J. N. Langston, and P. L. Fuller. 2014. *Pterois volitans/miles*. U.S. Geological Survey, Nonindigenous Aquatic Species Database, Gainesville, Florida. Available: http://nas.er.usgs.gov/queries/ FactSheet.aspx?speciesID=963. (September 2012).

Schultz, E. T. 1986. *Pterois volitans* and *Pterois miles*: two valid species. Copeia 1986:686–690.

Scyphers, S. B., S. P. Powers, J. L. Adkins, J. M. Drymon, C. W. Martin, Z. H. Schobernd, P. J. Schofield, R. L. Shipp, and T. S. Switzer. In press. The role of citizens in detecting and responding to a rapid marine invasion. Conservation Letters. DOI: 10.1111/conl.12127.

Sokal, R. R., and F. J. Rohlf. 2012. Biometry, 4th edition. Freeman, New York.

Sturges, W., and J. C. Evans. 1983. On the variability of the Loop Current in the Gulf of Mexico. Journal of Marine Research 41:639–653.

USGS (U.S. Geological Survey). 2014. Nonindigeinous Aquatic Species Database. USGS, Gainesville, Florida. Available: http://nas.er.usgs.gov. (October 2014).

Whitfield, P. E., J. A. Hare, A. W. David, S. L. Harter, R. C. Muñoz, and C. M. Addison. 2007. Abundance estimates of the Indo-Pacific lionfish *Pterois volitans/miles* complex in the western North Atlantic. Biological Invasions 9:53–64.

6

Spatial and Temporal Patterns of Black Sea Bass Sizes and Catches in the Southeastern United States Inferred from Spatially Explicit Nonlinear Models

Nathan M. Bacheler*

National Marine Fisheries Service, Southeast Fisheries Science Center, 101 Pivers Island Road, Beaufort, North Carolina 25887, USA

Joseph C. Ballenger

South Carolina Department of Natural Resources, Marine Resources Research Institute, 217 Fort Johnson Road, Post Office Box 12559, Charleston, South Carolina 29412, USA

Abstract

    Temporal and spatial variability in abundance often results from the effects of environmental and landscape variables interacting over multiple spatial scales, and understanding the complex interplay among these variables is key to elucidating the drivers of a species' population dynamics. We used a spatially explicit, variable-coefficient, generalized additive modeling approach with 24 years of fishery-independent trap data ($N = 11{,}726$ samples) to elucidate the spatiotemporal dynamics of size and size-specific CPUE of Black Sea Bass *Centropristis striata* along the southeastern Atlantic coast of the United States. Black Sea Bass catch exhibited complex spatial and temporal dynamics that were influenced by environmental, landscape, and sampling effects. Black Sea Bass were more commonly caught inshore than offshore, but were significantly smaller inshore and southward and larger offshore and northward in the study area. Moreover, the spatial distribution of Black Sea Bass changed as abundance varied within and among sampling seasons. Standardized mean length of Black Sea Bass also increased by more than 20% over the study period, from 230 mm TL in the early 1990s to 280 mm TL after 2010. These results elucidate the spatial and temporal dynamics of Black Sea Bass, inform population structure and indices of abundance, and provide an analytical framework that can be easily adapted to other species and systems.

All species exhibit spatial variability in abundance across a landscape, and elucidating the spatial patterns in size and abundance is key to understanding community dynamics, variation in life history traits, and temporal changes in abundance (Dunning et al. 1992; Jackson et al. 2001; Ciannelli et al. 2013). Spatial variability in abundance can result from larval or juvenile dispersal patterns, habitat patchiness, environmental variability, landscape features, or biotic interactions such as predation or competition (Levin 1992; Brown et al. 1995). While the focus of most historical studies has been on temporal variability in abundance or density, researchers now recognize that understanding spatial variability and dynamics is key to describing a species' ecology and explaining temporal abundance patterns

Subject editor: Patrick Sullivan, Cornell University, Ithaca, New York

*Corresponding author: nate.bacheler@noaa.gov

(Cadrin and Secor 2009; Bartolino et al. 2011; Ciannelli et al. 2012).

Recent analytical advances have helped us understand the complex interplay between environmental and landscape influences on the spatial dynamics of organisms (Lehmann et al. 2002; Bacheler et al. 2009). For instance, Ciannelli et al. (2012) used a spatially explicit nonlinear regression modeling approach to show that at low levels of abundance Arrowtooth Flounder *Atheresthes stomias* distribution was influenced solely by water temperature, but their distribution expanded into new habitats in a nonadditive fashion with increasing water temperature when abundance was high. Bacheler et al. (2012) used a similar modeling approach to document density-dependent estuarine habitat use of Red Drum *Sciaenops ocellatus* after removing variability in catches due to various environmental and landscape effects.

Another fish species for which spatially explicit modeling would be useful to explicate temporal and spatial dynamics is the Black Sea Bass *Centropristis striata*, a protogynous serranid that occurs in nearshore waters of the U.S. Atlantic Ocean and Gulf of Mexico (Lavenda 1949; Wenner et al. 1986; Hood et al. 1994). Two Black Sea Bass stocks have been identified along the U.S. Atlantic coast, separated at Cape Hatteras, North Carolina (Roy et al. 2012; McCartney et al. 2013). The northern stock is thought to migrate offshore and southward in colder winter months, then back inshore and northward when water warms in spring and summer (Musick and Mercer 1977; Moser and Shepherd 2009; Fabrizio et al. 2013). South of Cape Hatteras, however, Black Sea Bass movement rates appear to be lower, perhaps lacking seasonal migrations altogether.

Recreational and commercial fishers harvest Black Sea Bass throughout their range primarily using pots and hook and line (Coleman et al. 2000; McGovern et al. 2002). Along the southeastern U.S. Atlantic Coast (SEUS), size limits have increased from 203 mm TL in the 1980s and 1990s to 279 mm TL (commercial sector) or 330 mm TL (recreational sector) by 2015. While commercial harvest of Black Sea Bass is substantially greater in North Carolina than in states to its south, recreational harvest is similar among states in the SEUS, and historically, fishing by both sectors generally occurred year-round except when quotas had been met. However, there have been management changes to the fishing year of both the commercial and recreational sectors, gear restrictions implemented on the timing and use of certain gears of the Black Sea Bass fishery, and in-season closures of one or both sectors in recent years. These actions, coupled with regional differences in dominant weather patterns in the SEUS limiting access to Black Sea Bass habitat, have led to changes in the timing of peak landings in the calendar year across years and across subregions of the SEUS within a year.

We used long-term, spatially extensive monitoring data in a spatially explicit regression-modeling framework to determine the environmental, landscape, and temporal predictors of

Black Sea Bass size and CPUE along the SEUS. Black Sea Bass is an ideal species with which to use a spatially explicit modeling approach because they can be sampled efficiently (Bacheler et al. 2013c), catches generally reflect abundance (Bacheler et al. 2013b), and their spatial and temporal dynamics in the SEUS are poorly understood (Sedberry et al. 1998). There were two specific objectives of our work. Our first objective was to quantify the spatial and temporal patterns of Black Sea Bass size and size-specific CPUE throughout the SEUS after correcting for the influences of environmental variation and landscape features. Our second objective was to determine whether annual or seasonal variation in Black Sea Bass sizes or size-specific CPUE was spatially variable; in other words, whether temporal changes in Black Sea Bass sizes or catches occur more strongly in some locations than in others. Our aim was to improve our understanding of the seasonal movement patterns and temporal and spatial dynamics of Black Sea Bass in the SEUS to benefit their stock assessment by elucidating their population structure and improving annual indices of abundance. We also intended to provide an analytical framework that can be easily adapted to other species and systems.

METHODS

Study area.—We used long-term, fishery-independent, chevron-trap data to elucidate the spatial and temporal patterns of Black Sea Bass lengths and catches in the SEUS between North Carolina and Florida (Figure 1). In 1990, the Marine Resources Monitoring, Assessment, and Prediction (MARMAP) program began using chevron traps to index reef fish abundance in the SEUS. Since 2009, MARMAP funding has been supplemented by the Southeast Area Monitoring and Assessment Program–South Atlantic to allow for an expansion of coverage of the survey into historically undersampled areas. In 2010, the Southeast Fishery-Independent Survey began chevron-trap sampling cooperatively and identically in the region to increase sample sizes. Collectively these programs are now known as the Southeast Reef Fish Survey (SERFS); we used SERFS data from 1990 through 2013 in our analyses.

Sampling by SERFS targets hard substrates in continental shelf and shelf-break waters in the SEUS, the preferred habitat of Black Sea Bass (Powles and Barans 1980; Sedberry and Van Dolah 1984). The continental shelf and shelf-break in the SEUS are dominated by sand and mud substrates, but Black Sea Bass generally associate with the scattered patches of hard, rocky substrates ("hard bottom") that occur in the region (Kendall et al. 2008; Fautin et al. 2010). Hard-bottom habitats sampled in our study ranged in complexity from flat limestone pavement, sometimes covered with a sand or gravel veneer, to high-relief rocky ledges (Schobernd and Sedberry 2009; Glasgow 2010). Sampling in our study occurred between Cape Hatteras, North Carolina, and St. Lucie Inlet, Florida (Figure 1).

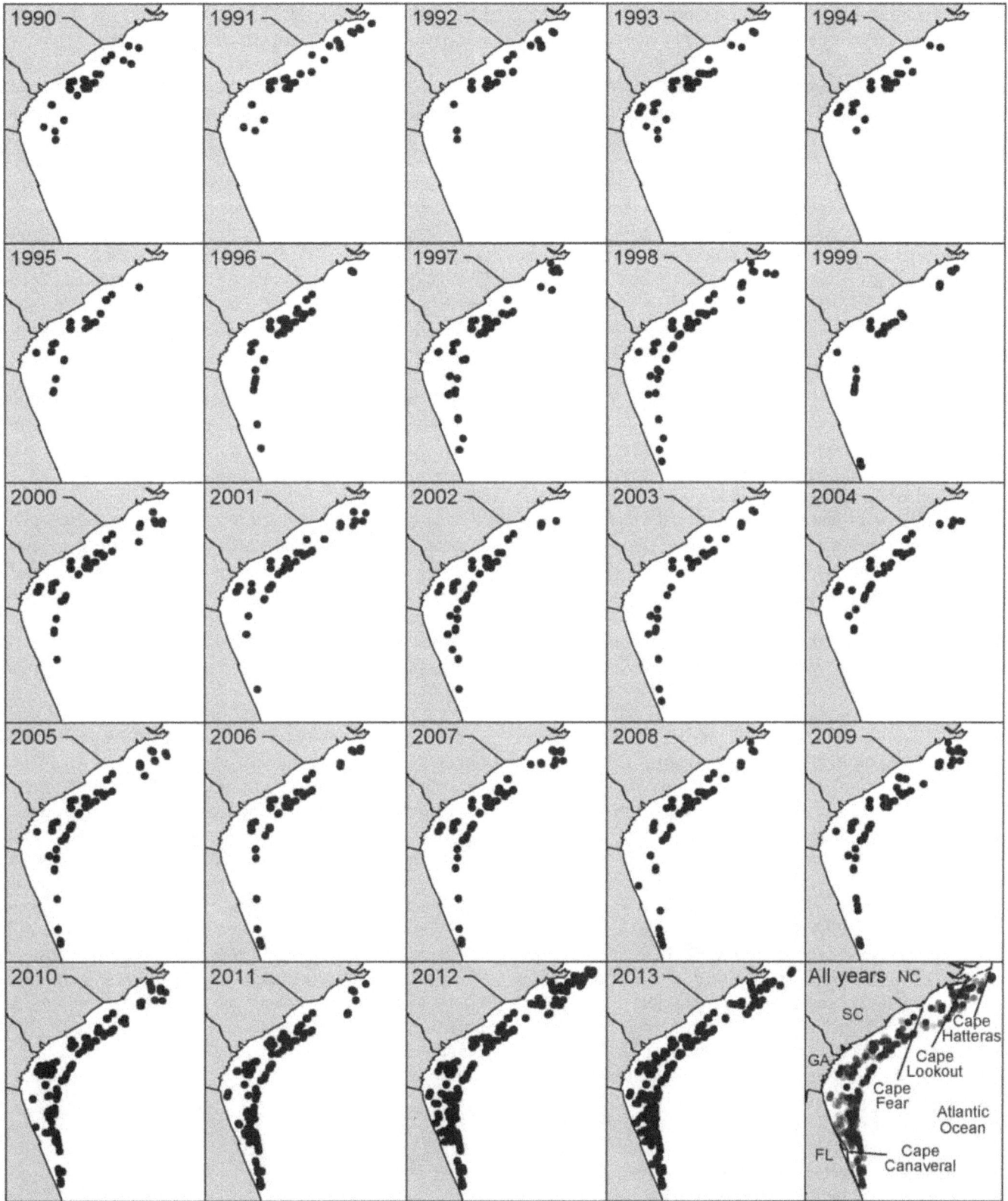

FIGURE 1.   Spatial distribution of chevron-trap sampling for Black Sea Bass by the Southeast Reef Fish Survey between North Carolina and Florida, 1990–2013. Each point represents a single chevron-trap deployment included in the analysis. Note that symbols often overlap. In the bottom right panel, all sampling from 1990 to 2013 is shown and the darker the symbol, the greater the overlap among points.

Sampling design and gear.—Hard-bottom sampling stations were selected for sampling in one of three ways. First, most sites were randomly selected from the SERFS sampling frame that consisted of approximately 1,000 sampling stations in the early 1990s to more than 3,000 sampling stations in the 2010s, all located on hard-bottom habitat. Second, some stations in the sampling frame were sampled opportunistically even though they were not randomly selected for sampling in a given year. Third, new hard-bottom stations were added during the study period through the use of information from fishers, charts, and historical surveys. These new locations were investigated using the vessel's echo sounder or drop cameras and sampled if hard bottom was detected. All sampling for this study occurred during daylight hours between late March and early November and was conducted on the RV *Palmetto* (1990–2013), RV *Savannah* (2010–2013), NOAA Ship *Nancy Foster* (2010), or NOAA Ship *Pisces* (2011–2013) using identical methods.

Chevron fish traps were deployed at each station sampled in our study. Black Sea Bass are caught very effectively in chevron traps (Bacheler et al. 2013c), and catches appear to be strongly and positively related to true abundance at a site (Bacheler et al. 2013b). Chevron traps were constructed from plastic-coated, galvanized, 2-mm-diameter wire (mesh size = 3.4 cm^2) and were shaped like an arrowhead that measured $1.7 \times 1.5 \times 0.6$ m with a total volume of 0.91 m^3 (Collins 1990). Trap-mouth openings were shaped like a teardrop and measured approximately 18 cm wide and 45 cm high. Each trap was baited with 24 menhaden *Brevoortia* spp. Traps were typically deployed in groups of six, and each trap in a set was deployed at least 200 m from all other traps in a given year to provide some measure of independence between traps. A soak time of 90 min was targeted for each trap deployed, and any trap not fishing properly (e.g., dragged due to current, damaged upon retrieval) was excluded from analysis. All Black Sea Bass caught in chevron traps were enumerated and measured for length (mm TL).

Data analysis.—We related mean length or catches of small or large (defined below) Black Sea Bass to various predictor variables using a spatially explicit, variable-coefficient, generalized additive model (Bacheler et al. 2009, 2010; Bartolino et al. 2011). A generalized additive model (GAM) is a nonlinear, nonparametric, regression model that does not require a priori specification of the functional relationship between the response and predictor variables (Venables and Dichmont 2004; Wood 2008). The GAMs extend traditional additive models by allowing for alternative error distributions, just as generalized linear models allow for alternative error distributions of linear models. The addition of variable-coefficient terms can be used to determine specific locations where Black Sea Bass sizes or catches are expected to increase or decrease with changes in any of the predictor variables in the model (Bacheler et al. 2009).

We developed two broad classes of spatially explicit GAMs to understand more about the spatial and temporal patterns of Black Sea Bass in the SEUS. The first model used mean Black Sea Bass TL (mm) in each trap as the response variable for the GAM (hereafter, "length model"). Here, chevron-trap samples were weighted in the model by the total number of Black Sea Bass caught in each trap, so that a mean length based on many fish in a trap was weighted more heavily than a mean length comprising only a single fish in a trap. Chevron-trap samples that failed to catch Black Sea Bass were excluded from this analysis. Mean lengths from the remaining, positive trap catches were log transformed to achieve normality. The second model type used the trap catch of small or large Black Sea Bass as the response variable (hereafter, "catch models"). Small Black Sea Bass were defined as <235 mm TL, and large Black Sea Bass were $\geq$235 mm TL, roughly the cutoff between age-2 and age-3 Black Sea Bass caught in traps (McGovern et al. 2002); 235 mm TL was also the modal size of Black Sea Bass caught in traps in our study (see Figure 2). While arbitrary, the 235-mm cutoff between small and large Black Sea Bass was chosen because sample sizes of these two groups were sufficient in chevron traps each year. Moreover, large Black Sea Bass have not been observed excluding small Black Sea Bass from entering traps, and asymptotic catch of Black Sea Bass is highly related to local abundance (Bacheler et al. 2013b, 2013c). Catch of small or large Black Sea Bass was fourth-root transformed, which resulted in better model fit than any other types of transformations or error distributions using standard model diagnostics (e.g., Bacheler et al. 2013a). Unlike the length model described above, chevron traps that did not catch any Black Sea Bass were included in the catch models.

We examined the influence of various predictor variables on the mean length or catch of Black Sea Bass. For the length model, six primary variables were considered for inclusion based on our hypotheses and previous knowledge: year (y) was included as a factor variable, and bottom temperature (°C; *temp*), day of the year (*doy*), and spatial position (latitude and longitude; *pos*) were included in the model as smoothed variables. In addition, two variable-coefficient terms were included: an interaction between spatial position and year and an interaction between spatial position and day of the year. The former allows for interaction between the spatial position smoother and year while the latter allows for the local effect of the spatial position smoother to vary seasonally (i.e., within the sampling season). Thus, the base length model ("Base$_{length}$") was formulated as:

$$z_{doy,y,pos} = f_1(y) + g_1(doy) + g_2(temp) + g_3(pos) + g_4(pos \cdot y)$$

$$+ g_5(pos \cdot doy) + e_{doy,y,pos},$$

where $z_{doy,y,pos}$ is the log-transformed mean TL of Black Sea Bass on day of the year *doy* in year y at spatial position *pos*, *temp* is bottom temperature, f_1 is a categorical function, g_{1-5} are nonparametric smoothing functions, and $e_{doy,y,pos}$ is the random error assumed to be normally distributed with a mean of zero and finite variance.

Catch models were coded similarly except five additional variables were included based on the results of Bacheler et al. (2013a). Station type was included as a factor variable and described any potential variation in catch between randomly selected stations and those newly found and sampled for the first time. Depth (m), time of day (Coordinated Universal Time), trap soak time (min), and moon phase were each included as smoothed variables. The base catch models ("Base$_{catch}$") were formulated as:

$$x_{doy,y,pos} = a + f_1(y) + f_2(type) + g_1(depth) + g_2(doy)$$

$$+ g_3(tod) + g_4(soak) + g_5(temp) + g_6(moon)$$

$$+ g_7(pos) + g_8(pos \cdot y) + g_9(pos \cdot doy) + e_{doy,y,pos},$$

where $x_{doy,y,pos}$ is the fourth-root transformed catch of small or large Black Sea Bass on day of the year *doy* in year *y* at spatial position *pos, type* is the station type, *depth* is bottom depth, *tod* is time of day, *soak* is the soak time of the trap, *moon* is the moon phase, f_{1-2} are categorical functions, g_{1-9} are non-parametric smoothing functions, and $e_{doy,y,pos}$ is the random error assumed to be normally distributed with a mean of zero and finite variance.

All base models were compared to reduced models using the Akaike information criterion (AIC; Burnham and Anderson 2002). The AIC approach balances the number of parameters of a model and its log-likelihood, and the model with the lowest AIC values is considered the best model out of the candidate models investigated given the data set used (Burnham and Anderson 2002). We used the mgcv library (Wood 2004, 2011) in R (R Development Core Team 2013) to construct and compare all models. For smoothed and variable coefficient terms, estimated degrees of freedom were chosen using automatic software selection. There was no significant multicolinearity among predictor variables given that the variance inflation factor was less than three for all variables (Neter et al. 1989). Also, there were no consistent patterns in the relationship between the semivariance of the model residuals and distance between sampling points, indicating negligible spatial autocorrelation in the residuals. Furthermore, there were no obvious trends in residuals over space, suggesting no spatial bias in model fit. Last, models only using data from 2010 to 2013, during which time the survey expanded spatially (see Table 1), were very similar to models incorporating the full data set (i.e., 1990–2013), as was a model that only examined Black Sea Bass caught in South Carolina and Georgia over the entire time series.

Three additional landscape variables were considered for inclusion in the catch models: rugosity (a measure of the roughness of the seafloor), slope of slope (a measure of the curvature or shape of the seafloor), and predicted hard bottom

(the likelihood of being hard bottom: Dunn and Halpin 2009). Since bottom-mapping information does not exist for most of our sampling locations, we calculated the first two metrics using data from the Coastal Relief Model (National Centers for Environmental Information, NOAA), which provided depth data for our entire study area at a resolution of approximately 90 m ($\sim$8,100-m^2 cells). Rugosity and slope of slope were calculated for each 90-m grid cell in our study area by comparing its depth to the depth of the eight surrounding grid cells using the Benthic Terrain Modeler in ArcGIS 10.2. Moreover, we obtained predicted hardbottom data from Dunn and Halpin (2009). For both small and large Black Sea Bass catch models, all three landscape variables were excluded based on AIC values, likely due to the weak habitat relationships displayed by Black Sea Bass (Kendall et al. 2008) and poor accuracy of the Coastal Relief Model (Dunn and Halpin 2009).

The overall influence of predictor variables on mean length or catch of Black Sea Bass was calculated using a bootstrapping approach described by Bacheler et al. (2013a). Briefly, we resampled the predictions ($N = 10,000$) for each model at mean values of all other predictor variables according to the pointwise estimates of error that were assumed to be normally distributed. Since average values did not exist for categorical variables (year and station type), the model predicted at all combinations of these categorical variables using the "expand. grid" function in R. Mean latitude and longitude values were not used because average values would be placed in the middle of the Atlantic Ocean (outside the range of our study area), so we instead used a latitude of 32°N and a longitude of 80°W (i.e., mid-continental shelf off southern South Carolina). All 95% CIs were estimated as the 0.025 and 0.975 quantiles of the 10,000 point estimates. To visualize spatial effects across the study area, a grid of 0.05° × 0.05° was created over the study area and mean Black Sea Bass length or catch was predicted for each cell given the depth (from the U.S. Coastal Relief Model, National Geophysical Data Center, NOAA), latitude, and longitude of each cell and mean values of all other model predictor variables. Black Sea Bass predictions only applied to hard bottom areas within each cell.

RESULTS

Overall, 11,726 chevron-trap deployments were included in the catch models, ranging from 250 samples in 1999 to 1,514 in 2013 (annual mean = 489, SD = 333; Table 1; Figure 1). Seasonality of sampling was relatively constant over the 24 years, and sampling commenced in late April through May and terminated in late September through October in most years (Table 1). Likewise, the range of depths sampled annually was relatively constant, ranging from approximately 15 to 95 m in most years. In contrast, the spatial extent of sampling increased from approximately 30–34°N in the early years of the survey to 27–35°N in later years; however, the long time series and large number of samples taken throughout the entire

TABLE 1. Sampling information for the 24 years included in the analysis of the Southeast Reef Fish Survey chevron-trap data. N = number of trap samples included each year, the number of small Black Sea Bass is the total number of small Black Sea Bass < 235 mm TL caught in traps in a given year, and the number of large Black Sea Bass is the total number of large Black Sea Bass ≥ 235 mm TL caught in traps in a given year.

Year	N	Dates sampled	Depth range (m)	Latitude range (°N)	Number of small Black Sea Bass	Number of large Black Sea Bass
1990	310	Apr 23–Aug 9	17–93	30.4–33.8	4,172	1,843
1991	268	Jun 11–Sep 24	17–95	30.8–34.6	2,824	1,079
1992	291	Mar 31–Aug 13	17–62	30.4–34.3	2,708	1,551
1993	412	May 10–Aug 13	16–94	30.4–34.3	2,122	1,138
1994	409	May 9–Oct 26	16–93	30.7–33.8	2,323	1,325
1995	376	Apr 17–Oct 26	16–60	29.9–33.7	2,170	777
1996	498	Apr 29–Oct 17	14–95	27.9–34.3	2,282	1,489
1997	476	Apr 21–Sep 29	15–97	27.9–34.6	2,300	1,513
1998	466	Mar 31–Oct 5	14–92	27.4–34.6	2,450	1,335
1999	250	May 18–Oct 6	15–79	27.3–34.4	2,278	1,215
2000	352	May 16–Oct 19	15–95	29.0–34.3	2,518	1,653
2001	276	May 23–Oct 24	14–91	27.9–34.3	1,838	1,455
2002	298	Jun 17–Nov 5	13–94	27.9–34.0	1,765	1,033
2003	276	Jun 3–Sep 22	16–92	27.4–34.3	941	854
2004	319	May 5–Oct 28	14–91	30.0–34.0	2,156	2,980
2005	338	May 3–Oct 20	15–69	27.3–34.3	2,121	2,235
2006	309	Jun 6–Oct 19	15–94	27.3–34.4	1,624	1,424
2007	361	May 21–Sep 24	15–92	27.3–34.3	1,811	1,455
2008	354	May 5–Sep 30	14–92	27.3–34.6	1,634	1,526
2009	458	Apr 23–Oct 8	14–91	27.3–34.6	1,501	2,196
2010	990	May 4–Oct 27	14–92	27.3–34.6	3,907	5,838
2011	817	May 3–Oct 26	14–93	27.2–34.5	7,444	5,501
2012	1,308	Apr 24–Oct 10	15–98	27.2–35.0	9,709	7,360
2013	1,514	Apr 24–Oct 4	15–92	27.2–35.0	10,565	8,766
Total	11,726	Mar 31–Oct 28	13–98	27.2–35.0	75,163	57,541

latitudinal extent in later years likely minimized any effects of inconsistent latitudinal sampling on GAM models (Table 1). Catch of small Black Sea Bass ranged from 0 to 137 individuals per trap, and catch of large Black Sea Bass ranged from 0 to 160 individuals per trap.

A total of 5,230 (44.6%) out of 11,726 chevron traps deployed in our study caught Black Sea Bass and were included in the length model. The annual percent frequency of occurrence of Black Sea Bass in chevron traps ranged from 27% in 2003 to 64% in 1990. Overall mean Black Sea Bass TL was 259 mm (SD = 47; range = 100–520 mm; Figure 2). The largest catches of Black Sea Bass had mean lengths of between 200 and 350 mm TL and were generally caught in depths of less than 45 m (Figure 3). Moreover, mean Black Sea Bass length appeared to increase with depth (Figure 3).

The best length model for Black Sea Bass excluded bottom temperature from the $Base_{length}$ model; the $Base_{length}$ model and all other reduced models had AIC scores of at least 23 points higher than the best model (Table 2). The best length model explained 54.4% of the deviance in Black Sea Bass mean length and included year, day of the year, position,

position × year, and position × day of the year (Table 2). The best catch models for small and large Black Sea Bass were the $Base_{catch}$ models, which explained 53.6% and 52.3% of the deviance in catch, respectively. Based on AIC scores, none of the reduced models compared favorably with the $Base_{catch}$ models. $Base_{catch}$ models for small and large Black Sea Bass included year, station type, depth, day of the year, time of the day, soak time, bottom temperature, moon phase, position, position × year, and position × day (Table 2).

Predicted mean TL of Black Sea Bass increased over the study period from approximately 230 mm in the early 1990s to approximately 280 mm after 2010 (Figure 4A). Increases in predicted mean length were gradual over the time series, perhaps increasing most dramatically between 2008 and 2009 (Figure 4A). The predicted catch of small Black Sea Bass was relatively constant between 1990 and 2010 at ~2–4 Black Sea Bass/trap, but increased in 2011–2013 to ~6–7 fish/trap (Figure 4B). The predicted catch of large Black Sea Bass increased throughout the study period from fewer than 1 fish/trap in the mid-1990s to ~4 fish/trap in 2011–2013 (Figure 4C).

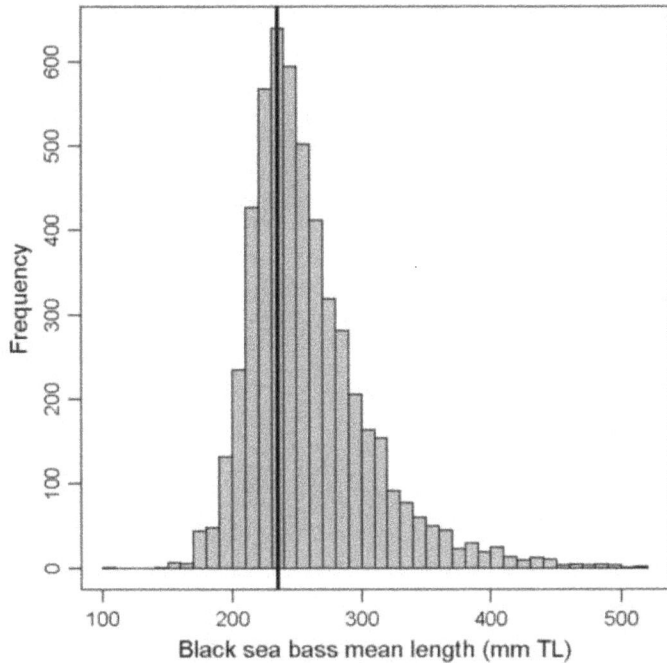

FIGURE 2. Mean length frequency distribution of Black Sea Bass (mm TL) caught in chevron traps by the Southeast Reef Fish Survey between North Carolina and Florida, 1990–2013. The black vertical line shows the cutoff used in this paper (i.e., 235 mm TL) between small and large Black Sea Bass.

Most of the effects of the smoothed predictor variables on Black Sea Bass mean length or catch were nonlinear. The predicted mean length of Black Sea Bass gradually decreased from approximately 260 mm TL on day of the year 100 (April 10) to 240 mm TL on day of the year 255 (September 12), but increased after that time to over 250 mm by day of the year 300 (October 27), although CIs were wide (Figure 5). Predicted catch of small and large Black Sea Bass was influenced very similarly by five of the six smoothed predictor variables. Mean catch of both small and large Black Sea Bass declined exponentially with increasing depth, declined throughout the sampling season, increased with time of day, displayed a dome-shaped response to bottom temperature, and appeared to be lowest during full moons (Figure 6). In contrast, the catch of large Black Sea Bass reached an asymptote beyond a soak time of 100 min, while the catch of small Black Sea Bass increased linearly with soak time over the range of soak times examined in this paper (Figure 6). The precision of estimates was highest for depth and day of the year and lowest for time of day and moon phase (Figure 6).

The predicted mean length of Black Sea Bass was spatially variable, generally smallest in shallower depths and off Cape Canaveral, Florida, and highest in deeper waters off North Carolina, South Carolina, Georgia, and northern Florida (Figure 7A). The predicted catch of small and large Black Sea Bass was also spatially variable, highest inshore in South Carolina, Georgia, and around Cape Canaveral and lowest in deep waters and off southern Georgia and northern Florida (Figure 7).

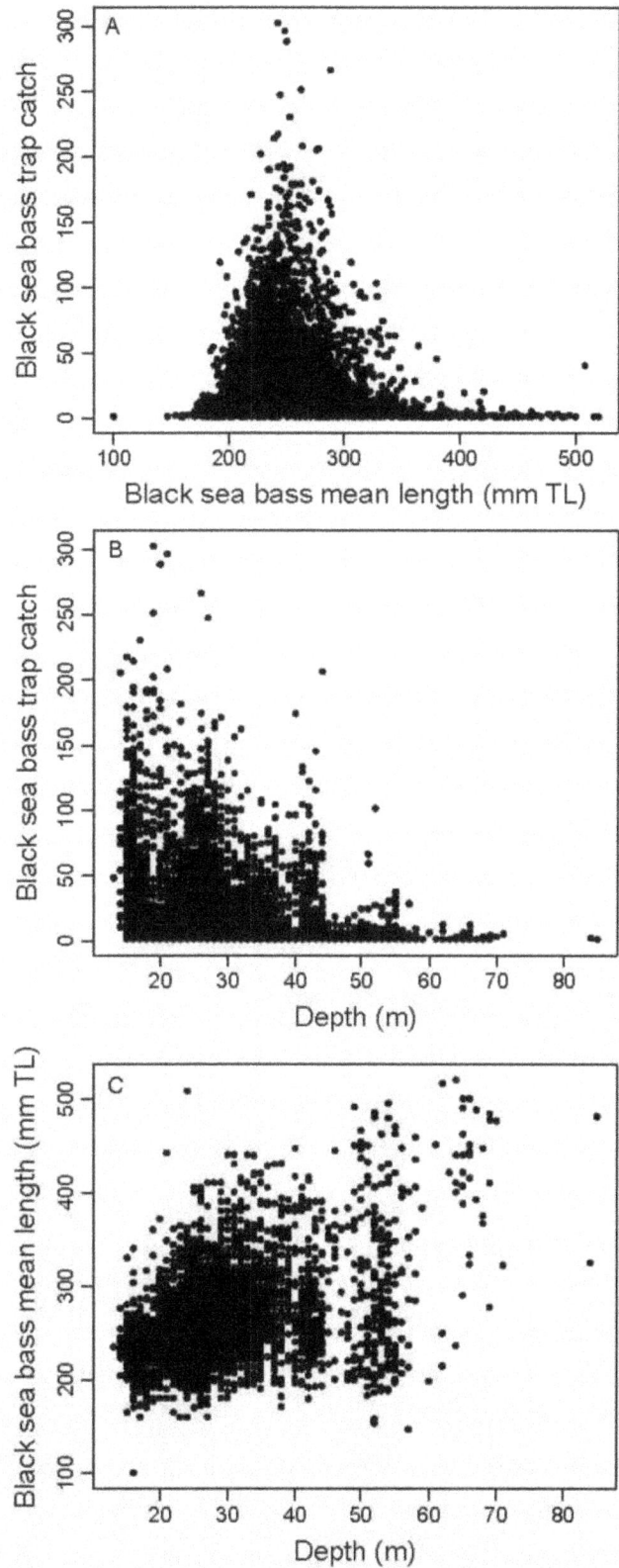

FIGURE 3. Relationship between (A) total Black Sea Bass catch (all sizes, number per trap) and Black Sea Bass mean length (mm TL), (B) total Black Sea Bass catch and depth (m), and (C) Black Sea Bass mean TL and depth in Southeast Reef Fish Survey chevron-trap sampling, 1990–2013.

TABLE 2. Model selection for the spatially explicit generalized additive models for mean length of Black Sea Bass, catch of small Black Sea Bass, or catch of large Black Sea Bass in chevron traps by the Southeast Reef Fish Survey, 1990–2013. Degrees of freedom are shown for factor (f) terms, and estimated degrees of freedom are shown for nonparametric, smoothed terms (g). Asterisks denote significance at the following alpha levels: *0.05, **0.01, ***0.001; Dev = deviance explained by the model; AIC = Akaike information criterion; y = year of the sample; $type$ = station type; $depth$ = bottom depth; t = day of the year; tod = Coordinated Universal Time; $soak$ = trap soak time; $temp$ = bottom temperature; $moon$ = moon phase; pos = position of the trap sample; NA = covariate was not applicable to that particular model. Only the four best models are shown for each response variable.

Model	Dev	AIC	y	$type$	$depth$	t	tod	$soak$	$temp$	$moon$	pos	$pos·y$	$pos·t$
Black Sea Bass mean length													
$Base_{length}$[a] − temp	54.4	−7,639	23***	NA	NA	7.1***	NA	NA	NA	NA	21.0***	28.8***	26.1***
$Base_{length}$ − doy	54.1	−7,616	23***	NA	NA	NA	NA	NA	5.9***	NA	20.1***	29.7***	25.9***
$Base_{length}$ − temp − doy	53.9	−7,598	23***	NA	NA	NA	NA	NA	NA	NA	20.9***	28.8***	26.8***
$Base_{length}$	53.8	−7,597	23***	NA	NA	6.3***	NA	NA	6.9***	NA	3.7***	29.8***	27.7***
Small Black Sea Bass catch													
$Base_{catch}$[b]	53.6	23,030	23***	1***	2.8***	3.6***	2.2***	1.9***	4.5***	4.0***	26.7***	29.5***	24.6***
$Base_{catch}$ − type	53.5	23,040	23***	NA	2.8***	3.9***	2.2***	2.2***	4.5***	4.0***	26.7***	29.6***	23.9***
$Base_{catch}$ − moon	53.5	23,048	23***	1***	2.8***	3.5***	2.2***	1.8***	4.5***	NA	26.7***	29.6***	25.1***
$Base_{catch}$ − temp	53.5	23,048	23***	1***	2.8***	3.3***	2.1***	1.9***	NA	4.1***	26.7***	29.6***	24.1***
Large Black Sea Bass catch													
$Base_{catch}$	52.3	21,681	23***	1***	2.7***	5.5***	1.0***	2.8***	4.7***	4.6***	26.9***	30.0***	28.0***
$Base_{catch}$ − tod	52.2	21,698	23***	1***	2.7***	5.4***	NA	2.8***	4.7***	4.6***	26.9***	29.9***	28.2***
$Base_{catch}$ − moon	52.1	21,714	23***	1***	2.7***	5.2***	1.0***	2.8***	4.7***	NA	26.9***	30.0***	28.3***
$Base_{catch}$ − doy	52.1	21,715	23***	1***	2.8***	NA	1.0***	2.8***	4.6***	4.6***	26.9***	29.9***	27.4***

[a] $Base_{length}$ is: $z_{doy,y,pos} = f_1(y) + g_1(doy) + g_2(temp) + g_3(pos) + g_4(pos·y) + g_5(pos·doy) + e_{doy,y,pos}$.

[b] $Base_{catch}$ is: $x_{doy,y,pos} = f_1(y) + f_2(type) + g_1(depth) + g_2(doy) + g_3(tod) + g_4(soak) + g_5(temp) + g_6(moon) + g_7(pos) + g_7(pos·y) + g_7(pos·doy) + e_{doy,y,pos}$.

There were significant spatially variable effects of year on mean Black Sea Bass length and the catch of small and large Black Sea Bass (Figure 8). Over the course of the study, mean Black Sea Bass length increased throughout the region (Figure 4A), but mean length increased disproportionately more off Cape Canaveral and inshore areas of South Carolina and Georgia and increased the least off Cape Lookout, North Carolina, and deep waters off Georgia (Figure 8A). Catches of small and large Black Sea Bass increased considerably in North Carolina, northern Florida, and inshore areas of South Carolina and Georgia, while they decreased disproportionately off Cape Canaveral and offshore areas of South Carolina and Georgia (Figure 8B, C).

The spatially variable effects of day of the year on mean Black Sea Bass length and the catch of small and large Black Sea Bass were also significant, but were not nearly as strong as the effects of year (Figure 8D–F). Mean Black Sea Bass length was more likely to decrease during the sampling season (spring–fall) off Cape Canaveral, inshore areas in South Carolina, and around Cape Lookout (Figure 8D). The catch of small Black Sea Bass tended to increase during the sampling season off Cape Lookout (Figure 8E) and increased off both Cape Lookout and Cape Canaveral for large Black Sea Bass (Figure 8F).

DISCUSSION

The spatial and temporal distributions of marine fish species can be ecologically complex and mediated by myriad interacting variables (Dingsør et al. 2007; Ciannelli et al. 2012). We found complex spatial and temporal dynamics of Black Sea Bass that were influenced by environmental, landscape, and sampling effects over the 24 years of the survey. Moreover, the spatial distribution of Black Sea Bass was not static among or within sampling seasons, but instead changed temporally over both seasonal and annual time scales. Standardized mean length of Black Sea Bass has also increased over time as a result of a higher proportional abundance of large Black Sea Bass in the SEUS. These results elucidate the spatial and temporal dynamics of Black Sea Bass, inform spatial management approaches, and provide an analytical framework that can be easily adapted to other species and systems.

The spatial dynamics of Black Sea Bass vary along the U.S. Atlantic coast. North of Cape Hatteras, Black Sea Bass undertake yearly migrations southward and offshore in fall and inshore and northward in the spring (Musick and Mercer 1977; Moser and Shepherd 2009; Fabrizio et al. 2013). However, Black Sea Bass do not appear to migrate across Cape Hatteras; Moser and Shepherd (2009) tagged over 16,000 Black Sea Bass north of Cape Hatteras, and not a single tag

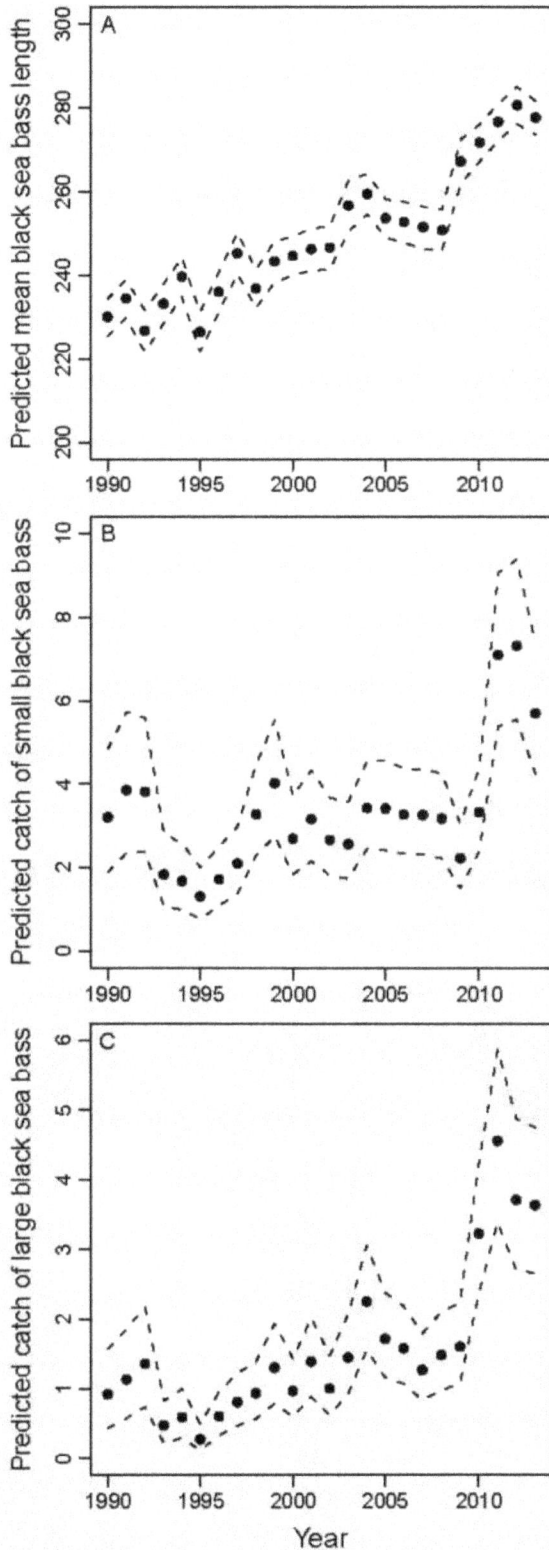

FIGURE 4. Predicted annual (**A**) mean Black Sea Bass length (mm TL), (**B**) catch of small Black Sea Bass, and (**C**) catch of large Black Sea Bass from spatially explicit generalized additive models built upon chevron-trap data, 1990–2013. Filled circles are mean predictions at average values of all other model covariates and dashed lines represent the 95% CI.

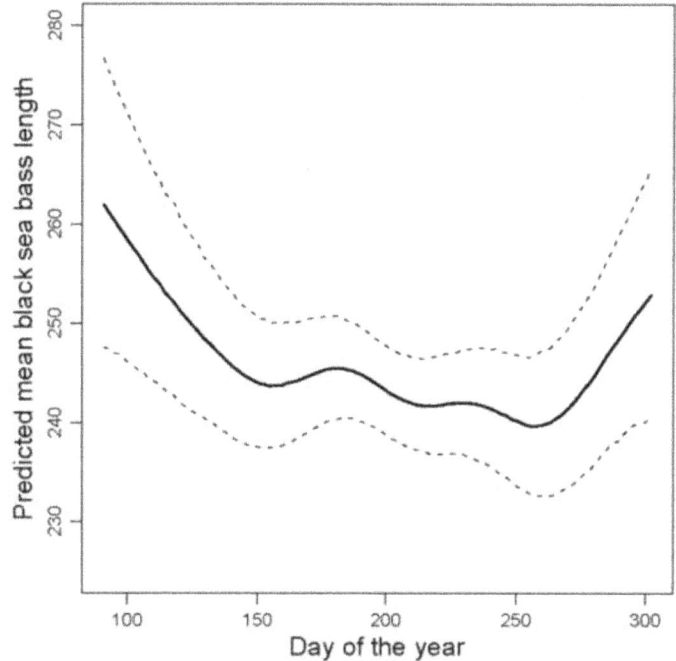

FIGURE 5. Predicted mean length of Black Sea Bass (mm TL) as a function of the day of the year using a spatially explicit generalized additive model built from Southeast Reef Fish Survey chevron-trap data, 1990–2013. The solid line is the predicted mean length of Black Sea Bass at average values of all other covariates, and dashed lines represent the 95% CI.

return (out of 2,800 returns) occurred south of Cape Hatteras. In the SEUS (i.e., south of Cape Hatteras), we found no clear evidence of inshore–offshore migration of Black Sea Bass over the seasonal time scale of this study (late March through early November). However, we did observe changes in the spatial distribution of sizes and catches of Black Sea Bass at smaller spatial scales within and among years. For instance, mean Black Sea Bass length tended to increase more in locations inshore in South Carolina, Georgia, and Florida over the course of the study than elsewhere, but seasonally showed modest declines off Cape Lookout, inshore off South Carolina, and off Cape Canaveral. Changes in the spatial structure of marine fishes may reflect migratory behaviors (Block et al. 2001), but habitat preferences (Gregory and Anderson 1997) and the spatial patterns of fishery harvest (Bartolino et al. 2012) can also influence observed spatial patterns. The potentially multiple, interacting forces causing Black Sea Bass to exhibit spatially variable sizes and catches are not known and deserve research attention.

The spatial patterns of Black Sea Bass were not consistent throughout the entire SEUS. Black Sea Bass tended to be smaller inshore and larger offshore, as has been previously noted (Sedberry et al. 1998; Steimle et al. 1999), but in our study small fish persisted into offshore waters around and south of Cape Canaveral. Black Sea Bass sizes and size at age can be positively related to latitude along the U.S. Atlantic coast, where individuals attain the largest body sizes north of

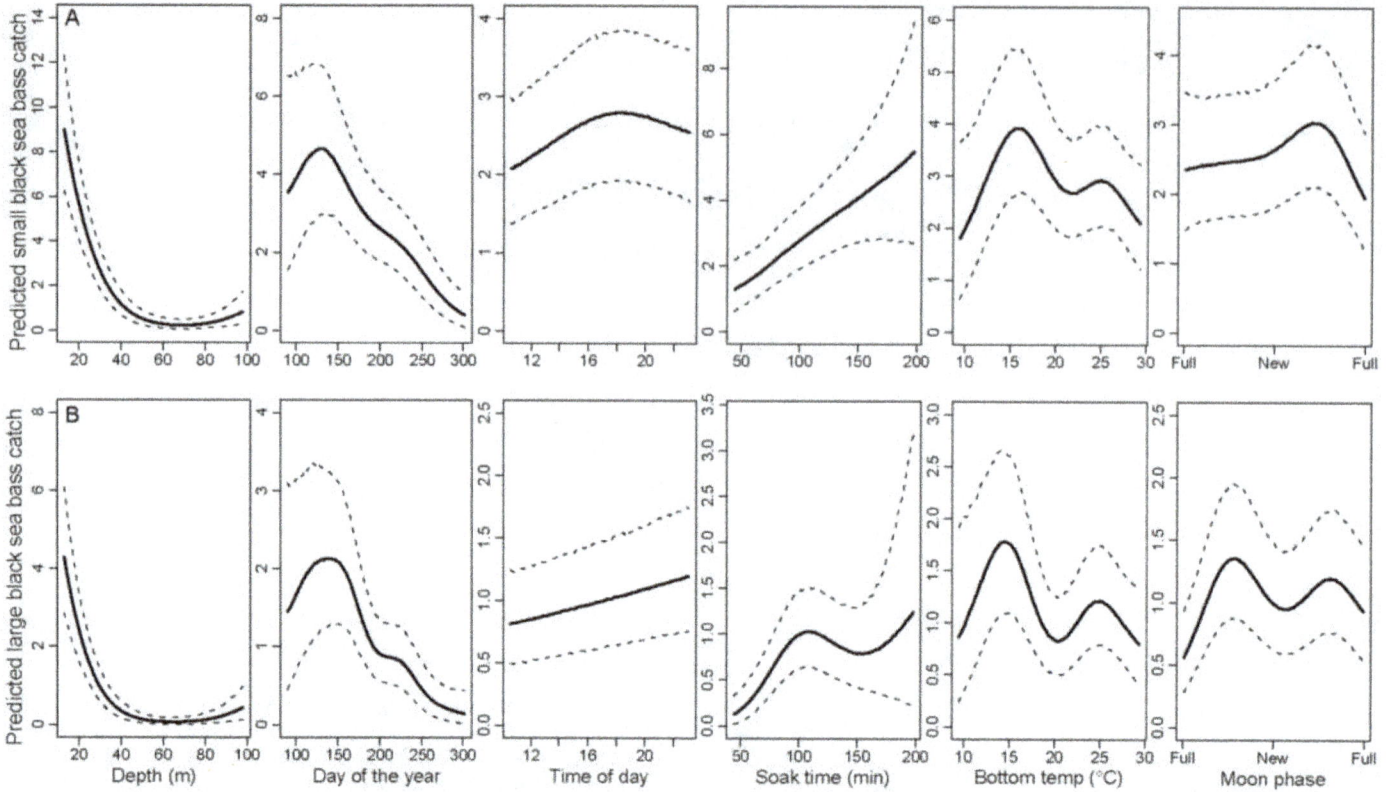

FIGURE 6.   Predicted catch of (**A**) small or (**B**) large Black Sea Bass as a function of depth (m), day of the year, time of day (Coordinated Universal Time), soak time (min), bottom temperature (°C), or moon phase using spatially explicit generalized additive models built using Southeast Reef Fish Survey chevron-traps data, 1990–2013. Solid lines are the predicted Black Sea Bass catch per trap at average values of all other covariates, and dashed lines represent the 95% CIs.

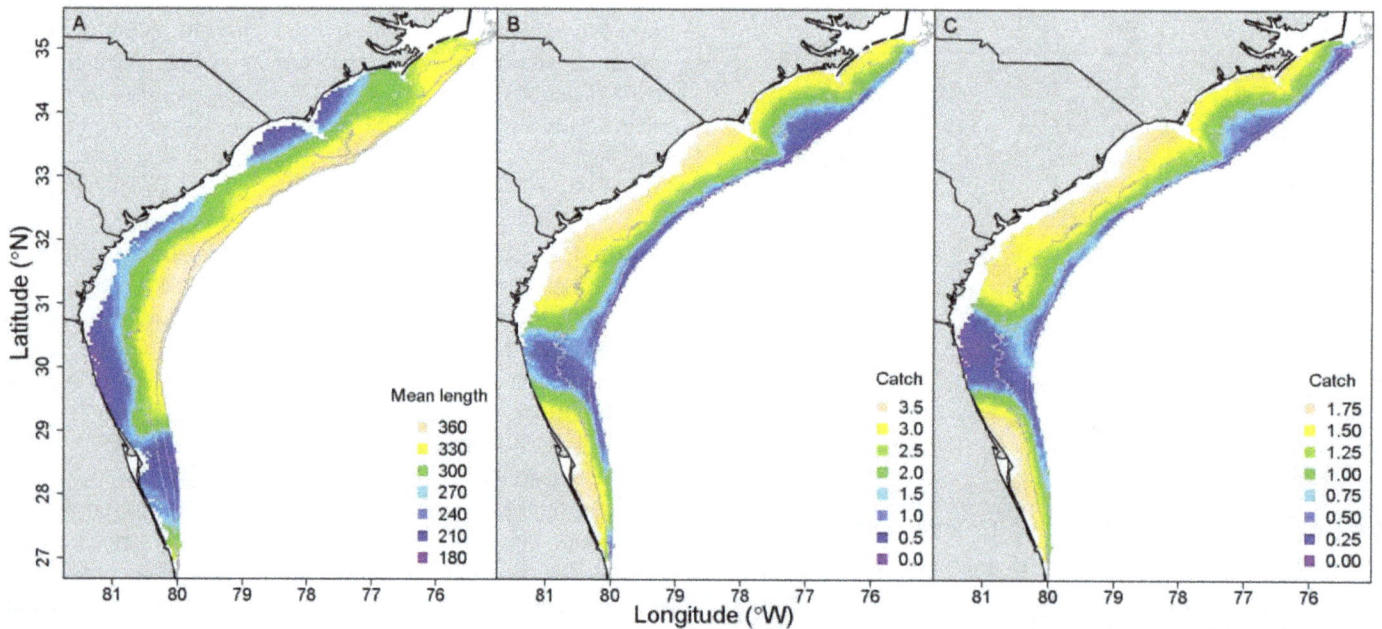

FIGURE 7.   Predicted (**A**) mean Black Sea Bass length (mm TL), (**B**) catch of small Black Sea Bass, and (**C**) catch of large Black Sea Bass across the study area given spatial position and depth of each cell and average values of all other covariates using spatially explicit generalized additive models built on chevron-trap data in 1990–2013. Gray lines indicate the 30-, 50-, and 100-m isobaths.

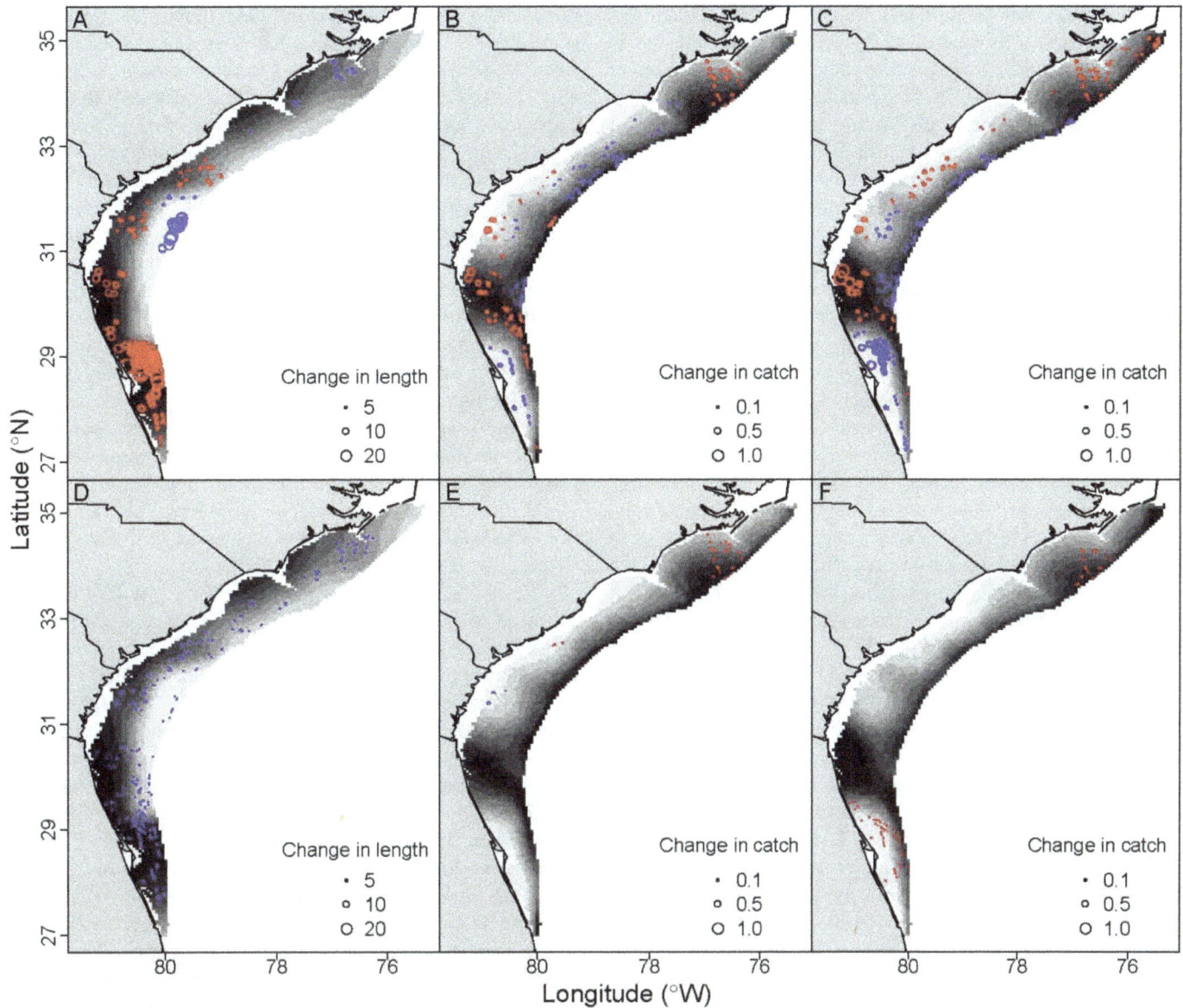

FIGURE 8. Spatially explicit variable-coefficient generalized additive model plots for the effects of (**A, B, C**) annual or (**D, E, F**) seasonal changes on the spatial distribution of (A, D) mean length of Black Sea Bass, (B, E) catch of small Black Sea Bass, and (C, F) catch of large Black Sea Bass in the Southeast Reef Fish Survey, 1990–2013. Light gray grid cells denote large mean size or catch, and black grid cells denote small mean size or catch. Overlaid on grid cells are red or blue circles, which indicate a disproportionate increase or decrease, respectively, in mean length or catch with an increase in the covariate. Size of the colored circles is scaled to the size of the positive or negative effect, and effects not significantly different from zero are excluded.

Cape Hatteras (Mercer 1978; Wenner et al. 1986). Even within the SEUS, Black Sea Bass can attain a larger size at age in North and South Carolina than those in Georgia and Florida (McGovern et al. 2002). But instead of a gradual change throughout the SEUS, we found that the depth distribution of Black Sea Bass sizes was similar over a broad range of latitudes (29.5–35°N), only changing qualitatively around Cape Canaveral.

Standardized mean length of Black Sea Bass increased significantly over the 24 years of the trap survey. During the 1970s and 1980s, harvest and fishery mortality rate of Black Sea Bass were high, which likely caused declines in mean length (Vaughan et al. 1995; McGovern et al. 2002). The stabilization of mean lengths and fishery-independent catch rates in the 1990s was attributed to the implementation of size limits, the prohibition of trawling, and reduced fishery harvests in the 1980s and 1990s (McGovern et al. 2002). When a population is heavily harvested, fish sizes tend to decline due to the selective removal of larger individuals from the population, both because of immediate effects of losing the largest individuals but also because of the potential long-term selection for early maturing and slower growing individuals (Conover and

Munch 2002; Olsen et al. 2004). Since the 1990s, Black Sea Bass mean length has increased by approximately 20 mm in the SEUS because of the relative increase in the number of large Black Sea Bass in the population, perhaps due to increased minimum size limits in the recreational fishery over the same time frame, and suggests there are few, persistent, selective effects on length for this population. This finding also hints at lower recent mortality rates of large Black Sea Bass in the SEUS.

The spatial dynamics and covariate effects were surprisingly similar between small and large Black Sea Bass in the SEUS. Most species tend to display ontogenetic changes in relation to environmental and habitat variables due to physiological and behavioral differences across ontogeny (Werner and Gilliam 1984; Mitchell et al. 2014). For Black Sea Bass in the SEUS, maturity occurs between 135 and 235 mm TL (McGovern et al. 2002), so some small and all large Black Sea Bass in our study were mature. We found that small and large Black Sea Bass varied similarly by depth, season, time of day, bottom temperature, moon phase, and spatial position. It is likely that the most significant ontogenetic changes for Black Sea Bass occur at fish sizes smaller than those caught in our survey, so that by the time they are caught in our survey, Black Sea Bass appear to behave similarly.

Small and large Black Sea Bass did differ significantly in their relationship to soak time in our trap survey. The catch of small Black Sea Bass increased linearly as soak time increased, whereas the catch of large Black Sea Bass reached an asymptote at longer soak times. Bacheler et al. (2013b) showed that the catch of Black Sea Bass in chevron traps reached an asymptote at a soak time of around 60 min and further suggested that catch at saturation is likely related to true abundance around the trap. Our results indicate that small and large Black Sea Bass may be responding differently to the trapping process, which was not accounted for in Bacheler et al. (2013b). One explanation for this difference in response of large and small Black Sea Bass to the trapping process is due to size-based selectivity of Black Sea Bass to the chevron-trap gear, which may not fully select for small Black Sea Bass.

There were some drawbacks of our study design. First, predictions of absolute mean lengths or catches of small and large Black Sea Bass were dependent upon the values or levels of the covariates chosen. In most cases we were able to use average values of covariates, but in some instances selection needed to occur manually. For instance, using mean latitude and longitude values would have resulted in unreasonable Black Sea Bass predictions from deep, offshore waters of the Atlantic Ocean; instead, values were chosen approximately in the center of the SEUS study area. Regardless, the covariate values or levels used only influence absolute but not relative values. Second, the latitudinal extent of sampling increased over time, which could have influenced study results. We believe bias is unlikely because mean length and catch rate

information from the core of the study area (i.e., South Carolina and Georgia), which has been sampled consistently since 1990, was nearly identical to that from the entire study area. However, mean length or catch information north of Cape Lookout, North Carolina, or south of Cape Canaveral, Florida, should be interpreted with caution due to low sample sizes. Third, spatial predictions were only developed for chevron traps deployed on hard-bottom sites, so inferences can only be made for hard-bottom (not sand or mud) substrates throughout the SEUS. Fourth, our regression models explained approximately 50–55% of the deviance in catch or length, suggesting that unmeasured variables, such as other characteristics of the site, interactions with other species, or fishery spatial structure, are important. Last, we assumed that the size-selectivity patterns of chevron traps were constant over space and time. Chevron-trap selectivity patterns would also need to be known in order to translate our length and catch predictions to make inferences about the population.

Despite these caveats, our study demonstrated that Black Sea Bass exhibited annual and seasonal variability in sizes and catches that had a unique spatial component unrelated to obvious inshore–offshore migrations. Black Sea Bass also appeared to have benefited from recent management actions by exhibiting increased mean length and fishery-independent trap catches; most significantly, large Black Sea Bass appeared to increase almost fourfold from the mid-1990s to the 2010s. The spatially explicit analytical approach we employed allowed us to quantify the spatial, temporal, environmental, and landscape correlates of Black Sea Bass size or catch in the SEUS, and provides important information if spatial management measures are considered in the future (Murphy and Jenkins 2010; Cardinale et al. 2011). Our results also suggest that little coastwide inshore–offshore or north–south movements by Black Sea Bass in the SEUS occurrs between spring and fall. The major benefits of a variable-coefficient modeling approach are that this approach is flexible and can be adapted to a variety of species and systems, it can test for myriad covariate effects simultaneously, and the results can be straightforward to interpret.

ACKNOWLEDGMENTS

We thank the captains and crews of the RV *Palmetto*, RV *Savannah*, NOAA Ship *Nancy Foster*, and NOAA Ship *Pisces*, all SERFS staff members, and the many volunteers for collection of field data. We thank J. Buckel and P. Rudershausen for many thoughtful discussions. We also thank L. Avens, A. Chester, A. Hohn, T. Kellison, P. Marraro, and K. Shertzer for providing comments on earlier versions of this manuscript. The use of trade, product, industry, or firm names, products, software, or models, whether commercially available or not, is for informative purposes only and does not constitute an endorsement by the U.S. Government or the National Oceanic and Atmospheric Administration.

REFERENCES

Bacheler, N. M., K. M. Bailey, L. Ciannelli, V. Bartolino, and K. S. Chan. 2009. Density-dependent, landscape, and climate effects on spawning distribution of Walleye Pollock *Theragra chalcogramma*. Marine Ecology Progress Series 391:1–12.

Bacheler, N. M., V. Bartolino, and M. J. M. Reichert. 2013a. Influence of soak time and fish accumulation on catches of reef fishes in a multispecies trap survey. U.S. National Marine Fisheries Service Fishery Bulletin 111:218–232.

Bacheler, N. M., J. A. Buckel, and L. M. Paramore. 2012. Density-dependent habitat use and growth of an estuarine fish. Canadian Journal of Fisheries and Aquatic Sciences 69:1734–1747.

Bacheler, N. M., L. Ciannelli, K. M. Bailey, and J. T. Duffy-Anderson. 2010. Spatial and temporal patterns of Walleye Pollock (*Theragra chalcogramma*) spawning in the eastern Bering Sea inferred from egg and larval distributions. Fisheries Oceanography 19:107–120.

Bacheler, N. M., Z. H. Schobernd, D. J. Berrane, C. M. Schobernd, W. A. Mitchell, and N. R. Geraldi. 2013b. When a trap is not a trap: converging entry and exit rates and their effect on trap saturation of Black Sea Bass (*Centropristis striata*). ICES Journal of Marine Science 70:873–882.

Bacheler, N. M., C. M. Schobernd, Z. H. Schobernd, W. A. Mitchell, D. J. Berrane, G. T. Kellison, and M. J. M. Reichert. 2013c. Comparison of trap and underwater video gears for indexing reef fish presence and abundance in the southeastern United States. Fisheries Research 143:81–88.

Bartolino, V., L. Ciannelli, N. M. Bacheler, and K. S. Chan. 2011. Ontogenetic and sex-specific differences in density-dependent habitat selection of a marine fish population. Ecology 92:189–200.

Bartolino, V., L. Ciannelli, P. Spencer, T. K. Wilderbuer, and K. S. Chan. 2012. Scale-dependent detection of the effects of harvesting a marine fish population. Marine Ecology Progress Series 444:251–261.

Block, B. A., H. Dewar, S. B. Blackwell, T. D. Williams, E. D. Prince, C. J. Farwell, A. Boustany, S. L. H. Teo, A. Seitz, A. Walli, and D. Fudge. 2001. Migratory movements, depth preferences, and thermal biology of Atlantic Bluefin Tuna. Science 293:1310–1314.

Brown, J. H., D. W. Mehlman, and G. C. Stevens. 1995. Spatial variation in abundance. Ecology 76:2028–2043.

Burnham, K. P., and D. R. Anderson. 2002. Model selection and multimodel inference: a practical information-theoretic approach, 2nd edition. Springer-Verlag, New York.

Cadrin, S. X., and D. H. Secor. 2009. Accounting for spatial population structure in stock assessment: past, present, and future. Pages 405–426 in R. J. Beamish and B. J. Rothschild, editors. The future of fisheries science in North America. Springer, Dordrecht, The Netherlands.

Cardinale, M., V. Bartolino, M. Llope, L. Maiorano, M. Sköld, and J. Hagberg. 2011. Historical spatial baselines in conservation and management of marine resources. Fish and Fisheries 12:289–298.

Ciannelli, L., V. Bartolino, and K. S. Chan. 2012. Non-additive and non-stationary properties in the spatial distribution of a large marine fish population. Proceedings of the Royal Society B 279:3635–3642.

Ciannelli, L., J. A. D. Fisher, M. Skern-Mauritzen, M. E. Hunsicker, M. Hildago, K. T. Frank, and K. M. Bailey. 2013. Theory, consequences and evidence of eroding population spatial structure in harvested fishes: a review. Marine Ecology Progress Series 480:227–243.

Coleman, F. C., C. C. Koenig, G. R. Huntsman, J. A. Musick, A. M. Eklund, J. C. McGovern, R. W. Chapman, G. R. Sedberry, and C. B. Grimes. 2000. Long-lived reef fishes: the grouper-snapper complex. Fisheries 25(3):14–20.

Collins, M. R. 1990. A comparison of 3 fish trap designs. Fisheries Research 9:325–332.

Conover, D. O., and S. B. Munch. 2002. Sustaining fisheries yields over evolutionary time scales. Science 297:94–96.

Dingsør, G. E., L. Ciannelli, K. S. Chan, G. Ottersen, and N. C. Stenseth. 2007. Density dependence and density independence during the early stages of four marine fish stocks. Ecology 88:625–634.

Dunn, D. C., and P. N. Halpin. 2009. Rugosity-based regional modeling of hard-bottom habitat. Marine Ecology Progress Series 377:1–12.

Dunning, J. B., B. J. Danielson, and R. H. Pulliam. 1992. Ecological processes that affect populations in complex landscapes. Oikos 65:169–175.

Fabrizio, M. C., J. P. Manderson, and J. P. Pessutti. 2013. Habitat associations and dispersal of Black Sea Bass from a mid-Atlantic Bight reef. Marine Ecology Progress Series 482:241–253.

Fautin, D., P. Dalton, L. S. Incze, J. C. Leong, C. Pautzke, A. Rosenberg, P. Sandifer, G. Sedberry, J. W. Tunnell, I. Abbott, R. E. Brainard, M. Brodeur, L. G. Eldredge, M. Feldman, F. Moretzsohn, P. S. Vroom, M. Wainstein, and N. Wolff. 2010. An overview of marine biodiversity in United States waters. PLoS (Public Library of Science) One [online serial] 5(8):e11914.

Glasgow, D. M. 2010. Photographic evidence of temporal and spatial variation in hardbottom habitat and associated biota of the southeastern U.S. Atlantic continental shelf. Master's thesis. College of Charleston, Charleston, South Carolina.

Gregory, R. S., and J. T. Anderson. 1997. Substrate selection and use of protective cover by juvenile Atlantic Cod *Gadus morhua* in inshore waters of Newfoundland. Marine Ecology Progress Series 146:9–20.

Hood, P. B., M. F. Godcharles, and R. S. Barco. 1994. Age, growth, reproduction, and the feeding ecology of Black Sea Bass, *Centropristis striata* (Pisces: Serranidae), in the eastern Gulf of Mexico. Bulletin of Marine Science 54:24–37.

Jackson, D. A., P. R. Peres-Neto, and J. D. Olden. 2001. What controls who is where in freshwater fish communities – the roles of biotic, abiotic, and spatial factors. Canadian Journal of Fisheries and Aquatic Sciences 58:157–170.

Kendall, M. S., L. J. Bauer, and C. F. G. Jeffrey. 2008. Influence of benthic features and fishing pressure on size and distribution of three exploited reef fishes from the southeastern United States. Transactions of the American Fisheries Society 137:1134–1146.

Lavenda, N. 1949. Sexual differences and normal protogynous hermaphroditism in the Atlantic sea bass, *Centropristis striatus*. Copeia 1949:185–194.

Lehmann, A., J. M. Overton, and J. R. Leathwick. 2002. GRASP: generalized regression analysis and spatial prediction. Ecological Modelling 157:189–207.

Levin, S. A. 1992. The problem of pattern and scale in ecology. Ecology 73:1943–1967.

McCartney, M. A., M. L. Burton, and T. G. Lima. 2013. Mitochondrial DNA differentiation between populations of Black Sea Bass (*Centropristis striata*) across Cape Hatteras, North Carolina (USA). Journal of Biogeography 40:1386–1398.

McGovern, J. C., M. R. Collins, O. Pashuk, and H. S. Meister. 2002. Temporal and spatial differences in life history parameters of Black Sea Bass in the southeastern United States. North American Journal of Fisheries Management 22:1151–1163.

Mercer, L. P. 1978. The reproductive biology and population dynamics of Black Sea Bass, *Centropristis striata*. Doctoral dissertation. College of William and Mary, Williamsburg, Virginia.

Mitchell, W. A., G. T. Kellison, N. M. Bacheler, J. C. Potts, C. M. Schobernd, and L. F. Hale. 2014. Depth-related distribution of postjuvenile Red Snapper in southeastern U.S. Atlantic Ocean waters: ontogenetic patterns and implications for management. Marine and Coastal Fisheries: Dynamics, Management, and Ecosystem Science [online serial] 6:142–155.

Moser, J., and G. R. Shepherd. 2009. Seasonal distribution and movement of Black Sea Bass (*Centropristis striata*) in the northwest Atlantic as determined from a mark–recapture experiment. Journal of Northwest Atlantic Fishery Science 40:17–28.

Murphy, H. M., and G. P. Jenkins. 2010. Observational methods used in marine spatial monitoring of fishes and associated habitats: a review. Marine and Freshwater Research 61:236–252.

Musick, J. A., and L. P. Mercer. 1977. Seasonal distribution of Black Sea Bass, *Centropristis striata*, in the Mid-Atlantic Bight with comments on the ecology and fisheries of the species. Transactions of the American Fisheries Society 106:12–25.

Neter, J., W. Wasserman, and M. H. Kutner. 1989. Applied linear regression models, 2nd edition. Irwin, Homewood, Illinois.

Olsen, E. M., M. Heino, G. R. Lilly, M. J. Morgan, J. Brattey, B. Ernande, and U. Dieckmann. 2004. Maturation trends indicative of rapid evolution preceded the collapse of northern cod. Nature 428:932–935.

Powles, H., and C. A. Barans. 1980. Groundfish monitoring in sponge-coral areas off the southeastern United States. Marine Fisheries Review 42:21–35.

R Development Core Team. 2013. R: a language and environment for statistical computing. R Foundation for Statistical Computing, Vienna. Available: http://cran.r-project.org/. (October 2014).

Roy, E. M., J. M. Quattro, and T. W. Greig. 2012. Genetic management of Black Sea Bass: influence of biogeographic barriers on population structure. Marine and Coastal Fisheries: Dynamics, Management, and Ecosystem Science [online serial] 4:391–402.

Schobernd, C. M., and G. R. Sedberry. 2009. Shelf-edge and upper-slope reef fish assemblages in the South Atlantic Bight: habitat characteristics, spatial variation, and reproductive behavior. Bulletin of Marine Science 84:67–92.

Sedberry, G. R., J. C. McGovern, and C. A. Barans. 1998. A comparison of fish populations in Gray's Reef National Marine Sanctuary to similar habitats off the southeastern U.S.: implications for reef fish and sanctuary management. Proceedings of the Gulf and Caribbean Fisheries Institute 57:463–514.

Sedberry, G. R., and R. F. Van Dolah. 1984. Demersal fish assemblages associated with hard bottom habitat in the South Atlantic Bight of the U.S.A. Environmental Biology of Fishes 11:241–258.

Steimle, F. W., C. A. Zetlin, P. L. Berrien, and S. Chang. 1999. Essential fish habitat source document: Black Sea Bass, Centropristis striata, life history and habitat characteristics. NOAA Technical Memorandum NMFS-NE-143.

Vaughan, D. S., M. R. Collins, and D. J. Schmidt. 1995. Population characteristics of the Black Sea Bass Centropristis striata from the southeastern United States. Bulletin of Marine Science 56: 250–267.

Venables, W. N., and C. M. Dichmont. 2004. GLMs, GAMs and GLMMs: an overview of theory for applications in fisheries research. Fisheries Research 70:319–337.

Wenner, C. A., W. A. Roumillat, and C. W. Waltz. 1986. Contributions to the life history of Black Sea Bass, Centropristis straita, off the southeastern United States. U.S. National Marine Fisheries Service Fishery Bulletin 84:723–741.

Werner, E. E., and J. F. Gilliam. 1984. The ontogenetic niche and species interactions in size-structured populations. Annual Review of Ecology and Systematics 15:393–425.

Wood, S. N. 2004. Stable and efficient multiple smoothing parameter estimation for generalized additive models. Journal of the American Statistical Association. 99:673–686.

Wood, S. N. 2008. Generalized additive models: an introduction with R. Chapman and Hall/CRC Press, Boca Raton, Florida.

Wood, S. N. 2011. Fast stable restricted maximum likelihood and marginal likelihood estimation of semiparametric generalized linear models. Journal of the Royal Statistical Society B 73:3–36.

7

Spatial and Temporal Patterns in Smolt Survival of Wild and Hatchery Coho Salmon in the Salish Sea

Mara S. Zimmerman*
Washington Department of Fish and Wildlife, 600 Capitol Way North, Olympia, Washington 98501, USA

James R. Irvine
Fisheries and Oceans Canada, Pacific Biological Station, 3190 Hammond Bay Road, Nanaimo, British Columbia V9T 6N7, Canada

Meghan O'Neill
407-960 Inverness Road, Victoria, British Columbia V8X 2R9, Canada

Joseph H. Anderson
Washington Department of Fish and Wildlife, 600 Capitol Way North, Olympia, Washington 98501, USA

Correigh M. Greene
National Oceanic and Atmospheric Administration, National Marine Fisheries Service, Northwest Fisheries Science Center, Fish Ecology Division, 2725 Montlake Boulevard East, Seattle, Washington 98112, USA

Joshua Weinheimer
Washington Department of Fish and Wildlife, 600 Capitol Way North, Olympia, Washington 98501, USA

Marc Trudel
Fisheries and Oceans Canada, Pacific Biological Station, 3190 Hammond Bay Road, Nanaimo, British Columbia V9T 6N7, Canada; and Department of Biology, University of Victoria, Post Office Box 1700, Station CSC, Victoria, British Columbia V8W 3N5, Canada

Kit Rawson
Swan Ridge Consulting, 3601 Carol Place, Mount Vernon, Washington 98273-8583, USA

Abstract
    Understanding the factors contributing to declining smolt-to-adult survival (hereafter "smolt survival") of Coho Salmon *Oncorhynchus kisutch* originating in the Salish Sea of southwestern British Columbia and Washington State is a high priority for fish management agencies. Uncertainty regarding the relative importance of mortality

Subject editor: Carl Walters, University of British Columbia, Canada

*Corresponding author: mara.zimmerman@dfw.wa.gov

operating at different spatial scales hinders the prioritization of science and management activities. We therefore examined spatial and temporal coherence in smolt survivals for Coho Salmon based on a decision tree framework organized by spatial hierarchy. Smolt survival patterns of populations that entered marine waters within the Salish Sea were analyzed and compared with Pacific coast reference populations at similar latitudes. In all areas, wild Coho Salmon had higher survival than hatchery Coho Salmon. Coherence in Coho Salmon smolt survival occurred at multiple spatial scales during ocean entry years 1977–2010. The primary pattern within the Salish Sea was a declining smolt survival trend over this period. In comparison, smolt survival of Pacific coast reference populations was low in the 1990s but subsequently increased. Within the Salish Sea, smolt survival in the Strait of Georgia declined faster than it did in Puget Sound. Spatial synchrony was stronger among neighboring Salish Sea populations and occurred at a broader spatial scale immediately following the 1989 ecosystem regime shift in the North Pacific Ocean than before or after. Smolt survival of Coho Salmon was synchronized at a more local scale than reported by other researchers for Chinook Salmon *O. tshawytscha*, Pink Salmon *O. gorbuscha*, Chum Salmon *O. keta*, and Sockeye Salmon *O. nerka*, suggesting that early marine conditions are especially important for Coho Salmon in the Salish Sea. Further exploration of ecosystem variables at multiple spatial scales is needed to effectively address linkages between the marine ecosystem and Coho Salmon smolt survival within the Salish Sea. Since the relative importance of particular variables may have changed during our period of record, researchers will need to carefully match spatial and temporal scales to their questions of interest.

During their ocean residence, Pacific salmon *Oncorhynchus* spp. travel thousands of kilometers through spatially and temporally dynamic environments. Survival in the marine environment plays a major role in determining the numbers of adult Pacific salmon recruiting to fisheries and returning to freshwater to spawn (Pearcy 1992). Most Pacific salmon "marine survival" estimates cover the period from when smolts leave their spawning stream to when they return and, therefore, include some freshwater effects; hereafter we refer to these as "smolt survival" estimates. Understanding factors that influence smolt survival and using this information to predict Pacific salmon run sizes has proven to be a challenging undertaking (Dorner et al. 2013; Irvine and Akenhead 2013). When smolt survival patterns are compared among populations and species, results can provide important insight into the factors contributing to survival. When smolt survival patterns are similar across broad geographic regions, this suggests that broad-scale climate conditions are affecting survival (Mantua et al. 1997; Beamish et al. 1999b; Mueter et al. 2007; Peterman and Dorner 2012). In comparison, synchronous smolt survival patterns found only for populations within a limited geographic region suggest that local factors, disconnected from broad-scale drivers, are having the greatest influence on survival (Pyper et al. 2005; Beamish et al. 2012). Furthermore, when smolt survival patterns differ between species or regions, these contrasts can inform hypotheses and future study of the key factors affecting survival.

The marine environment is linked to the survival of Pacific salmon, such as Coho Salmon *Oncorhynchus kisutch*, at many spatial scales. For example, at the broadest scale, Coho Salmon populations entering the Gulf of Alaska Current north of Vancouver Island have had different smolt survival patterns than those entering the California Current south of this location (Coronado and Hilborn 1998a; Hobday and Boehlert 2001; Teo et al. 2009). At the scale of the California Current, smolt survival among Coho Salmon populations exhibit some synchrony (Beamish et al. 2000; Botsford and Lawrence 2002); however, several studies have noted different smolt survival patterns for populations that enter the California Current but originate inside versus outside the Salish Sea (Coronado and Hilborn 1998a; Hobday and Boehlert 2001; Beetz 2009). The Salish Sea encompasses the network of inland marine waters from the northern extent of the Strait of Georgia within British Columbia to the southern extent of Puget Sound in Washington State and includes the Strait of Juan de Fuca that connects the Salish Sea to the California Current (Figure 1A). The Strait of Georgia basin can be divided into northern, central, and southern subbasins based on physical factors, such as water depth, salinity, turbidity, and currents (Thomson 2014). Similarly, Puget Sound basin is comprised of four major subbasins, Whidbey Basin, Central Puget Sound, South Puget Sound, and Hood Canal (Babson et al. 2006; Moore et al. 2008b). Smolt survival differences at this subbasin scale have not been demonstrated.

In recent decades, several salmonid species that spawn in watersheds draining into the Salish Sea (e.g., Coho Salmon, Chinook Salmon *O. tshawytscha*, and steelhead *O. mykiss*) have returned in increasingly low numbers (Coronado and Hilborn 1998b; Scott and Gill 2008; Beamish et al. 2010), whereas other species (e.g., odd-year returning Pink Salmon *O. gorbuscha*) have returned in unprecedented high numbers (Irvine et al. 2014). Understanding factors contributing to these declines is a high priority for fish management agencies in the Pacific Northwest. Uncertainty in the relative importance of mortality operating at different spatial scales hinders the prioritization of science and management activities. This paper focuses on smolt survival of Coho Salmon in the Salish Sea and adds more than a decade of information to previous analyses for this species (Coronado and Hilborn 1998a; Beamish et al. 2000), explores heterogeneity within the Salish Sea, and provides a framework for future investigation of explanatory factors.

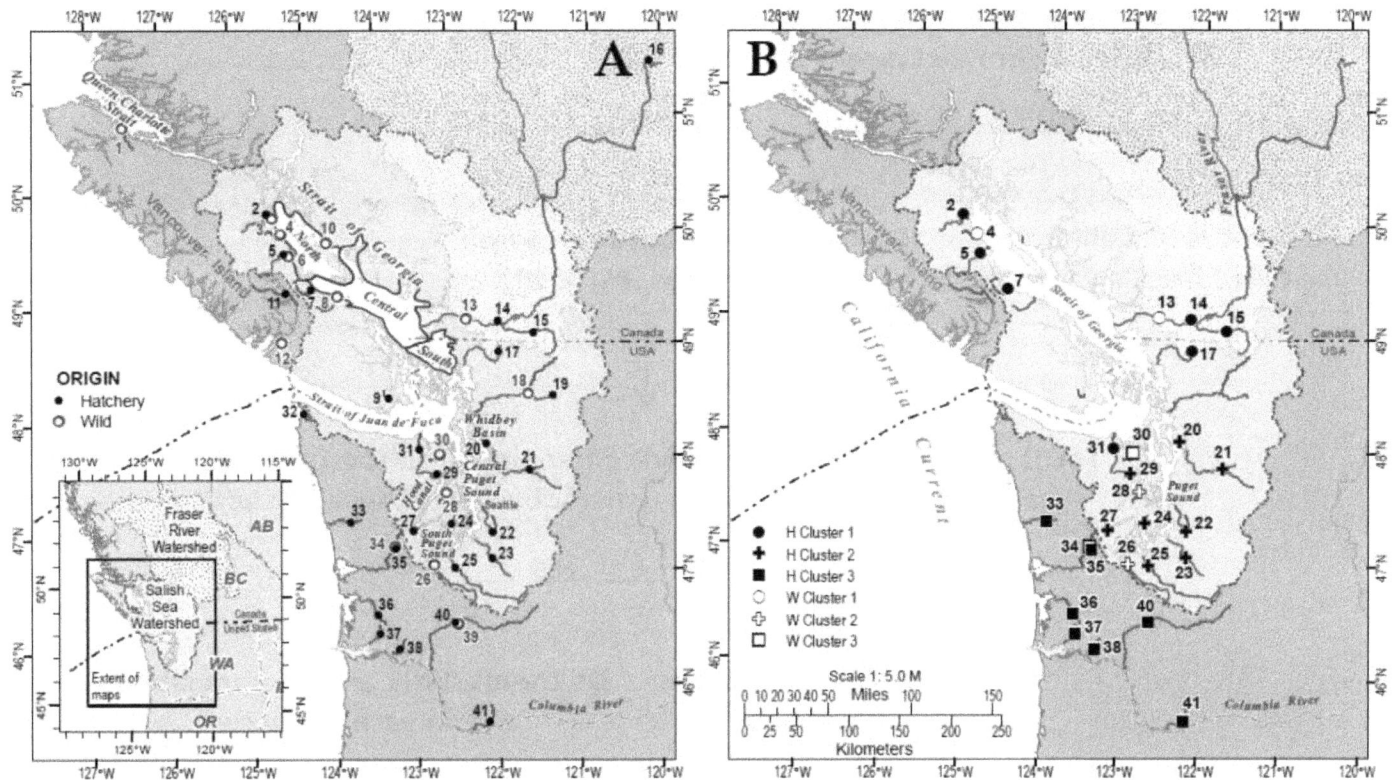

FIGURE 1. Study area showing **(A)** the major oceanographic areas and locations of Coho Salmon populations (open circle = wild, filled circle = hatchery) and **(B)** the spatial organization of cluster analysis groupings for the Strait of Georgia (circles), Puget Sound (crosses), and the Pacific coast (squares). Numbers correspond to population information provided in Table 1.

The primary objectives of this study were to examine patterns of spatial and temporal coherence in smolt survival for wild and hatchery Coho Salmon populations within the Salish Sea and to identify appropriate spatial scales for the subsequent identification of key ecosystem variables. These objectives were addressed using a decision tree framework (Figure 2) that connects the scale of survival patterns within multiple populations entering the California Current to the scale of ecosystem variables. In addition to populations within the Salish Sea, we also examined smolt survivals for Pacific coast reference populations from the Columbia River, Washington coast, and western and northeastern coasts of Vancouver Island. Since only populations originating from the Salish Sea spend time rearing and growing in the inland sea environment, survival differences between Salish Sea and Pacific coast reference populations were likely attributable to their early marine behavior and ecology. Coho Salmon smolt survival was also compared between and within the two primary basins of the Salish Sea (Puget Sound versus Strait of Georgia) and the subbasins within each primary basin.

Five separate analyses were used in combination to examine the patterns of spatial and temporal coherence in smolt survival of hatchery and wild Coho Salmon. The first two analyses investigated the smolt survival patterns without prior geographic assignment. A cluster analysis identified population groupings based on smolt survival. An exponential decay model examined the strength of survival correlations as a function of distance for the entire time series and for three preassigned time periods corresponding to major climatic regimes in the North Pacific Ocean. The next three analyses investigated smolt survival patterns at preassigned spatial scales identified in the decision tree (Figure 2). A regional mixed-effects model tested the contribution of geographic regions (Salish Sea versus outside the Salish Sea, Strait of Georgia versus Puget Sound) to Coho Salmon smolt survival. Linear regression and structural change analysis identified temporal trends and whether breakpoints occurred at these spatial scales. Residuals from this regional model were used in a second mixed-effects model to test whether smolt survival of Salish Sea populations covaried with Pacific coast populations (California Current scale) and whether survival differed at the subbasin scale. Finally, we calculate effect sizes to compare the relative contributions of patterns at multiple scales with the overall variation in survival.

METHODS

Populations.—Stream populations, identified by their stream of origin and the location of monitoring activities, were

Do trends differ between Salish Sea and non-Salish Sea California Current populations?

NO | YES

Focus on California Current scale variables

Do trends differ between Salish Sea basins (e.g., PS, SoG)?

NO | YES

Focus on Salish Sea scale variables

Do trends differ among subbasins that are within major basins of the Salish Sea?

NO | YES

Focus on basin scale variables

Do trends differ among populations within each subbasin?

NO | YES

Focus on subbasin scale variables

Focus on local scale variables

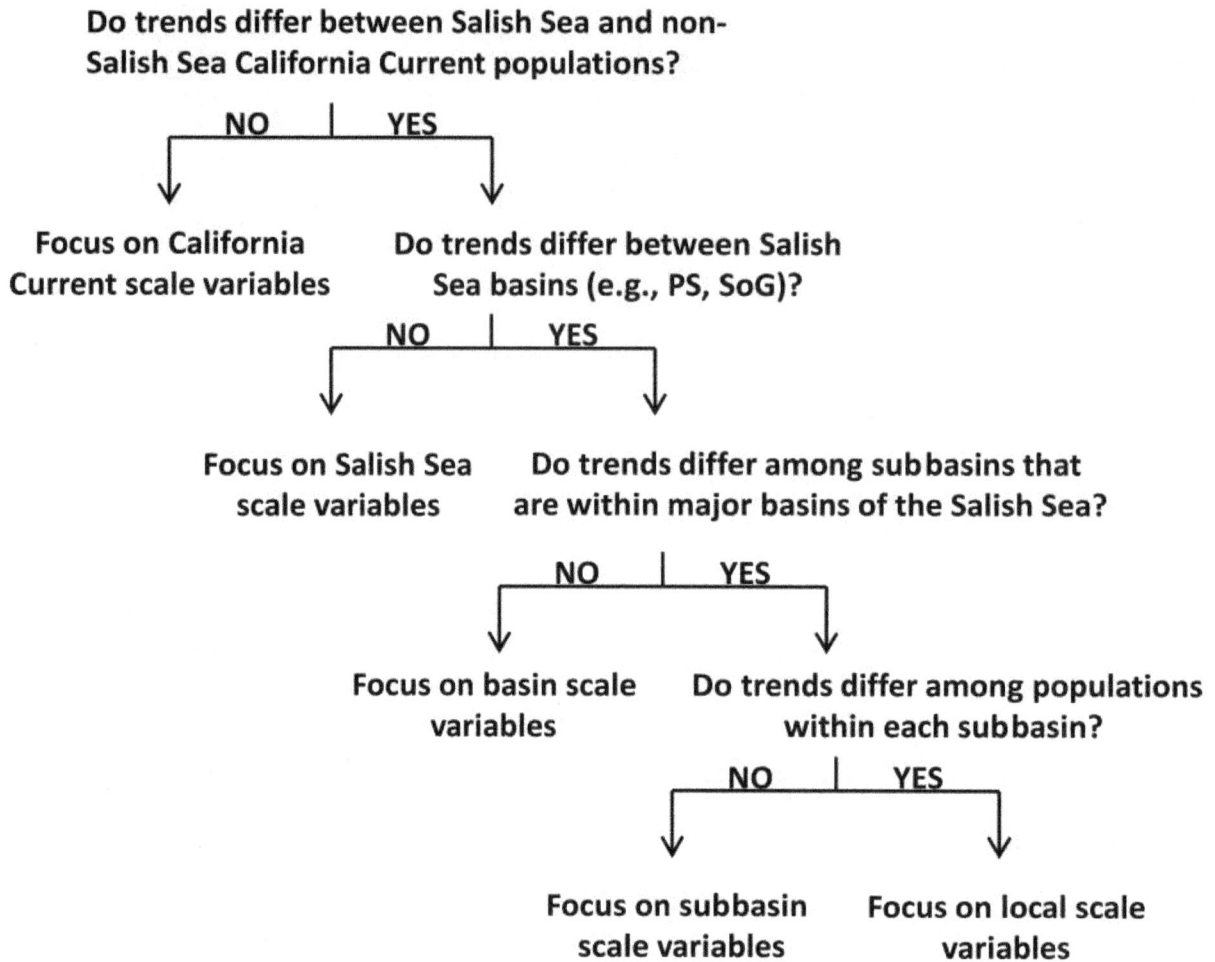

FIGURE 2.   Decision tree showing the series of dichotomous questions regarding survival patterns that are used to infer an appropriate scale for considering environmental variables associated with the observed patterns (SoG = Strait of Georgia, PS = Puget Sound).

groups of fish produced either by natural spawning ("wild") or in hatcheries and represented the major areas within the Salish Sea, as well as locations on the western and northeastern coasts of Vancouver Island, the Olympic Peninsula, southwestern Washington, and the lower Columbia River (Figure 1; Table 1). One population (Louis Creek) in the Thompson River (tributary to the Fraser River) watershed originates outside the terrestrial geographic boundaries of the Salish Sea but migrates into and through the Salish Sea and was therefore included with the Strait of Georgia populations.

Smolt survival estimates.—Our measures of smolt survival included survival from the release location in freshwater, which was often upstream of the marine entry point; therefore, we chose not to use the term marine survival. Smolt survival was the estimated number of 3-year-old Coho Salmon caught in all fisheries plus the number of 3-year-old Coho Salmon escaping fisheries to return to the rivers or hatchery to spawn divided by the number of smolts that produced these adults. Age 3 is the typical age of returning spawners in the Salish Sea region (Sandercock 1991; Labelle et al. 1997) and

represented the majority of the adult returns in most of our datasets. Jacks (precocious 2-year male Coho Salmon) were not included in the analysis because they rarely contribute to fisheries; their small size makes them difficult to enumerate (Irvine et al. 1992), resulting in estimates of unknown accuracy and precision.

The primary method to estimate smolt survival was based on the release and subsequent recovery of coded-wire-tagged (CWT) Coho Salmon smolts (Table 1). Tag data for U.S. populations were retrieved from the coastwide CWT database (www.rmpc.org). Tag data for Canadian populations were retrieved from the Mark Recovery Program database (Kuhn et al. 1988) and regional agency datasets (S. Baillie, Fisheries and Oceans Canada in Nanaimo, British Columbia, personal communication). A second estimation method relied on estimates of smolts leaving a system during the spring and adults returning during the fall and winter 18 months later. Smolt survival estimates using this second approach will be biased high if there are significant numbers of Coho Salmon subyearlings leaving the system in fall that survive to adulthood, as was found for some streams in our study area (Craig et al. 2014; Bennett et al. 2015). For this second method, spawner escapement estimates were expanded by exploitation rate, either modeled or calculated using tag recoveries of a nearby population, in order to include the fish retained in fisheries in the survival estimates. Modeled exploitation rates (CoTC 2013) were estimated using either a mixed-stock model based on annual CWT recoveries or using backwards runs of the Fishery Regulation Assessment Model. The Fishery Regulation Assessment Model is a comprehensive fishery model used to plan and assess the impacts of mixed-stock ocean fisheries on Coho Salmon stocks from Alaska to California (PFMC 2008).

Selection criteria were developed in order to remove the CWT codes that were likely to be affected by year-specific factors other than marine conditions. Tag codes from Puget Sound populations were ignored prior to 1977 releases due to concerns about incomplete fishery and escapement sampling. Additional criteria for excluding CWT codes included no sampling data from major fisheries impacting that population, missing escapement recoveries for age-3 returns, incorrectly cut tag lengths, known disease noted for smolts upon release, poor environmental conditions noted at release, fish that were part of a sterilization experiment and genetically altered (e.g., triploid), or release at the fry rather than smolt life stage. Net pen or sea pen releases were not used, as fish escaping terminal fisheries were not likely to be well documented (one exception was three tag codes from the Big Qualicum Estuary sea pen releases in 1989 because estimated smolt survival was similar to the releases from Big Qualicum Hatchery). Eighty-four percent of the tag codes met these criteria and were summed by year and watershed (hatchery and wild separately).

Groupings in smolt survival without prior assignment.—A cluster analysis identified population groups with similar patterns in smolt survival. Because this analysis did not use prior information about geographic regions, the results could be used to objectively examine whether patterns in smolt survival were associated with population characteristics, including geography and origin (i.e., hatchery versus wild). The time series used in this analysis included 28 populations (including 7 wild) that spanned ocean entry years (OEYs) 1977–2010 (Table 1). Datasets had a minimum of 19 years that met the selection criteria and were not truncated on either extreme (e.g., a time series with 19 years beginning in 1992 would not be included in this analysis, whereas a time series with 19 years of data beginning in 1980 would be included).

Ward's hierarchical cluster analysis, which minimizes within-group variance among clusters, described associations among populations based on Euclidian distances between annual smolt survival estimates (Legendre and Legendre 2012). Prior to analysis, all data were logit transformed and scaled to a mean of 0 and standard deviation of 1. As a result of these transformations, the cluster analysis identified similarities in data trends (e.g., high and low years), removing among-population differences in the magnitude of smolt survival. Statistical

support for the cluster groupings was examined with a multiscale bootstrap resampling technique using the pvclust package in R version 3.1 (Suzuki and Shimodaira 2006). The approximately unbiased *P*-value from this analysis represented the statistical support for each branch of the dendrogram based on *N* = 10,000 bootstraps. A *P*-value greater than or equal to 95% was considered to be strong support.

Smolt survival correlations over distance and time.—Correlograms and correlation-by-distance analyses were conducted on the entire data series and the data series divided into three OEY time periods (1977–1988, 1989–1997, and 1998–2010). These time periods were selected because they have been identified as regimes in the physical climate of the North Pacific Ocean (Overland et al. 2008). The end of the last time period (1998–2010) represented the end of the time series, not an identified regime break. These analyses tested whether time periods associated with broad-scale (i.e., climate) environmental changes could be linked to different scales of smolt survival correlations among Coho Salmon populations.

Correlograms visually displayed the strength and direction of the correlations (Pearson's product-moment correlation coefficients) among populations for the entire data set and among the three time periods. Smolt survival data were logit transformed and scaled prior to analysis. Correlations among a single set of populations with overlapping data in all three time periods were used so that patterns could easily be compared across time periods. At least 5 years of overlapping data within each time period were required, resulting in 17 populations being included in the display.

The rate of decay in pairwise correlations was modeled as a function of distance between populations. All populations with overlapping smolt survival estimates in a given time period were used. A geographic information system (GIS) was used to create a series of path distance arrays and to calculate distance over sea between points of entry into the marine environment. Data were fit with an exponential decay model (Myers et al. 1997; Pyper et al. 2005):

$$\rho(d) = \rho_o e^{-d/v},$$

where $\rho(d)$ was the survival correlation measure and d the distance between marine entry locations. The intercept, ρ_o, represented the correlation between neighboring populations, and the parameter v (*e*-folding scale) was the distance at which the population correlations were reduced by 37% (i.e., $e^{-1} \times 100\%$). The fit of the exponential decay model was weighted by the number of years of data for each population, and parameters were estimated with nonparametric bootstrap sampling with replacement (1,000 iterations) using the stats package in R version 3.1 (R Development Core Team 2014). In order to compare results from this study with those previously published (Pyper et al. 2005; Teo et al. 2009; Kilduff et al. 2014), the analysis was first conducted allowing the intercept (ρ_o) to vary and then conducted while constraining the intercept to a constant value of 1. The fit of the unconstrained and constrained models to the data was compared using Akaike information criteria corrected for small sample size (AIC$_c$; Burnham and Anderson 2002). For all analyses, models were considered to have strong support when the difference in AIC$_c$ values between the best (lowest AIC$_c$) and a given model was less than three.

Smolt survival patterns at Salish Sea and basin scales.—A model selection process was used to evaluate the differences in smolt survival among geographic regions. Linear mixed-effect models evaluated the fixed effects explicitly accounted for in the study design, while taking into account the random effects that were not strictly controlled for in the analysis (Stroup 2012). The model evaluated whether geographic region (Table 1; Pacific coast, Strait of Georgia, or Puget Sound), origin (wild or hatchery origin), and year explained variation in smolt survival; population was a random effect in all analyses. All populations with available data were included in the regional analysis, except for three populations as explained below. This analysis was restricted to OEYs ≥ 1977 because this was the first year of data for many populations.

Several of the populations did not fall neatly into any of the three regions in the first analysis. The Keogh River (number 1 in Table 1; Figure 1A) was assigned to the Pacific coast region, as this population enters marine waters (Queen Charlotte Strait) at the north end of Vancouver Island that connect most directly with the Pacific Ocean. Goldstream River (number 9), Snow Creek (number 30), and Dungeness River (number 31) populations enter the Strait of Juan de Fuca, which has oceanographic characteristics more similar to the Pacific Ocean shelf than the inland waters of the Strait of Georgia (Johannessen et al. 2006; Masson and Pena 2009). Because the Strait of Juan de Fuca lies within the geographic boundaries of the Salish Sea but did not obviously assign to either the Puget Sound or Strait of Georgia region, these populations were excluded in the mixed-effects model analysis.

A series of models examined whether geographic region, origin (hatchery or wild), year, and possible two-way interactions predicted variation in smolt survival. The AIC was used to compare the model probabilities given the data. Akaike information criteria values were not corrected for small sample size as the number of samples (independent survival estimates) were well above the threshold where small sample size issues are a problem. (Burnham and Anderson 2002). The strength of support for each model was based on model weights. Year was modeled as a linear covariate to determine trend. An autoregressive error structure with a 3-year lag was used for all analyses, as this structure matched the dominant 3-year Coho Salmon life cycle and resulted in the lowest AIC values when compared with the same model with other possible autoregressive and autoregressive moving-average variance structures. Unlike the cluster analysis, in which data were logit transformed and scaled (mean of 0, standard deviation of 1), the mixed-effects model analyses were performed on logit-transformed but nonscaled

data. Consequently, the two analyses explored smolt survival patterns in different ways; the mixed-effects model focused on differences in the magnitude of smolt survival among explanatory variables, whereas the cluster analysis focused on similarities in smolt survival trends (high years versus low years).

The regional model evaluated whether smolt survival differed among the three geographic regions (Pacific coast, Strait of Georgia, and Puget Sound) but did not determine how smolt survival differed at these two scales in our decision tree (Figure 2). Furthermore, the use of year as a covariate in the regional model assumed smolt survival trends were consistent over time. A temporal shift in smolt survival trends might be expected given the reported shifts in climate regimes during the time series (Overland et al. 2008; Perry and Masson 2013).

In order to simultaneously evaluate the differences between the Salish Sea and the Pacific coast and between the Strait of Georgia and Puget Sound, linear regressions were applied to smolt survival time series from the Pacific coast, Strait of Georgia, and Puget Sound and structural changes in these time series were investigated using the strucchange package in R (Zeileis et al. 2002). Smolt survival data used in this analysis were the annual predicted smolt survival values (hatchery and wild combined) from the regional mixed-effects model for the Pacific coast, the Strait of Georgia, and Puget Sound. This analysis allowed both the intercept and slope of the time series segments to vary and tested whether a "shift," or a change in trend, was supported by the data. A comparison of AIC$_c$ values was used to determine whether one breakpoint was more likely than no breakpoints in the data set. The fit of a breakpoint to the data set was determined with an *F*-test calculated from the ordinary least-squares residuals of segmented and nonsegmented models (Zeileis et al. 2003). Analysis was conducted on logit-transformed smolt survival data. Differences between Pacific coast, Strait of Georgia, and Puget Sound time series were supported if the slope for one regression was outside the 95% confidence interval of the regression slope for the contrasting region. For time series in which more than one linear trend was identified, the breakpoint year and associated confidence intervals were calculated.

Regional coherence and smolt survival patterns at a subbasin scale.—The results from the mixed-effects model of regional variation and structural change analysis did not address whether smolt survival differed among subbasins within the Salish Sea or whether covariation in survival between Salish Sea and Pacific coast reference populations was significant. In order to investigate smolt survival patterns at both these spatial scales, primary trends identified in the regional model were removed from the survival data. Residual annual smolt survival for each Salish Sea population was the response variable in a mixed effects model and was the difference between the annual population estimates and annual predictive smolt survival from the best regional model. Subbasin and Pacific coast residual survivals were the fixed effects and population was the random effect. Subbasins included seven subbasin delineations within Puget Sound

TABLE 1. List of Coho Salmon populations used for the analysis of smolt survival patterns. Location is listed by biogeographic region (PC = Pacific coast, SS = Salish Sea), Salish Sea basin (SoG = Strait of Georgia, PS = Puget Sound, JDF = Strait of Juan de Fuca), Salish Sea subbasin, and watershed. The watershed number refers to the map number in Figure 1. The origin is either hatchery (H) or wild (W). Ocean entry years (OEYs) include the first and last year of data, the number in parentheses is the number of years of data (excluding gaps) within the range. Method describes the way in which smolt survival was calculated. Method 1 was based on a coded wire tag reconstruction. Method 2 was smolt and spawner counts from weirs expanded by an exploitation rate that was either modeled (2.1) or based on coded wire tag recoveries from a neighboring population (2.2).

Region	Salish Sea basin	Salish Sea subbasin	Watershed	Origin	OEY	Method
PC			1. Keogh River	W	1997–2010 (14)	2.1
SS		Northern SoG	2. Quinsam River	H	1976–2010 (35)	1
SS	SoG	Northern SoG	3. Simms Creek	W	1998–2008 (11)	2.1
SS	SoG	Northern SoG	4. Black Creek	W	1978–2010 (27)	1
SS	SoG	Northern SoG	5. Puntledge River	H	1978–2004 (25)	1
SS	SoG	Northern SoG	6. Millard Creek	W	1999–2006 (8)	2.1
SS	SoG	Central SoG	7. Big Qualicum River	H	1973–2010 (37)	1
SS	SoG	Central SoG	8. Englishman River	W	1998–2010 (9)	2.1
SS	JDF		9. Goldstream River	W	1998–2010 (12)	1
SS	SoG	Northern SoG	10. Myrtle Creek	W	2001–2010 (9)	1
PC			11. Robertson Creek	H	1998–2010 (13)	1
PC			12. Carnation Creek	W	1998–2010 (13)	2.2
SS	SoG	Central SoG	13. Salmon River	W	1986–2007 (17)	1
SS	SoG	Central SoG	14. Inch Creek	H	1984–2010 (27)	1
SS	SoG	Central SoG	15. Chilliwack River	H	1982–2004 (20)	1
SS	SoG	Central SoG	16. Louis Creek	H	1990–2007 (14)	1
SS	SoG	Southern SoG	17. Nooksack River	H	1982–2009 (28)	1
SS	PS	Whidbey	18. Baker River (Skagit)	W	1991–2010 (18)	1
SS	PS	Whidbey	19. Skagit River	H	1995–2010 (16)	1
SS	PS	Whidbey	20. Tulalip Bay	H	1980–2010 (30)	1
SS	PS	Whidbey	21. Skykomish River	H	1983–2010 (27)	1
SS	PS	Central PS	22. Green River	H	1977–2010 (31)	1
SS	PS	Central PS	23. Puyallup River	H	1979–2010 (32)	1
SS	PS	South PS	24. Minter Creek	H	1979–2009 (14)	1
SS	PS	South PS	25. Kalama Creek	H	1979–2010 (20)	1
SS	PS	South PS	26. Deschutes River	W	1977–2008 (24)	1
SS	PS	Hood Canal	27. Skokomish River	H	1979–2010 (32)	1
SS	PS	Hood Canal	28. Big Beef Creek	W	1977–2010 (34)	1
SS	PS	Hood Canal	29. Quilcene River	H	1979–2010 (26)	1
SS	JDF		30. Snow Creek	W	1985–2010 (22)	2.1
SS	JDF		31. Dungeness River	H	1977–2010 (17)	1
PC			32. Sooes River	H	1982–2010 (23)	1
PC			33. Quinault River	H	1977–2010 (34)	1
PC			34. Bingham Creek	W	1982–2010 (29)	1
PC			35. Satsop River	H	1973–2010 (34)	1
PC			36. Willapa River	H	1973–2010 (24)	1
PC			37. Grays River	H	1977–2010 (29)	1
PC			38. Elochoman River	H	1985–2008 (22)	1
PC			39. Upper Cowlitz River	W	2001–2010 (8)	2.1
PC			40. Cowlitz River	H	1982–2010 (26)	1
PC			41. Washougal River	H	1977–2010 (28)	1

and the Strait of Georgia (Table 1; Figure 1A) and tested whether there were differences in smolt survival patterns among subbasins within the Salish Sea. Subbasin differences were further examined using Bonferonni multiple comparisons and were considered different when the *P*-value was less than an alpha of 0.05. Residuals of smolt survival for Pacific coast reference populations were detrended values of annual survival outside the Salish Sea and were calculated as the mean residual survival among Pacific coast populations for each year. Residuals for each population were calculated from the best regional model. Inclusion of this factor tested whether similar trends could be detected between the California Current populations in the Salish Sea and those outside the Salish Sea (top bifurcation in decision tree; see Figure 2). This analysis was conducted for the entire time series, as well as for subdivisions of the full time series as identified from the results of the structural change analysis.

Effect sizes of multiple spatial scales.—Effect sizes from the two sets of mixed-effects models were computed in order to compare the relative importance of factors explaining variation in smolt survival. Effect sizes were the parameter estimate divided by its standard error, akin to Cohen's D effect-size calculation. For a given factor, these values were summed for multiple comparisons (e.g., multiple subbasin contrasts). Comparing effect size between the regional mixed-effects model and the subsequent analysis of residual survival was complicated by the fact that the second analysis started with a substantial amount of variation explained by factors in the regional mixed-effects model, including variation from Pacific coast populations. Therefore, we adjusted the effect sizes of parameters in each model by the initial variation in the Salish Sea. For the regional mixed-effects model, this adjustment used residuals of a model that

included only a constant and the process error generated by random effects and correlation structure. For the residual survival model, this adjustment used residuals of the best regional mixed-effects model. In both adjustments, we focused on the variance generated by Salish Sea populations only. We then weighted relative effect sizes by the amount of variance each model sought to explain.

RESULTS

Groupings of Smolt Survival According to Cluster Analyses

Populations clustered into two major geographical groupings, (mostly) Salish Sea (Puget Sound and the Strait of Georgia) and (mostly) Pacific coast (Figures 1B, 3). Wild populations clustered with hatchery populations from their respective geographic regions. The Salish Sea cluster included a Puget Sound cluster and a Strait of Georgia cluster (bootstrap support for these groupings was 100%). The Puget Sound cluster included populations from all four subbasins (Table 1; Figure 3). The Strait of Georgia cluster included populations from all three subbasins of the Strait of Georgia and the Dungeness River (Strait of Juan de Fuca). The Pacific coast cluster was split into two groups (lower Columbia River and coastal

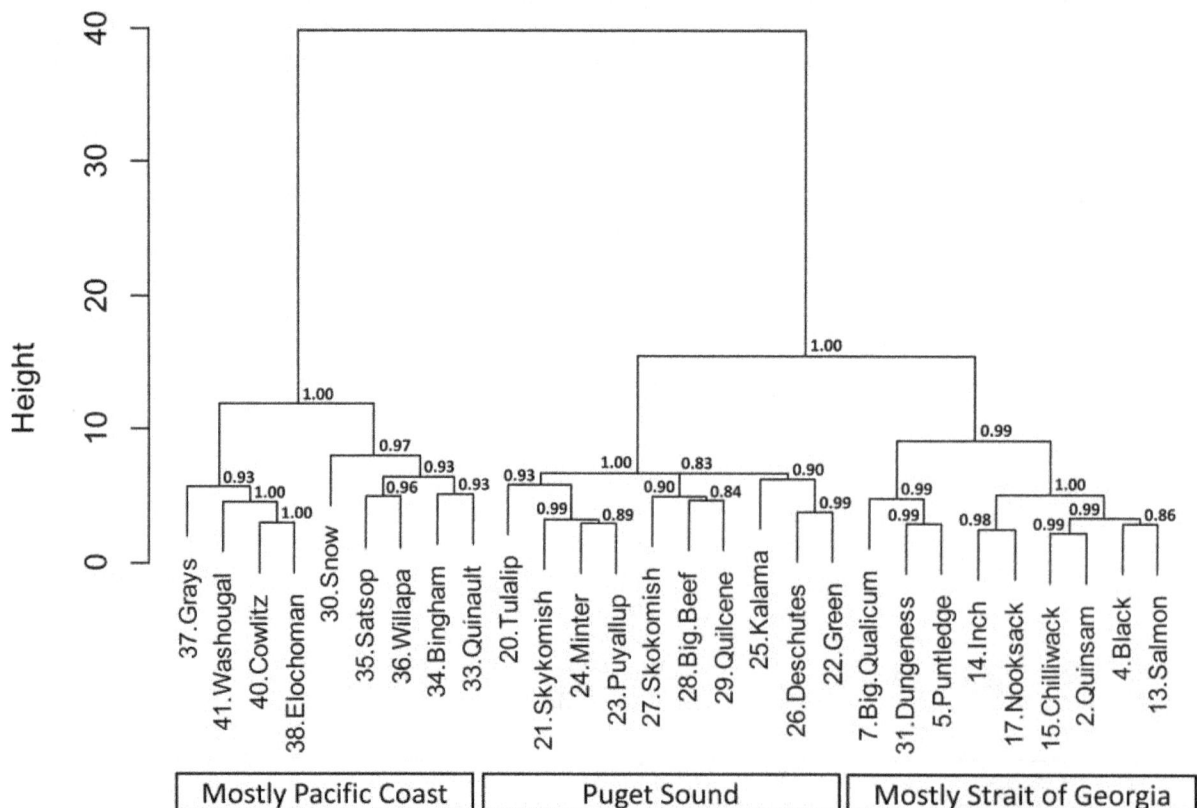

FIGURE 3. Dendrogram showing the results of Ward's hierarchical cluster analysis of smolt survival estimates for Coho Salmon populations in the Salish Sea and neighboring watersheds, ocean entry years 1977–2010. The vertical line height represents the Euclidean distance (magnitude of difference) between pairs of populations. Bootstrap support for each cluster is provided as an approximately unbiased *P*-value (significance, $P \geq 0.95$).

Washington) that were also strongly supported by the bootstrap analysis (bootstrap support for these groupings was 100%). The one geographic exception to membership in the Pacific coast cluster was the Snow Creek population, which enters the marine environment in the Strait of Juan de Fuca.

Smolt Survival Correlations over Distance and Time

Correlograms showed that smolt survival correlations were spatially structured among the Pacific coast, Strait of Georgia, and Puget Sound regions but that the strength of the correlations changed over time (Figure 4). For the entire time series,

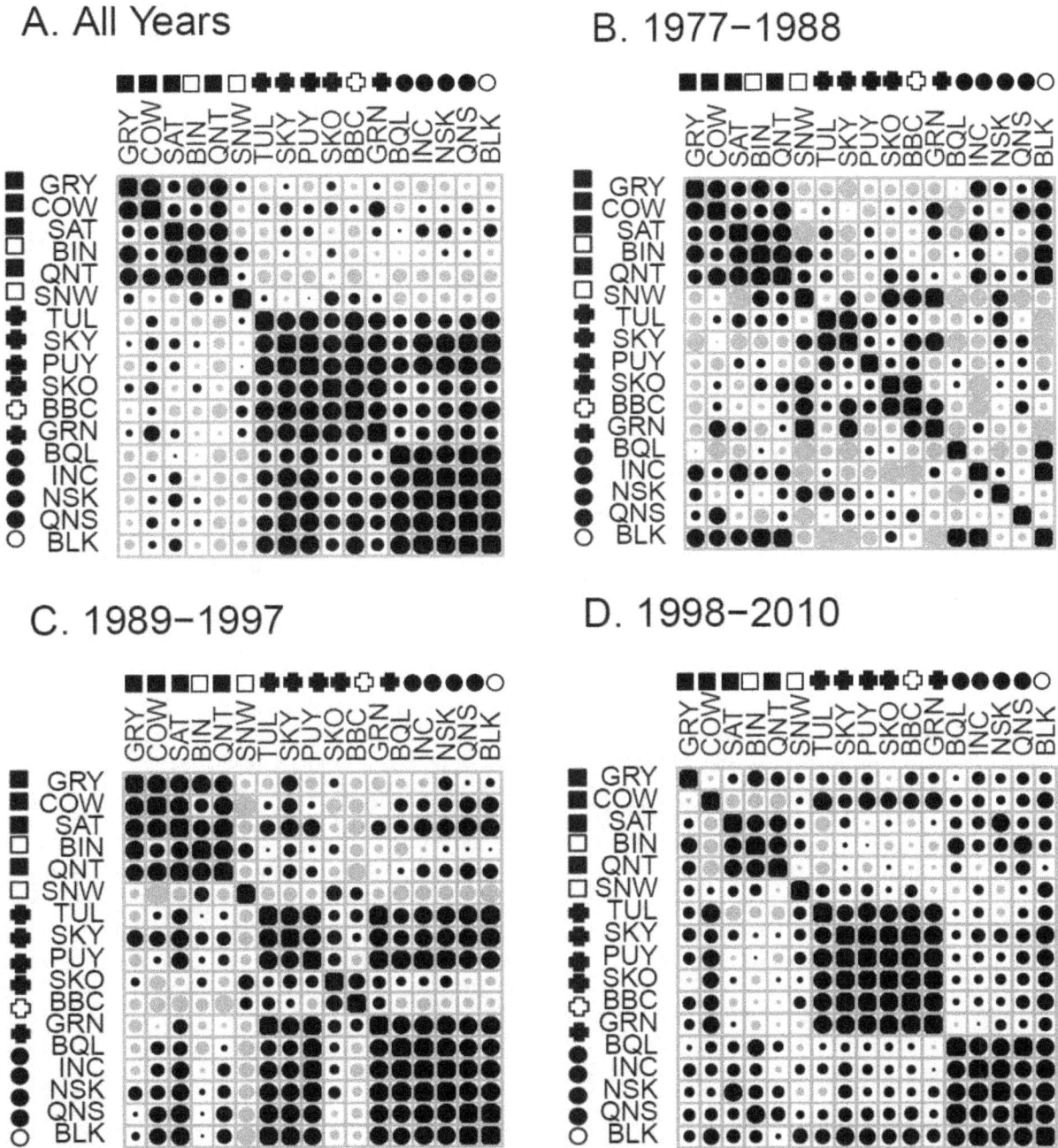

FIGURE 4. Pairwise correlations in Coho Salmon smolt survival over different ocean entry year time periods shown as a correlogram for (A) the entire time series, (B) 1977–1988, (C) 1989–1997, and (D) 1998–2010. Correlogram results are mirrored above and below the diagonal, with larger circles representing stronger correlations than smaller circles. Within the correlogram, black circles are positive correlations and gray circles are negative correlations. The first six populations listed are from the Pacific coast (squares): Grays (GRY), Cowlitz (COW), Satstop (SAT), Bingham (BIN), Quinault (QNT), and Snow (SNW). The second six populations are from Puget Sound (crosses): Tulalip (TUL), Skykomish (SKY), Puyallup (PUY), Skokomish (SKO), Big Beef (BBC), and Green (GRN). The final five populations are from the Strait of Georgia (circles): Big Qualicum (BQL), Inch (INC), Nooksack (NSK), Quinsam (QNS), and Black (BLK). For the population labels, filled symbols are hatchery fish and open symbols are wild populations.

correlations in Coho Salmon smolt survival were strong and positive within the Strait of Georgia and the Salish Sea (Puget Sound and Pacific coast), as well as between the Strait of Georgia and Puget Sound. However, they were generally weak or negative between the Pacific coast and the Strait of Georgia or Puget Sound (Figure 4A). During the first period (OEYs 1977–1988), smolt survival was weakly correlated among populations in the Salish Sea region but positively correlated among populations in the Pacific coast region (Figure 4B). Two populations from the Strait of Georgia (Inch Creek [number 14] and Black Creek [number 4]) were positively correlated with Pacific coast populations during the first time period. During the second period (OEYs 1989–1997), smolt survival was positively correlated within the Salish Sea and Pacific coast regions but not between them (Figure 4C). Exceptions during this period were the Skokomish River and Big Beef Creek (numbers 27 and 28, respectively, in the Hood Canal subbasin of Puget Sound), which were positively correlated with each other but weakly correlated with the rest of the Salish Sea populations. During the third period (OEYs 1998–2010), smolt survival was positively correlated within the Strait of Georgia and Puget Sound basins. However, between these basins, only Black Creek in the Strait of Georgia was positively correlated with Puget Sound populations (Figure 4D). Populations in the Pacific coast region were weakly correlated with Salish Sea populations during this third time period.

The correlations among populations declined with distance, and the shape of the exponential decay curve fit to the data differed among time periods (Figure 5). The e-folding scale (v), representing the extent of spatial synchrony, was three to four times greater for the second two time periods (OEYs 1989–1997 and 1998–2010) than the earliest time period (OEYs 1977–1988; Table 2). The intercept parameter (ρ_o), representing the correlation between neighboring populations at a theoretical separation distance of 0, was nearly two times higher (stronger) in the second (OEYs 1989–1997) time period than the first (OEYs 1977–1988) or third (OEYs 1998–2010) time periods (Table 2). For each time series, the exponential decay model that included an intercept parameter was a better fit to the data than when the intercept was fixed at a value of 1 (full time series: $\Delta\text{AIC}_c = 5.5$; OEYs 1977–1988: $\Delta\text{AIC}_c = 10$; OEYs 1989–1997: $\Delta\text{AIC}_c = 5.8$; and OEYs 1998–2010: $\Delta\text{AIC}_c = 91.4$).

Regional Differences in Smolt Survival

The mixed-effects model including region, origin (hatchery or wild), year, and an interaction between region and year was overwhelmingly supported by the data (model weight = 0.977, Model 9; Table 3). Two other potentially contending models (Models 2 and 3) added an additional interaction term to these four effects but did not reduce the negative log likelihood, suggesting these additional terms were predictively

neutral. All parameter values in the best model differed from 0. The region × year interaction indicated that annual trends in smolt survival differed among regions. Across all regions, annual smolt survival was consistently higher for wild than hatchery Coho Salmon.

Recognizing that highly variable smolt survival is the norm in Pacific salmon populations (e.g., Teo et al. 2009), we visualized predictions for each year by using the structure of the best model but including year as a categorical variable rather than a covariate (Figure 6). Predictions were back-calculated to nontransformed values for ease of interpretation. Modeled predictions of annual smolt survival of hatchery Coho Salmon ranged from 0.6% to 16.6% inside the Strait of Georgia, 0.6% to 15.7% in Puget Sound, and 0.1% to 4.6% on the Pacific coast. In comparison, annual smolt survival of wild Coho Salmon ranged from 1.1% to 30.8% for the Strait of Georgia, 1.3% to 25.7% for Puget Sound, and 0.2% to 10.4% for the Pacific coast.

Smolt Survival Patterns at Salish Sea and Basin Scales

Survival trends differed within regions of the California Current (in the Salish Sea versus outside the Salish Sea) and between basins of the Salish Sea (Table 4; Figure 7). The Strait of Georgia and Puget Sound time series were best fit with single regression models, whereas two separate linear trends were supported for the Pacific coast time series, with a break occurring in OEY 1991 (95% confidence interval = 1986–1992). A lack of structural change in the two Salish Sea time series was supported by AIC_c model comparisons of one versus two linear models ($\Delta\text{AIC}_c < 3$) and by F-tests ($P > 0.05$). However, smolt survival trends differed between Salish Sea basins, with survival declining more rapidly in the Strait of Georgia than in Puget Sound (regression slopes were outside the 95% confidence intervals calculated for the other basin). The break in the Pacific coast time series was supported by AIC_c model comparisons of one versus two linear models ($\Delta\text{AIC}_c = 13.6$) and the F-test ($P < 0.05$) used to test goodness of fit for the model. For OEYs 1977–1991, the temporal trend for the Pacific coast region did not differ from a slope of 0. Following a decrease in smolt survival in the early 1990s, survival of Pacific coast populations increased through 2010 (Table 4; Figures 6, 7).

Regional Coherence and Smolt Survival Patterns at a Subbasin Scale

Based on the entire time series, residual smolt survival of Salish Sea populations covaried with Pacific coast reference populations ($F = 7.51$, $P = 0.006$) but did not differ among subbasins ($F = 1.68$, $P = 0.18$). Overall, residual smolt survival values were predominantly negative in the 1990s and again in the mid-2000s and positive during other portions of the time series (Figures 7, 8). Results differed when the time

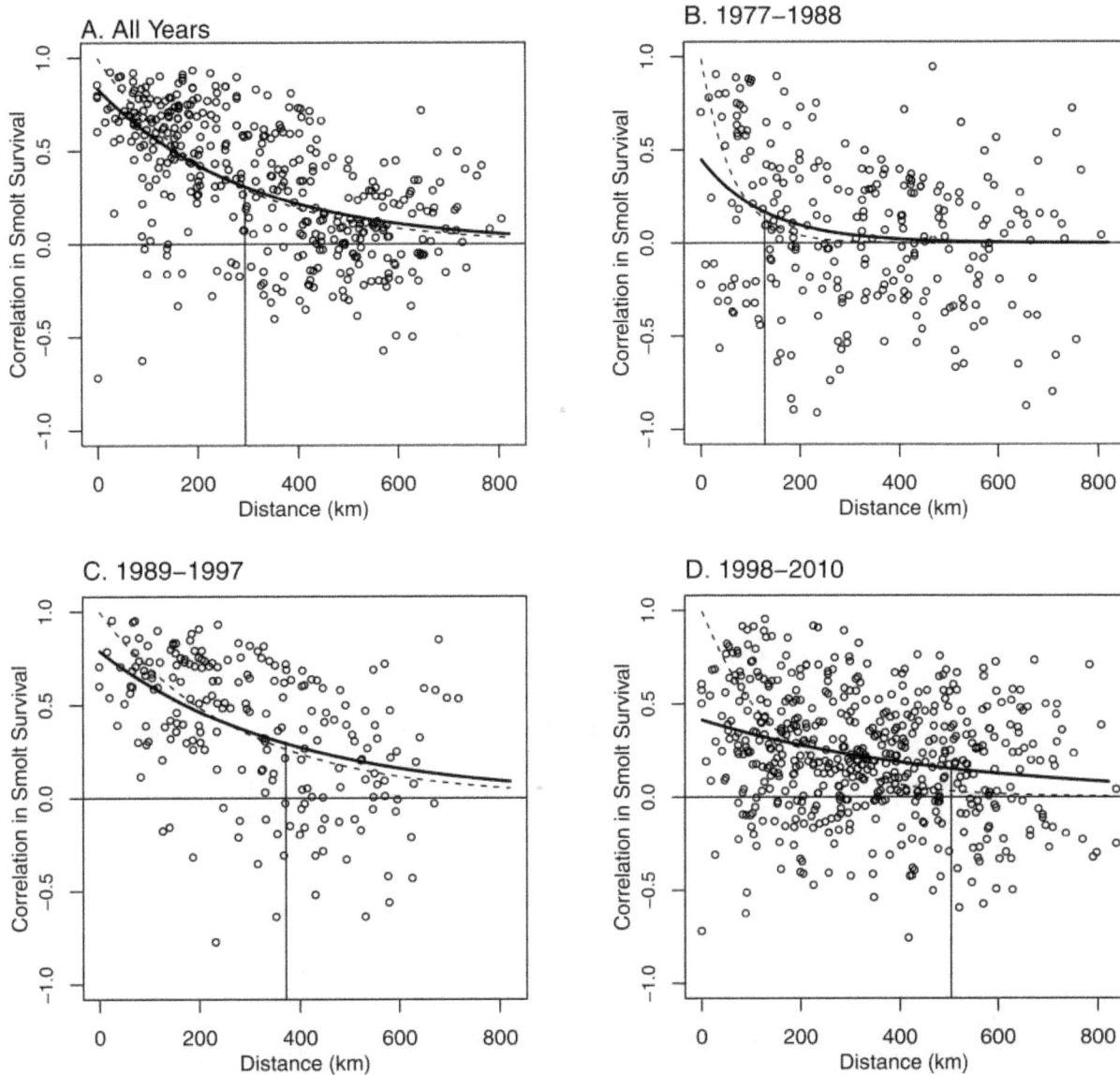

FIGURE 5. Pairwise correlations of Coho Salmon smolt survival as a function of distance between populations. Panels represent (A) the entire time series dataset and three time periods: (B) 1977–1988, (C) 1989–1997, and (D) 1998–2010. Each point represents an individual pairwise correlation. The curvilinear lines represent an exponential decay function fit to the data with an estimated intercept (solid line) and an intercept fixed at a value of 1 (dashed line). The vertical thin line represents the distance at which pairwise correlations decrease by 37% (e-folding scale) for the nonlinear function with an estimated intercept.

TABLE 2. Parameter values for exponential decay models, $\rho(d) = \rho_o e^{-d/v}$, which described the relationship between the strength of the survival correlation, $\rho(d)$, and the ocean entry distance (d) among populations.

Time period	ρ_o (Intercept)		v (e-folding scale)	
	Estimate	95% CI	Estimate	95% CI
All years	0.84	0.75–0.93	294 km	246–354 km
1977–1988	0.46	0.22–0.73	129 km	64–307 km
1989–1997	0.79	0.65–0.94	372 km	287–518 km
1998–2010	0.41	0.34–0.50	506 km	367–747 km

series was divided into two periods based on identified breakpoints for the Pacific coast region (1977–1991, 1992–2010). In the first time period (1977–1991), neither Pacific coast smolt survival nor subbasin was a good predictor of residual smolt survival of Salish Sea populations ($F < 0.8$, $P > 0.5$, Figure 9A). However, between 1992 and 2010 the residual smolt survival of Salish Sea populations covaried with the residual smolt survival of Pacific coast populations and differed among subbasins ($F = 3.91$ and $F = 3.71$, respectively, $P < 0.05$; Figure 9b). In the 1977–1991 time period, Bonferroni multiple comparisons indicated contrast between residual smolt survival of just two subbasins (Hood Canal and Whidbey). In the

TABLE 3. Factors contributing to Coho Salmon smolt survival from ocean entry years 1977 to 2010. An "X" indicates fixed-effect variables included in linear mixed-effects models predicting variation in smolt survival. Factors include region (Strait of Georgia, Puget Sound, or Pacific coast), origin (hatchery or wild), and year (ocean entry year). Population (not shown) was included in the model as a random effect. The differences in AIC values from the model with the lowest AIC score (ΔAIC) were used to compare models. An asterisk indicates the best model.

Model	Intercept	Region	Origin	Year	Region × year	Origin × year	Origin × region	Log likelihood	ΔAIC	Model weight
1	X	X						1,110.2	48.9	0.000
2	X		X					1,112.0	50.5	0.000
3	X			X				1,102.1	30.8	0.000
4	X	X		X				1,101.2	32.9	0.000
5	X		X	X				1,099.7	27.9	0.000
6	X	X	X					1,109.0	48.6	0.000
7	X	X	X	X				1,097.7	28.1	0.000
8	X	X	X	X		X		1,101.5	37.6	0.000
9*	X	X	X	X	X			1,081.7	0.0	0.977
10	X	X	X	X			X	1,098.3	33.2	0.000
11	X	X	X	X	X		X	1,084.3	9.1	0.010
12	X	X	X	X	X	X		1,085.1	8.7	0.012
13	X	X	X	X		X	X	1,102.1	42.7	0.000

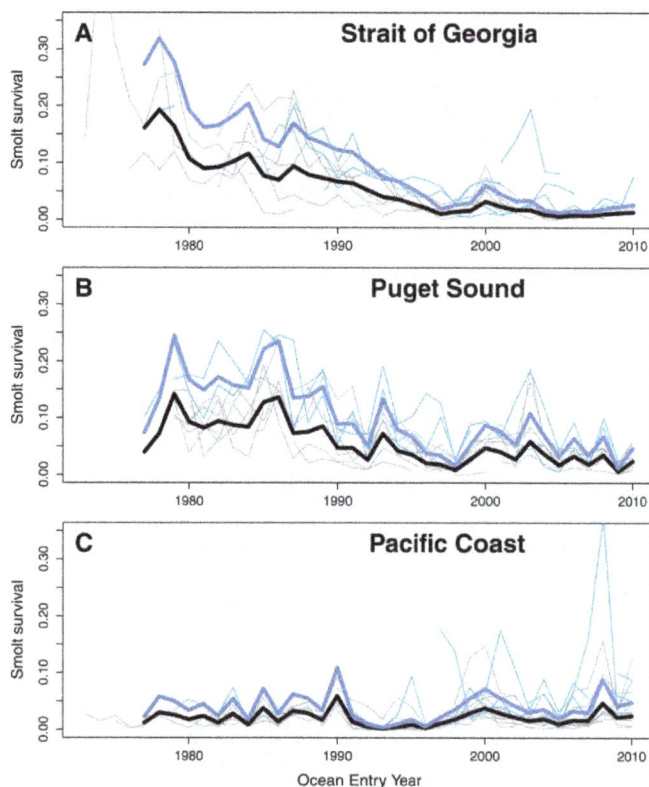

FIGURE 6. Smolt survival time series for Coho Salmon from ocean entry years 1977 to 2010. The values shown are not transformed (logit transformation was used in the analysis to account for heterogeneous variance among years). Panels represent the three geographic regions: **(A)** the Strait of Georgia, **(B)** Puget Sound, and **(C)** the Pacific coast. The thin lines represent individual populations: the thin gray lines are hatchery populations and the thin blue lines are wild populations. The thick lines show the predicted survival for hatchery (black) and wild (blue) populations in each region from the best mixed-effects regional model (Model 9 in Table 3), with year incorporated as a categorical variable to highlight annual variation in smolt survival.

1992–2010 time period, five strong contrasts existed, all between South Puget Sound and the five subbasins with above-average residual smolt survival. Across the two time periods, the South Puget Sound subbasin went from having high smolt survival relative to other subbasins to the lowest survival relative to other subbasins. The reverse occurred for Hood Canal and southern Strait of Georgia subbasins.

Effect Sizes of Multiple Spatial Scales

The comparison of relative effect sizes (Figure 10) revealed how Coho Salmon smolt survival was explained at different spatial scales (Figure 2). The largest spatial scale was regionwide coherence of Salish Sea and Pacific coast populations. This source of variation had a much lower relative effect for Salish Sea populations than any other parameter. The next smaller spatial scale was the region-specific scale, i.e., the differences between Salish Sea (Strait of Georgia and Puget Sound) populations and Pacific coast populations. These patterns, represented by the Region and Region × Year terms, exhibited the largest effect size. The next smaller spatial scale was the basin scale (Strait of Georgia versus Puget Sound). These patterns, represented by the Year and Region × Year terms, exhibited the second largest effect sizes. Variation at the subbasin scale was modest compared with these larger spatial scales. The finest spatial scale includes population-specific factors, including hatchery or wild origin, as well as within-population temporal variation. This latter source of variation could not be assigned a relative effect size but includes much of the variation not explained by any model. Given the relative effect sizes of origin alone, population-specific variation may be on the same order of magnitude as subbasin effects.

TABLE 4. Linear model parameters (95% confidence intervals in parentheses) and breakpoints identified for Coho Salmon smolt survival time series from Strait of Georgia, Puget Sound, and Pacific coast regions. Analysis was conducted using predicted smolt survival (hatchery and wild combined) based on the best regional mixed-effects model. Slope parameters correspond to logit-transformed survival time series. Strait of Georgia and Puget Sound time series did not have a significant breakpoint (only one segment).

Region	Breakpoint years	First segment slope	Second segment slope
Strait of Georgia		−0.09 (−0.08 to −0.10)	
Puget Sound		−0.06 (−0.04 to −0.08)	
Pacific coast	1991 (1986 to 1992)	0.03 (−0.03 to 0.10)	0.13 (0.07 to 0.20)

DISCUSSION

This study provides evidence for coherence in Coho Salmon smolt survival at multiple spatial scales for OEYs 1977–2010. Covariation at the Salish Sea scale (versus Pacific coast) was supported by the cluster analysis, which did not rely on a priori geographic assignment, and the mixed-effects model, which tested differences among preassigned geographic regions. The primary Salish Sea scale pattern was declining smolt survival over the entire period of record. This pattern contrasted with Pacific coast reference populations, which had lower smolt survival than the Salish Sea populations in the 1970s and 1980s, but similar smolt survival in recent years. Smolt survival for Pacific coast populations was consistently low in the early 1990s, but subsequently increased. Within the Salish Sea, major basins shared a similar

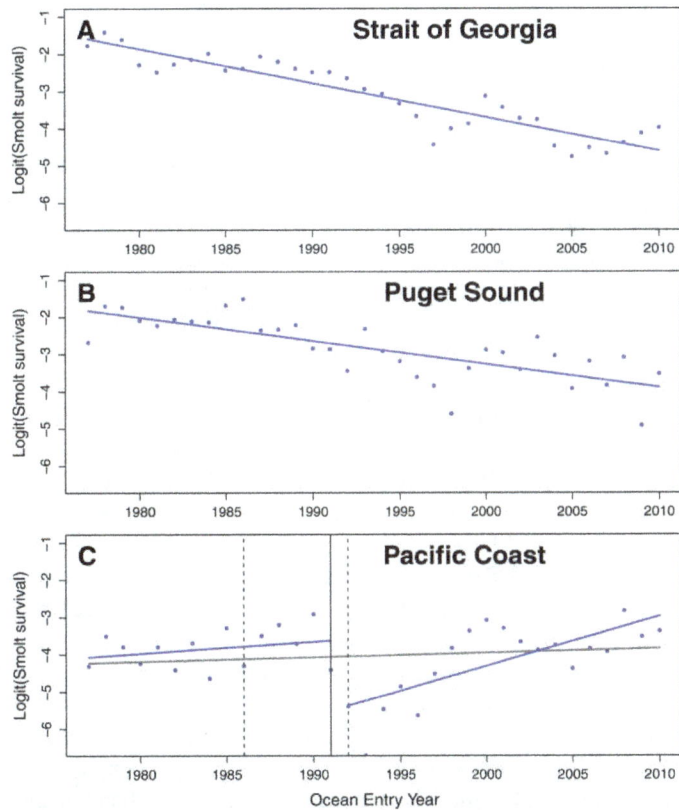

FIGURE 7.   Trends in smolt survival for Coho Salmon from ocean entry years 1977 to 2010. The values shown are logit transformed. The panels represent the three geographic regions: (A) the Strait of Georgia, (B) Puget Sound, and (C) the Pacific coast. Points represent the average smolt survival each year for each region (hatchery and wild combined). The thick blue lines are best-fit linear regressions based on breakpoint analysis. The vertical lines are the breakpoint and 95% confidence intervals (solid and dashed lines, respectively). No breakpoints were identified for the Strait of Georgia or Puget Sound. The thick gray line in panel (C) is the best-fit linear regression model with no breakpoints and is provided for comparison.

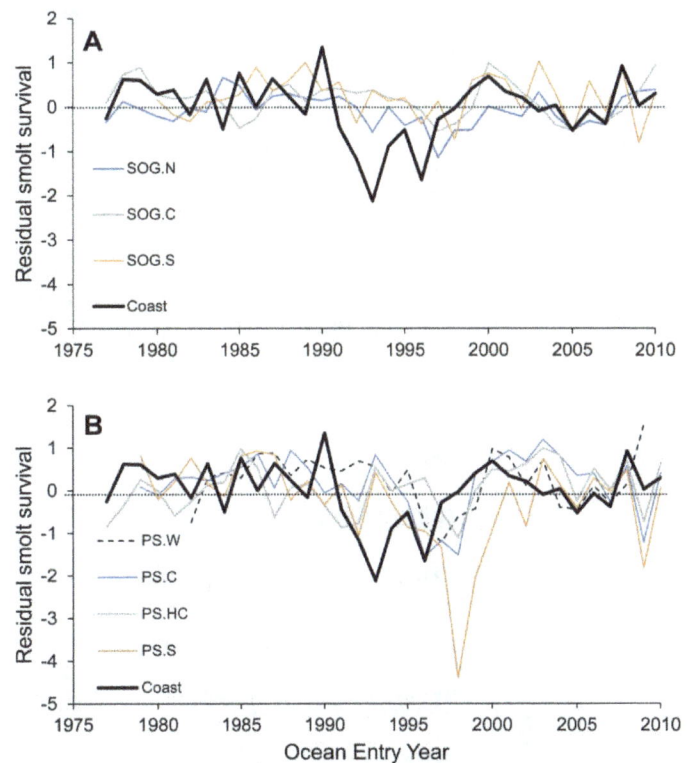

FIGURE 8.   Smolt survival residuals from ocean entry years 1977 to 2010. Residuals were calculated from the best regional model in Table 3. The values shown were averaged by subbasin within (A) the Strait of Georgia (SOG; solid colored lines in panel A) and (B) Puget Sound (PS; dashed and colored lines in panel B) and along the Pacific coast (thick solid black line in both panels). Subbasin abbreviations are as follows: N = North, C = Central, S = South, HC = Hood Canal, and W = Whidbey.

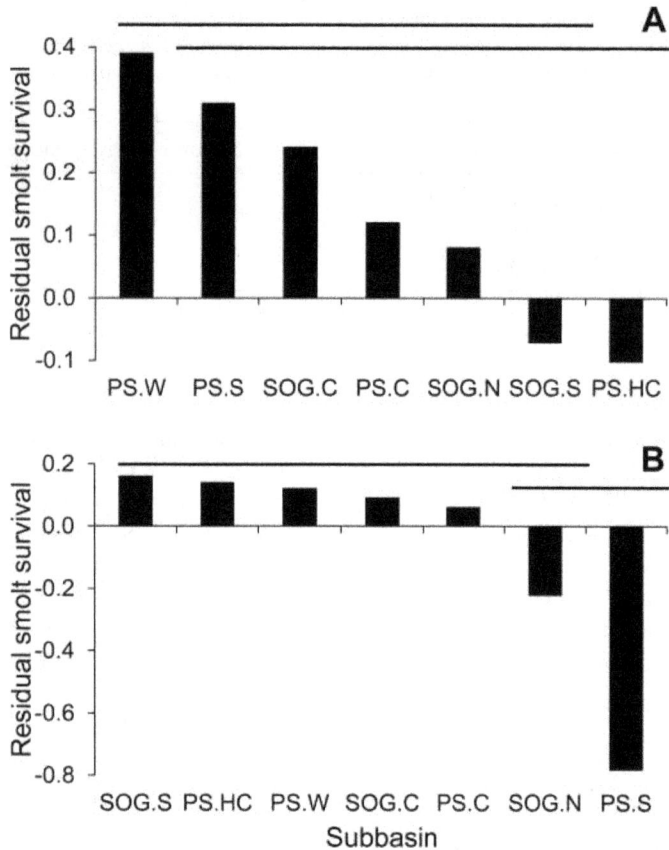

FIGURE 9. Mean smolt survival residuals for each subbasin of the Salish Sea for two time periods: ocean entry years (A) 1977–1991 and (B) 1992–2010. Residuals were calculated from the best regional model (Table 3) and represent variation in survival after main effects in the time series were removed. Subbasins are shown for the Strait of Georgia (SOG) and Puget Sound (PS). Horizontal lines link subbasins lacking significant differences as determined by Bonferroni multiple comparisons. Subbasin abbreviations are as follows: N = North, C = Central, S = South, HC = Hood Canal, and W = Whidbey. Note that the order of the subbasins and the range of the y-axis changes between panels.

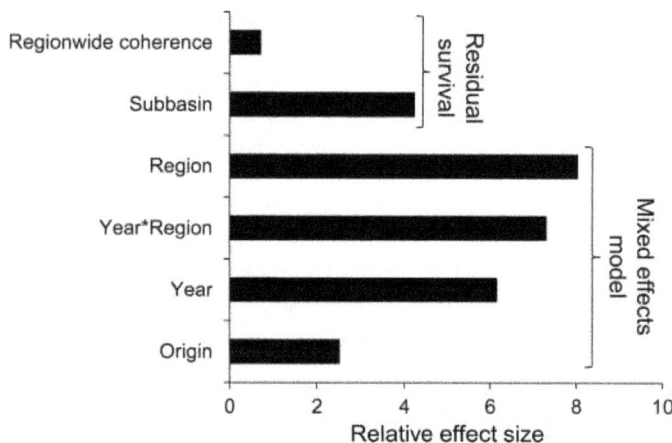

FIGURE 10. Relative effect sizes of multiple factors contributing to smolt survival of Coho Salmon in the Salish Sea. The effect size was calculated from the best regional mixed-effects model and subsequent analysis of residual survival.

declining survival trend but the decline was steeper in the Strait of Georgia than Puget Sound. Smolt survival differences at the subbasin scale were apparent only after the primary Salish Sea scale trend was removed from the data and were distinct only for the Whidbey and Hood Canal subbasins early in the time series and the South Puget Sound subbasin later in the time series. Although differences between populations in the Salish Sea and outside the Salish Sea were significant, both regions had low smolt survival in the early 1990s.

Our results are consistent with previous studies that demonstrated Coho Salmon smolt survival patterns at the scale of the California Current (Beamish et al. 2000; Botsford and Lawrence 2002; Beamish et al. 2004b; Teo et al. 2009), the Salish Sea (Coronado and Hilborn 1998a; Hobday and Boehlert 2001; Beetz 2009) and between basins of the Salish Sea (Coronado and Hilborn 1998a). However, these previous studies did not resolve the issue of multiple scales and arrived at seemingly conflicting results with respect to the importance of factors affecting smolt survival at different scales. For example, how could patterns differ between the Salish Sea and Pacific coast (Coronado and Hilborn 1998a; Hobday and Boehlert 2001) if a shared climate regime was affecting both patterns (Beamish et al. 2000; Beamish et al. 2004b)?

The time frames for each study have contributed to different conclusions regarding spatial scale. Coronado and Hilborn (1998a) investigated Coho Salmon smolt survival for OEYs 1971–1990 and concluded that Salish Sea populations (termed "BC South Coast" and "Puget Sound") had declining survival trends compared with a lack of temporal trend observed for Washington coast or Columbia River populations. With an additional 4 years of information, Beamish et al. (2000) demonstrated a synchronous and declining trend for California Current populations originating both inside and outside the Salish Sea. This result was driven by the synchronous low smolt survival that occurred in the 1990s (Figure 7C), data not available at the time Coronado and Hilborn (1998a) conducted their analysis. The breakpoint (OEY 1991) identified in our analysis for the Pacific coast time series was consistent with the low smolt survival period during the early 1990s (Beamish et al. 2000; Peterson et al. 2006). However, with an additional 15 years of information, our results show that smolt survival of Pacific coast populations has increased since the 1990s, whereas there is no evidence for increasing smolt survival in the Salish Sea through OEY 2010. This perspective helps to reconcile the identified differences between the Salish Sea and Pacific coast with the identified coastwide similarities in smolt survival.

Spatial Scale is Time Dependent

The temporal scale determined what spatial pattern in smolt survival was identified. For example, when the entire time series was examined, smolt survival patterns were similar within the Salish Sea but differed from Pacific coast reference populations (Figures 3, 4A). However, when this same time series was subdivided, spatial coherence within the Salish Sea was weak in the first time period (Figure 4B; OEYs 1977–1988), synchronous in the second time period (Figure 4C; OEYs 1989–1997), and stronger within basins than across the entire Salish Sea in the third time period (Figure 4D; OEYs 1998–2010).

The e-folding parameter (v) described the scale, and the intercept (ρ_o) of the exponential decay model described the strength of spatial synchrony. The e-folding scale value itself has no specific biological relevance but rather provides a common metric to compare scales of spatial synchrony in smolt survival among species. The intercept can be interpreted as shared environmental effects among neighboring populations (Pyper et al. 2005). A comparison among time periods revealed broader spatial synchrony in smolt survival since OEY 1989 and stronger shared environmental effects for OEYs 1989–1997 than in the other two time periods.

In addition, our results suggest that spatial synchrony among Coho Salmon populations occurs at a more localized scale than for other Pacific salmon species, implying that local marine conditions are particularly important for Coho Salmon within the Salish Sea. The e-folding scale estimated for the entire time series was 294 km (95% CI = 246–354 km), approximately one-third the value for Chinook Salmon (1,069 km; Kilduchumff et al. 2014) and one-half to three-quarters the values for Pink Salmon (431 km; Pyper et al. 2001) and Chum Salmon O. keta (564 km; Pyper et al. 2002). Calculation of an e-folding scale for Sockeye Salmon O. nerka does not appear to have been made; however, Pyper et al. (2005) found that the 50% scale for Sockeye Salmon was slightly larger than that for Pink and Chum salmons. Some caution is needed when comparing estimates for Chum, Sockeye, and Pink salmons with those of Coho and Chinook salmons, as the former were based on spawner-recruit residuals that incorporated a greater degree of freshwater influence than the smolt survival estimates for Coho and Chinook salmons.

The e-folding scale in this study was slightly larger than that reported in a previous coastwide study of Coho Salmon populations (217 km; Teo et al. 2009). This difference can be explained by how the intercept (ρ_o) was selected for analysis. Teo et al. (2009) found that model fit was not improved by estimating the intercept and therefore constrained the intercept (ρ_o) to a value of 1. In our analysis, model fit was substantially improved by estimating the intercept; allowing the intercept to be < 1 changed the shape of the exponential decay function and increased the estimated e-folding scale.

Importance of Early Marine Environment

Different smolt survival patterns between Salish Sea and Pacific coast populations suggest that the early marine environment determines interannual differences in smolt survival. By itself, this result cannot distinguish between effects of the early marine environment versus effects of marine environments encountered during population-specific ocean migration routes (Weitkamp and Neely 2002). However, the importance of the early marine environment to Coho Salmon smolt survival is broadly supported by other studies. For example, across the range of Coho Salmon, smolt survival was better explained by ocean conditions following ocean entry than by those at later life stages (Hobday and Boehlert 2001). On the outer coastal continental shelf adjacent to the Columbia River, catches of juvenile Coho Salmon in the first month of ocean residence were strongly correlated with hatchery jack returns, which were, in turn, strongly correlated with hatchery adult returns (Pearcy 1988). Other studies have pointed to the importance of growth in the early marine environment. Early marine growth was correlated with smolt survival of wild Coho Salmon from Carnation Creek on the western coast of Vancouver Island (Holtby et al. 1990), and Beamish et al. (2004a) proposed that growth to a critical size during the early marine rearing period determines the overall survival trajectory for Coho Salmon in the Strait of Georgia.

Growth and survival are integrated responses to ecosystem conditions. In order to explain survival patterns in the Salish Sea, we must answer the following question: what has changed during the last 30 years and what drove those changes? Ultimately, this question may require an ecosystem research program, such as that implemented off the coast of Washington and Oregon (Brodeur et al. 2000) and recently extended from the Strait of Georgia (Masson and Perry 2013) to include Puget Sound in the Salish Sea (Riddell et al. 2009; U.S. Salish Sea Technical Team 2012). Retrospective analyses such as ours provide valuable insights on changing conditions. Based on the survival patterns identified in this study, we briefly summarize below the ecosystem processes recognized at spatial scales identified in the decision tree framework (Figure 2). These preliminary hypotheses should provide guidance for future ecosystem research within the Salish Sea, as well as for future development of predictive ecosystem indicators for these populations.

North Pacific Ocean and California Current Scales

The geographic scope of this study did not allow us to contrast patterns within regions of the North Pacific Ocean, but these have been described by other researchers (Coronado and Hilborn 1998a; Hare et al. 1999; Teo et al. 2009). For the purpose of this discussion, environmental factors at the North Pacific Ocean scale will be considered as they are manifested in the northern California Current, which was the geographic scope of our study populations.

Our study encompassed three climate regimes reported in the North Pacific Ocean (Hare and Mantua 2000; Overland et al. 2008). Coho Salmon smolt survival in the Salish Sea declined throughout this period. Some similarities within and outside the Salish Sea were detected in the detrended time series due to lower-than-average smolt survival in all regions during the 1990s, corresponding with an extended El Nino period from 1990 to 1996 (Peterson et al. 2006). North Pacific Ocean climate indices useful in predicting Coho Salmon smolt survival for coastal populations include the Pacific Decadal Oscillation Index, North Pacific Gyre Oscillation Index, and Multivariate El Nino Southern Oscillation Index (Rupp et al. 2012). Ocean climate changes may be linked to the Salish Sea ecosystem through tidal mixing and the introduction of seasonally varying nutrient- and oxygen-rich waters, as well as changes in stream flows (Mantua et al. 1997), temperature (Ebbesmeyer et al. 1989; Moore et al. 2008b), and wind patterns (Beamish et al. 1999a).

Salish Sea Scale

The inland sea environment of the Salish Sea is unique with respect to oceanography and has historically supported higher smolt survival rates of Coho Salmon than the outer coast continental shelf. On the outer coast continental shelf, increased productivity is associated with a deeper mixed layer and delivery of nutrients to the photic zone (Povolina et al. 1995). In the Strait of Georgia, increased productivity is associated with increased stratification and light exposure (Yin et al. 1997; Collins et al. 2009; Masson and Pena 2009) and nutrients are not thought to be limiting (Mackas and Harrison 1997; Johannessen et al. 2014). In addition to the physical environment, zooplankton communities in the Strait of Georgia are composed of large lipid-rich taxa belonging to a subarctic zoogeographic range (Mackas et al. 2013) and are mostly devoid of the southern-origin copepods associated with low smolt survival of Coho Salmon on the outer coast continental shelf (Peterson 2009).

The Salish Sea has undergone significant changes during the period of study that are related to Coho Salmon smolt survival in ways that are not fully understood. At present, temporal trends in the Strait of Georgia ecosystem have been better studied than those in Puget Sound and the Strait of Juan de Fuca. Over multiple decades in the Strait of Georgia, seawater and river water temperatures have increased, deep water oxygen has declined, sea level has risen, and timing of the Fraser River freshets has changed (Riche et al. 2013). These changes can be organized into major regime shifts occurring in the late 1970s and mid-1990s (Perry and Masson 2013), with a major shift in the zooplankton community in 1998–1999 (Li et al. 2013) and the mean vertebrate trophic level decreasing since the 1980s (Preikshot et al. 2013). After 1990, the biomass of zooplankton, calanoid copepods, and Pacific Herring *Clupea pallasii* were the best indicators of early smolt survival for Coho Salmon in the Strait of Georgia (Araujo et al. 2013).

Based on current understanding, freshwater inflows and wind speed, which are associated with stratification depths and primary productivity (Denman and Gargett 1983; Yin et al. 1997; Moore et al. 2008b; Preikshot et al. 2013), and

zooplankton biomass and composition, which are indicators of smolt survival (Peterson 2009; Araujo et al. 2013), may serve as important indicators of early marine conditions at the scale of the Salish Sea.

Basin Scale

The major difference observed between basins of the Salish Sea was the rate at which smolt survival declined over the past three decades. Interpreting these differences with respect to ecosystem processes is limited by a lack of detailed studies comparing the oceanography of the Strait of Georgia with that of Puget Sound. In general, the Strait of Georgia is deep, greatly influenced by freshwater input from the Fraser River, and connected to the Pacific Ocean at both its southern and northern boundaries. In comparison, Puget Sound is relatively shallow, influenced primarily by freshwater from the Skagit River, and connected to the Pacific Ocean only at its northern boundary.

One potential hypothesis to explain differences in Coho Salmon smolt survival in the Strait of Georgia versus Puget Sound is that populations in the two basins have different abilities to respond to changing ecosystem states based on genetically predisposed migration behavior or innate growth potential. The migration behavior of Coho Salmon and the duration of time spent within the Salish Sea ecosystem are better described for the Strait of Georgia than for Puget Sound. Migrations out of the Salish Sea have been observed for populations originating in both basins (Weitkamp and Neely 2002; Morris et al. 2007), but trawl surveys suggest that Coho Salmon from Puget Sound leave the Salish Sea earlier that those from the Strait of Georgia (R. J. Beamish, Fisheries and Oceans Canada, personal communication). In addition, a large portion of Coho Salmon historically overwintered within the Strait of Georgia and contributed to active recreational and troll fisheries in this basin. In 1991, for the first time on record, returning Coho Salmon were rarely caught in active recreational and troll fisheries within the Strait of Georgia (Simpson et al. 2000, especially Figure 9), and reduced catches were concluded to be the result of shifts in distribution to predominantly outside the Salish Sea (Beamish et al. 1999a). Residency of Coho Salmon has also been observed in the Puget Sound basin (Allen 1959; Rhode et al. 2013); however, a detailed comparison between the two basins has not been conducted.

Subbasin Scale

The ability to understand Coho Salmon smolt survival at a subbasin scale is critical to managing terminal fisheries in these areas. Subbasin scale variation in the Salish Sea ecosystem has been identified for temperature and salinity (Babson et al. 2006; Moore et al. 2008b), oxygen (Johannessen et al. 2014), nitrogen (Sutton et al. 2013), and chlorophyll (Masson and Pena 2009), which may be linked with early marine growth and survival. Moore et al. (2008a) found that subbasin scale oceanographic properties were better predicted by local air temperatures and river flows than climate-scale indices. In addition to oceanographic properties, the influence of urbanization, toxic chemicals (e.g., polychlorinated biphenyls), harmful algal blooms (e.g., *Heterosigma*; Rensel et al. 2010), and disease (e.g., *Nanophyetus*) are most likely to vary at a subbasin scale. Heterogeneity among the Salish Sea subbasins provides a natural laboratory to examine the relative impacts of these factors on Coho Salmon smolt survival and will require investment in a research program to collect the needed time series.

Our results indicate that subbasin scale differences in smolt survival were small in magnitude compared with basin scale or Salish Sea scale patterns. However, the ability to detect subbasin differences suffered from low statistical power (few populations per subbasin, few long-term smolt survival time series). The South Sound subbasin of Puget Sound was the most distinctive in that smolt survival in this subbasin was relatively high in the early portion of the time series and very low in the latter portion of the time series. South Sound, which is a series of shallow inlets separated from the remainder of Puget Sound by a sill at the Tacoma Narrows, is distinctive from other Puget Sound subbasins by its large temperature ranges (Moore et al. 2008b), low stratification (Moore et al. 2008a), and salinity that is more strongly influenced by freshwater inputs than is the salinity in the Strait of Juan de Fuca (Babson et al. 2006).

Conclusions

A desired outcome of this study was to guide the development of ecosystem indicators of Coho Salmon smolt survival, with specific reference to survival within the Salish Sea. We found a more localized scale of synchronized survival than reported by other researchers for Chinook, Pink, Chum, and Sockeye salmons, suggesting that early marine conditions are especially important for Coho Salmon within the Salish Sea. Within the geographic scope of our study, the dominant pattern in smolt survival occurred at the scale of the Salish Sea. Explaining this pattern will require future study on mechanistic links between the observed ecosystem changes and Coho Salmon growth and survival. We highlight the importance of the early marine environment to help focus future investigation. The ability to predict secondary patterns from the scale of the California Current to the scale of individual subbasins will be equally relevant to contemporary management concerns, such as climate change and run-size forecasting. For example, in an era characterized by low smolt survival, average smolt survival in Puget Sound over the past decade ranged by an order of magnitude, greater than that observed within the Strait of Georgia. Further exploration of ecosystem variables at multiple spatial scales is needed to effectively address linkages between the marine ecosystem and Coho Salmon

smolt survival within the Salish Sea. Since the relative importance of particular variables may have changed during our period of record, researchers will need to carefully match spatial and temporal scales to their questions of interest.

ACKNOWLEDGMENTS

This is Publication Number 1 from the Salish Sea Marine Survival Project, an international research collaboration designed to determine the primary factors affecting the survival of juvenile salmon and steelhead in the Salish Sea. Funding was received through a grant from the Pacific Salmon Commission Southern Endowment Fund sponsored by Long Live the Kings and Pacific Salmon Foundation. Andrew Weiss and Dale Gombert (Washington Department of Fish and Wildlife) produced the maps and calculated distances between river mouths, Steve Baillie (Fisheries and Oceans Canada) provided valuable input including updated time series of smolt releases and escapement estimates for Strait of Georgia streams, Peter Tschaplinski (British Columbia Forestry) provided data on Carnation Creek Coho Salmon, Nick Komick (Fisheries and Oceans Canada) generated exploitation rate estimates from the Canadian Mark Recovery Program database, and Jeff Haymes and Thomas Buehrens (Washington Department of Fish and Wildlife) contributed to discussions on U.S. data sets. The paper benefited from constructive comments provided by two anonymous reviewers.

REFERENCES

Allen, G. H. 1959. Growth of marked Silver Salmon (*Oncorhynchus kisutch*) of the 1950 brood in Puget Sound. Transactions of the American Fisheries Society 88:310–318.

Araujo, H. A., C. Holt, J. M. R. Curtis, R. I. Perry, J. R. Irvine, and C. G. J. Michielsens. 2013. Building an ecosystem model using mismatched and fragmented data: a probabilistic network of early marine survival for Coho Salmon *Oncorhynchus kisutch* in the Strait of Georgia. Progress in Oceanography 115:41–52.

Babson, A. L., M. Kawase, and P. MacCready. 2006. Seasonal and interannual variability in the circulation of Puget Sound, Washington: a box model study. Atmosphere-Ocean 44:29–45.

Beamish, R. J., C. Mahnken, and C. M. Neville. 2004a. Evidence that reduced early marine growth is associated with lower marine survival of Coho Salmon. Transactions of the American Fisheries Society 133:26–33.

Beamish, R. J., G. A. McFarlane, and R. E. Thomson. 1999a. Recent declines in the recreational catch of Coho Salmon (*Oncorhynchus kisutch*) in the Strait of Georgia are related to climate. Canadian Journal of Fisheries and Aquatic Sciences 56:506–515.

Beamish, R. J., C. M. Neville, R. Sweeting, and K. L. Lange. 2012. The synchronous failure of juvenile Pacific salmon and herring production in the Strait of Georgia in 2007 and the poor return of Sockeye Salmon to the Fraser River in 2009. Marine and Coastal Fisheries: Dynamics, Management, and Ecosystem Science [online serial] 4:403–414.

Beamish, R. J., D. A. Noakes, G. A. McFarlane, L. Klyoshtorin, V. V. Ivanov, and V. Kurashov. 1999b. The regime concept and natural trends in the production of Pacific salmon. Canadian Journal of Fisheries and Aquatic Sciences 56:516–526.

Beamish, R. J., D. J. Noakes, G. A. McFarlane, W. Pinnix, R. Sweeting, and J. King. 2000. Trends in coho marine survival in relation to the regime concept. Fisheries Oceanography 9:114–119.

Beamish, R. J., R. M. Sweeting, K. L. Lange, and D. J. Noakes. 2010. Early marine survival of Coho Salmon in the Strait of Georgia declines to very low levels. Marine and Coastal Fisheries: Dynamics, Management, and Ecosystem Science [online serial] 2:424–439.

Beamish, R. J., R. M. Sweeting, and C. M. Neville. 2004b. Improvement of juvenile Pacific salmon production in a regional ecosystem after the 1998 climatic regime shift. Transactions of the American Fisheries Society 133:1163–1175.

Beetz, J. L. 2009. Marine survival of Coho Salmon (*Oncorhynchus kisutch*) in Washington State: characteristics patterns and their relationship to environmental and biological factors. Master's thesis. University of Washington, Seattle.

Bennett, T. R., P. Roni, K. Denton, M. McHenry, and R. Moses. 2015. Nomads no more: early juvenile Coho Salmon migrants contribute to the adult return. Ecology of Freshwater Fish 24:264–275.

Botsford, L. W., and C. A. Lawrence. 2002. Patterns of co-variability among California Current Chinook Salmon, Coho Salmon, Dungeness crab, and physical oceanographic conditions. Progress in Oceanography 53:283–305.

Brodeur, R. D., G. W. Bouehlert, E. Casillas, M. B. Eldridhe, J. H. Helle, W. T. Peterson, W. R. Heard, S. T. Lindley, and M. H. Schiewe. 2000. A coordinated research plan for estuarine and ocean research on Pacific salmon. Fisheries 25(6):7–16.

Burnham, K. P., and D. R. Anderson. 2002. Model selection and multi-model inference: a practical information-theoretic approach. Springer-Verlag, New York.

Collins, A. K., S. E. Allen, and R. Pawlowicz. 2009. The role of wind in determining the timing of the spring bloom in the Strait of Georgia. Canadian Journal of Fisheries and Aquatic Sciences 66:1597–1616.

Coronado, C., and R. Hilborn. 1998a. Spatial and temporal factors affect survival in Coho Salmon (*Oncorhynchus kisutch*) in the Pacific Northwest. Canadian Journal of Fisheries and Aquatic Sciences 55:2067–2077.

Coronado, C., and R. Hilborn. 1998b. Spatial and temporal factors affecting survival in Coho and fall Chinook salmon in the Pacific Northwest. Bulletin of Marine Science 62:409–425.

CoTC (Coho Joint Technical Committee). 2013. 1986–2009 Periodic report revised. Pacific Salmon Commission, CoTC, Report TCCOHO (13)-1. Available: http://www.psc.org/pubs/TCCOHO13-1.pdf. (March 2015).

Craig, B. E., C. A. Simenstad, and D. L. Bottom. 2014. Rearing in natural and recovering tidal wetlands enhanced growth and life-history diversity of Columbia estuary tributary Coho Salmon *Oncorhynchus kisutch* population. Journal of Fish Biology 85(Special Issue):31–51.

Denman, K. L., and A. E. Gargett. 1983. Time and space scales of vertical mixing and advection of phytoplankton in the upper ocean. Limnology and Oceanography 28:801–815.

Dorner, B., K. R. Holt, R. M. Peterman, C. Jordan, D. P. Larsen, A. R. Olsen, and O. I. Abdul-Aziz. 2013. Evaluating alternative methods for monitoring and estimating responses of salmon productivity in the North Pacific to future climatic change and other processes: a simulation study. Fisheries Research 147:10–23.

Ebbesmeyer, C. C., C. A. Coomes, G. A. Cannon, and D. E. Bretschneider. 1989. Linkage of ocean and fjord dynamics at decadal period. Geophysical Monograph 55: 399–417.

Hare, S. R., and N. J. Mantua. 2000. Empirical evidence for North Pacific regime shifts in 1977 and 1989. Progress in Oceanography 47:103–145.

Hare, S. R., N. J. Mantua, and R. C. Francis. 1999. Inverse production regimes: Alaska and West Coast Pacific salmon. Fisheries 24(1):6–14.

Hobday, A. J., and G. W. Boehlert. 2001. The role of coastal ocean variation in spatial and temporal patterns in survival and size of Coho Salmon (*Oncorhynchus kisutch*). Canadian Journal of Fisheries and Aquatic Sciences 58:2021–2036.

Holtby, L. B., B. C. Andersen, and R. K. Kadowaki. 1990. Importance of smolt size and early ocean growth to interannual variability in marine survival of Coho Salmon (*Oncorhynchus kisutch*). Canadian Journal of Fisheries and Aquatic Sciences 47:2181–2194.

Irvine, J. R., and S. A. Akenhead. 2013. Understanding smolt survival trends in Sockeye Salmon. Marine and Coastal Fisheries: Dynamics, Management, and Ecosystem Science [online serial] 5:303–328.

Irvine, J. R., R. C. Bocking, K. K. English, and M. Labelle. 1992. Estimating Coho Salmon (*Oncorhynchus kisutch*) spawning escapement by conducting visual surveys in areas selected using stratified random and stratified index sampling designs. Canadian Journal of Fisheries and Aquatic Sciences 49:1972–1981.

Irvine, J. R., C. G. J. Michielsens, M. O'Brien, B. A. White, and M. Folkes. 2014. Increasing dominance of odd-year returning Pink Salmon. Transactions of the American Fisheries Society 143:939–956.

Johannessen, S. C., D. Masson, and R. Macdonald. 2006. Distribution and cycling of suspended particles inferred from transmissivity in the Strait of Georgia, Haro Strait, and Juan de Fuca Strait. Atmosphere-Ocean 44:17–27.

Johannessen, S. C., D. Masson, and R. W. Macdonald. 2014. Oxygen in the deep Strait of Georgia, 1951–2009: the roles of mixing, deep-water renewal, and remineralization of organic carbon. Limnology and Oceanography 59:211–222.

Kilduff, D. P., L. W. Botsford, and S. L. H. Teo. 2014. Spatial and temporal covariability in early ocean survival of Chinook Salmon (*Oncorhynchus tshawytscha*) along the west coast of North America. ICES Journal of Marine Science 71:1671–1682.

Kuhn, B., L. Lapi, and J. M. Hamer. 1988. An introduction to the Canadian database on marked Pacific salmonids. Canadian Technical Report of Fisheries and Aquatic Sciences 1949.

Labelle, M., C. J. Walters, and B. Riddell. 1997. Ocean survival and exploitation of Coho Salmon (*Oncorhynchus kisutch*) stocks from the east coast of Vancouver Island, British Columbia. Canadian Journal of Fisheries and Aquatic Sciences 54:1433–1449.

Legendre, P., and L. Legendre. 2012. Numerical ecology, 3rd English edition. Elsevier, Amsterdam.

Li, L., D. Mackas, B. Hunt, J. Schweigert, E. Pakhomov, R. I. Perry, M. Galbraith, and T. J. Pitcher. 2013. Zooplankton communities in the Strait of Georgia, British Columbia track large-scale climate forcing over the Pacific Ocean. Progress in Oceanography 115:90–102.

Mackas, D., M. Galbraith, D. Faust, D. Masson, K. Young, W. Shaw, S. Romaine, M. Trudel, J. Dower, R. Campbell, A. Sastri, E. A. Bornhold, Pechter, E. Pakhomov, and R. El-Sabaawi. 2013. Zooplankton time series from the Strait of Georgia: results from year-round sampling at deep water locations, 1990–2010. Progress in Oceanography 115:129–159.

Mackas, D. L., and P. J. Harrison. 1997. Nitrogenous nutrient sources and sinks in the Juan de Fuca/Strait of Georgia/Puget Sound estuarine system: assessing the potential for eutrophication. Estuarine, Coastal, and Shelf Science 44:1–21.

Mantua, N. J., S. R. Hare, Y. Zhang, J. M. Wallace, and R. C. Francis. 1997. A Pacific decadal climate oscillation with impacts on salmon. Bulletin of the American Meteorological Society 78:1069–1079.

Masson, D., and A. Pena. 2009. Chlorophyll distribution in a temperate estuary: the Strait of Georgia and Juan de Fuca Strait. Estuarine, Coastal, and Shelf Science 82:19–28.

Masson, D., and R. I. Perry. 2013. The Strait of Georgia ecosystem research initiative: an overview. Progress in Oceanography 115:1–5.

Moore, S. K., N. J. Mantua, J. P. Kellog, and J. A. Newton. 2008a. Local and large-scale climate forcing of Puget Sound oceanographic properties on seasonal to interdecadal time scales. Limnology and Oceanography 53:1746–1758.

Moore, S. K., N. J. Mantua, J. A. Newton, M. Kawase, M. J. Warner, and J. P. Kellog. 2008b. A descriptive analysis of temporal and spatial patterns of variability in Puget Sound oceanographic properties. Estuarine, Coastal, and Shelf Science 80:545–554.

Morris, J. F. T., M. Trudel, M. E. Thiess, R. M. Sweeting, J. Fisher, S. A. Hinton, E. A. Fergusson, J. A. Orsi, E. V. Farley, and D. W. Welch. 2007. Stock-specific migrations of juvenile Coho Salmon serviced from coded-wire tag recoveries on continental shelf of western North America. Pages 81–104 in C. B. Grimes, R. D. Brodeur, L. J. Haldorson, and S. M. McKinnell, editors. The ecology of juvenile salmon in the northeast Pacific Ocean. American Fisheries Society, Symposium 57, Bethesda, Maryland.

Mueter, F. J., J. L. Boldt, B. A. Megrey, and R. M. Peterman. 2007. Recruitment and survival of mortheast Pacific Ocean fish stocks: temporal trends, covariation, and regime shifts. Canadian Journal of Fisheries and Aquatic Sciences 64:911–927.

Myers, R. A., G. Mertz, and J. Bridson. 1997. Spatial scales of interannual recruitment variations of marine, anadromous, and freshwater fish. Canadian Journal of Fisheries and Aquatic Sciences 54:1400–1407.

Overland, J., S. Rodionov, S. Minobe, and N. Bond. 2008. North Pacific regimes shifts: definitions, issues and recent transitions. Progress in Oceanography 77:92–102.

Pearcy, W. G. 1988. Factors affecting survival of Coho Salmon off Oregon and Washington. Pages 67–73 in W. J. McNeil, editor. Salmon production, management, and allocation. Oregon State University Press, Corvallis.

Pearcy, W. G. 1992. Ocean ecology of north Pacific salmonids. University of Washington Press, Washington Sea Grant Program, Seattle.

Perry, R. I., and D. Masson. 2013. An integrated analysis of the marine social-ecological system of the Strait of Georgia, Canada, over the past four decades, and development of a regime shift index. Progress in Oceanography 115:14–27.

Peterman, R. M., and B. Dorner. 2012. A widespread decrease in productivity of Sockeye Salmon (Oncorhynchus nerka) populations in western North America. Canadian Journal of Fisheries and Aquatic Sciences 69:1255–1260.

Peterson, W. T. 2009. Copepod species richness as an indicator of long-term changes in the coastal ecosystem of the northern California Current. California Cooperative Oceanic Fisheries Investigations Reports 50:73–81.

Peterson, W. T., R. C. Hoof, C. A. Morgan, K. L. Hunter, E. Casillas, and J. W. Ferguson. 2006. Ocean conditions and salmon survival in the northern California Current. Northwest Fisheries Science Center, Seattle.

PFMC (Pacific Fishery Management Council). 2008. Fisheries regulation assessment model (FRAM) an overview for coho and Chinook v. 3.0. PFMC, Portland, Oregon.

Povolina, J. J., G. T. Mitchum, and G. T. Evans. 1995. Decadal and basin-scale variation in mixed layer depth and the impact on biological production in the Central and North Pacific, 1960–1988. Deep Sea Research Part I: Oceanographic Research Papers 42:1701–1716.

Preikshot, D., R. J. Beamish, and C. M. Neville. 2013. A dynamic model describing ecosystem-level changes in the Strait of Georgia from 1960 to 2010. Progress in Oceanography 115:28–40.

Pyper, B. J., F. J. Mueter, and R. M. Peterman. 2002. Spatial covariation in survival rates of northeast Pacific Chum Salmon. Transactions of the American Fisheries Society 131:343–363.

Pyper, B. J., F. J. Mueter, and R. M. Peterman. 2005. Across-species comparisons of spatial scales of environmental effects on survival rates of northeast Pacific salmon. Transactions of the American Fisheries Society 134:86–104.

Pyper, B. J., F. J. Mueter, R. M. Peterman, D. J. Blackbourn, and C. C. Wood. 2001. Spatial covariation in survival rates of northeast Pacific Pink Salmon (Oncorhynchus gorbuscha). Canadian Journal of Fisheries and Aquatic Sciences 58:1501–1515.

R Development Core Team. 2014. R: a language and environment for statistical computing. R Foundation for Statistical Computing, Vienna. Available: http://www.R-project.org. (March 2015).

Rensel, J. R., N. Haigh, and T. J. Tynan. 2010. Fraser River Sockeye Salmon marine survival decline and harmful blooms of Heterosigma akashiwo. Harmful Algae 10:98–115.

Rhode, J., A. N. Kagley, K. L. Fresh, F. A. Goetz, and T. P. Quinn. 2013. Partial migration and diel movement patterns in Puget Sound Coho Salmon. Transactions of the American Fisheries Society 142:1615–1628.

Riche, O., S. C. Johannessen, and R. W. Macdonald. 2013. Why timing matters in a coastal sea: trends, variability and tipping points in the Strait of Georgia, Canada. Journal of Marine Systems 131:36–53.

Riddell, B., I. Pearsall, R. J. Beamish, B. Devlin, A. P. Farrell, S. McFarlane, K. Miller-Saunders, A. Tautz, A. Trites, and C. Walters. 2009. Strait of Georgia Chinook and coho proposal. Pacific Salmon Foundation, Vancouver.

Rupp, D. E., T. C. Wainwright, P. W. Lawson, and W. T. Peterson. 2012. Marine environment-based forecasting of Coho Salmon (Oncorhynchus kisutch) adult recruitment. Fisheries Oceanography 21:1–19.

Sandercock, F. K. 1991. Life history of Coho Salmon (Oncorhynchus kisutch). Pages 395–446 in C. Groot and L. Margolis, editors. Pacific salmon life histories. University of British Columbia Press, Vancouver.

Scott, J. B., and W. T. Gill, editors. 2008. Onchorhynchus mykiss: assessment of Washington State's steelhead populations and programs. Washington Department of Fish and Wildlife, Olympia.

Simpson, K., R. Semple, D. Dobson, J. Irvine, S. Lehmann, and S. Baillie. 2000. Status in 1999 of coho stocks adjacent to the Strait of Georgia. Canadian Stock Assessment Secretariat Research Document 2000/158.

Stroup, W. W. 2012. Generalized linear mixed models: modern concepts, methods, and applications. CRC Press, Boca Raton, Florida.

Sutton, J. N., S. C. Johannessen, and R. W. Macdonald. 2013. A nitrogen budget for the Strait of Georgia, British Columbia, with emphasis on particulate nitrogen and dissolved inorganic nitrogen. Biogeosciences 10:7179–7194.

Suzuki, R., and H. Shimodaira. 2006. Pvclust: an R package for assessing the uncertainty in hierarchical clustering. Bioinformatics Applications 22:1540–1542.

Teo, S. L. H., L. W. Botsford, and A. Hastings. 2009. Spatio-temporal covariability in Coho Salmon (Oncorhynchus kisutch) survival, from California to southwest Alaska. Deep Sea Research Part II: Topical Studies in Oceanography 56:2570–2578.

Thomson, R. 2014. The physical ocean. Pages 13–40 in R. J. Beamish and G. A. McFarlane, editors. The sea around us: the amazing Strait of Georgia. Harbour Publishing, Madeira Park, British Columbia.

U.S. Salish Sea Technical Team. 2012. Marine survival of salmon and steelhead in the Salish Sea: hypootheses and preliminary research recommendations for Puget Sound. Available: http://marinesurvivalproject.com/wp-content/uploads/Puget-Sound-Hypotheses-and-Prelminary-Recs-SSMSP-2012-2.pdf. (May 2015).

Weitkamp, L. A., and K. Neely. 2002. Coho Salmon (Oncorhynchus kisutch) ocean migration patterns: insight from marine coded-wire tag recoveries. Canadian Journal of Fisheries and Aquatic Sciences 59:1100–1115.

Yin, K., R. H. Goldblatt, P. J. Harrison, M. A. St. John, P. J. Clifford, and R. J. Beamish. 1997. Importance of wind and river discharge in influencing nutrient dynamics and phytoplankton production in summer in the central Strait of Georgia. Marine Ecology Progress Series 161:173–183.

Zeileis, A., C. Kleiber, W. Kraemer, and K. Hornik. 2003. Testing and dating of structural changes in practice. Computational Statistics and Data Analysis 44:109–123.

Zeileis, A., F. Leisch, K. Hornik, and C. Kleiber. 2002. strucchange: an R package for testing for structural change in linear regression models. Journal of Statistical Software 7:1–38.

8

Simulating the Trophic Impacts of Fishery Policy Options on the West Florida Shelf Using Ecopath with Ecosim

David D. Chagaris*
Florida Fish and Wildlife Conservation Commission, Fish and Wildlife Research Institute,
100 8th Avenue Southeast, St. Petersburg, Florida 33701, USA; and Department of Fisheries and
Aquatic Sciences, University of Florida, 7922 Northwest 71st Street, Gainesville, Florida 32653, USA

Behzad Mahmoudi
Florida Fish and Wildlife Conservation Commission, Fish and Wildlife Research Institute,
100 8th Avenue Southeast, St. Petersburg, Florida 33701, USA

Carl J. Walters
Department of Fisheries and Aquatic Sciences, University of Florida, 7922 Northwest 71st Street,
Gainesville, Florida 32653, USA; and Fisheries Centre, University of British Columbia,
2202 Main Mall, Vancouver, British Columbia V6T 1Z4, Canada

Micheal S. Allen
Department of Fisheries and Aquatic Sciences, University of Florida, 7922 Northwest 71st Street,
Gainesville, Florida 32653, USA

Abstract
    The recovery of several top predators in the Gulf of Mexico is likely to increase predation on and competition with other target and nontarget species, possibly causing the abundance of those species to decline. While changes are taking place at the upper trophic levels, exploitation of prey species and climate change are altering productivity at the lower levels. An Ecopath with Ecosim model was developed to simulate the ecosystem impacts of Reef Fish Fishery Management Plan Amendment 30B (which aims to rebuild Gag *Mycteroperca microlepis*) and Amendment 31 (which reduces effort in the longline fishery). We also evaluated the impact of a hypothetical increase in the exploitation of baitfish and future changes to phytoplankton productivity. The model predicted that rebuilding Gag will cause the biomass of Black Sea Bass *Centropristis striata* to be 20% lower than it is now and those of Black Grouper *M. bonaci*, King Mackerel *Scomberomorus cavalla*, and other shallow-water groupers to be 5–10% lower. Reducing effort in the longline fishery will lead to biomass declines for Black Sea Bass (13%) and Vermilion Snapper *Rhomboplites aurorubens* (7%). Harvesting baitfish at historically high levels caused the biomass of Red Snapper *Lutjanus campechanus*, Vermilion Snapper, Greater Amberjack *Seriola dumerili*, King Mackerel, and numerous species of dolphins and seabirds to be 5–12% lower after 20 years, while biomass increased for species whose diet consists of benthic-associated prey. This paper demonstrates that ecosystem models can be used to quantify the potential ecological impacts of management goals and that the predictions of such models should be considered alongside stock projections from single-species models that assume a constant environment. We intend for this research effort to lead to a more focused and coherent strategy for ecosystem-based fishery management in the Gulf of Mexico.

Subject editor: Anthony Overton, East Carolina University, Greenville, North Carolina

*Corresponding author: dave.chagaris@myfwc.com

A basic tenet of ecosystem-based fisheries management (EBFM) is that species are interconnected and that fishing, along with other human and natural perturbations, has the potential to impact entire ecosystems (Link 2010). Ecosystem impacts, whether induced by fishing or environmental change, typically arise through predator–prey interactions. Removing predators can cause an increase in the abundance of their prey and a decline in species two trophic levels below them, a phenomenon known as a trophic cascade (Carpenter et al. 1985; Frank et al. 2005; Steneck 2012). Harvesting prey, even at sustainable rates, can impact the growth and reproductive success of predators, ultimately causing their populations to decline (Walters and Martell 2004; Walters et al. 2005; Smith et al. 2011; Pikitch et al. 2012). Competition also plays a structuring role in ecosystems (Pianka 1974). In regards to trophic-dynamic models, competition requires that a change in the abundance of one species cause reciprocal changes in the abundance of other species that utilize the same resource (Hollowed et al. 2000). Simulation models have shown that competitive interactions were important in structuring demersal fish communities, especially during periods when predator abundances were high (Overholtz and Tyler 1986; Collie and DeLong 1999). Because of predation and competition, rebuilding plans for depleted predator species are likely to have consequences for other members of the community (Hartman 2003; Andersen and Rice 2010).

Reef fish such as groupers (Epinephelidae) and snappers (Lutjanidae) support some of the most valuable recreational and commercial fisheries in the southeastern United States and Gulf of Mexico. In 2009, the commercial fishery landed over 6,400 metric tons of reef fishes on the west coast of Florida with a dockside value of nearly US$32 million, and recreational anglers captured an estimated 3,400 metric tons of reef fish (National Marine Fisheries Service, Office of Science and Technology; http://www.st.nmfs.noaa.gov/index). Over the last 50 years, several reef fish species have been severely depleted. In 2009, Red Snapper *Lutjanus campechanus*, Greater Amberjack *Seriola dumerili*, and Gag *Mycteroperca microlepis* were all determined to be overfished and undergoing overfishing, and rebuilding plans are currently in place (NMFS 2011a). Alternatively, the stock of Red Grouper *Epinephelus morio* in the Gulf of Mexico has been increasing since the mid-1990s (SEDAR 2009a). The simultaneous increase in the stock sizes of a suite of top predators will increase predation on and competition with other target and nontarget species, possibly causing their abundances to decline. While these changes are taking place at the upper trophic levels, the exploitation of prey species and climate change are altering productivity at the lower levels. Despite these impending changes to the ecosystem, management goals are still based on model projections that assume no change in ecological circumstances.

In the Gulf of Mexico, reef fish are managed by the Gulf of Mexico Fishery Management Council (GMFMC) using a combination of recreational bag limits, minimum size limits, commercial trip limits, gear restrictions, annual catch limits, seasonal closures, area closures, and individual fishing quotas. Rule changes proposed in Reef Fish Fishery Management Plan Amendment 30B aim to end the overfishing of Gag and respond to the improved status of Red Grouper (GMFMC 2008). To limit the bycatch of the endangered loggerhead sea turtle *Caretta caretta*, Amendment 31 establishes an endorsement requirement, seasonal area closure, and hook limit in the bottom longline fishery that is expected to reduce overall effort in that fishery by 48–67% (GMFMC 2009). These two regulations are expected to increase the biomass of Gag and other reef fish captured by bottom longlines. However, the effect that this will have on other species in the system through predation and competition has not been evaluated. Moreover, the response of predator populations to variability in the abundance of their prey, whether induced by fishing or climate change, is not well known.

Ecological forecasting has become a common goal for EBFM because it can provide resource managers a comprehensive picture of how the ecosystem will respond to a diverse set of policy options (Clark et al. 2001; Valette-Silver and Scavia 2003). Ecosystem models are increasingly being utilized as ecological prediction tools because they provide the capability to simulate the entire ecosystem from primary producers to top predators and fisheries. Ecopath with Ecosim (EwE) is an ecosystem modeling package that simulates population dynamics and explicitly accounts for trophic interactions, fisheries, and environmental forcing (Christensen and Walters 2004). Ecosim has been used to simulate the ecosystem response to climate change (Ainsworth et al. 2011), fisheries (Heymans et al. 2009), bycatch (Walters et al. 2008), invasive species (Pinnegar et al. 2014), marine aquaculture (Forrestal et al. 2012), organic pollution (Libralato and Solidoro 2009), and bioaccumulation of toxins (Booth and Zeller 2005). Because Ecosim is a biomass-dynamic model with only coarse age and size representation, it is not capable of simulating tactical management measures such as bag limits and size limits. Despite their widespread use, Ecosim and other ecosystem models have played only a limited role in actual fisheries management decisions because of their large data requirements and high levels of uncertainty (Plaganyi and Butterworth 2004).

To date, there have been few attempts at using ecosystem models to evaluate the impacts of harvest policies and environmental change on the fisheries and ecosystems in the Gulf of Mexico (Okey et al. 2004; Walters et al. 2008). Limited by data requirements, ecosystem modeling in the Gulf of Mexico has lagged behind that in regions such as Alaska and the northeastern United States, which have a long history of data collection programs and especially food web investigations (Link and Almeida 2000; Aydin et al. 2007; Link et al. 2010; Boldt et al. 2012). In this study, we developed an EwE model of the West Florida Shelf (WFS) to predict the biomass changes caused by Reef Fish Fishery Management Plan Amendments

30B and 31, a hypothetical increase in the exploitation of bait-fish, and changes to primary production. Like all ecosystem models, this model is a simplified representation of a far more complex system. To make it useful to management, we attempted to strike a balance between capturing what we believe to be the major ecological processes and keeping the model flexible, functional, and interpretable. This research serves as a case study for EBFM in the Gulf of Mexico and demonstrates that ecosystem models can provide quantitative and predictive information that is useful for fisheries assessment and management in this region.

METHODS

Model description.—The EwE model that we developed centered on regulated species on the WFS, including reef fishes, coastal migratory pelagic species, and highly migratory pelagic species as defined by the GMFMC and the National Marine Fisheries Service. The area modeled is approximately 170,000 km^2 and extends from the Florida Panhandle south to a boundary that excludes the Florida Keys and out to the 250-m isobaths contour. Particular emphasis was given to groupers and snappers that inhabit reefs on the WFS and support valuable commercial and recreational fisheries. Gag, Red Grouper, Black Grouper *M. bonaci*, and Yellowedge Grouper *Hyportho-dus flavolimbatus* were represented in the model by three age stanzas (0–1, 1–3, and 3+ years) to capture basic ontogenetic changes in diet, habitat, and fishery selectivity. Red Snapper, Spanish Mackerel *Scomberomorus maculatus*, and King Mackerel *S. cavalla* were divided into juveniles (0–1 years) and adults (1+ years). Other reef fishes and pelagic fishes were included either as single-species biomass groups or aggregated into groups of similar species. Coastal and inshore species were included because they interact with reef fish juveniles yet to migrate offshore. Aggregate groups of nontarget fishes, invertebrates, zooplankton, and primary producers were necessary for a complete food web. The resulting model consisted of 70 biomass pools, including one each for dolphins and seabirds, 43 fish groups (of which 11 are nonadult life stages), 18 invertebrate groups, 4 primary producers, and 3 detritus groups (Table 1).

Biomass (*B*; metric tons/km^2) values were taken from single-species stock assessments, estimated by dividing observed catches by assumed fishing mortality (*B* = *C/F*) or derived from survey data. The production rate (*P/B*) or instantaneous total mortality (*Z*) was calculated by adding an assumed natural mortality to the fishing mortality from stock assessments or by using empirical equations for mortality (Pauly 1980; Ralston 1987). Estimates of consumption (*Q*) were derived empirically using equations that incorporate data on morphometrics, ambient water temperature, and diet (Pauly 1989; Palomares and Pauly 1998). The diet compositions of fish were estimated by combining data from the Florida Fish and Wildlife Conservation Commission (FWC) fisheries-independent monitoring

program's trophic database with information available in the literature using weighted averages that account for the number of nonempty stomachs, the locations at which they were collected, and the quality of the data (see Supplement A for details). Much effort was put into the derivation of parameters for invertebrates in an earlier WFS model (Okey and Mahmoudi 2002; Okey et al. 2004), and we used those values as initial input for this reef fish–centric model.

The fishery included four recreational (shore-based, private boat, charter boat, and headboat) and nine commercial (vertical line, bottom longline, pelagic longline, pelagic troll, gill/trammel net, cast net, purse seine, trawl, fish trap, and crab trap) fishing "fleets." Commercial landings were obtained from trip tickets in the Florida Marine Resources Information System, and discards were based on bycatch reports (NMFS 2011b) and information from other observer programs (Pierce et al. 1998; Passerotti et al. 2010; NMFS–Southeast Fisheries Science Center, personal communication; FWC, Fish and Wildlife Research Institute, personal communication). Recreational landings and discards were made available by the Marine Recreational Fisheries Statistics Survey, and headboat landings were obtained from the Southeast Fisheries Science Center's headboat survey. After entering the required input data, the Ecopath model was "mass-balanced" by making small adjustments to diet, mortality, and biomass so that fishing and predation mortality rates did not exceed total mortality.

Model calibration.—Before being used to make predictions, the Ecosim model was calibrated to time series of observed trends in abundance and catch over the period 1950–2009. Reference time series were obtained directly from stock assessments or taken from fisheries-independent and other survey data. Fleet-specific fishing effort from the Vessel Operating Units database (Jason Rueter, National Oceanic and Atmospheric Administration, Southeast Regional Office, personal communication), and species-specific fishing mortality rates from Southeast Data Assessment and Review stock assessments (www.sefsc.noaa.gov/sedar) were used as forcing time series. Chlorophyll-*a* production along the WFS is dependent on a variety of factors, including the outflow from the Mississippi River (Gilbes et al. 1996, 2002; Castillo et al. 2001). Therefore, we used nutrient loads from the Mississippi River as a proxy for phytoplankton production on the WFS (Goolsby and Battaglin 2000; Aulenbach et al. 2007). Because the calibration simulation began in 1950, the biomass, catch, and total mortality parameters from the 2009 Ecopath model were first rescaled to represent a historical (1950s) condition (Table 1). This involved increasing biomass, reducing catch, and reducing total mortality to a level closer to natural mortality. In most cases, the stock assessment or time series data provided the information necessary to make such adjustments. The diet matrices were the same in the 2009 and 1950 models except in a few cases in which minor adjustments were required for mass balance.

TABLE 1. Biomass, catch (including dead discards), instantaneous total mortality (Z), and instantaneous fishing mortality (F) representing historical (1950) and present-day (2009) Ecopath models. Biomass and catch are in thousands of metric tons; Z and F are per year.

Taxon	1950 Biomass	Catch	Z	F	2009 Biomass	Catch	Z	F
Dolphins	2.89	0.00	0.16	0.00	2.89	0.00	0.16	0.00
Seabirds	0.85	0.00	0.30	0.00	2.04	0.00	0.30	0.00
Large coastal sharks[a]	15.30	0.83	0.15	0.05	6.89	1.65	0.41	0.24
Small coastal sharks[a]	18.34	0.13	0.30	0.01	11.37	0.62	0.54	0.05
Rays and skates[a]	40.63	1.35	0.50	0.03	40.63	1.35	0.85	0.03
Billfish and tunas[a]	7.90	0.29	0.34	0.04	5.50	0.59	0.68	0.11
Oceanic small pelagics[a]	68.00	0.42	1.74	0.01	68.00	0.42	1.36	0.01
Cobia *Rachycentron canadum*	6.19	0.06	0.50	0.01	2.11	0.42	0.70	0.20
King Mackerel (juvenile)	0.08	0.00	2.00	0.00	0.33	0.01	2.00	0.04
King Mackerel (adult)	13.55	0.94	0.40	0.07	8.78	1.85	0.80	0.21
Spanish Mackerel (juvenile)	0.35	0.00	2.00	0.00	1.31	0.03	2.00	0.02
Spanish Mackerel (adult)	29.42	1.10	0.40	0.04	13.16	2.10	1.08	0.16
Jacks, dolphins, and tunnies[a]	30.76	0.31	0.50	0.01	21.53	3.14	0.72	0.15
Red Snapper (juvenile)	0.04	0.00	2.00	0.01	0.23	0.01	1.50	0.03
Red Snapper (adult)	17.58	0.67	0.25	0.04	7.87	1.57	0.74	0.20
Vermilion Snapper *Rhomboplites aurorubens*	2.84	0.03	0.40	0.01	1.53	0.70	0.86	0.46
Other snappers[a]	41.07	0.41	0.60	0.01	32.86	9.96	0.63	0.30
Tilefish *Lopholatilus chamaeleonticeps*	1.27	0.00	0.30	0.01	0.94	0.16	0.50	0.17
Yellowedge Grouper (0–1 years)	0.00	0.00	6.00	0.00	0.02	0.00	2.50	0.00
Yellowedge Grouper (1–3 years)	0.03	0.00	0.80	0.00	0.33	0.00	0.80	0.00
Yellowedge Grouper (3+ years)	12.21	0.12	0.08	0.01	4.03	0.28	0.40	0.07
Other deepwater groupers	5.37	0.15	0.20	0.03	1.77	0.15	0.40	0.08
Gag (0–1 years)	0.07	0.00	3.00	0.01	0.17	0.01	1.48	0.04
Gag (1–3 years)	0.62	0.01	0.90	0.02	2.66	0.63	1.06	0.23
Gag (3+ years)	22.88	0.60	0.15	0.03	7.55	3.85	0.80	0.51
Red Grouper (0–1 years)	0.29	0.00	2.00	0.00	0.50	0.00	2.00	0.00
Red Grouper (1–3 years)	3.92	0.05	0.80	0.01	6.40	0.13	0.80	0.02
Red Grouper (3+ years)	55.33	5.60	0.25	0.10	32.66	5.19	0.40	0.16
Black Grouper (0–1 years)	0.05	0.00	2.00	0.00	0.04	0.00	2.00	0.02
Black Grouper (1–3 years)	0.79	0.00	0.80	0.00	0.50	0.04	0.80	0.08
Black Grouper (3+ years)	9.70	0.24	0.30	0.02	3.13	0.34	0.40	0.11
Other shallow-water groupers	8.87	0.17	0.25	0.02	3.40	0.17	0.40	0.05
Atlantic Goliath Grouper *Epinephelus itajara*	1.25	0.04	0.13	0.03	0.65	0.04	0.40	0.05
Gray Triggerfish *Balistes capriscus*	3.03	0.00	0.50	0.00	0.94	0.39	0.89	0.42
Greater Amberjack	4.29	0.04	0.25	0.01	1.09	0.62	1.06	0.57
Black Sea Bass	4.95	0.08	1.00	0.02	2.04	0.15	1.12	0.08
Reef carnivores[a]	229.50	1.59	1.32	0.01	153.00	1.59	1.32	0.01
Reef omnivores[a]	100.98	0.00	1.98	0.00	68.00	0.00	1.66	0.00
Coastal piscivores[a]	35.36	3.66	0.71	0.10	17.70	3.66	0.71	0.21
Large coastal carnivores[a]	130.36	5.10	0.92	0.04	72.42	5.10	0.92	0.07
Small coastal carnivores[a]	130.56	0.79	1.76	0.01	97.92	0.79	1.76	0.01
Coastal omnivores[a]	158.67	1.25	1.98	0.01	119.00	1.25	1.98	0.01
Sardines, herrings, and scads[a]	289.01	5.77	2.31	0.02	289.01	7.53	2.20	0.03
Anchovies and silversides[a]	132.20	0.02	2.67	0.00	100.73	0.02	2.67	0.00
Mullets[a]	31.74	13.45	1.20	0.42	26.06	6.66	1.42	0.26

(*Continued on next page*)

TABLE 1. Continued.

Taxon	1950				2009			
	Biomass	Catch	Z	F	Biomass	Catch	Z	F
Squid	54.47	0.02	2.67	0.00	54.47	0.02	2.67	0.00
Shrimp	116.11	3.52	3.66	0.03	154.72	3.52	3.66	0.02
Lobsters	11.90	0.68	0.90	0.06	5.95	0.68	0.90	0.11
Large crabs[a]	151.30	3.49	1.69	0.02	87.00	3.50	1.69	0.04
Octopods	21.74	0.00	3.10	0.00	17.36	0.00	3.10	0.00
Stomatopods	168.98	0.00	1.50	0.00	168.98	0.00	1.34	0.00
Echinoderms and gastropods[a]	3,271.88	0.55	2.60	0.00	3,271.88	0.55	2.60	0.00
Bivalves	8,261.80	0.00	5.35	0.00	8,261.80	0.00	5.35	0.00
Sessile epibenthos	3,723.06	0.00	1.62	0.00	3,723.06	0.00	1.62	0.00
Small infauna	3,235.49	0.00	4.02	0.00	3,235.49	0.00	4.02	0.00
Small mobile epifauna	2,109.23	0.25	4.76	0.00	2,109.23	0.25	4.76	0.00
Meiofauna	2,210.04	0.00	6.20	0.00	2,210.04	0.00	6.20	0.00
Small copepods	1,411.02	0.00	10.60	0.00	1,411.02	0.00	10.60	0.00
Mesozooplankton	1,139.02	0.00	10.60	0.00	1,139.02	0.00	10.60	0.00
Carnivorous zooplankton	1,836.03	0.00	8.70	0.00	1,836.03	0.00	8.70	0.00
Ichthyoplankton	32.30	0.00	50.45	0.00	32.30	0.00	50.45	0.00
Carnivorous jellyfish	37.57	0.24	20.08	0.01	37.57	0.24	20.08	0.01
Microbes	10,200.17	0.00	100.00	0.00	10,200.17	0.00	100.00	0.00
Macroalgae	6,128.60	0.00	4.00	0.00	6,128.60	0.00	4.00	0.00
Microphytobenthos	5,062.68	0.00	23.73	0.00	5,062.68	0.00	23.73	0.00
Phytoplankton	2,232.14	0.00	182.13	0.00	4,250.07	0.00	182.13	0.00
Sea grasses	29,855.90	0.00	9.00	0.00	29,855.90	0.00	9.00	0.00
Water column detritus	21,250.36	0.00	0.00	0.00	21,250.36	0.00	0.00	0.00
Sediment detritus	59,501.00	0.00	0.00	0.00	66,301.12	0.00	0.00	0.00
Dead discards	0.17	0.00	0.00	0.00	0.17	0.00	0.00	0.00

[a]See Supplement A for a detailed breakdown.

The most important parameters when calibrating Ecosim models are the vulnerability exchange rates (v_{ij}) between prey i and predator j. These vulnerability parameters represent the rates at which prey move from an invulnerable state to a vulnerable state, and there is one parameter for each predator–prey interaction (Ahrens et al. 2012). At very high values of v_{ij} (>100) prey become vulnerable to predators at faster rates and the invulnerable pools are quickly depleted. This essentially implies a linear relationship between predator biomass and predation mortality and can lead to unstable Lotka–Volterra dynamics. At low values of v_{ij} (<2), predation mortality rates remain relatively constant at their Ecopath base values when predator abundances change. To fit the model to time series, manual adjustments were made to the foraging arena parameters, especially the v_{ij}s, to correct for any gross divergence from the data. For example, groups for which biomass declined to zero required that the v_{ij}s be reduced or that feeding time adjustment be turned on to generate stronger compensatory improvements in survival at low stock sizes. After correcting for obvious errors, we executed an automated search that adjusts the v_{ij}s to minimize the sum of squared deviations (SS) between predicted and observed biomass and catch data. This process was repeated iteratively, focusing on the group with the highest SS, until the model was able to reproduce the major patterns in biomass and catch for all groups over the entire time period.

As a further diagnostic, we evaluated how groups responded under no fishing and very high fishing mortality. We also compared the values of fishing mortality at the maximum sustainable yield (F_{msy}) from Ecosim with those estimated by single-species stock assessment models. In Ecosim, F_{msy} was estimated using the "MSY Search" interface that runs the model to equilibrium under a range of fishing mortality rates while holding all other groups stationary. These diagnostics were performed to correct for spurious parameter estimates obtained during the calibration process for reasons such as lack of adequate contrast in the historical biomass trend data.

To conduct forward-projecting policy simulations with the 2009 present-day model, we rescaled the v_{ij}s from the calibrated historical model so that the maximum possible predation mortalities were the same in both the historical and 2009 models. This was done by multiplying each v_{ij} from the historical model by the ratio of historical to present-day predation mortality rates for the same predator–prey interaction, i.e.,

$$\hat{v}_{ij} = v_{ij} \times {}^{M2_{ij}} \Big/ {}_{\widehat{M2}_{ij}},$$

where $\hat{v}_{ij}$ is the rescaled vulnerability for the present-day model, $M2_{ij}$ is the predation mortality from the 1950s model, and $\widehat{M2}_{ij}$ is the predation mortality from the 2009 model. Biomass accumulation was added to the 2009 model to account for the initial rate of biomass change occurring during the first year of forecasting (2009); this was calculated as the change in biomass during the last year of the historical simulation. All input parameter values estimated for the historical (calibrated) and 2009 models are available in Supplements B and C and can be examined and changed using the EwE 6.4 user interface. Comments describing the source and derivation of input values are embedded in each cell in which data were entered.

Policy screening.—We prescribed two actual management actions (Reef Fish Fishery Management Plan Amendments 30B and 31), a hypothetical expansion in the baitfish fishery, and two alternative scenarios of future phytoplankton productivity. Although we simulated the impact of each scenario on the biomass of all of the species in the model, in the results that we present we focus on recreationally and commercially valuable species. In each case, we conducted a 20-year projection using the present-day model with vulnerability exchange rates rescaled as described above. Fishing mortality, fishing effort, and phytoplankton production were held constant at either the prescribed test values or the 2009 Ecopath base values throughout each simulation.

To determine the impacts on species biomass, we compared the change in biomass ($\Delta B = B_{end}/B_{start}$) for each scenario with that of the status quo scenario. The percent change in biomass from status quo ($\%\Delta B$) was calculated as $100 \cdot ([\Delta B_{scenario}/\Delta B_{status\ quo}] - 1)$. In the status quo scenario the model was projected forward 20 years using the baseline fishing mortality rates and the fishing effort in the 2009 Ecopath model. Thus, the status quo scenario simulated a continued increase in biomass for species that were recovering as of 2009 (e.g., Greater Amberjack, Red Grouper, and Red Snapper) and declining stock sizes for species whose fishing mortality rates had not yet been reduced below the overfishing limit (Gag).

For each scenario, we conducted a deterministic run over 20 years using the base parameterization with the 2009 biomass, catch, and fishing mortalities. We also performed 100 Monte Carlo simulations to establish the sensitivity of the model predictions to uncertainty in Ecopath biomass values. The Monte Carlo simulations randomly selected a biomass value for each species from a uniform distribution, where the mean was the 2009 base value and the upper and lower limits were based on knowledge about the uncertainty in the source data and the estimates derived from them. If the random draws of biomass did not violate mass balance in Ecopath, then a

Monte Carlo trial was conducted in Ecosim. Otherwise, another draw was made. For each 20-year simulation, $\%\Delta B$ values were calculated along with standard errors and 95% confidence intervals from the 100 Monte Carlo simulations.

Rebuilding Gag stocks.—The 2009 stock assessment for Gulf of Mexico Gag determined the stock to be overfished and undergoing overfishing (SEDAR 2009b). Amendment 30B was adopted to address the overfished status and develop a stock rebuilding plan for Gag (GMFMC 2008). The rebuilding plan called for reducing fishing mortality to about a third of the 2009 level, to be achieved through a combination of larger size limits, smaller bag limits, and/or shorter seasons (SEDAR 2009b). The baseline (Ecopath) fishing mortality rate of Gag was 0.52 in 2009, so the fishing mortality on Gag was set at $F_{rebuild} = 0.16$ for the duration of the simulation.

Longline effort reduction.—A 2008 report indicated that bottom longline gear took between 339 and 1,884 loggerhead sea turtles over an 18-month period, far exceeding the allowable take of 85 turtles (NMFS 2005, 2008). To reduce the frequency of interactions between bottom longlines and sea turtles, Amendment 31 prohibited the use of bottom longline gear in depths shallower than 64 m (i.e., the 35-fathom depth contour) from June through August; reduced the number of longline vessels to those with annual average landings of at least 18 metric tons during 1999–2007; and restricted the number of hooks per vessel to 1,000, of which only 750 may be fished at a time (GMFMC 2009). The overall reduction in effective effort based on these regulations is expected to be between 48% and 67%. For this scenario, we chose a middle value of 60% by which to reduce effort in the longline fishery.

Increased exploitation of baitfishes.—On the WFS, Scaled Sardine *Harengula jaguana*, Spanish Sardine *Sardinella aurita*, Atlantic Thread Herring *Opisthonema oglinum*, Round Scad *Decapterus punctatus*, and other less dominant clupeids and small carangids are commonly referred to as "baitfish." These small pelagic planktivores make up an important forage base on the WFS and support a commercial baitfish fishery with average annual landings of 2,500 metric tons from 2006 to 2010 (Florida Marine Resources Information System; http://myfwc.com/research/saltwater/fishstats/commercial-fisheries/landings-in-florida/). While the current and historical stock sizes and fishing mortality rates of baitfish are not well known for the WFS, we assumed that a low fishing mortality rate of 0.02 in 2009 was reasonable given the magnitude of the biomass and the small size and scale of the fishery. During 1989, effort and catch in the baitfish fishery were both almost 20 times higher than they were from 2006 to 2010. To evaluate the impact of harvesting baitfish at historically high levels, we simulated a 20-fold increase in effort in the purse seine fishery, which generated a fishing mortality rate of about 0.40 on the baitfish complex. For comparison, the average annual landings in the Gulf Menhaden *Brevoortia patronus* reduction fishery from 2006 to 2010 were 436,160 metric tons, with an average fishing morality rate of 0.43 (SEDAR 2011).

Changes in primary production.—How complex marine ecosystems will respond to climate change is uncertain, and there are plausible hypotheses for both lower and higher overall productivity in the future. Warming sea temperatures have been shown to decrease phytoplankton productivity by reducing mixing throughout the water column and lowering nutrient supply (Behrenfeld et al. 2006; Doney 2006). Severe and prolonged droughts will reduce the delivery of nutrients from freshwater sources and lower the productivity in coastal and estuarine areas (Wiseman et al. 1999; Wetz et al. 2011). On the other hand, it has been hypothesized that increased productivity will occur in warmer, more stratified waters due to enhanced atmospheric nitrogen fixation at the surface (Karl et al. 1997). To investigate how the WFS might respond to broad changes in productivity driven by climate change, we considered two scenarios. In one scenario phytoplankton productivity increased 1% each year for 20 years; in the other, productivity decreased 1% each year. Linear forcing functions without seasonality or random variation were used to simplify the analysis and interpretation of results.

RESULTS

The model was capable of reproducing historical trends in abundance and catch for the period 1950–2010, with a total SS of 223.24 (Figures 1, 2). The nonstationary behavior in the status quo simulation (solid black lines in Figure 3) was a result of biomass accumulation rates calculated from the historical model. The biomass accumulation rates for most species were positive, leading to increasing stock sizes reflecting generally more conservative management in recent years. Gag, which was overfished and undergoing overfishing in 2009, was predicted to continue declining to approximately 50% of its 2009 stock size with a biomass of 3,773 metric tons (Table 2). The biomass of Red Grouper was predicted to be approximately 9,000 metric tons higher in year 20, a 28% increase from 2009. Under the status quo, Red Snapper (27% increase) continue to recover because of the reduced fishing mortality achieved by their rebuilding plans. Biomass was predicted to increase for Black Grouper (16%), Greater Amberjack (8%), Yellowedge Grouper (14%), Atlantic Goliath Grouper (40%), Vermilion Snapper (7%), and Gray Triggerfish (14%), while Tilefish were predicted to decline by 5%. The other shallow-water grouper (SWG) and deepwater grouper (DWG) species also increased over the 20-year simulation. All of the species in the coastal migratory pelagic group increased in the status quo simulation, King Mackerel by 22%, Spanish Mackerel by 18%, and Cobia by 13%. Dolphins and seabirds showed only a little change in biomass (±2%), while the sardine–herring–scad baitfish complex declined 9% as predators recovered.

Rebuilding Gag Stocks

Under an $F_{rebuild}$ of 0.16, Gag increased 70% over its 2009 biomass to 12,835 metric tons in year 20, which was 240% larger than the biomass predicted under the overfishing scenario in the status quo simulation (Figures 3, 4). One hundred Monte Carlo trials produced stock biomass estimates ranging from about 9,000 to 17,000 metric tons (Figure 4), with 95% confidence intervals between 12,000 and 13,000 metric tons (Table 2). For reference, the single-species stock assessment biomass projections for Gag were 10,000 and 13,500 metric tons after 10 years under $F_{rebuild}$ values of 0.19 and 0.14, respectively (SEDAR 2009b). The biomass of Black Sea Bass was predicted to be 20% lower than the status quo in this scenario. The contribution of Black Sea Bass to the diet of Gags was just 1%, and the baseline predation mortality rate was 0.15/year. The predation mortality rate was more than twice as high under rebuilding (0.19/year) than under the status quo (0.09/year). The total predation mortality rates of Black Sea Bass increased from 0.55/year in 2009 to 0.58/year in the status quo simulation and 0.64/year in this scenario, an increase of about 10% after 20 years. Black Grouper, SWG, DWG, Vermilion Snapper, Greater Amberjack, Black Sea Bass, King Mackerel, and the sardine–herring–scad complex all had lower 95% confidence intervals that were more than 5% below the status quo (Table 2).

Longline Effort Reduction

The impact of a 60% reduction in bottom longline effort had direct positive effects on the biomass of several reef fish species, including Gag (5%), Red Grouper (20%), Yellowedge Grouper (65%), SWG (6%), DWG (22%), and Tilefish (74%). Biomass declined for Vermilion Snapper (7%) and Black Sea Bass (13%). The biomass of the baitfish complex was 6% lower than under the status quo, and stochastic sensitivity runs (Monte Carlo trials with randomly varying Ecopath biomasses) indicated that it could be as much as 11% lower. Consequently, the impact on pelagic species was negative but within 5% for the base scenario. The 95% confidence intervals from stochastic runs were centered on zero for Spanish Mackerel and Cobia, indicating that the impact, though small, could be in either direction depending on the starting biomasses of other species.

Increased Exploitation of Baitfish

Increasing effort in the purse seine fishery 20-fold, to a historical high, reduced the biomass of baitfish by 23% from the status quo (Table 2). The biomasses of SWG, Red Snapper, Vermilion Snapper, Amberjack, King Mackerel, dolphins, and seabirds were between 5% and 12% lower after 20 years of harvesting baitfish at $F = 0.40$ than under status quo conditions. Monte Carlo simulations predicted the loss in biomass to be no more than 15% for a given predator. The base model predicted

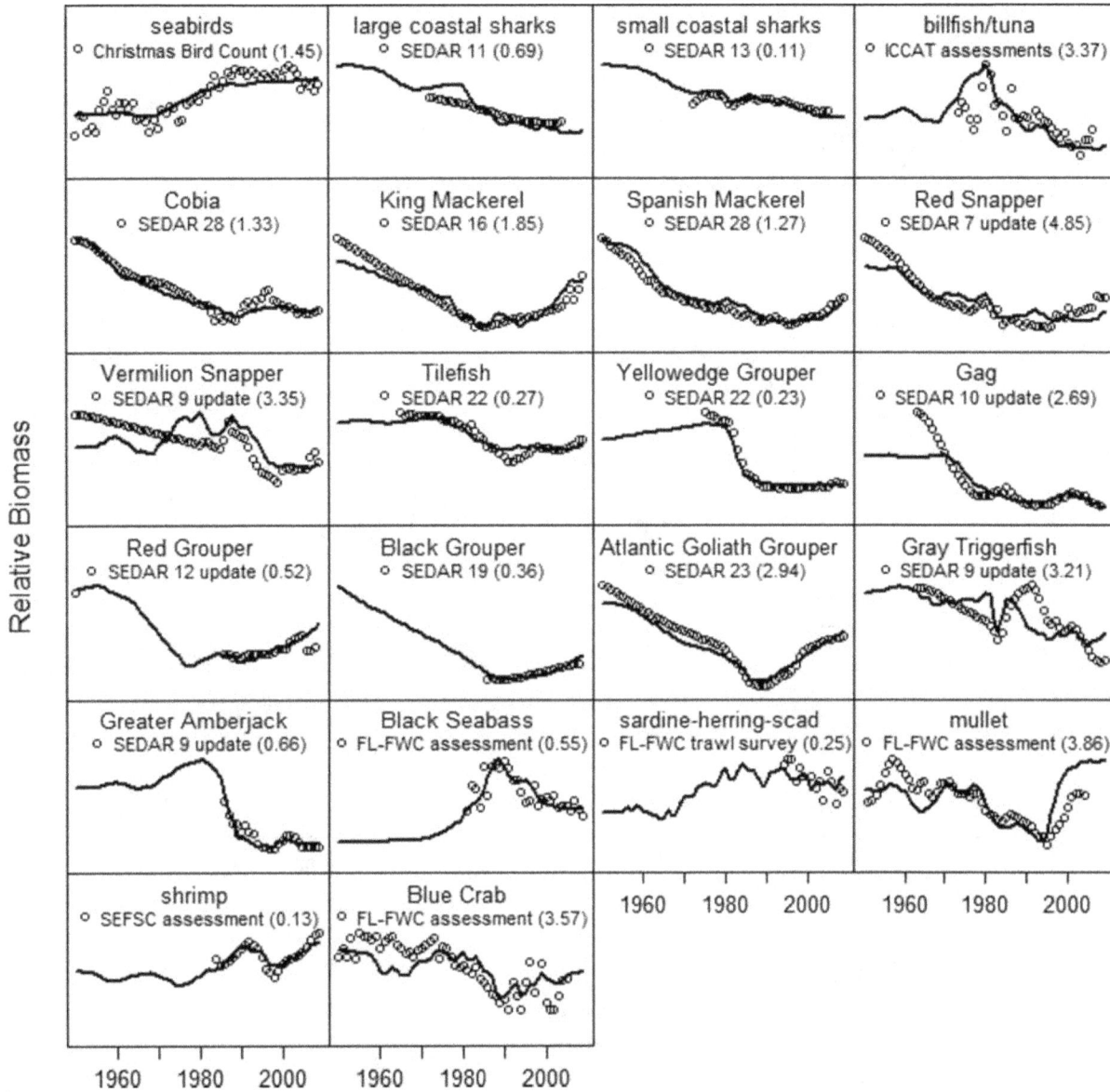

FIGURE 1.   Predicted biomass (solid lines) from the Ecosim model and observed trends in biomass (circles) for selected species, with the associated sums of squares in parentheses. Observed trends in abundance were obtained from stock assessments by the Southeast Data Assessment and Review, the Florida Fish and Wildlife Conservation Commission, NOAA's Southeast Fisheries Science Center, and the International Commission for the Conservation of Atlantic Tunas.

a greater than 5% increase in biomass for DWG and Gray Triggerfish and produced mostly positive Markov chain–Monte Carlo values for Atlantic Goliath Grouper, Tilefish, and Cobia.

Changes in Primary Production

As expected, the model predicted widespread reductions in fish biomass as productivity declined and increases in biomass when it increased. Under a low-production regime, the biomasses of Red Grouper, Red Snapper, Vermilion Snapper, Amberjack, and Gray Triggerfish were all at least 5% less than

under the status quo (Table 2). More severe impacts on biomass were predicted for Red Snapper (-19%), Vermilion Snapper (-16%), King Mackerel (-12%), and Spanish Mackerel (-17%). The impacts of reduced productivity on the pelagic baitfish group were between -18% and -28%, whereas the impacts on benthic-associated prey species such as reef carnivores, small coastal carnivores, coastal omnivores, shrimp, and crabs were between 0% and -12%. Monte Carlo simulation trials for Atlantic Goliath Grouper, Tilefish, Black Sea Bass, and Cobia all showed improvements in biomass under lower primary production.

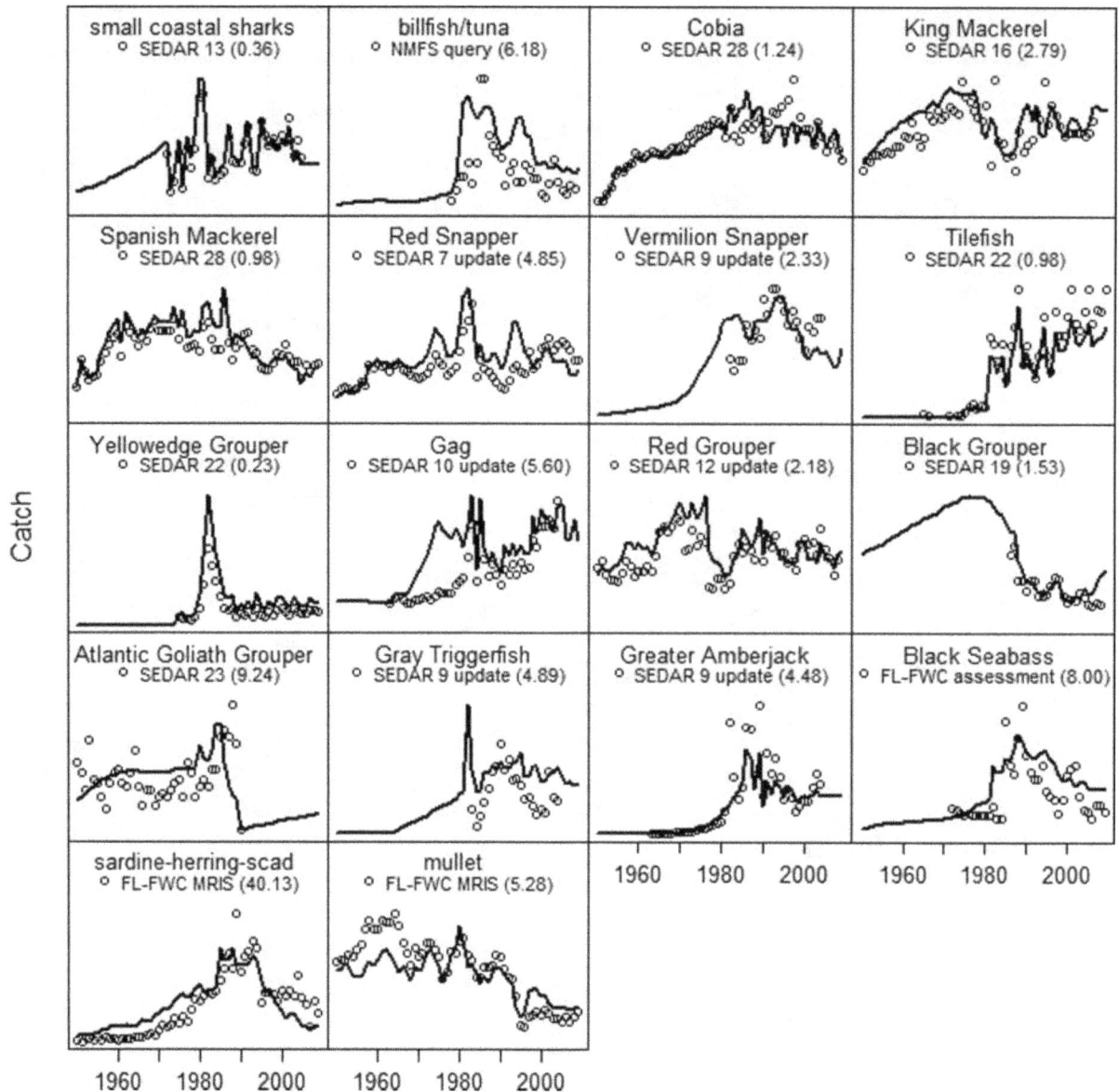

FIGURE 2. Predicted (solid lines) and observed catch (circles) for selected species from 1950 to 2009, with the associated sums of squares in parentheses. For visualization purposes, the scales of the *y*-axes are not shown. Observed catch was taken from stock assessments by the Southeast Data Assessment and Review and the Florida Fish and Wildlife Conservation Commission (FL-FWC) or obtained from trip tickets in the FL-FWC Marine Resources Information System.

DISCUSSION

The overall conclusion from these simulations is that there are winners and losers in all policy options. Management options oriented toward a single species, such as rebuilding an overfished stock, had less widespread and more modest (±5%) impacts on biomass than policies affecting a suite of species. Simulations which involved perturbations to the middle of the food web or changes to primary production had more drastic impacts over a broader set of species. None of the harvest policies or environmental conditions that we considered was predicted to cause any species to collapse.

Differential use of resources (i.e., resource partitioning) may partly explain why competition caused the biomass of some species to decrease and that of others to increase. For example, the high utilization of anchovies by Spanish Mackerel (22% of diet) and of crabs by Cobia (32% of diet) likely provided some relief in the longline scenario from competition with groupers, whose diets are dominated by sardines, herrings, and smaller reef fishes (see Supplement A). Pelagic species such as the sardine–herring–scad complex and anchovies are more tightly coupled to changes in phytoplankton abundance than benthic prey items. In general, reef fish diets are

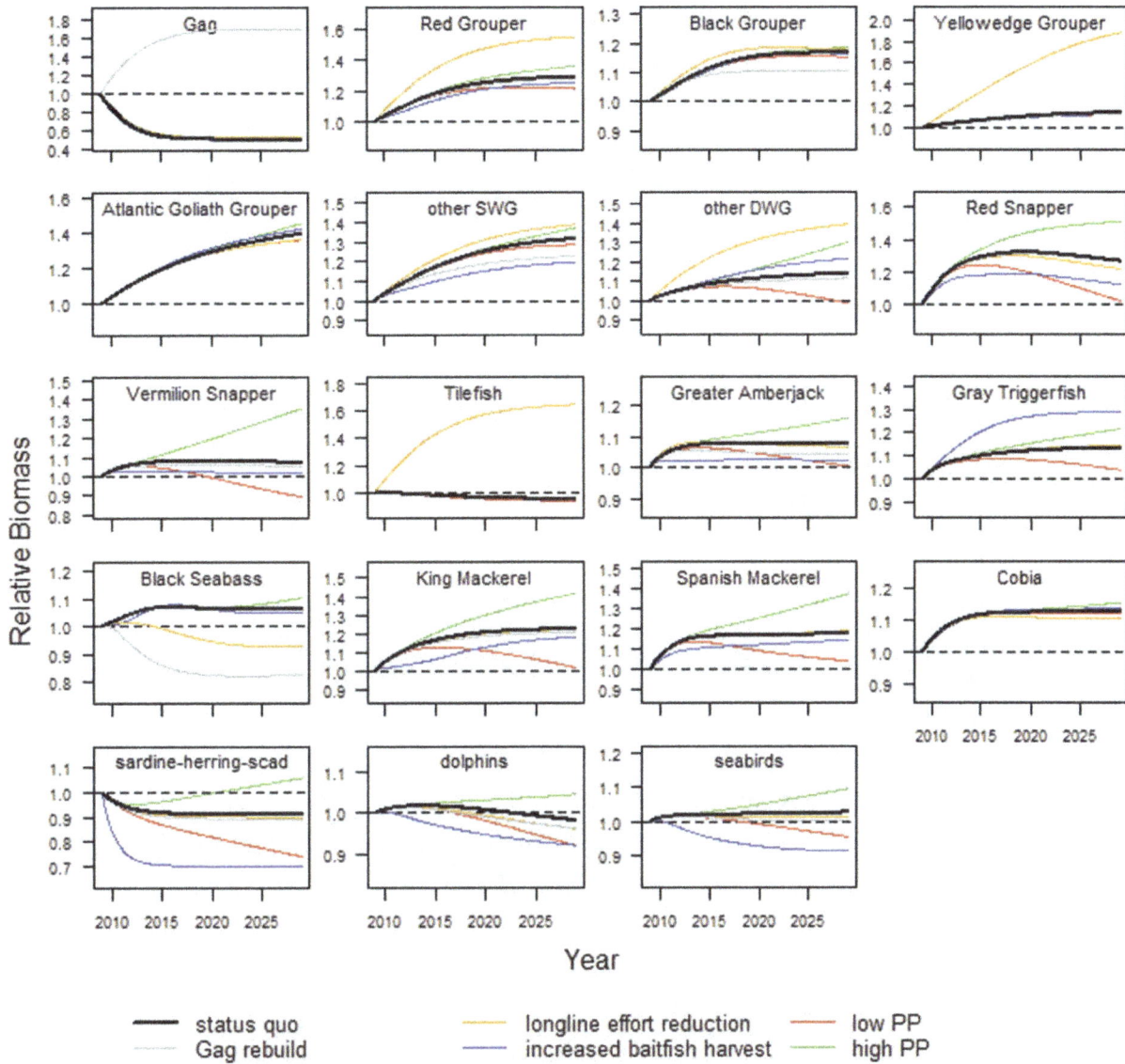

FIGURE 3. Future biomass trajectories simulated by the Ecosim model. Scenarios that caused an increase or decrease in biomass from the status quo are indicated by lines above or below the solid black lines. In some cases there was little change, and those scenarios may be obscured by the status quo line. The dotted line represents the Ecopath base 2009 biomass level.

partially composed of benthic and demersal prey items (e.g., shrimps, crabs, and grunts). The ability of reef fish to access benthic energy channels may stabilize their biomass when pelagic forage fish are removed or phytoplankton production causes changes in their abundance (Rooney et al. 2006).

Competitive interactions are believed to influence reef fish communities in the eastern Gulf of Mexico (Smith 1979). The biomasses of Vermilion Snapper and Black Sea Bass were predicted to decline in response to an increase of other predators, suggesting that these two species are at a competitive disadvantage. Black Sea Bass were observed in higher densities on experimental reefs where Gag were excluded (Lindberg et al. 2006), and Vermilion Snapper became more abundant after a

decrease in resident piscivores (groupers) on artificial reefs (Dance et al. 2011). These observations at artificial and experimental reefs in the northeastern Gulf of Mexico support the predictions made by the Ecosim model. In another case, divers in the northeastern Gulf of Mexico observed a school of Greater Amberjacks driving prey downward toward the reef where Gag were waiting to feed (Stallings and Dingeldein 2012). This illustrates fine-scale competition for food between these two predators and that separate foraging arenas can exist for multiple predators over a single prey resource, water column, and reef. It also raises the possibility that vulnerability exchange rates can be mediated by multiple species that pursue the same prey in different microhabitats (e.g., at different depths).

TABLE 2. Impact of three policy options and two primary productivity scenarios on select taxa. The 2009 Ecopath baseline biomass and the predicted biomass in 2029 under the status quo are expressed in thousands of metric tons; the other values are percent changes from the 2029 status quo biomass. The values in parentheses are the 95% confidence limits estimated from 100 Monte Carlo simulations.

Taxon	2009 Ecopath base biomass	2029 status quo biomass	Policy options			Primary productivity	
			Gag stock rebuilding	Longline effort	Baitfish $F = 0.4$	Low	High
Gag	7.55	3.77	240 (230, 246)	5 (3, 6)	−1 (−3, 0)	−1 (−4, −1)	5 (3, 6)
Red Grouper	32.67	42.10	−1 (−2, 0)	20 (20, 23)	−3 (−3, −1)	−6 (−7, −5)	6 (5, 7)
Black Grouper	3.13	3.65	−5 (−5, −3)	1 (0, 3)	0 (−1, 2)	−1 (−4, −1)	2 (0, 2)
Yellowedge Grouper	4.03	4.60	−1 (−1, 1)	65 (62, 65)	−2 (−2, 0)	−2 (−4, −2)	2 (1, 2)
Atlantic Goliath Grouper	0.65	0.91	0 (−2, 13)	−2 (−6, 12)	2 (5, 22)	−2 (−2, 15)	4 (4, 22)
Other shallow-water groupers	3.40	4.47	−7 (−11, −2)	6 (1, 13)	−9 (−16, −6)	−2 (−8, 2)	4 (3, 14)
Other deepwater groupers	1.77	2.02	−2 (−5, 12)	22 (28, 49)	7 (−4, 12)	−13 (−15, 0)	14 (5, 23)
Red Snapper	7.87	9.98	0 (−2, 7)	−4 (−3, 7)	−7 (−11, −3)	−19 (−22, −15)	19 (16, 27)
Vermilion Snapper	1.53	1.64	−2 (−9, −2)	−7 (−10, −3)	−5 (−15, −9)	−16 (−24, −18)	17 (12, 21)
Tilefish	0.94	0.89	4 (−3, 10)	74 (69, 89)	1 (−1, 13)	−1 (−6, 6)	0 (0, 14)
Greater Amberjack	1.09	1.17	−3 (−5, −1)	−2 (−4, 0)	−5 (−7, −2)	−7 (−9, −5)	8 (4, 9)
Gray Triggerfish	0.94	1.07	2 (−1, 4)	1 (−1, 3)	14 (13, 18)	−8 (−7, −2)	7 (8, 13)
Black Sea Bass	2.04	2.17	−20 (−22, −18)	−13 (−18, −13)	−1 (−4, 2)	0 (−3, 3)	4 (4, 11)
King Mackerel	8.79	10.76	−6 (−8, −4)	−3 (−5, −1)	−10 (−12, −8)	−17 (−20, −17)	16 (11, 15)
Spanish Mackerel	13.16	15.50	0 (−4, 0)	1 (−3, 2)	−3 (−6, −2)	−12 (−15, −11)	17 (14, 20)
Cobia	2.11	2.39	0 (−2, 4)	−2 (−3, 3)	1 (1, 7)	−1 (−3, 2)	2 (2, 9)
Sardines, herrings, and scads	289.13	262.68	−2 (−11, −2)	−6 (−11, −2)	−23 (−32, −23)	−18 (−25, −18)	16 (14, 25)
Dolphins	2.89	2.83	−2 (1, 13)	−2 (7, 19)	−6 (−12, 1)	−6 (−6, 6)	7 (10, 23)
Seabirds	2.04	2.09	−2 (16, 26)	−1 (14, 23)	−11 (−11, 1)	−7 (6, 14)	7 (26, 37)

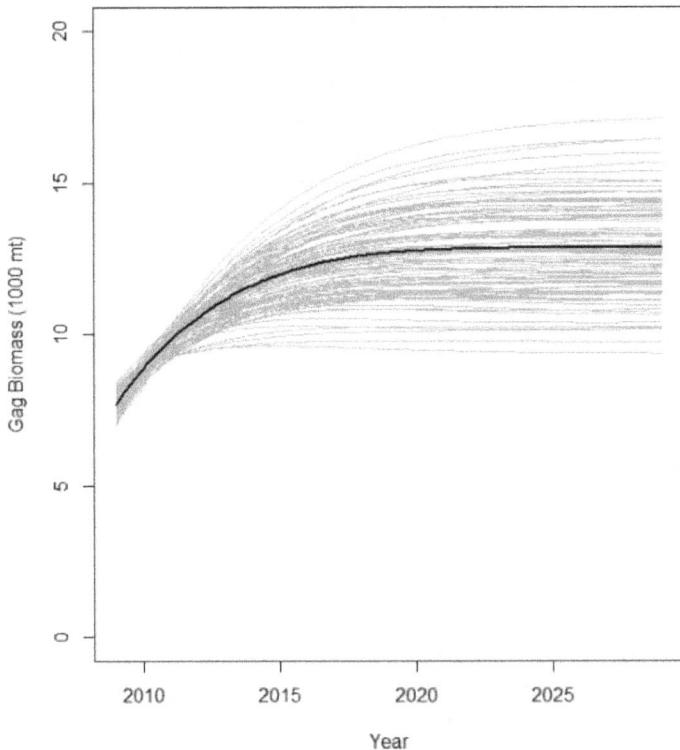

FIGURE 4. Gag projections under $F_{rebuild} = 0.16$, where the solid black line represents the base run and the gray lines the Monte Carlo simulation trials in which the biomasses of all species were randomly chosen from uniform distributions.

Ecosim only predicts the impacts due to trophic interactions and ignores any competition for habitat. There are opposing hypotheses about the role of competition for habitat in structuring reef fish communities (Sale and Williams 1982), and there is conflicting evidence for habitat limitation in the Gulf of Mexico (Bohnsack 1989; Grossman et al. 1997; Shipp and Bortone 2009). Assuming that Vermilion Snapper and Black Sea Bass do compete with other species for habitat and that they are at a disadvantage in those interactions, we would expect the impacts to be greater than predicted from trophic interactions alone. Other species such as Gag and Red Snapper, for which the model predicted little to no impact because of a small overlap in diet or low predation mortalities, could in fact be affected by competition for space, a process not accounted for by the Ecosim model.

Fish abundance trajectories were expected to vary among species but were similar to projections made in the stock assessments for most species (SEDAR stock assessments, available at www.sefsc.noaa.gov/sedar/). The status quo biomasses were similar in magnitude and direction of change to projections made with single-species models for Gag, Red Grouper, Red Snapper, Greater Amberjack, Yellowedge Grouper, Atlantic Goliath Grouper, and Vermillion Snapper. The similarity between the status quo forecasts made by Ecosim and those of the stock assessment models should not come as a surprise because many groups were calibrated using historical

data generated by the stock assessment and are therefore expected to have similar biomass dynamics. While this does not validate the model, it does facilitate direct comparison between predictions made by Ecosim and those of single-species models and allows us to characterize the environmental uncertainty not captured by the single-species models. Divergences between the predictions made by Ecosim and those of the single-species models (i.e., for Cobia and King Mackerel) could be due to incorrect parameter estimates (especially the vulnerability exchange and biomass accumulation rates) or the failure of the single-species models to capture some important environmental process.

The predicted responses of predators to forage fish depletion are consistent with—and perhaps slightly more conservative than—those made by other ecosystem models. When half of a predator's diet is composed of forage fish, ecosystem models tend to predict a 20–40% loss of predator biomass when forage fishes are reduced to between 80% and 40% of their virgin stock sizes (Pikitch et al. 2012). The model described here predicted predator biomass to decline by at most 15% when baitfish were harvested at the high rates of the 1980s. A diverse prey resource, such as that available in the Gulf of Mexico, would likely lessen the impact because predators can switch to prey that are more abundant and opportunistic prey species can replace niches left behind by those targeted in the fishery.

Changes in primary production were predicted to have rather large effects on the biomass of fish on the WFS. The magnitude of the impacts predicted in the low and high primary production scenarios is consistent with those predicted by a suite of Ecosim models from Australia (Brown et al. 2010). These models showed that consumer biomass is proportional to the primary production rate and that a 20% change in primary production would lead to a similar change in biomass. It is likely that primary production will be highly variable in the future, making the ecosystem response far less predictable. For example, an increase in phytoplankton biomass could lead to declines in submerged aquatic vegetation (Greening and Janicki 2006) or cause widespread hypoxic zones (Breitburg 2002; Diaz and Rosenberg 2008; Justic et al. 2002) that will have negative, nonlinear effects on marine organisms. Nevertheless, the simple linear simulations explored here offer some insight into how bottom-up processes impact the entire ecosystem and provide a framework for forecasting more detailed climate change scenarios.

There are several caveats and limitations associated with the EwE approach, which are described elsewhere (Christensen and Walters 2004). A few of the more important caveats to consider when interpreting such projections include the lack of spatial representation, the lack of a response in fishing effort to changing biomass, and the absence of management feedback as stocks recover (or decline). For instance, in our model effort reduction in the longline fishery was achieved through a combination of regulations, including spatial closures and depth restrictions. Prohibiting longlines in depths less than 35 fathoms, as is the rule in Amendment 31, will shift effort farther offshore and therefore not benefit the deepwater species (Yellowedge Grouper, Tilefish, and other DWG) nearly as much as predicted in the nonspatial Ecosim model. Preliminary simulations conducted using Ecospace, a spatially explicit component of EwE, predict biomass to decline for deepwater species under Amendment 31.

There are two basic types of uncertainty in ecosystem models, that associated with the data used to derive the input parameters and that associated with model structure. Structural uncertainty manifests itself in the definition of biomass pools and functional relationships and the choice of environmental drivers (Pauly et al. 2000). This type of uncertainty was addressed early in model development through a series of internal reviews, iterative improvements, and a workshop at which the model was reviewed by a group of scientists familiar with the WFS and reef fish species.

Regarding data, the two most critical sources of uncertainty are the diet compositions for large-bodied, offshore predators and the biomass of the forage species. Quality stomach contents from deepwater reef species are difficult to obtain due to barotrauma (which can lead to stomach eversion) and the inability to sample with active, nonbaited gear. Much of the data used to establish the diet compositions of adult reef fish were outdated and obtained using baited gear, or were only available from small samples. Sampling beyond that conducted in traditional fisheries-independent surveys is needed to more adequately describe the diet compositions of these ecologically and economically important predator species.

Baitfish are one of the most ecologically important groups in this system, yet there is considerable uncertainty about their biomass. Houde (1976) estimated the biomass of baitfish (Atlantic Thread Herring, Scaled Sardine, and Spanish Sardine) from egg and larval surveys in the eastern Gulf of Mexico to be nearly 1 million metric tons during the early 1970s. Recent estimates based on the FWC baitfish trawl and an acoustic survey conducted offshore of west-central Florida are 50,000 and 800,000 metric tons, respectively (Keith Fischer, FWC, personal communication). Nearly every commercially and recreationally important species utilizes the baitfish resource to some extent. Therefore, it is critical to gain a better understanding of their abundance, productivity, and contribution to the diets of predator species.

In general, the model does indicate that trophic impacts are potentially strong and can lead to ecological trade-offs that will trigger management actions, especially for species currently near a threshold or under a rebuilding plan. For instance, simultaneously rebuilding the stocks of multiple species with similar diets may require lower target catch limits or fishing mortality rates than those estimated using single-species models that do not account for the rebuilding of competing species. While ecosystem-based fisheries science and modeling has grown greatly over the last decade, agencies have had difficulty incorporating it into the management process. Even with the caveats and uncertainties, there is great utility in food web models that can quantify the changes in biomass and mortality arising from trophic interactions and environmental change. Ecosystem models such as ours are intended to complement single-species stock assessment and management. For example, they can generate vectors of time-varying natural mortality as an input to single-species models or be used to simultaneously evaluate the performance of several management options. Because large-scale, long-term ecological experiments are impractical, if not impossible, scientists will continue to rely on simulation models to predict the impacts of broad-scale management actions and environmental change on complex ecosystems. The work presented here demonstrates how the West Florida Shelf may respond to natural and anthropogenic perturbations, and we hope this effort will lead to a more focused and coherent strategy for EBFM throughout the Gulf of Mexico.

ACKNOWLEDGMENTS

This research was supported by Florida Sea Grant and the Florida Fish and Wildlife Conservation Commission with funds from Sport Fish Restoration. We also acknowledge Sherry Larkin, Bill Pine, Bill Lindberg, and the Florida FWC Stock Assessment group for their participation and critique at various stages of model development. Sergio Alvarez, Ed Camp, Jake Tetzlaff, the FWC Fisheries Independent Monitoring Program, and the National Marine Fisheries Service assisted by providing support and necessary data. Lastly, we thank all those who participated in the science and stakeholder workshops.

REFERENCES

Ahrens, R. N. M., C. J. Walters, and V. Christensen. 2012. Foraging arena theory. Fish and Fisheries 13:41–59.

Ainsworth, C. H., J. F. Samhouri, D. S. Busch, W. W. L. Cheung, J. Dunne, and T. A. Okey. 2011. Potential impacts of climate change on northeast Pacific marine food webs and fisheries. ICES Journal of Marine Science 68:1217–1229.

Andersen, K. H., and J. C. Rice. 2010. Direct and indirect community effects of rebuilding plans. ICES Journal of Marine Science 67:1980–1988.

Aulenbach, B. T., H. T. Buxton, W. T. Battaglin, and R. H. Coupe. 2007. Streamflow and nutrient fluxes of the Mississippi–Atchafalaya River basin and subbasins for the period of record through 2005. U.S. Geological Survey, Open-File Report 2007–1080, Reston, Virginia.

Aydin, K., S. Gaichas, I. Ortiz, D. Kinzey, and N. Friday. 2007. A comparison of the Bering Sea, Gulf of Alaska, and Aleutian Islands large marine ecosystems through food web modeling. NOAA Technical Memorandum NMFS-AFSC-178.

Behrenfeld, M. J., R. T. O'Malley, D. A. Siegel, C. R. McClain, J. L. Sarmiento, G. C. Feldman, A. J. Milligan, P. G. Falkowski, R. M. Letelier, and E. S. Boss. 2006. Climate-driven trends in contemporary ocean productivity. Nature 444:752–755.

Bohnsack, J. A. 1989. Are high densities of fishes at artificial reefs the result of habitat limitation or behavioral preference? Bulletin of Marine Science 44:631–645.

Boldt, J. L., T. W. Buckley, C. N. Rooper, and K. Aydin. 2012. Factors influencing cannibalism and abundance of Walleye Pollock (*Theragra chalcogramma*) on the eastern Bering Sea shelf, 1982–2006. U.S. National Marine Fisheries Service Fishery Bulletin 110:293–306.

Booth, S., and D. Zeller. 2005. Mercury, food webs, and marine mammals: implications of diet and climate change for human health. Environmental Health Perspectives 113:521–526.

Breitburg, D. 2002. Effects of hypoxia, and the balance between hypoxia and enrichment, on coastal fishes and fisheries. Estuaries 25:767–781.

Brown, C. J., E. A. Fulton, A. J. Hobday, R. J. Matear, H. P. Possingham, C. Bulman, V. Christensen, R. E. Forrest, P. C. Gehrke, N. A. Gribble, S. P. Griffiths, H. Lozano-Montes, J. M. Martin, S. Metcalf, T. A. Okey, R. Watson, and A. J. Richardson. 2010. Effects of climate-driven primary production change on marine food webs: implications for fisheries and conservation. Global Change Biology 16:1194–1212.

Carpenter, S. R., J. F. Kitchell, and J. R. Hodgson. 1985. Cascading trophic interactions and lake productivity. BioScience 35:634–639.

Castillo, C. E. D., P. Coble, R. Conmy, F. Muller-Karger, L. Vanderbloemen, and G. Vargo. 2001. Multispectral in situ measurements of organic matter and chlorophyll fluorescence in seawater: documenting the intrusion of the Mississippi River plume in the West Florida Shelf. Limnology and Oceanography 46:1836–1843.

Christensen, V., and C. Walters. 2004. Ecopath with Ecosim: methods, capabilities, and limitations. Ecological Modelling 172:109–139.

Clark, J. S., S. R. Carpenter, M. Barber, S. Collins, A. Dobson, J. Foley, D. M. Lodge, M. Pascual, R. Pielke Jr., W. Pizer, C. Pringle, W. Reid, K. A. Rose, O. Sala, W. H. Schlesinger, D. H. Wall, and D. Wear. 2001. Ecological forecasts: an emerging imperative. Science 293:657–659.

Collie, J. S., and A. K. DeLong. 1999. Multispecies interactions in the Georges Bank fish community: ecosystem approaches for fisheries management. Alaska Sea Grant College Program, AK-SG-99-01, Fairbanks.

Dance, M. A., W. F. Patterson III, and D. T. Addis. 2011. Fish community and trophic structure at artificial reef sites in the northeastern Gulf of Mexico. Bulletin of Marine Science 87:301–324.

Diaz, R. J., and R. Rosenberg. 2008. Spreading dead zones and consequences for marine ecosystems. Science 321:926–929.

Doney, S. C. 2006. Oceanography: plankton in a warmer world. Nature 444:695–696.

Forrestal, F., M. Coll, D. J. Die, and V. Christensen. 2012. Ecosystem effects of Bluefin Tuna, *Thunnus thynnus*, aquaculture in the NW Mediterranean Sea. Marine Ecology Progress Series 456:215–231.

Frank, K. T., B. Petrie, J. S. Choi, and W. C. Leggett. 2005. Trophic cascades in a formerly cod-dominated ecosystem. Science 308:1621–1623.

Gilbes, F., F. E. Muller-Karger, and C. E. Del Castillo. 2002. New evidence for the West Florida Shelf Plume. Continental Shelf Research 22:2479–2496.

Gilbes, F., C. Tomas, J. J. Walsh, and F. E. Muller-Karger. 1996. An episodic chlorophyll plume on the West Florida shelf. Continental Shelf Research 16:1201–1224.

GMFMC (Gulf of Mexico Fishery Management Council). 2008. Final Reef Fish Amendment 30B. National Marine Fisheries Service, NA05NMF4410003, Tampa, Florida.

GMFMC (Gulf of Mexico Fishery Management Council). 2009. Final Reef Fish Amendment 31. National Marine Fisheries Service, NA05NMF4410003, Tampa, Florida.

Goolsby, D. A., and W. A. Battaglin. 2000. Nitrogen in the Mississippi River basin: estimating sources and predicting flux to the Gulf of Mexico. U.S. Geological Survey, Fact Sheet 135–00, Reston, Virginia.

Greening, H., and A. Janicki. 2006. Toward reversal of eutrophic conditions in a subtropical estuary: water quality and seagrass response to nitrogen loading reductions in Tampa Bay, Florida, USA. Environmental Management 38:163–178.

Grossman, G. D., G. P. Jones, and W. J. Seaman. 1997. Do artificial reefs increase regional fish production? A review of existing data. Fisheries 22 (4):17–23.

Hartman, K. J. 2003. Population-level consumption by Atlantic coastal Striped Bass and the influence of population recovery upon prey communities. Fisheries Management and Ecology 10:281–288.

Heymans, J. J., U. R. Sumaila, and V. Christensen. 2009. Policy options for the northern Benguela ecosystem using a multispecies, multifleet ecosystem model. Progress in Oceanography 83:417–425.

Hollowed, A. B., N. Bax, R. Beamish, J. Collie, M. Fogarty, P. Livingston, J. Pope, and J. C. Rice. 2000. Are multispecies models an improvement on single-species models for measuring fishing impacts on marine ecosystems? ICES Journal of Marine Science 57:707–719.

Houde, E. D. 1976. Abundance and potential for fisheries development of some sardine-like fishes in the eastern Gulf of Mexico. Pages 73–83 *in* J. B. Higman, editor. Proceedings of the 28th Annual Session of the Gulf and Caribbean Fisheries Institute. Gulf and Caribbean Fisheries Institute, Miami.

Justic, D., N. N. Rabalais, and R. E. Turner. 2002. Modeling the impacts of decadal changes in riverine nutrient fluxes on coastal eutrophication near the Mississippi River delta. Ecological Modelling 152:33–46.

Karl, D., R. Letelier, L. Tupas, J. Dore, J. Christian, and D. Hebel. 1997. The role of nitrogen fixation in biogeochemical cycling in the subtropical North Pacific Ocean. Nature 388:533–538.

Libralato, S., and C. Solidoro. 2009. Bridging biogeochemical and food web models for an end-to-end representation of marine ecosystem dynamics: the Venice Lagoon case study. Ecological Modelling 220:2960–2971.

Lindberg, W. J., T. Frazer, K. Portier, F. Vose, J. Loftin, D. Murie, D. Mason, B. Nagy, and M. Hart. 2006. Density-dependent habitat selection and performance by a large mobile reef fish. Ecological Applications 16:731–746.

Link, J. S. 2010. Ecosystem-based fisheries management: confronting trade-offs. Cambridge University Press, New York.

Link, J. S., and F. P. Almeida. 2000. An overview and history of the food web dynamics program of the Northeast Fisheries Science Center, Woods Hole, Massachusetts. NOAA Technical Memoradum NMFS-NE-159.

Link, J. S., E. A. Fulton, and R. J. Gamble. 2010. The northeast US application of ATLANTIS: a full system model exploring marine ecosystem dynamics in a living marine resource management context. Progress in Oceanography 87:214–234.

NMFS (National Marine Fisheries Service). 2005. Endangered Species Act: Section 7 consultation on the continued authorization of reef fish fishing under the Gulf of Mexico Reef Fish Fishery Management Plan and Proposed Amendment 23. NMFS, Southeast Region Biological Opinion (February 15), St. Petersburg, Florida.

NMFS (National Marine Fisheries Service). 2008. Estimated takes of sea turtles in the bottom longline portion of the Gulf of Mexico reef fish fishery July 2006 through 2007 based on observer data. NMFS, Southeast Fisheries Science Center Contribution PRD-07/08-15, Miami.

NMFS (National Marine Fisheries Service). 2011a. Annual report to Congress on the status of U.S. fisheries: 2010. National Marine Fisheries Service, Silver Spring, Maryland.

NMFS (National Marine Fisheries Service). 2011b. U.S. national bycatch report. NOAA Technical Memorandum NMFS-F/SPO-117E.

Okey, T. A., and B. Mahmoudi. 2002. An ecosystem model of the West Florida Shelf for use in fisheries management and ecological research, volume II. Model construction. Florida Marine Research Publications 163.

Okey, T. A., G. A. Vargo, S. Mackinson, M. Vasconcellos, B. Mahmoudi, and C. A. Meyer. 2004. Simulating community effects of sea floor shading by plankton blooms over the West Florida Shelf. Ecological Modelling 172:339–359.

Overholtz, W. J., and A. V. Tyler. 1986. An exploratory simulation model of competition and predation in a demersal fish assemblage on Georges Bank. Transactions of the American Fisheries Society 115:805–817.

Palomares, M. L. D., and D. Pauly. 1998. Predicting food consumption of fish populations as functions of mortality, food type, morphometrics, temperature, and salinity. Marine and Freshwater Research 49:447–453.

Passerotti, M. S., J. K. Carlson, and S. J. B. Gulak. 2010. Catch and bycatch in U.S. southeast gill-net fisheries, 2009. NOAA Technical Memorandum NMFS-SEFSC-600.

Pauly, D. 1980. On the interrelationship between natural mortality, growth parameters, and mean environmental temperature in 175 fish stocks. ICES Journal of Marine Science 39:175–192.

Pauly, D. 1989. Food consumption by tropical and temperate fish populations: some generalizations. Journal of Fish Biology 35:11–20.

Pauly, D., V. Christensen, and C. Walters. 2000. Ecopath, Ecosim, and Ecospace as tools for evaluating ecosystem impact of fisheries. ICES Journal of Marine Science 57:1–10.

Pianka, E. R. 1974. Evolutionary ecology. Harper and Row, New York.

Pierce, D. J., J. E. Wallin, and B. Mahmoudi. 1998. Spatial and temporal variations in the species composition of bycatch collected during a Striped Mullet (*Mugil cephalus*) survey. Gulf of Mexico Science 16:15–27.

Pikitch, E. K., P. D. Boersma, I. L. Boyd, D. O. Conover, P. Cury, T. Essington, S. S. Heppell, E. D. Houde, M. Mangel, D. Pauly, E. Plagányi, K. Sainsbury, and R. S. Steneck. 2012. Little fish, big impact: managing a crucial link in ocean food webs. Lenfest Ocean Program, Washington, D.C.

Pinnegar, J. K., M. T. Tomczak, and J. S. Link. 2014. How to determine the likely indirect food web consequences of a newly introduced nonnative species: a worked example. Ecological Modelling 272:379–387.

Plaganyi, E. E., and D. S. Butterworth. 2004. A critical look at the potential of Ecopath with Ecosim to assist in practical fisheries management. African Journal of Marine Science 26:261–287.

Ralston, S. 1987. Mortality rates of snappers and groupers. Pages 375–404 *in* J. J. Polovina and S. Ralston, editors. Tropical snappers and groupers: biology and fisheries management. Westview Press, Boulder, Colorado.

Rooney, N., K. McCann, G. Gellner, and J. C. Moore. 2006. Structural asymmetry and the stability of diverse food webs. Nature 442:265–269.

Sale, P. F., and D. M. Williams. 1982. Community structure of coral reef fishes: are the patterns more than those expected by chance? American Naturalist 120:121–127.

SEDAR (Southeast Data Assessment and Review). 2009a. SEDAR 12 update: Gulf of Mexico Red Grouper stock assessment report. SEDAR, North Charleston, South Carolina.

SEDAR (Southeast Data Assessment and Review). 2009b. SEDAR 10 update: Gulf of Mexico Gag stock assessment report. SEDAR, North Charleston, South Carolina.

SEDAR (Southeast Data Assessment and Review). 2011. SEDAR 27 Gulf Menhaden stock assessment report. SEDAR, North Charleston, South Carolina.

Shipp, R. L., and S. A. Bortone. 2009. A perspective of the importance of artificial habitat on the management of Red Snapper in the Gulf of Mexico. Reviews in Fisheries Science 17:41–47.

Smith, G. B. 1979. Relationship of eastern Gulf of Mexico reef–fish communities to the species equilibrium theory of insular biogeography. Journal of Biogeography 6:49–61.

Smith, A. D. M., C. J. Brown, C. M. Bulman, E. A. Fulton, P. Johnson, I. C. Kaplan, H. Lozano-Montes, S. Mackinson, M. Marzloff, L. J. Shannon, Y. Shin, and J. Tam. 2011. Impacts of fishing low–trophic level species on marine ecosystems. Science 333:1147–1150.

Stallings, C. D., and A. L. Dingeldein. 2012. Intraspecific cooperation facilitates synergistic predation. Bulletin of Marine Science 88:317–318.

Steneck, R. S. 2012. Apex predators and trophic cascades in large marine ecosystems: learning from serendipity. Proceedings of the National Academy of Sciences of the USA 109:7953–7954.

Valette-Silver, N. J., and D. Scavia. 2003. Ecological forecasting: new tools for coastal and ecosystem management. NOAA Technical Memorandum NOS NCCOS 1.

Walters, C., V. Christensen, S. J. D. Martell, and J. F. Kitchell. 2005. Possible ecosystem impacts of applying MSY policies from single-species assessment. ICES Journal of Marine Science 62:558–568.

Walters, C., S. J. D. Martell, V. Christensen, and B. Mahmoudi. 2008. An Ecosim model for exploring Gulf of Mexico ecosystem management options: implications of including multistanza life history models for policy predictions. Bulletin of Marine Science 83:251–271.

Walters, C. J., and S. J. D. Martell. 2004. Fisheries ecology and management. Princeton University Press, Princeton, New Jersey.

Wetz, M. S., E. A. Hutchinson, R. S. Lunetta, H. W. Paerl, and J. C. Taylor. 2011. Severe droughts reduce estuarine primary productivity with cascading effects on higher trophic levels. Limnology and Oceanography 56:627–638.

Wiseman, W. J. Jr., N. N. Rabalais, M. J. Dagg, and T. E. Whitledge, editors. 1999. Nutrient-enhanced coastal ocean productivity in the northern Gulf of Mexico: understanding the effects of nutrients on a coastal ecosystem. National Oceanic and Atmospheric Administration, Coastal Ocean Program, Decision Analysis Series 14, Silver Spring, Maryland.

9

The Impact of the Second Seasonal Spawn on the Nantucket Population of the Northern Bay Scallop

Valerie A. Hall*

Maria Mitchell Association, Department of Natural Sciences, 4 Vestal Street, Nantucket, Massachusetts 02554, USA; and Department of Fisheries Oceanography, School for Marine Science and Technology, University of Massachusetts Dartmouth, 200 Mill Street Suite 325, Fairhaven, Massachusetts 02719, USA

Chang Liu and Steven X. Cadrin

Department of Fisheries Oceanography, School for Marine Science and Technology, University of Massachusetts Dartmouth, 200 Mill Street Suite 325, Fairhaven, Massachusetts 02719

Abstract

Nantucket, Massachusetts, has one of the last remaining commercial fisheries of the bay scallop *Argopecten irradians*, which is based largely on natural recruitment. Though previously thought to spawn only once in early summer at age 1, individuals of the northern subspecies often spawn again in late summer or fall, and recruits from this second spawning can survive to reproduce again in their second summer. We formulated an age-based Leslie matrix model and estimated population growth rate with and without a second spawn based on data from 5 years of life history research. Elasticity analysis revealed that the population growth rate was most sensitive to juvenile survival, the major factor in recruitment rate, and year-1 adult fertility was a close second. We varied those two rates randomly in a stochastic matrix model, which represented the effect of environmental fluctuations on population growth. A life history modeled with a second spawn had a negligible effect on the deterministic population growth rate under constant conditions, but under variable conditions the second spawn increased the mean of the stochastic growth rates up to 58.3% over that of a single early-spawning life history. These results suggest that the second spawn is a successful bet-hedging strategy. The northern bay scallop increases its chances for successful recruitment in a variable environment by spreading reproductive effort over more than one period in a season. This strategy appears to have sustained the Nantucket scallop population in spite of severe annual fluctuations and the eventual collapse seen in other locations.

The bay scallop *Argopecten irradians*, a short-lived bivalve mollusk, has three subspecies that occupy estuarine and coastal environments along the U.S. Atlantic and Gulf of Mexico coasts, usually in or near beds of submerged aquatic plants such as eelgrass *Zostera marina* (Brand 2006). The northern subspecies, the northern bay scallop *A. irradians irradians*, extends from Massachusetts to New Jersey, the southern subspecies, *A. irradians concentricus*, is found intermittently from North Carolina (under dispute) and again along the west coast of Florida to Louisiana, and the third subspecies,

Subject editor: Kenneth Rose, Louisiana State University, Baton Rouge

*Corresponding author: scalloplady@gmail.com

A. irradians amplicostatus, exists sparsely in the Gulf of Mexico along the coasts of Texas and Mexico (Clarke 1965). Partially because of the bay scallop's short lifespan, population size and commercial landings have had extreme yearly fluctuations, and their vulnerability to environmental and anthropomorphic change has led to a general decline throughout the species' range (MacFarlane 1991; MacKenzie 2008; Tettelbach and Smith 2009). Nantucket, Massachusetts, is one of the few remaining locations where a wild fishery of bay scallops persists, but Nantucket landings from 1966 to 2010 showed similar decreasing trends.

In addition to an early spawn in late May through early July, a second seasonal spawn occurs frequently in the Nantucket Island (Massachusetts) northern bay scallop population. The early spawn results in a "classic" juvenile scallop that reaches a shell height of 40–55 mm before overwintering. It then forms a growth ring at that position in the spring, spawns twice during the next summer at year 1, and enters the fishery in the fall. Classic scallops will almost always die before entering another spawning season. The second spawn in late summer or fall produces a scallop known to islanders as a "nub," with its growth ring close to the hinge. Juvenile scallops produced by the second spawn reach shell heights averaging 10 mm during their first autumn before growth ceases for the winter (Tettelbach et al. 2001). Growth then resumes in the spring and is marked by the formation of the first (nub) growth ring ≤ 10 mm from the umbo. This nub ring is the only growth ring found in late-spawned scallops until they form another one after surviving their second winter. While less than 40% of nub scallops spawn in their first summer, at least 50% are able to survive and spawn during their second summer before dying late in the season (Hall 2014). Nubs in Nantucket scallop population are subject to harvest in their second winter if they exceed 63 mm in shell height; however, they rarely survive into a third commercial season.

The relative frequency and existence of late-spawned northern bay scallops appear to vary greatly. Belding (1910) estimated that 10–20% of Massachusetts bay scallops survive to spawn a second year, suggesting that those living longer were born late in the first spawning season. Kelley (1981) reported adult Nantucket bay scallops spawning into the month of September. MacFarlane (1991) found a 9% occurrence of "ring-at-hinge" scallops in Pleasant Bay, Massachusetts, during fall 1979. Taylor and Capuzzo (1983) inferred fall spawning in bay scallops in Waquoit Bay, Cape Cod, Massachusetts, as did Tettelbach (1991) in Groton, Connecticut. Juvenile scallops < 20 mm averaged 2.59% of the populations in some parts of the Peconic Bays, Long Island, New York, during the winter of 1990–1991 after a nonbrown tide year (Tettelbach et al. 2001), but by October 1992, after a brown tide bloom, adults with growth rings 2–7 mm from the hinge approached 100% in one bay (Tettelbach et al. 1999). These may have been the result of the fall spawn of 1991. Bologna et al. (2001) observed settling juveniles of less than 15 mm shell height in October of 1998 and 1999, which strongly suggests there is a fall spawn in New Jersey scallops.

Leslie (1945) first introduced matrix population models to ecologists. While scalar population models rely only on counts or estimates of total population size, matrix models can describe populations in terms of ages, size-classes, or stages of development (Caswell 2001). They are now readily implementable to computer programs such as MATLAB. The matrix elements, or vital rates, of Leslie's model are derived from a life table of an organism and a table of age-specific fertility rates, and consist of fertility rates across the top row and survival probabilities from one age to another along the principle subdiagonal. The population growth rate (λ) can be determined from the principle eigenvalue of the Leslie matrix. The λ term is related to the more familiar r (intrinsic rate of natural increase) by the equations $\lambda = e^r$ and $r = \ln\lambda$. The right and left eigenvectors of the Leslie matrix represent the stable age distribution and the reproductive value of each age-class, respectively. An age-specific Leslie matrix can be multiplied by a column vector of the initial age distribution of the population to achieve a projection of future population structure. (Caswell 2001).

While Leslie (1945) constructed his matrices based upon age-classes, Leftovitch (1965) realized that some species could not easily be divided into age-groups. Rather, stages are often more recognizable. He added the vital rate of growth of one stage to another to the vital rates of fertility and survival. Two studies that used stages are Brault and Caswell (1993) and Smith et al. (2005). The former researchers designed a matrix model with four stages (yearlings, juveniles, reproductive adults, postreproductive adults) to study the demography of killer whales *Orcinus orca*. The latter evaluated variability in flood and precipitation as they affected the demography of the threatened floodplain plant, *Boltonia decurrens*, along the Illinois river, using seeds, seedlings, and various reproductive stages. Many others have applied Leftovitch's model to sizes rather than stages. Werner and Caswell (1977) found that size was a better predictor of population dynamics than age in the teasel *Dipsacus sylvestris*. Doak et al. (1994) established size-classes of the desert tortoise *Gopherus agassizii* in the Mojave Desert to evaluate various proposed management regimes for this threatened species. Although Nakaoka (1997) had previously used an age-based matrix model to study the demography of the infaunal clam, *Yoldia notabilis*, in Japan, he constructed a size-classified stochastic matrix model to assess the effect of fluctuating recruitment rates on the population. He did so because survivorship and reproduction in this species is more dependent upon size than age; it is easier to incorporate annual growth fluctuations into the model, and older age-classes with smaller sample sizes can be combined into one large size-class.

"Multistate" matrix models, either age- or size-classified, can be used to classify individuals and to follow their transitions either by geographic regions or by demographic group (Caswell 2001). They have often been applied to metapopulation models, where subpopulations are connected through larval dispersal, diffusion, or migration, or multiregional models when applied to human demography. For example, Wootton and Bell (1992) analyzed the response of the endangered California peregrine falcon *Falco peregrines* population to different management strategies by developing a model with two subpopulations linked only by migration, Barbeau and Caswell (1999) applied a multistate model to evaluate the effectiveness of seeding strategies of juvenile sea scallops *Placopecten magellanicus*, and Strasser (2008) applied the model to a theoretical two-patch metapopulation of the soft-shell clam *Mya arenaria* to evaluate the relative importance of "sources" and "sinks" in fishery management. When the models are used to combine multiple demographic classifications, they are specifically called multistate or multidimensional models, and the same mathematical principles apply to both (Caswell 2001).

Matrix population models were first applied to a marine mollusk, *M. arenaria*, in Ipswich Bay, Cape Ann, Massachusetts (Brousseau 1978a, 1978b; Brousseau et al. 1982; Brousseau and Baglivo 1984). After determining the clam's spawning cycle and gathering size-specific fecundity and mortality data, an age-based matrix model was formulated. Survival probability was divided into two factors: settlement rate (r_s) and survival from settlement to year 1 (b_1). Population growth rate in that study was most sensitive to changes in settlement rate, which is mainly composed of survival. The second most important influence on population growth rate was survival to age 1. Ripley (1998) investigated another population of *M. arenaria* in relatively pristine Barnstable Harbor, Cape Cod, Massachusetts, and compared it with one growing in the contaminated sediments of Boston Harbor. Ripley (1998) found that variability in settlement rate in this long-live species, due to the occurrence of occasional years with exceptionally high recruitment, was responsible for the persistence of the population. This conclusion was confirmed by Ripley and Caswell (2006) using stochasticity analysis of clams in Barnstable Harbor. Sporadic recruitment success in a species with broadcasting larvae becomes more important as environmental variability increases.

Other population studies of marine mollusks that have successfully employed matrix models are those of Malinowski and Whitlatch (1988), Weinberg (1989), Noda and Nakao (1996), and Barbeau and Caswell (1999). Malinowski and Whitlatch (1988) used matrix models at different larval settlement rates to analyze life history tactics of *M. arenaria*, the oyster, *C. virginica*, and the hard clam, *Mercenaria mercenaria*, then assessed the relative benefit of different management strategies on various stages of their life cycles. Since population growth rate was most sensitive to the survival of larval and juvenile stages (mainly due to predation), Malinowski and Whitlatch (1988) concluded that the best

management practices were those that increased the survival of those stages. Weinberg (1989) used cohort-specific growth rates for a population of infaunal clams, *Gemma gemma*, in Little Narragansett Bay, Rhode Island, to make annual forecasts of population size from an age-based matrix model. Weinberg (1989) found that the model was realistic enough to describe the demography of this population, particularly because the species has little migration and no dispersing larval stage. Noda and Nakao (1996) reported the first documented case of recruitment limitation in a whole population of marine benthic animals using the Japanese snail, *Umbonium costatum*. Those investigators varied recruitment rate only in eight independent Leslie matrices that described the transition in age from age 0 to over 6 years, combining them to simulate long-term population dynamics. Barbeau and Caswell (1999) divided sea scallops into 10 size-classes, each with a range of sizes from juveniles to harvestable adults, to elucidate the best strategy by which to seed juveniles. They compared the contributions of predation and dispersal to the growth and survival of the juveniles, which had been seeded in four locations along the coast of Nova Scotia, and identified possible management strategies. Barbeau and Caswell (1999) concluded that reductions in predator densities would have the most dramatic effect on scallop survival, especially if the size of seeded scallops was increased.

In this study we addressed the effects of environmental stochasticity on the bay scallop, with particular interest in the role that its short life span plays in its population dynamics. Some studies have found that shorter life histories are most often affected by environmental stochasticity (Benton and Grant 1996; Tettelbach and Smith 2009). However, such long-lived mollusk species as the gastropod, *U. costatum*, (Noda and Nakao 1996), the protobranch bivalve, *Y. notabilis*, (Nakaoka 1996), and the sea scallop (Barbeau and Caswell 1999) are also vulnerable to environmental variation, especially in certain life history stages. Random environmental fluctuations alter the vital rates in stochastic matrix models compared with the constant rates in deterministic models. Information attained from a deterministic matrix model (growth rate λ, a measure of fitness, reproductive values, stable age distribution, elasticity, and future projections) assumes a constant environment, but this assumption is often unrealistic for natural populations (Gourley and Lawrence 1977). Orzack and Tuljapurkar (1989) examined 25 life histories with differing patterns of variability and found a lognormal distribution of random environmental variation. They assumed that variability in vital rates depends upon the way in which organisms experience these changes.

Most mollusk species have life spans longer than that of the bay scallop's 2 years: *Arctica islandica* (up to 400 years); *Panopea abrupta* (up to 200 years); *M. mercenaria* (up to 40 years); *Placopecten magellanicus*, *C. virginica*, and *Y. notabilis* (up to 20 years); *Mya arenaria* (10–12 years); *C. islandica* and *U. costatum* (17 years); *G.*

gemma (8 years) (Abele et al. 2009). Constructing and analyzing a matrix population model for such a short-lived species as the bay scallop is a challenging task that has not been undertaken, except for an initial attempt by the U.S. Environmental Protection Agency (Hinchey et al. 2004). The further addition of the demographic effects of two different bay scallop life history patterns, classic (spawned early in season) and nub (spawned late in season), has never been attempted.

We constructed a multistate, age-based, Leslie matrix population model to investigate the contribution of late summer and fall spawning to population dynamics of the bay scallop. Projection matrices were constructed with and without the contribution of nubs, and then using initial conditions for both good and poor nub years (based on population estimates from September surveys conducted from 2006 to 2010). Population growth rates, stable age distributions, and reproductive values were determined from each projection. Total population estimates for 10 years at 1-year intervals were compared between life histories with and without a second seasonal spawn. Elasticity analysis revealed that population growth rate was sensitive to changes in vital rates in the projections, and stochasticity analysis modeled the effects of a randomly changing estuarine environment on bay scallop population dynamics. If our analysis shows that spreading reproduction out to include a second seasonal spawn does enhance bay scallop population growth, our matrix population models can be used as powerful tools to assist in the management of this critically important shellfish resource.

METHODS

Model formation and analysis.—We used a Leslie age-based matrix model with the addition of multistate and stochastic analyses. The six stages used in the bay scallop model were: (1) classic (early spawned) newly settled juvenile, (2) year-1 classic adult, (3) year-2 classic adult, (4) nub (late spawned) newly settled juvenile, (5) year-1 nub adult, and (6) year-2 nub adult.

In our multistate matrix model, reproductive products did not move physically from one region to another, but from one life history strategy to another. Our multistate matrix (**M**) consisted of two "regions," which represented classic (early spawning only) and nub life histories (late spawning only), with four submatrices: (**A₁**) simple classic life history giving rise only to classics (thus remaining in that region), (**A₂**) simple nub life history giving rise only to nubs (thus remaining in that region), (**M₁→₂**) year-1 classic late spawning that gives rise to nubs, and (**M₂→₁**) year-2 nub early spawning that gives rise to classics (matrix, Figure 1a; life cycle graph, Figure 2). We compared a simple age-classified Leslie matrix model for classic bay scallops with a multistate one, basing its construction on that of Barbeau and Caswell (1999) for sea scallops in Lunenburg Bay, Nova Scotia. Initial deterministic Leslie

a.

	1	2	3	4	5	6	
A₁	0	F_{2E}	F_{3E}	0	0	F_{6E}	$M_{2 \to 1}$ Nub contrib
Basic Classic	P_1	0	0	0	0	0	
	0	P_2	0	0	0	0	
M₁→₂	0	F_{2L}	0	0	F_{5L}	F_{6L}	A_2
Classic contrib.	0	0	0	P_4	0	0	**Basic Nub**
	0	0	0	0	P_5	0	

b.

	1	2	3	4	5	6	
	0	7.2	0.6	0	0	0.9	Nub contrib
Basic Classic	0.4	0	0	0	0	0	
	0	0.02	0	0	0	0	
	0	7.2	0	0	2.8	3.2	
Classic contrib.	0	0	0	0.4	0	0	**Basic Nub**
	0	0	0	0	0.13	0	

FIGURE 1. (**a**) Generalized multistate population matrix (**M**) of the Nantucket bay scallop population. **A₁** is the basic classic life history matrix; F_{2E} and F_{3E} are fertilities of stages 2 and 3 during the early spawn only. **A₂** is basic nub life history matrix; F_{5L} and F_{6L} are fertilities of stages 5 and 6 during the late spawn only. **M₁→₂** is the contribution of classics to the nub life history; F_{2L} is the fertility of stage 2 during the late spawn only. **M₂→₁** is the contribution of nubs to the classic life history; F_{6E} is the fertility of stage 6 during the early spawn only. (**b**) Multistate matrix with proposed values for vital rates substituted into the matrix described in (a).

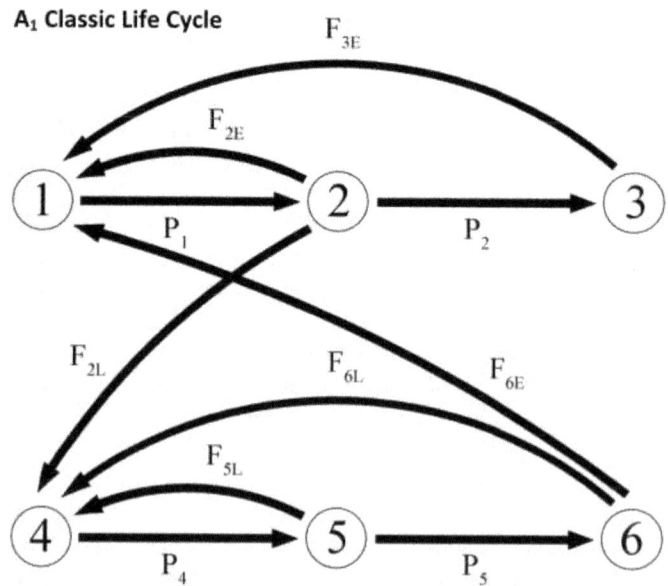

FIGURE 2. Life cycle graph for the Nantucket Harbor bay scallop population. **A₁** is the classic life cycle; stages 1, 2, and 3 are year-0, year-1, and year-2 classics, respectively. **A₂** is the nub life cycle; stages 4, 5, and 6 are year-0, year-1, and year-2 nubs, respectively. Crossovers between life cycles represent contributions of one life cycle to the other: F_{2L} is the second (late) spawn of year-1 classics; F_{6E} is the early spawn of year-2 nubs.

TABLE 1. Vital rates chosen for the matrix population model for the northern bay scallop. Fecundity is number of eggs per individual per spawning event. Fertilization rate, larval survival, and settlement rate, expressed as percent, are multiplied with fecundity to yield Fertility (F) for the top row of the matrix. Probability (P) of juvenile and adult survival is expressed as percent for the major subdiagonal of the matrix.

Subcategory	Life stage	Early versus late	Fecundity	F	P (%)
Fecundity	1 (classic juvenile)	Neither	0	0	40
	2 (year-1 classic)	Both same	2.0×10^6	7.2	2
	3 (year-2 classic)	Early only	1.8×10^5	0.6	0
	4 (nub juvenile)	Neither	0	0	40
	5 (year-1 nub)	Late only	7.8×10^5	2.8[a]	13
	6 (year-2 nub)	Early	2.4×10^5	0.9[b]	0
		Late	8.8×10^5	3.2[c]	
Fertilization	All	Both	10.3%		
Larval survival	All	Both	0.35%		
Settlement	All	Both	1%		

[a] Based on 39% observed spawning rate.
[b] Based on 6% observed spawning rate.
[c] Based on 44% observed spawning rate.

matrices for early spawned classics only (submatrix $\mathbf{A_1}$), late-spawned nubs only (submatrix $\mathbf{A_2}$), and the multistate two-spawn model ($\mathbf{M}$), not taking into account environmental variation or uncertainty in assumed parameters, were used to estimate eigenvalues. The principal eigenvalue, i.e. the largest real positive root, represents the population growth rate (λ) (Caswell 2001). The right eigenvector (w_i), indicates stable age distribution at time $t + 1$, and the left eigenviector (v^*_i), indicates reproductive value of each age-class.

Vital rates.—Values chosen for the vital rates of fertility (F) and probability of survival (P) in our model are described below, summarized in Table 1, and diagrammed in Figure 1b.

To calculate F for the model, we multiplied our estimated values for fecundity, fertilization rate, larval survival, and settlement rate and applied them to each age-class of early spawned classics and late-spawned nubs (juvenile, year-1, and year-2) (Table 1).

The first stage in our bay scallop model (age-0) began with recently settled spat (juveniles), because this was the first stage when scallops could be quantified. Fertility thus involved several component processes spanning from initial spawning through larval settlement. Because of the difficulty in assessing bivalves at an earlier time, matrix models often begin with this stage (Ripley 1998; Ripley and Caswell 2006).

Fecundity is defined as the number of eggs produced by one individual per spawning event (Llodra 2002; Barber and Blake 2006). Values in the literature for the bay scallop range from 2×10^6 (Belding 1910) to 23.7×10^6 eggs (Bricelj et al. 1987) per spawning. We chose the more conservative rate of 2×10^6 (Belding 1910) for year-1 classics because it is still considered a valid indicator of egg production in the wild (Tettelbach et al. 2011). There was no difference in fecundity during the first and second spawning periods of year-1 classics during 2010, the only year tested (ANOVA: $P = 0.64$, $F =$ 0.2167, $n = 40$) (Hall 2014), so we used the same value for both. We assumed that 100% of year-1 classics released eggs during each period. Only 39% of year-1 nubs released eggs and then only in the late spawning period (Hall 2014), so we calculated their fecundity for the model as zero for the early spawn and 7.8×10^5 (39% of classic fecundity) for the late spawn. Bricelj et al. (1987) calculated the average fecundity of a year-2 scallop as 8.6×10^6. However, we observed that only 9% (SD, ±4.3%) of ovaries of year-2 classics examined histologically showed any spawning and only in the early spawn (Hall 2014), so we calculated their fecundity as 1.8×10^5 (9% of classic fecundity) in the early spawn and zero in the late. Year-2 nubs were at least twice as fecund as year-1 classics during the early spawn, but only 4% spawned in the early period (thus, a fecundity of 4% of 4×10^6, or 2.4×10^5), but equal in fecundity to classics during their second spawn, though only 44% of them spawned (thus, 44% of 2×10^6, or 8.8×10^5).

The mean percentage of eggs fertilized for our model was based upon the method of Tettelbach et al. (2013), who determined the regression equation from data provided by Lundquist and Botsford (2004) Figure 7:

$$y = 2.6667x^3 - 10.28x^2 + 15.905x + 0.0571, \quad (1)$$

where y = mean percent of eggs fertilized and x = density of spawners per square meter.

We calculated the density of spawners by using the population density of the survey year with the largest percent of adult scallops (0.89 scallops/m² in 2007: Hall 2014) and the mean survival of adult classic and nub scallops from two unpublished caged studies (54%). The resulting spawning density was 1.65 scallops/m², determined by dividing the mean density of adults from the 2007 September survey (0.89 scallops/m^2) by the mean survival through the summer spawning period (0.54).

In broadcast spawners such as bay scallops, fertilization success (percent of eggs released that actually become fertilized) also depends upon tidal current flow. Pennington (1985) showed that percent fertilization in green sea urchins *Strongylocentrotus droebachiensis* was lower in fast currents (>0.2 m/s) than in slow ones (<0.2 m/s). Lundquist and Botsford (2004) used that study to develop two simulated negative exponential sperm distributions for their model: a broad exponential distribution for slow current speeds and a narrow exponential distribution for high current speeds (see their Figure 7). Using Lundquist and Botsford's (2004) model, Tettelbach et al. (2013) employed the broad exponential sperm distribution based on current speeds of <0.2 m/s recorded in the Peconic Bays to determine percent of eggs fertilized (i.e., fertilization success). We used the narrow exponential sperm distribution after determining that mean current speeds in Nantucket Harbor (0.43 m/s) exceeded 0.2 m/s (P. Boyce, Maria Mitchell Association, unpublished data). We also assumed a slightly aggregated spawner distribution as did Tettelbach et al. (2013). Using a spawner density of 1.65 scallops/m^2 for Nantucket Harbor, we calculated a mean fertilization rate of 10.29% for Nantucket Harbor (equation 1).

Larval survival has not been directly estimated for bay scallops, but is thought to be very low. Vance (1973) predicted that any invertebrate with a pelagic larval stage will experience survival rates below 1% during that time. Previous studies found that the average for four scallop species was only 0.54% (Vahl 1981; LePennec et al. 1998; Martinez et al. 2007; Soria et al. 2010) and that for seven other bivalve species was 0.26% (Strathmann 1985; Hines 1986; Malinowski and Whitlatch 1988; Rumrill 1990). We chose a larval survival rate of 0.35% for the model because that was the rate calculated by Soria et al. (2010) for the rock scallop *Spondylus calcifer*, whose 14-day larval period is similar to that of the bay scallop. We chose a settlement rate of 1%, which is an estimated maximum for marine invertebrates with planktotrophic larvae to survive metamorphosis and settlement (Thorson 1966).

Probability of survival (P) was calculated for two postsettlement stages: (1) juvenile to age 1, and (2) age 1 to age 2. Survival of nubs and classics was based on both the literature and on results of earlier unpublished studies.

Juvenile survival varies widely depending upon whether or not the scallops are caged. Caged juvenile classics and nubs in Nantucket Harbor had a similar survival of ~75% from late fall (early postsettlement period) through late spring (prespawning period) (Hall 2014). In order to obtain an estimate of survival to be used in the population model, we compared those results with ones found in the literature. Wild bay scallop juvenile survival ranges from 25% (Irlandi et al. 1995) to 70% (Tettelbach 1990) for classics and 16.5% for nubs (Tettelbach

et al. 2001). The latter figure represents an average of two values reported from two locations in the Peconic Bays after the winter of 1990–1991. Survival values of other juvenile bivalves range from 1% in *Mya arenaria* (Ayers 1956) to 50% in *Mercenaria mercenaria* (Zarnoch and Schreibman 2008). Brousseau et al. (1982) surveyed the literature for 11 species and found that the average juvenile survival in the wild was 17%. We assessed the wild, unfished population of juvenile scallops, so we assumed 40% survival for classics, an average of the wild bivalve data mentioned above. Although Tettelbach et al. (2001) observed 16.5% mortality for nubs, we assumed, based on our caged studies, that their survival was equal to that of classic juveniles (40%). We also assumed no fishing mortality in juveniles since no fishing occurs until after spawning at year 1.

Adult survival from year 1 to year 2 was assessed in both caged and wild scallops in Nantucket (Hall 2014), with caged nub adults exceeding survivorship of classic adults (39% versus 6%) over 2 years. Based upon our survey results, mean survival of wild scallops from age 1 to age 2 in Nantucket Harbor was 2% for classics and 13% for nubs (Hall 2014), the difference presumably was due to fishing pressure on classics. Extremely low salinities and burial in sediments in winter can further reduce wild scallop survival at times (Tettelbach et al. 1985; Tettelbach 1990). For the population model, the probability of year-1 classics surviving to year 2 was assumed to be 2%, while the same probability for nubs was assumed to be 13%, both based only on our field data. It was not possible to separate fishing and natural mortality in making these assumptions.

Population projections were first made from a simple 3 × 3 Leslie matrix of early spawned classic scallops only (submatrix $\mathbf{A_1}$) then from the 6 × 6 multistate matrix ($\mathbf{M}$), adding the effect of late spawning. For comparison, matrix $\mathbf{A_2}$ (late spawning only) was also analyzed. Initial abundance vectors shown in Table 2 for classics only, nubs only, and multistate with both contributions were multiplied by their corresponding matrices to produce population projections 10 years into the future at 1-year intervals, using the equation:

$$n_{(t+1)} = \mathbf{A}n_t, \qquad (2)$$

where $\mathbf{A}$ is the projection matrix and n_t is the initial population vector.

Initial vectors were based on estimations for each age-class in the harbor, determined from population surveys (authors' unpublished data). The year 2008 was chosen as the "low-classic–high-nub year," when 7,921,177 (39.3%) of the estimated scallop population in Nantucket Harbor were year-1 nubs, while year-1 classics made up only 3.6% of the population. The "high-classic–low-nub year" was 2009 when an estimated 1,495,072 (10.3%) were year-1 nubs, while year-1 classics made up 73.9% of the population.

TABLE 2. Values for initial vectors (n_t) which were multiplied by the Leslie matrices (**A**) to produce population projections using the equation $n_{(t+1)} = \mathbf{A}n_t$ for 2008, a "low-classic–high-nub year," and 2009, a "high-classic–low-nub year."

Age-class	Description	% in survey	Estimated number in population
	2008 (estimated total population size, ~ 28,402,000)		
n_1	Year-0 classic spat	50.6	10,198,767
n_2	Year-1 classics	3.6	725,604
n_3	Year-2 classics	4.5	907,005
n_4	Year-0 nub spat		8,225,931[a]
n_5	Year-1 nubs	39.3	7,921,177
n_6	Year-2 nubs	2.1	423,269
	2009 (estimated total population size, ~ 37,808,000)		
n_1	Year-0 classic spat	10.1	1,466,042
n_2	Year-1 classics	73.9	10,726,782
n_3	Year-2 classics	0.1	14,515
n_4	Year-0 nub spat		23,292,358[a]
n_5	Year-1 nubs	10.3	1,495,072
n_6	Year-2 nubs	5.6	812,855

[a]Year-0 nubs were not surveyed but estimated using the equation: $n_1 = (n_5 \times F_5L) + (n_6 \times F_6L) + (n_2 \times F_2L)$ where n_i = age-class, F_i = fertility of age-class, and L = late spawn (spawn that produces nubs).

We made an approximation of the carrying capacity of Nantucket Harbor by comparing the commercial scallop landings (in bushels) during a survey year where the vast majority of scallops were adults (2007) with those of a previous year (1980), which had the maximum recorded landings for the harbor. We estimated the total population of bay scallops in the harbor in 1980 was 110 million individuals by setting up an equation comparing the ratio of the 2007 estimated scallop population to that year's commercial landings with the landings in 1980 (assuming an average of 400 scallops/bushel). The two adjacent years, 2008 and 2009, were compared with or without a second spawn to see whether its addition made a significant difference to the population growth rate and to the time during which each would reach an approximation of carrying capacity of the harbor.

The matrices were then subjected to sensitivity analysis to test the effects on population growth rate of changing the vital rates. The sensitivity matrix is calculated by multiplying the right and left eigenvectors together and reveals the sensitivity of the population growth rate (λ) to changes in rates of fertility and survival. Sensitivity of λ to both vital rates tends to decline with age (Caswell 2001). It is more straightforward to express sensitivities in an elasticity matrix rather than by using absolute numbers, because its elements are proportional (ranging from zero to one). Elasticity can be calculated by multiplying the corresponding elements of the sensitivity matrix by those in each original projection matrix and dividing their product by λ (Caswell 2001). We prepared elasticity matrices for classics only (submatrix $\mathbf{A_1}$), nubs only (submatrix $\mathbf{A_2}$), and the full multistate model with both classics and nubs (matrix **M**).

Stochasticity analysis tested the effects of a varying random environment on the population growth rate. The vital rates P_1 (juvenile survival) and F_2 (classic adult fertility) were first each varied alone in a stochastic model and then varied together. The stochastic growth rate (λ_S) is the dominant eigenvalue calculated for each stochastic matrix. The mean of the stochastic growth rates ($\overline{\lambda_S}$) resulting from 2,000 iterations (and its CI) was calculated for each case ($\mathbf{A_1}$ [classics only] and **M** [multistate with both classics and nubs]). The stochastic growth rate is a function of the intrinsic rate of increase ($\lambda = e^r$) and the mean generation time, and represents fitness in a randomly varying environment (Orzack and Tuljapurkar 1989; Caswell 2001). One should use a gamma (γ) or lognormal distribution for fecundity and a beta (β) distribution for survival in order to randomize the vital rates for stochastic simulations (Benton and Grant 1996; Caswell 2001). We drew 2,000 vital rates for stochasticity analysis randomly from a (1) β distribution of possible values of the vital rate to which population growth rate was most sensitive (P_1, early juvenile survival) and (2) lognormal distribution of the next highest rate (F_2, year-1 classic fertility) (Benton and Grant 1996; Caswell 2001).

Values of P_1 following the β distribution ranged from a minimum of zero to a maximum of 1.0. This early juvenile survival covered the period from early postsettlement to early summer of the following year, when the scallops would enter the reproductive population. It was thus equivalent to the bay scallop's recruitment rate. Although the observed survival rate of juvenile bay scallops maintained in published field studies and subject to predation, but not to fishing, are 44% to 93% (Pohle et al. 1991; Irlandi et al. 1995), there have been reports of nearly unsuccessful recruitment of early spawned (classic) juveniles in situations such as during the 1995 brown tides caused by

Aureococcus anophagefferens in the Peconic Bays (Tettelbach et al. 1999) and the 2009 rust tide outbreak caused by *Cochlodium polykrikoides* in Nantucket Harbor, Massachusetts (Hall et al. 2011).

Values for F_2 following the lognormal distribution ranged from a minimum of 0.777 to a maximum of 2,150 eggs/individual per spawn. These values were determined by multiplying together the lowest figures for fecundity, fertilization rate, larval survival, and settlement rate obtained from the literature (Belding 1910; Brousseau 1978b; Soria et al. 2010; Tettelbach et al. 2011) and then doing the same with the highest figures (Thorson 1966; Bricelj et al. 1987; LePennec et al. 1998; Tettelbach et al. 2011).

The following equations were used to determine the parameters of the two distributions:

(1) P_1 β distribution

$$a = [(1 - \mu)/\sigma^2 - (1/\mu)]\mu^2$$
$$b = \alpha[(1/\mu) - 1] \tag{3}$$

where $\mu = 0.4$ is the mean of P_1 (Table 1) and $\sigma = 0.25$ is the SD, taken as one-quarter of the range of P_1.

(2) F_2 lognormal distribution

$$\mu = \ln\left(m^2 / \sqrt{\upsilon + m^2}\right)$$
$$\sigma = \sqrt{\ln(\upsilon/m^2 + 1)} \tag{4}$$

where $m = 7.2$ is the mean of F_2 (Table 1), υ is the variance of F_2, and $\sigma = 1.98$ is the SD of $\ln(F_2)$ taken as one-quarter of the range of $\ln(F_2)$. By rearranging equation (4), parameter μ was calculated using the formula

$$\mu = \ln(m) - \sigma^2/2. \tag{5}$$

RESULTS

Deterministic Model

The upper left quadrant (A_1) of the matrix (Figure 1a) signifies the contribution of the classic life history alone to the bay scallop population. Its principal eigenvalue (λ) indicated that the population growth rate was 1.6979, implying that the population was expected to increase even with only one seasonal spawn. When the entire multistate matrix M was used (with all four quadrants of Figure 1a), the principal eigenvalue ($\lambda = 1.7114$) increased by only 0.81% over that of the classic life history alone. Thus, the population was not expected to grow significantly more in a constant environment with the addition of the second seasonal spawn. The lower right quadrant (A_2) of the multistate matrix (Figure 1a) signifies the contribution of the nub life history alone to the bay scallop

TABLE 3. Left and right eigenvectors (v_1, w_1) for matrices analyzed in the Nantucket bay scallop population indicating reproductive values and stable age distributions, respectively, for classic life history with only a single early spawn (matrix A_1), nub life history with only a late spawn (matrix A_2), and full life history (multistate matrix M) with two spawns per season.

Stage	Reproductive value	Stable age distribution
Classic (A_1)		
Year-0 classic	0.5134	0.8075
Year-1 classic	2.1794	0.1902
Year-2 classic	0.1814	0.0022
Nub (A_2)		
Year-0 nub	0.5018	0.7162
Year-1 nub	1.4124	0.2544
Year-2 nub	1.4260	0.0294
Full (M)		
Year-0 classic	0.9706	0.2992
Year-1 classic	4.1525	0.0699
Year-2 classic	0.3403	0.0008
Year-0 nub	0.0155	0.5034
Year-1 nub	0.0663	0.1177
Year-2 nub	0.5394	0.0089

population. In this case $\lambda = 1.1260$, a population growth rate 34% less than either a classic life history alone or one with two seasonal spawns.

In the "classics only" matrix (A_1), reproductive values (left eigenvector) were lowest for year-2 classic scallops (0.1814) and highest for year-1 adults (2.1794), whereas year-0 juveniles had an intermediate value of 0.5134. In the "nubs only" matrix (A_2), reproductive values were lowest for year-0 juveniles (0.5018) and almost the same for year-1 and year-2 nubs (1.4124 and 1.4260, respectively). When the entire multistate matrix model (M) was used, allowing the two life histories to interact, the reproductive values of year-0 classics and nubs were very different (0.9706 and 0.0155, respectively) and those of year-1 classics and nubs were 4.1525 and 0.0663, respectively. Within year-classes of the multistate model, the closest reproductive values were between those of year-2 nubs (0.5394) and year-2 classics (0.34030) (Table 3).

In the "classics only" matrix (A_1), year-0 juveniles made up by far the largest portion of the population (81%), whereas year-1 classics comprised 19% and year-2 classics only 0.22% of the population at the stable age distribution (right eigenvector). In the "nubs only" matrix (A_2), year-0 juveniles made up the largest portion (72%), whereas year-1 nubs comprised 25% and year-2 nubs contributed 3% to the population at the stable age distribution. When the entire multistate matrix (M) was used, showing the contribution of each life history to the other, year-0 juvenile classics and nubs were found in the greatest quantities (29% and 50%, respectively). Year-1 classics made up 7% of the population, while year-1 nubs were 12%. Year-2 classics and nubs made up 0.08% and 0.9%, respectively, of the population at the stable age distribution (Table 3).

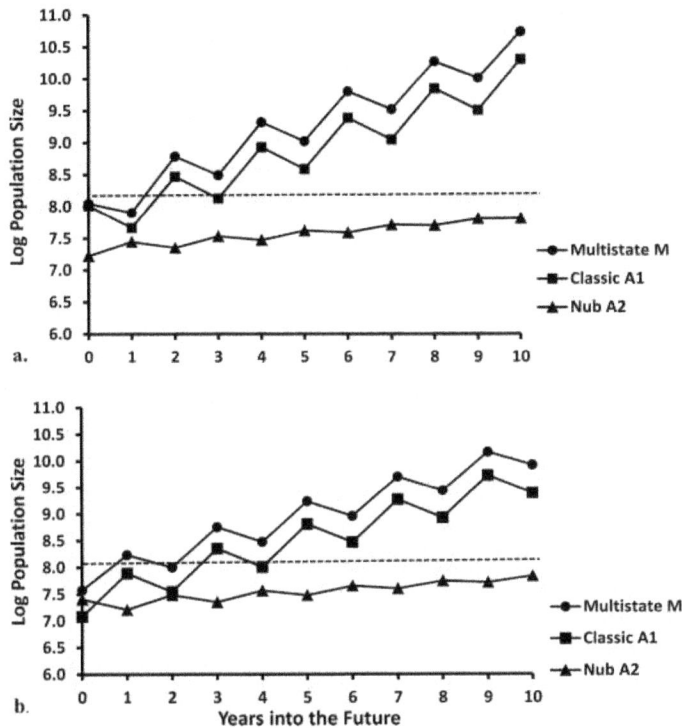

FIGURE 3. Future projections of total Nantucket bay scallop population based on initial conditions in two survey years (2008 and 2009) and comparing three models (full multistate matrix M, classics only submatrix A_1, nubs only submatrix A_2) for (a) 2008 with low-percent classics–high-percent nubs, and (b) 2009 with high-percent classics–low-percent nubs. Initial conditions were based on September population survey data. Population size is expressed in log_{10} values. The dashed line is a rough approximation of Nantucket Harbor's carrying capacity (110,000,000 scallops; $log_{10} = 8.05$) based on 1980 commercial landings.

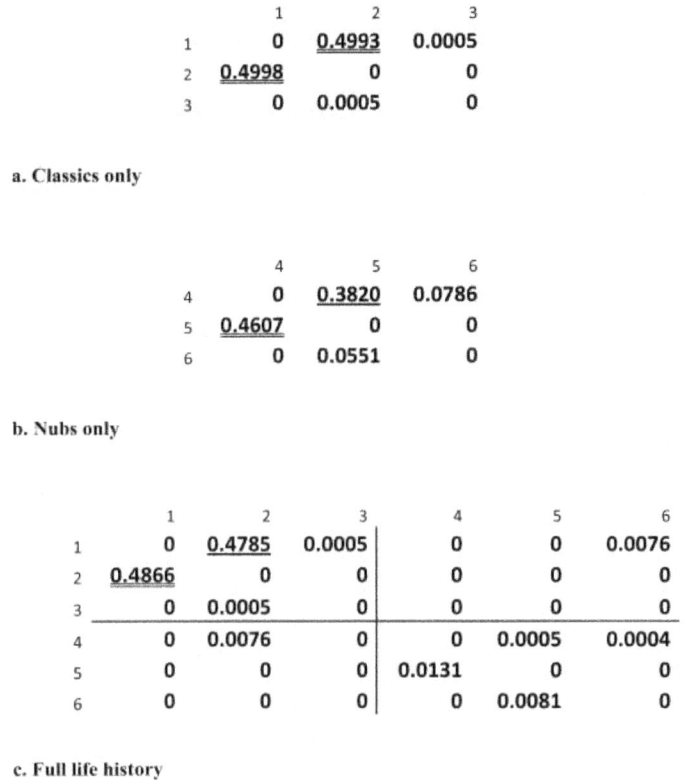

FIGURE 4. Elasticity matrices for (a) A_1 classics only submatrix, (b) A_2 nubs only submatrix, and (c) M full multistate matrix. Elasticity values range from zero to one; the larger the value, the more sensitive λ is to changes in that vital rate. Values underlined with double line are highest; those underlined with a single line are next highest in elasticity.

The multistate matrix, allowing contribution of nubs to the classic life history and of classics to the nub life history, allowed the population to increase by nearly three orders of magnitude in 10 years, regardless of which of the two initial conditions was used (Figure 3). The classic life history alone, with no second spawn, allowed the population to increase by two orders of magnitude in 10 years, while the nub life history, with only a late spawn, increased less than one order of magnitude regardless of which of the two initial conditions was used. Population growth with two seasonal spawns was greater beginning with the initial conditions for 2008 (low classic–low nub) than with those for 2009 (high classic–low nub). These population projections, however, represent the Nantucket bay scallop's biotic potential for exponential growth, are density independent, and do not take into account the carrying capacity of its estuarine environment. When a rough approximation of carrying capacity was considered (dashed line in Figure 3), the population reached carrying capacity with two seasonal spawns (multistate M model) in 1 to 2 years and in the classics alone model (A_1) with one early spawn in 3 to 4 years (depending on initial

conditions), but in the nubs alone model (A_2) with one late spawn carrying capacity was not reached within 10 years.

The elasticity matrices indicated that, for the "classics only" (A_1) life history, λ was essentially equally sensitive to juvenile survival (0.4998) (P_1 in Figure 4a) and to the fertility of year-1 adults (0.4993) (F_2 in Figure 4a). In the "nubs only" (A_2) life history, λ was more sensitive to the survival of juveniles (0.4607) than to the fertility of year-1 adults (0.3820). In the two matrices, the population growth rate was not especially sensitive to changes in either fertility or survival of year-2 scallops. When both life histories were combined in the multistate matrix, M (Figure 4c), the population growth rate was most sensitive to changes in P_1 (classic juvenile survival, 0.4866) and F_2 (year-1 classic fertility, 0.4785) was a close second. The population growth rate was relatively insensitive to juvenile and year-1 nub survival (0.0131 and 0.0081, respectively) and particularly insensitive to changes in nub fertility (0.0005 and 0.0004 for year-1 and year-2, respectively).

Stochastic Model

When survival rates (P_1) for 2,000 juveniles were randomly selected from a β distribution, the mean of the stochastic

TABLE 4. Population growth rates (λ) of classic ($\mathbf{A_1}$) versus full multistate ($\mathbf{M}$) bay scallop life histories using the deterministic model and then stochastic models varying vital rates P_1 (classic juvenile survival) and F_2 (classic year-1 fertility). First P_1 and F_2 varied alone and subsequently both vital rates varied together. Stochastic growth rates calculated from 2,000 iterations of vital rate(s), thus expressed as mean and lower and upper limits to 95% CI (in parentheses). The percent increase from $\mathbf{A_1}$ classic (one spawn) to $\mathbf{M}$ multistate (two spawns) is shown in the right column.

Model	λ $\mathbf{A_1}$ classic	λ $\mathbf{M}$ multistate	Percent increase
Deterministic	1.6979	1.7114	0.81
Stochastic (P_1 only)	1.6232 (0.4665, 2.5468)	1.7154 (1.1295, 2.5512)	5.7
Stochastic (F_2 only)	1.0053 (0.1848, 4.1548)	1.4652 (1.1261, 4.1562)	45.8
Stochastic (P_1 and F_2)	0.8762 (0.1180, 3.5747)	1.3866 (1.1260, 3.5767)	58.3

population growth rates ($\overline{\lambda_S}$) of the resulting "classics only" matrices ($\mathbf{A_1}$) was 1.6232. Using the same P_1 values, the mean of the stochastic growth rates of the full multistate matrices ($\mathbf{M}$) increased to 1.7154. The mean of the stochastic growth rates of the full matrix was 5.7% higher than that of classics alone matrix, but nearly the same as the deterministic growth rate (1.7114). When fertility rates (F_2) for 2,000 classic adults were randomly selected from a lognormal distribution, the mean of the stochastic growth rates of the "classics only"

matrices was 1.0053 and the mean of the stochastic growth rates ($\overline{\lambda_S}$) of the multistate matrices was 1.4652, an increase of 45.8%. When juvenile survival (P_1) was allowed to fluctuate together with classic adult fertility (F_2), $\overline{\lambda_S}$ increased by 58.3% from the classic matrices (0.8762) to the multistate matrices with two spawns (1.3866). All $\overline{\lambda_S}$ values with their 95% CIs are shown in Table 4. Box plots in Figure 5 compare population growth rates in all matrix models with and without the second spawn.

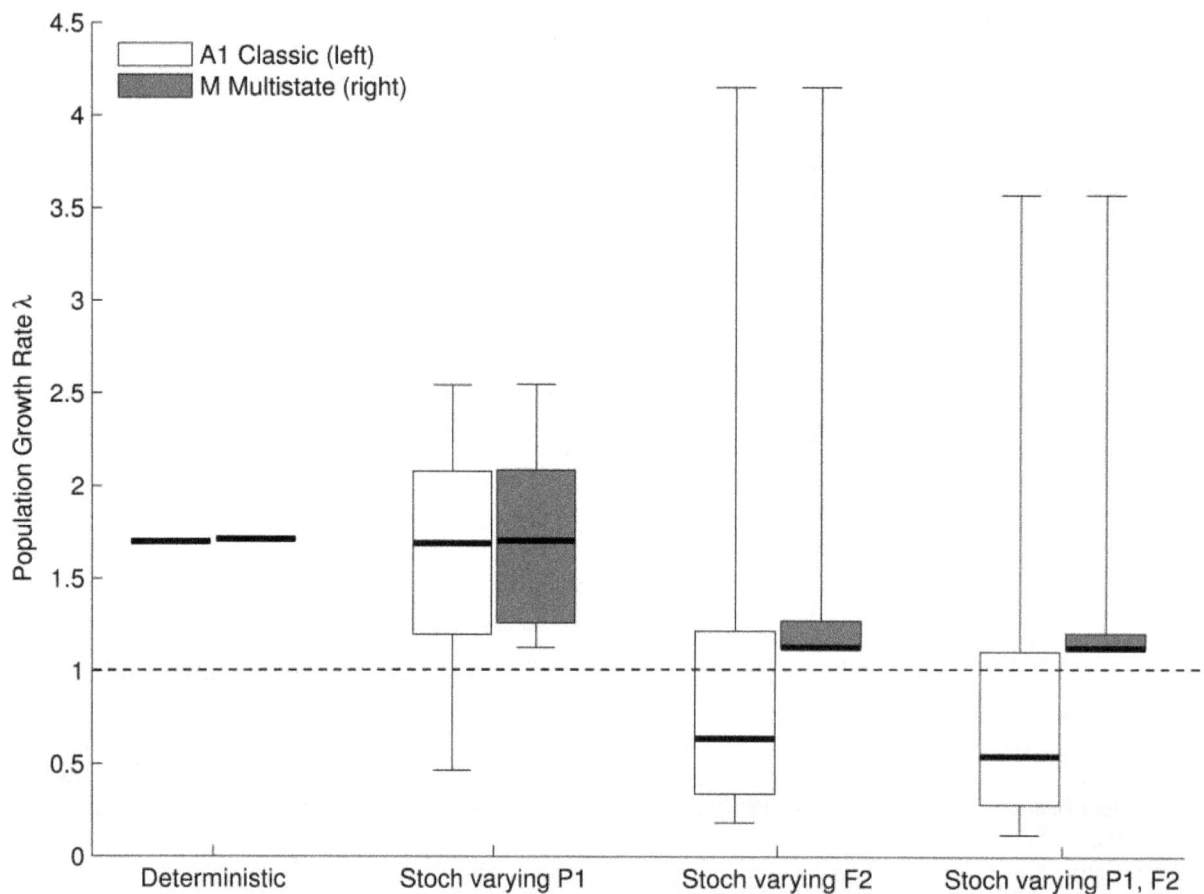

FIGURE 5. Comparison of mean population growth rates (λ) of bay scallop populations in Nantucket Harbor with ($\mathbf{M}$ multistate) and without ($\mathbf{A_1}$ classic) the second spawn of (1) the deterministic model, (2) a stochastic model (Stoch) using a beta distribution of juvenile classic survival (P_1), (3) a stochastic model using a lognormal distribution of year-1 classic adult fertility (F_2), and (4) a stochastic model varying P_1 and F_2 together. All vital rates for which growth rates were calculated were drawn randomly from a distribution of 2,000 possible values. Box plots show median values (solid horizontal line), 25% and 75% percentile values (box outline), and 2.5% and 97.5% percentile values (whiskers). The dashed line represents the growth rate at which the population is neither increasing nor declining ($\lambda = 1$).

DISCUSSION

When a deterministic matrix population model was employed, the presence of a second seasonal spawn in the Nantucket bay scallop increased the population growth by less than 1% over that of the classic early spawning life history. When stochasticity was added to the model, varying the two vital rates to which population growth rate was most sensitive, the bay scallop population growth increased by 58.3% when the second spawn was added. The lower limits of stochastic $\overline{\lambda_S}$ estimates (Figure 5) were much greater when a second spawn was included, suggesting there is a much lower frequency of population decrease caused by environmental variability.

Elasticity analysis revealed that the population growth rate was most sensitive to the survival of classic juveniles (P_1), and the fertility of year-1 classic adults (F_2) was a close second. We were surprised that adding a second spawn increased λ when fertility was varied by so much greater a percentage than when juvenile survival was varied alone (45.8% versus 5.7%, respectively) (Table 4). However, since the first age-class in our model was newly settled spat, fertility had to be calculated with a combination of fecundity, fertilization and settlement rates, and the probability of larval survival, almost all of which are subject to wide fluctuations. Malinowski and Whitlatch (1988) found the highest sensitivity to be survival of the youngest age-classes in three commercially important bivalves (*Mya arenaria, Mercenaria mercenaria, C. virginica*). In general, sensitivity of survivorship values in those species is two orders of magnitude greater than those for fecundity. Doak et al. (1994) found a similar difference in elasticity values between survival and fecundity, but in the case of the desert tortoise, elasticity of survival rates was highest in the largest size-class. Nakaoka (1997) also found the highest sensitivity for survivorship of the largest size-class of the bivalve, *Y. notabilis*. Greater sensitivities in older age- or size-classes seem to be the norm in such especially long-lived species as the desert tortoise and *Y. notabilis*.

Because classic juvenile bay scallop survival rate (P_1) and classic age-1 adult fertility (F_2) had the highest calculated elasticities in our study, we chose to vary only those vital rates in performing the stochastic analyses: first P_1 and F_2 each alone and then P_1 and F_2 together. Adding a second spawn increased the population growth rate by 5.7% under conditions of stochastic juvenile survival. There was an increase of 58.3% when the means of the stochastic growth rates were compared between classic and full multistate life histories, using random distributions of both juvenile survival and adult fertility. Although the variability within the estimates of $\overline{\lambda_S}$ for the two life histories was greater than the difference between the two estimates, there are some meaningful implications. The distribution of population growth rates for each revealed that, while the majority of values were below one (indicating population decline) in the classics only ($\mathbf{A_1}$) stochastic model, no values below one were found in the multistate ($\mathbf{M}$) stochastic model

with two seasonal spawns. In all life histories (classic alone, nub alone, and multistate), variability in stochastic growth rates, as shown by CIs, decreased substantially when a second spawn was added (Figure 5).

Juvenile bay scallop survival in this study represented survival from settlement to recruitment into the reproductive population at age 1, which is therefore analogous to recruitment rate in other studies. Marine bivalves, most of which have planktotrophic larval stages, experience considerable recruitment fluctuations (Brousseau 1978a; Strathmann 1985; Mann 1988; Ambrose and Lin 1991; LePennec et al. 2003). Bay scallop populations also have large variations in recruitment; thus, the stochastic estimate of population growth seems to be more relevant. Variable recruitment rates were modeled in stochastic matrix analysis of the subtidal snail, *Umbonium costatus*, (Noda and Nakao 1996), the bivalve, *Y. notabilis*, (Nakaoka 1997), and the soft-shell clam, *Mya arenaria* (Ripley 1998; Ripley and Caswell 2006). In the first study (Noda and Nakao 1996), an unstable age distribution indicated sporadic recruitment, and recruitment densities showed yearly fluctuations. The second study (Nakaoka 1997) also noted large interannual recruitment fluctuations and rates that were highly skewed towards smaller values (thus, lognormally distributed). In the last study, Ripley and Caswell (2006) found that the great variability in recruitment rate actually increased the stochastic growth rate of *Mya arenaria* in Barnstable Harbor, Massachusetts, when only those rates are varied. The great persistence of the population may be due to the ability of this soft-shell clam to live long enough to experience at least one good recruitment event in a lifetime.

The deterministic population growth rate (λ) for bay scallops in Nantucket Harbor was 1.7114 with a second seasonal spawn and 1.6979 without. However, it is not uncommon for species with variable reproduction or recruitment to have deterministic estimates of growth rate that are less than or just slightly over one. Population growth rates in the teasel, a biennial plant found at later stages of old-field succession, ranged from 0.275 to 2.605 depending upon location (Werner and Caswell 1977). The population growth rate of *G. gemma*, a small clam living in Rhode Island sandflats, fluctuated on either side of one because this species does not undergo reproduction each year (Weinberg 1989). Noda and Nakao (1996) calculated an average λ of 0.9268 for the subtidal snail, *U. costatum*, studied for 8 years when the year with extremely high recruitment (1982) was removed. With that year included, the average population growth rate rose to 1.078. In Nakaoka's (1996) review of population growth rates of 10 studies using matrix population models, four species had λ values less than one: Striped Bass *Morone saxatilis*, desert tortoise, and the marine mollusks, *Y. notabilis* and *U. costatum* (cited above). The other species (jack-in-the pulpit *Arisaema triphylllum*, savanna grasses *Andropogon semiberbis* and *A. brevifolis*, seaweed *Ascophyllum nodosum*, gorgonian coral *Leptogorgia virgulata*, and red deer *Cervus elephas*) had λ values equaling or slightly above one. None of the latter populations are increasing substantially; rather, they are either stable or just holding on.

When the bay scallop population was projected into the future at the growth rate calculated by our deterministic matrix model, adding a second spawn made a difference in the time it took for the population to increase to the carrying capacity. In our stochastic models, population growth rates without a second spawn often fell below one, indicating a declining population (Figure 5). However, adding a second spawn while varying the vital rates of juvenile survival and year-1 adult fertility allowed the population to increase by over 50% from a single-spawn to a two-spawn life history, a much greater increase than that seen in the deterministic model. Although the stochastic growth rate with a two-spawn life history was not as great as the deterministic estimate, we believe that it was more realistic, considering the scallop's variable environment. In the bay scallop, a second spawn could thus be a mechanism to prevent the population from declining to extinction.

Reproductive values of classic early spawned bay scallops in Nantucket followed the typical pattern of being lowest in juveniles, reaching their highest values in adults at first age of reproduction, and then declining with increasing age (Caswell 2001). In contrast, late-spawned nubs, which can spawn twice in their second summer, had their highest reproductive value as age-2 adults, slightly surpassing year-1 nubs, less than 40% of which could spawn once but only late in their first summer (Hall 2014), and both had greater reproductive value than did juveniles. It is possible that juveniles in both cases were underrepresented in our model because we were not able to assess stages earlier than newly settled spat. Malinowski and Whitlatch (1988) found that all three bivalves (*Mya arenaria, Mercenaria mercenaria*, and *C. virginica*) showed increasing reproductive values between age 0 and age 2 (age of first reproduction), followed by a slow decline until death.

The stable age distribution represents the constant age structure achieved when the population reaches an asymptote at the given growth rate (λ) (Caswell 2001). This distribution is ergodic (independent of initial conditions). The stable age distribution of Nantucket bay scallops for both classic and nub life histories revealed that juveniles less than 1 year old were found in the highest abundance (81% and 72% for classics and nubs, respectively). Year-1 adults were next in abundance (19% and 25%, respectively), while year-2 adults were lowest (0.2% and 2.9%, respectively). In the multistate matrix, representing the combined contributions of classic and nub life histories, juvenile classics and nubs also made up the largest percentage of the population (30% and 50%, respectively) (Table 3). Relative numbers of scallops in each cohort differed widely from year to year in the September survey. For example, year-1 classics made up 4% and year-1 nubs 39% of surveyed scallops in 2008, while year-1 classics made up 74% and year-1 nubs 10% of surveyed scallops in 2009 (Table 2).

Thus the modeled and observed age-class distributions were not the same.

Nantucket Harbor is an unpredictable environment and experiences sudden changes in temperature and salinity as well as episodic blooms of harmful algae and population explosions of predators. Bay scallops are especially sensitive to these changes during larval and juvenile stages when they experience wide fluctuations in recruitment. When only one vital rate (P_1) was varied, representing classic juvenile survival, stochastic population growth rate increased 5.7% from a classics-only life history to one incorporating a second seasonal spawn. When two vital rates were varied together (P_1 and classic adult fertility F_2), the difference in the mean of the population growth rates ($\overline{\lambda_S}$) between the two life histories was 58.3% (Table 4). The substantial larval survival component of F_2 must have interacted with juvenile survival (P_1) to produce the great increase in population growth rate between life histories with and without a second spawn.

The second spawn of the bay scallop population in Nantucket Harbor appears to illustrate the life history strategy of bet-hedging (Slatkin 1974). In bet-hedging, there is a tradeoff between current reproductive effort (and fecundity) and future survivorship (Caswell 1980; Goodman 1982; Strathmann 1985; MacDonald et al. 1987). Some examples of bet-hedging are spreading reproduction over multiple years in biennial plants (Klinkhamer and DeJong 1983), prolonged dormancy in annual desert plants (Philippi 1993), delayed hatching in anostracans living in vernal pools (Simovich and Hathaway 1997; Philippi et al. 2001), hatching asynchrony in nesting birds (Laaksonen 2004), and extended hatching of larval king crabs (Stevens 2002). Bet-hedging appears to be particularly advantageous in stochastic environments, where adaptations to unpredictable environmental conditions ensure that at least some offspring will find suitable conditions in which to survive (Menu et al. 2000; Krug 2009). Bet-hedging has been observed in several scallop species. "Dribble spawning" (partial spawning over a long period of time) in the sea scallop ensures that at least some larvae will survive in an uncertain environment (Langton et al. 1987). The Australian saucer scallop *Amusium balloti* also employs partial spawning to provide multiple opportunities for larval development and recruitment success (Joll and Caputi 1995). Tettelbach et al. (2001) hypothesized that the ability of late-spawned New York bay scallops to survive and spawn a second year suggests that this is important to ensure the population persists when early recruitment has failed. Bishop et al. (2005) stated that the adoption of an iteroparous (more than one spawning event) life style by North Carolina bay scallops is an example of bet-hedging. In that case, it is the early spawn that persists in spite of a large fall spawn because of the latter's risk of recruitment failure from autumn events such as harmful algal blooms and hurricanes. Multiple spawnings in Florida bay scallops throughout the year could allow for a greater reproductive output over their lifetimes, another example of bet-hedging (Geiger et al. 2010).

Bimodal spawning in northern bay scallop populations thus appears to be a strategy for coping in a stochastic environment. Fluctuations in recruitment, combined with a short lifespan, have caused the bay scallop to be particularly vulnerable to environmental change. The progeny of the late spawn may be exposed to a different set of environmental conditions than that of the early spawn and may experience greater growth and survival. By spreading out reproductive effort over more than one spawning event in a season, the northern bay scallop may be capable of increasing its chances for successful recruitment. This life cycle diversity also gives the Nantucket bay scallop population its resilience, or the ability to recover quickly from environmental perturbations. The importance of the second spawn to the persistence of northern bay scallop populations has significant management implications. The greater the survival of the second spawn's progeny (nub scallops), the more they can contribute to the population in their second spawning season. If recreational and commercial harvest of first-year nubs is again prohibited, a major source of their mortality could be eliminated. With that accomplished, a successful second spawn would be able to buffer the effects of the bay scallop's highly variable recruitment rate and prevent possible population collapse.

ACKNOWLEDGMENTS

We are grateful for the close collaboration of R. Kennedy and P. Boyce on research leading to this model, mentoring of B. Stevens, advice on modeling from M. Chintala, S. Tettelbach, H. Caswell, and M. Barbeau, local fishery information from K. Kelley, T. Riley, and C. Sjolund, and helpful comments from two anonymous reviewers. The life cycle illustrations were drawn by Sean Donnan. Support was provided by the Nantucket Shellfish Association, Nantucket Maria Mitchell Association, Nancy Sayles Day Foundation, T Theory Foundation, Quebec–Labrador Foundation, and the University of Massachusetts Dartmouth.

REFERENCES

Abele, D., T. Brey, and E. Phillipp. 2009. Bivalve models of aging and the determination of molluscan lifespans. Experimental Gerontology 44:307–315.

Ambrose, W. G., and J. Lin. 1991. Settlement preference of *Argopecten irradians* (Lamarck 1819) for artificial substrata. Pages 16–20 *in* S. E. Shumway and P. A. Sandifer, editors. An international compendium of scallop biology and culture. World Aquaculture Society, Baton Rouge, Louisiana.

Ayers, J. C. 1956. Population dynamics of the marine clam, *Mya arenaria*. Limnology and Oceanography 1:26–34.

Barbeau, M. A., and H. Caswell. 1999. A matrix model for short-term dynamics of seeded populations of sea scallops. Ecological Applications 9:266–287.

Barber, B. J., and N. J. Blake. 2006. Reproductive physiology. Pages 357–416 *in* S. E. Shumway and G. J. Parsons, editors. Scallops: biology, ecology and aquaculture, 2nd edition. Elsevier, Amsterdam.

Belding, D. L. 1910. The scallop fisheries of Massachusetts: including an account of natural history of the common scallop. Commonwealth of Massachusetts Commission on Fisheries and Game, Marine Fisheries 3, Boston.

Benton, T. G., and A. Grant. 1996. How to keep fit in a real world: elasticity analysis and selection pressures on life histories in a variable environment. American Naturalist 147:115–139.

Bishop, M. J., E. A. Rivera, W. G. Irlandi, W. G. Ambrose, and C. H. Peterson. 2005. Spatio-temporal patterns in the mortality of bay scallop recruits in North Carolina: investigation of a life history anomaly. Journal of Experimental Marine Biology and Ecology 315:127–146.

Bologna, P. X., A. E. Wilbur, and K. E. Able. 2001. Reproduction, population structure, and recruitment limitation in a bay scallop (*Argopecten irradians* Lamarck) population from New Jersey, USA. Journal of Shellfish Research 20:89–96.

Brand, A. R. 2006. Scallop ecology: distributions and behavior. Pages 651–744 *in* S. E. Shumway and G. J. Parsons, editors. Scallops: biology, ecology and aquaculture, 2nd edition. Elsevier, Amsterdam.

Brault, S., and H. Caswell. 1993. Pod-specific demography of killer whales (*Orcinus orca*). Ecology 74:1444–1454.

Bricelj, V. M., J. Epp, and R. E. Malouf. 1987. Intraspecific variation in reproductive and somatic growth cycles of bay scallops *Argopecten irradians*. Marine Ecology Progress Series 36:123–137.

Brousseau, D. J. 1978a. Spawning cycle, fecundity, and recruitment in a population of soft- shell clam, *Mya arenaria*, from Cape Ann, Massachusetts. U.S. National Marine Fisheries Service Fishery Bulletin 76:155–166.

Brousseau, D. J. 1978b. Population dynamics of the soft-shell clam *Mya arenaria*. Marine Biology 50:63–71.

Brousseau, D. J., and J. A. Baglivo. 1984. Sensitivity of the population growth rate to changes in single life history parameters: its application to *Mya arenaria* (Mollusca: Pelecypoda). U.S. National Marine Fisheries Service Fishery Bulletin 82:537–541.

Brousseau, D. J., J. A. Baglivo, and G. E. Lang. 1982. Estimation of equilibrium settlement rates for benthic marine invertebrates: its application to *Mya arenaria* (Mollusca: Pelecypoda). U.S. National Marine Fisheries Service Fishery Bulletin 82:642–644.

Caswell, H. 1980. On the equivalence of maximizing reproductive value and maximizing fitness. Ecology 61:19–24.

Caswell, H. 2001. Matrix population models: construction, analysis, and interpretation, 2nd edition. Sinauer, Sunderland, Massachusetts.

Clarke, H. A. Jr. 1965. The scallop superspecies *Aequipecten irradians* (Lamarck). Malacologia 2:161–188.

Doak, D., P. Kareiva, and B. Klepetka. 1994. Modeling population viability for the desert tortoise in the western Mojave Desert. Ecological Applications 4:446–460.

Geiger, S. P., S. P. Stephenson, and W. S. Arnold. 2010. Protracted recruitment in the bay scallop *Argopecten irradians* in a west Florida estuary. Journal of Shellfish Research 29:809–817.

Goodman, D. 1982. Optimal life histories, optimal notation, and the value of reproductive value. American Naturalist 199:813–823.

Gourley, R. S., and C. E. Lawrence. 1977. Stable population analysis in periodic environments. Theoretical Population Biology 11:49–59.

Hall, V. A. 2014. Impact of the second seasonal spawn on reproduction, recruitment, population, and life history of the northern bay scallop, *Argopecten irradians irradians* (Lamarck, 1819). Doctoral dissertation. University of Massachusetts, Dartmouth.

Hall, V. A., T. A. Riley, P. B. Boyce, and R. S. Kennedy. 2011. Possible impacts of the harmful alga *Cochlodinium polykrikoides* on bay scallops (*Argopecten irradians irradians*) in Nantucket Harbor. Journal of Shellfish Research 30:448–449.

Hinchey, E. K., M. M. Chintala, and T. R. Gleason. 2004. A stage-based population model for bay scallops *Argopecten irradians irradians* and implications for population-level effects of habitat alteration. Journal of Shellfish Research 12:295–296.

Hines, A. H. 1986. Larval problems and perspectives in life histories of marine invertebrates. Bulletin of Marine Science 39:506–525.

Irlandi, E. A., W. G. Ambrose, and B. A. Orlando. 1995. Landscape ecology and the marine environment: how spatial configuration of seagrass habitat influences growth and survival in the bay scallop. Oikos 72:307–313.

Joll, L. M., and N. Caputi. 1995. Geographic variation in the reproductive cycle of the saucer scallop, *Amusium balloti* (Bernardi, 1861) (Mollusca: Pectinidae), along the western Australian coast. Marine and Freshwater Research 46:770–792.

Kelley, K. M. 1981. The Nantucket bay scallop fishery: the resource and its management. Report to the Town of Nantucket Marine Department, Nantucket, Massachusetts.

Klinkhamer, P. G. L., and T. J. DeJong. 1983. Is it profitable for biennials to live longer than two years? Ecological Modelling 20:223–232.

Krug, P. J. 2009. Not my "type": larval dispersal bimorphisms and bet-hedging in Opisthobranch life histories. Biological Bulletin 216:355–372.

Laaksonen, T. 2004. Hatching asynchrony as a bet-hedging strategy – an offspring diversity hypothesis. Oikos 10:616–620.

Langton, R. W., W. E. Robinson, and D. Schick. 1987. Fecundity and reproductive effort of sea scallops *Placopecten magellanicus* from the Gulf of Maine. Marine Ecology Progress Series 37:19–25.

Leftovitch, L. P. 1965. The study of population growth in organisms grouped by stages. Biometrika 21:1–8.

LePennec, M., A. Paugam, and G. LePennec. 2003. The pelagic life of the pectinid *Pecten maximus*—a review. ICES Journal of Marine Science 60:211–223.

LePennec, M., R. Robert, and M. Avendano. 1998. The importance of gonadal development on larval production in pectinids. Journal of Shellfish Research 17:97–101.

Leslie, P. H. 1945. On the use of matrices in certain population mathematics. Biometrika 35:83–212.

Llodra, E. R. 2002. Fecundity and life-history strategies in marine invertebrates. Pages 87–170 *in* A. J. Southward, P. A. Tyler, C. M. Young, and L. A. Fuiman, editors. Advances in marine biology. Elsevier and Academic Press, London.

Lundquist, C. J., and L. W. Botsford. 2004. Model projections of the fishery implications of the Allee effect in broadcast spawners. Ecological Applications 14:929–941.

MacDonald, B. A., R. J. Thompson, and B. L. Bayne. 1987. Influence of temperature and food availability on the ecological energetics of the giant scallop *Placopecten magellanicus* IV. Reproductive effort, value, and cost. Oecologia 72:440–556.

MacFarlane, S. L. 1991. Managing scallops *Argopecten irradians irradians* (Lamarck 1819) in Pleasant Bay, Massachusetts; large is not always legal. Pages 264–272 *in* S. E. Shumway and P. A. Sandifer, editors. An international compendium of scallop biology and culture. World Aquaculture Society, Baton Rouge, Louisiana.

MacKenzie, C. L. 2008. The bay scallop *Argopecten irradians*, Massachusetts through North Carolina: its biology and the history of its habitats and fisheries. Marine Fisheries Review 70:6–79.

Malinowski, S., and R. B. Whitlatch. 1988. A theoretical evaluation of shellfish resource management. Journal of Shellfish Research 7:95–100.

Mann, R. 1988. Field studies of bivalve larvae and their recruitment to the benthos: a commentary. Journal of Shellfish Research 7:7–10.

Martinez, G., L. Mettifogo, M. A. Perez, and C. Callijas. 2007. A method to eliminate self- fertilization in a simultaneous hermaphroditic scallop. 1. Effects on growth and survival of larvae and juveniles. Aquaculture 273:459–469.

Menu, F., J.-P. Roebuck, and M.Viala. 2000. Bet-hedging diapause strategies in stochastic environments. American Naturalist 155:724–734.

Nakaoka, M. 1996. Dynamics of age- and size-structured populations in fluctuating environments: applications of stochastic matrix models to natural populations. Researches on Population Ecology 38:141–152.

Nakaoka, M. 1997. Demography of the marine bivalve *Yoldia notabilis* in fluctuating environments: an analysis using a stochastic matrix model. Oikos 79:59–68.

Noda, T., and S. Nakao. 1996. Dynamics of an entire population of the subtidal snail *Umbonium costatum*: the importance of annual recruitment limitation. Journal of Animal Ecology 65:196–204.

Orzack, S. H., and S.Tuljapurkar. 1989. Population dynamics in variable environments. VII. The demography and evolution of iteroparity. American Naturalist 133:901–923.

Pennington, J. T. 1985. The ecology of fertilization of echinoid eggs: the consequences of sperm dilution, adult aggregation, and synchronous spawning. Biological Bulletin 169:417–430.

Philippi, T. 1993. Bet-hedging germination of desert annuals beyond the first year. American Naturalist 142:474–487.

Philippi, T., M. A. Simovitch, E. T. Bauder, and J. A. Moorad. 2001. Habitat ephemerality and hatching fractions of a diapausing Anostracan (Crustacean: Branchipoda). Israel Journal of Zoology 47:387–395.

Pohle, D. G., V. M., Bricelj, and Z. Garcia-Esquivel. 1991. The eelgrass canopy: an above-bottom refuge from benthic predators for juvenile bay scallops *Argopecten irradians*. Marine Ecology Progress Series 74:47–59.

Ripley, B. J. 1998. Life history traits and population processes in marine bivalve molluscs. Doctoral dissertation. Massachusetts Institute of Technology/Woods Hole Oceanographic Institution, Woods Hole.

Ripley, B. J., and H. Caswell. 2006. Recruitment variability and stochastic population growth of the soft-shell clam, *Mya arenaria*. Ecological Modelling 193:517–530.

Rumrill, S. S. 1990. Natural mortality of marine invertebrate larvae. Ophelia 32:163–198.

Simovich, M. A., and S. A. Hathaway. 1997. Diversified bet-hedging as a reproductive strategy of some ephemeral pool anostracans (Branchiopoda). Journal of Crustacean Biology 17:38–44.

Slatkin, M. 1974. Hedging one's evolutionary bets. Nature 250:704–705.

Smith, M., H. Caswell, and P. Mettler-Cherry. 2005. Stochastic flood and precipitation regimes and the population dynamics of a threatened floodplain plant. Ecological Applications 15:1036–1052.

Soria, T., J. Tordechilla-Guillen, R. Cudney-Bueno, and W. Shaw. 2010. Spawning induction, fecundity estimation, and larval culture of *Spondylus calcifer* (Carpenter, 1857) (Bivalvia: Spondylidae). Journal of Shellfish Research 29:143–149.

Stevens, B. G. 2002. Survival of tanner crabs tagged with floy tags in the laboratory. Pages 551–560 *in* A. J. Paul, E. G. Dawe, R. Elner, G. S. Jamieson, G. H. Kruse, R. S. Otto, B. S. Sainte-Marie, T. C. Shirley, and D. Woodby, editors. Crabs in cold water regions: biology, management, and economics. University of Alaska Sea Grant, AK-SG-02-01, Fairbanks.

Strasser, C. A. 2008. Metapopulation dynamics of the softshell clam, *Mya arenaria*. Doctoral dissertation. Massachusetts Institute of Technology/Woods Hole Institution of Oceanography Joint Program, Woods Hole.

Strathmann, R. R. 1985. Feeding and nonfeeding larval development and life-history evolution in marine invertebrates. Annual Review of Ecology, Evolution, and Systematics 16:339–361.

Taylor, R. E., and J. M. Capuzzo. 1983. The reproductive cycle of the bay scallop, *Argopecten irradians irradians* (Lamarck), in a small coastal embayment of Cape Cod, Massachusetts. Estuaries 6:431–435.

Tettelbach, S. T. 1990. Burial of transplanted bay scallops *Argopecten irradians irradians* (Lamarck, 1819) in winter. Journal of Shellfish Research 9:127–134.

Tettelbach, S. T. 1991. Seasonal changes in a population of northern bay scallops, *Argopecten irradians irradians* (Lamarck, 1819). Pages 164–175 *in* S. E. Shumway and P. A. Sandifer, editors. An international compendium of scallop biology and culture. World Aquaculture Society, Baton Rouge, Louisiana.

Tettelbach, S. T., P. J. Auster, E. W. Rhodes, and J. Widman. 1985. A mass mortality of northern bay scallops, *Argopecten irradians irradians*, following a severe spring rainstorm. Veliger 27:381–385.

Tettelbach, S. T., D. Barnes, J. Aldred, G. Rivara, D. Bonal, A.Weinstock, C. Fitzsimmons-Diaz, J. Thiel, M. C. Cammarota, A. Stark, K. Wejnert, R. Ames, and J. Carroll. 2011. Utility of high-density plantings in bay scallop, *Argopecten irradians irradians*, restoration. Aquaculture International 19:715–739.

Tettelbach, S. T., and C. F. Smith. 2009. Bay scallop restoration in New York. Ecological Restoration 27:20–22.

Tettelbach, S. T., B. J. Peterson, J. M. Carroll, S. W. T. Hughes, D. M. Bonar, A. J. Weinstock, J. R. Europe, B. T. Furman, and C. F. Smith. 2013. Priming the larval pump: resurgence of bay scallop recruitment following initiation of intensive restoration efforts. Marine Ecology Progress Series 478:153–172.

Tettelbach, S. T., C. F. Smith, R. Smolowitz, K. Tetreault, and S. Dumais. 1999. Evidence for fall spawning of northern bay scallops *Argopecten irradians irradians* (Lamarck 1819) in New York. Journal of Shellfish Research 18:47–58.

Tettelbach, S. T., P. Wenczel, and S. W. Hughes. 2001. Size variability of juvenile (0 + yr) bay scallops *Argopecten irradians irradians* (Lamarck, 1819) at eight sites in eastern Long Island, New York. Veliger 44:389–397170.

Thorson, G. 1966. Some factors influencing the recruitment and establishment of marine benthic communities. Netherlands Journal of Sea Research 3:267–293.

Vahl, O. 1981. Age-specific residual reproductive value and reproductive effort in the Iceland scallop, *Chlamys islandica* (O.F. Müller). Oecologia 51:53–56.

Vance, R. R. 1973. On reproductive strategies in marine benthic invertebrates. American Naturalist 107:339–352.

Weinberg, J. R. 1989. Predicting population abundance and age structure: testing theory with field data. Marine Ecology Progress Series 53:59–64.

Werner, P. A., and H. Caswell. 1977. Population growth rates and age versus stage distribution models for teasel (*Dipsacus sylvestris* Huds.). Ecology 58:1103–1111.

Wootton, T. J., and D. A. Bell. 1992. A metapopulation model of the peregrine falcon in California: viability and management strategies. Ecological Applications 2:307–321.

Zarnoch, C. B., and M. P. Schreibman. 2008. Influence of temperature and food availability on the biochemical composition and mortality of juvenile *Mercenaria mercenaria* (L.) during the over-winter period. Aquaculture 274:281–291.

10

Stock-Specific Size and Timing at Ocean Entry of Columbia River Juvenile Chinook Salmon and Steelhead: Implications for Early Ocean Growth

Laurie A. Weitkamp*

National Oceanic and Atmospheric Administration, National Marine Fisheries Service,
Northwest Fisheries Science Center, Conservation Biology Division, Newport Field Station,
2032 Marine Science Drive, Newport, Oregon 97365, USA

David J. Teel

National Oceanic and Atmospheric Administration, National Marine Fisheries Service,
Northwest Fisheries Science Center, Conservation Biology Division, Manchester Field Station,
Post Office Box 130, Manchester, Washington 98353, USA

Martin Liermann

National Oceanic and Atmospheric Administration, National Marine Fisheries Service,
Northwest Fisheries Science Center, Fish Ecology Division, 2725 Montlake Boulevard East,
Seattle, Washington 98112, USA

Susan A. Hinton

National Oceanic and Atmospheric Administration, National Marine Fisheries Service,
Northwest Fisheries Science Center, Fish Ecology Division, Point Adams Field Station,
520 Heceta Place, Hammond, Oregon 97121, USA

Donald M. Van Doornik

National Oceanic and Atmospheric Administration, National Marine Fisheries Service,
Northwest Fisheries Science Center, Conservation Biology Division, Manchester Field Station,
Post Office Box 130, Manchester, Washington 98353, USA

Paul J. Bentley

National Oceanic and Atmospheric Administration, National Marine Fisheries Service,
Northwest Fisheries Science Center, Fish Ecology Division, Point Adams Field Station,
520 Heceta Place, Hammond, Oregon 97121, USA

Abstract

   Juvenile salmon transitioning from freshwater to marine environments experience high variation in growth and survival, yet the specific causes of this variation are poorly understood. Size at and timing of ocean entry may contribute to this variation because they influence both the availability of prey and vulnerability to predators. To

Subject editor: Carl Walters, University of British Columbia, Canada

*Corresponding author: laurie.weitkamp@noaa.gov

explore this issue, we used stock assignments based on genetic stock identification and internal tags to document the stock-specific size and timing of juvenile hatchery and presumed wild Columbia River Chinook Salmon *Oncorhynchus tshawytscha* and steelhead *O. mykiss* at ocean entry during 2007–2011. We found that juvenile salmon and steelhead had consistent stock-specific capture dates, with lower-river stocks typically having earlier timing than those originating farther upstream. Mean size also varied among stocks and was related to hatchery practices. Hatchery yearling Chinook Salmon and steelhead were consistently larger than wild fish from the same stocks, although timing in the estuary was similar. In contrast, hatchery subyearling Chinook Salmon were of similar size to wild fish but entered the ocean up to a month earlier. We evaluated the potential importance of these traits on early marine growth by estimating stock-specific growth rates for Chinook Salmon caught in estuarine and ocean habitats. Growth rates were related to relative ocean entry timing, with lower growth rates for stocks that had only recently arrived in marine waters. Our results demonstrate that stocks within a single basin can differ in their size and timing of ocean entry, life history traits that contribute to early marine growth and potentially to the survival of juvenile salmon. Our results also highlight the necessity of considering stock-specific variation in life history traits to understand salmon ecology and survival across the entire life cycle.

The movement of juvenile salmon from freshwater to marine habitats is a poorly understood but critical transition (Pearcy 1992; Pearcy and McKinnell 2007). During this transition, fish must not only physiologically adapt to salt water, but also contend with entirely new prey, predators, and habitats (Spence and Hall 2010). The size at and timing of ocean entry have been identified as important factors during this period. Minor variation in timing can have major consequences for survival (Holtby et al. 1990; Scheuerell et al. 2009; Chittenden et al. 2010; Beamish et al. 2013), while size affects growth and survival via vulnerability to predators and the availability of appropriately sized prey (Ivlev 1961; Mittelbach and Persson 1998). Several recent studies of Pacific salmon *Oncorhynchus* spp. have shown that individuals that survive to adulthood were often larger than average as juveniles (Beamish et al. 2004; Zabel and Achord 2004; Moss et al. 2005; Claiborne et al. 2011; Thompson and Beauchamp 2014), and growth rates during initial marine residence are often correlated with survival in both Atlantic Salmon *Salmo salar* and Pacific salmon (Holtby et al. 1990; Jonsson et al. 2003; Miller et al. 2014) and marine fish in general (Sogard 1997). However, while variation in timing and size at ocean entry is well documented between species or populations occupying independent river basins (e.g., Groot and Margolis 1991; Quinn 2005; Spence and Hall 2010), far less is known about the variation among populations within basins that enter the ocean at a common location (Beamish et al. 2013).

A first step to understanding the influence of size and timing of ocean entry is to document whether these traits vary between stocks, species, or production types (hatchery versus wild) occupying common environments. The Columbia River basin is ideal for this because its ecologically diverse subbasins support numerous populations of Chinook Salmon *O. tshawytscha* and steelhead *O. mykiss* (hereafter referred to collectively as "salmon") that are genetically and phenotypically distinct yet that all enter the ocean at a common location (Rich 1920; Busby et al. 1996; Waples et al. 2004). Although a variety of factors likely influence migration timing (Whalen et al. 1999; Beckman et al. 2000; Achord et al. 2007; Sykes et al.

2009), our fundamental hypothesis was that ocean entry timing would largely be a function of distance to the ocean, i.e., that stocks lower in the basin (closer to the ocean) would enter the ocean earlier than those farther upstream.

The Columbia River also provides an opportunity to document differences between hatchery and wild salmon because of its extensive hatchery production and—in sharp contrast—numerous wild populations that receive protection under the U.S. Endangered Species Act (ESA). Approximately 140 million hatchery salmon are released into the basin each year (Fish Passage Center Web site [www.fpc.org]), while five evolutionarily significant units (ESUs) of Chinook Salmon and five distinct population segments (DPSs) of steelhead are listed under the ESA (Table 1), in large part because of severely depressed population sizes (Ford 2011). We expected that juvenile hatchery salmon would be larger than wild fish, as has been shown in many studies (Quinn 2005; Tatara and Berejikian 2012). However, very little is known about differences in ocean timing between hatchery and wild stocks. We also expected that wild fish would have more variable timing than hatchery fish (Teel et al. 2014) because wild fish respond to environmental cues to initiate migration (Beckman et al. 2000) whereas the downstream movements of hatchery fish are restricted by hatchery release dates. However, it is uncertain whether hatchery fish would tend to have earlier or later timing than wild fish.

While documenting stock-specific variation in the size and timing of hatchery and wild salmon at ocean entry is important to understand the estuarine and marine ecology of salmon, it also provides insight into other processes that may affect depressed wild populations. In particular, there are concerns about potential behavioral interactions between hatchery and wild fish in estuarine or marine habitats where populations that are segregated in freshwater may co-occur (Naish et al. 2008; Rand et al. 2012). The extent of such spatial and temporal overlap and potential size differences between hatchery and wild fish has not been well documented in the Columbia River estuary. In addition, far more is known about hatchery than wild salmon in the Columbia River because of their

TABLE 1. Columbia River evolutionarily significant units (ESUs) for Chinook Salmon and distinct population segments (DPSs) for steelhead as well as genetic stocks and typical hatchery smolt ages (years) for the juvenile salmon used in the analysis. In the first column, the stock's status under the federal Endangered Species Act is indicated by the following abbreviations: N = not warranted, T = threatened, and E = endangered.

ESU/DPS (status)	Genetic stock	Smolt age (years)
	Chinook Salmon	
Lower Columbia River (T)	West Cascade, fall	0
	West Cascade, spring	1
	Spring Creek Group, fall	0
Mid Columbia River, spring (N)	Mid Columbia River, spring	1
Upper Columbia River, spring (E)	Upper Columbia River, spring	1
Upper Columbia River, summer/fall (N)	Upper Columbia River, summer	0, 1
	Upper Columbia River, fall	0
Snake River, fall (T)	Snake River, fall	0, 1
Snake River, spring/summer (T)	Snake River, summer	1
	Snake River, spring	1
Upper Willamette River (T)	Willamette River, spring	1
	Steelhead	
Lower Columbia River (T)	Lower Columbia River, summer–winter	1–3
Mid Columbia River (T)	Mid Columbia River, summer–winter	1–3
Upper Columbia River (T)	Upper Columbia River, summer	1–3
Snake River (T)	Snake River, summer	1–3
Upper Willamette River (T)	Upper Willamette River, winter	1–3

numerical dominance as both out-migrating juveniles (e.g., Roegner et al. 2012; Weitkamp et al. 2012) and returning adults (NRC 1996; Fish Passage Center Web site). Consequently, it is unclear whether the life history traits documented for hatchery fish can serve as valid proxies for those of relatively scarce wild fish.

In this article we document stock-specific variation in size and timing of ocean entry for juvenile hatchery and presumed wild Chinook Salmon and steelhead in the Columbia River estuary, one of the first studies of its kind in a large river basin. We also use data from sampling these same stocks in marine waters to examine the influence of timing on ocean growth. Despite numerous anthropogenic alterations to both salmon and salmon habitats throughout the Columbia River basin (NRC 1996; Williams 2006), our results show that there is variation between stocks and production types that is associated with growth opportunities in marine waters.

METHODS

Our primary objective was to determine whether there were differences in the size and date of capture of different groups of juvenile salmon caught in the Columbia River estuary immediately before ocean entry. These groups included both different stocks (defined by geographic origin, genetic distinctiveness, and life history type) and production types (known hatchery versus presumed wild). A secondary objective was to estimate early ocean growth rates for the Chinook Salmon stocks based on differences in size and timing among the individuals collected in the estuary and those collected during our ocean surveys.

Collection of Fish

The juvenile salmon used in this analysis came from two studies conducted by NOAA Fisheries' Northwest Fisheries Science Center and Oregon State University during 2007–2011. The objective of both studies was to sample juvenile Columbia River salmon, either in the open waters of the lower Columbia River estuary or in marine waters off the Washington and Oregon coasts. We considered the fish collected by the estuary study to represent fish at ocean entry because the study area is close to the mouth of the Columbia River (a passive particle released at the site would exit the estuary within 3 h during a typical ebb tide). Our methodologies are described in detail in Weitkamp et al. (2012) for the estuary study and Teel et al. (2015) for the ocean study and summarized here.

In the Columbia River estuary, juvenile salmon were sampled during daylight hours at two stations, North Channel (46°14.2′N, 123°54.2′W) and Trestle Bay (46°12.9′N, 123°57.7′W) (Figure 1). These stations are located in the lower estuary 17 and 13 km, respectively, from the river's entrance (rkm 0 is the seaward end of the jetties) and adjacent to the deep north and south channels of the lower estuary. Sampling was conducted every 2 weeks from mid-April until late June or early July during 2007–2011. In 2007 and 2008 we also made a single sampling trip in September, while in

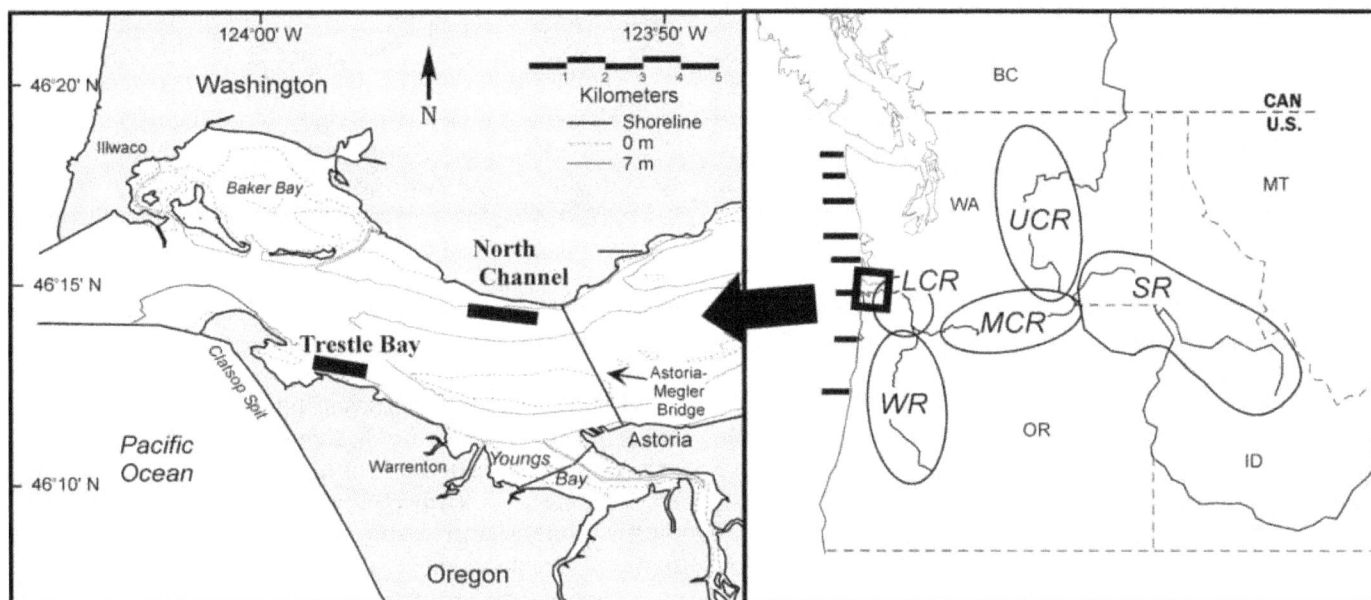

FIGURE 1. Maps showing the locations of the two sampling stations (North Channel and Trestle Bay) in the lower Columbia River estuary (left panel) and the ocean sampling transects (black horizontal lines) and freshwater regions (circles or polygons) within the Columbia River basin (right panel). Region abbreviations are as follows: LCR = Lower Columbia River, WR = Willamette River, MCR = Mid Columbia River, UCR = Upper Columbia River, and SR = Snake River.

2009–2011 we sampled at roughly monthly intervals during July–October.

Estuarine sampling used a fine-mesh purse seine (10.6 m deep × 155 m long; stretched mesh opening, 1.7 cm; knotless bunt mesh, 1.5 cm) set in water 8–10 m deep. All of the juvenile salmon captured were kept in running water until processed, then anesthetized with MS-222, identified to species, and measured (FL [mm]). The fish were checked for the presence of internal tags (passive integrated transponder [PIT] tags and coded wire tags [CWTs]) and fin clips, which are indicative of hatchery origin (see below). We randomly collected (i.e., lethally sampled) 50 individuals each of juvenile steelhead and yearling (age-1) and subyearling (age-0) Chinook Salmon on each cruise. These fish were given a lethal dose of MS-222, processed as above, individually labeled and bagged, and then frozen. Juvenile salmon that were not needed for laboratory analyses were allowed to fully recover and then released. In some cases (e.g., steelhead caught in 2007 and 2008), fin clips were collected from fish prior to release for genetic analysis; these tissues were immediately placed in labeled containers filled with 95% ethanol.

In marine waters, juvenile salmon were caught with a pelagic rope trawl along a series of seven or eight east-west-oriented transects from 48°14′N (Cape Flattery, Washington) to 44°40′N (Newport, Oregon; Figure 1) in late May, late June, and late September. Each transect had 6–7 stations where fishing was conducted with a Nordic 264 rope trawl (mouth opening, 30 m wide × 20 m high). The trawl was towed at the surface at 6 km/h for 30 min at each station, sampling approximately 3 km or 90,000 m^2 of water. The shipboard processing of the juvenile salmon collected in marine waters was similar to that of fish collected in the estuary: all salmon were identified to species, measured (FL [mm]), checked for fin clips, individually labeled and bagged, and immediately frozen.

In the laboratory, we reconfirmed the species of the fish from both studies, remeasured them (FL and total wet weight [g]), and rechecked them for tags and clipped fins. Fin tissue was collected for genetic stock identification (GSI) analysis and placed in 95% ethanol. Snouts were removed from fish with CWTs (see below), and PIT tags were read electronically.

Stock Assignments

Our analysis was restricted to juvenile salmon for which we were able to identify the stock. Stock information came from GSI analysis and PIT or CWT tags. The genetic stocks for both Chinook Salmon and steelhead correspond to ESUs and DPSs, respectively (Busby et al. 1996; Myers et al. 1998; Table 1); by definition, each represents an important component of the evolutionary legacy of the species (Waples 1991). We also caught large numbers Coho Salmon *O. kisutch*, but their limited genetic population structure in the Columbia River does not allow for the genetic differentiation of stocks (Van Doornik et al. 2007) and too few tagged fish were available for analysis.

Our GSI analysis used microsatellite DNA loci and standard DNA preparation techniques (e.g., Teel et al. 2014). We used 13 loci described by Seeb et al. (2007) for Chinook Salmon and the 13 described by Blankenship et al. (2011) for steelhead. Individual fish were assigned to regional genetic stocks using the likelihood model of Rannala and Mountain

(1997) as implemented in the GSI computer program ONCOR (Kalinowski et al. 2007). The origins of the Chinook Salmon sampled in the estuary were estimated using genetic stocks described by Teel et al. (2014), while the genetic baseline for the ocean-caught Chinook Salmon included the same baseline supplemented with populations ranging from California to southern British Columbia (Teel et al. 2015). Genetic stock assignment of the steelhead sampled in the estuary was made using baseline population data reported by Stephenson et al. (2009) and Blankenship et al. (2011). Steelhead caught in marine waters were not genetically analyzed because regional standardized DNA baselines do not exist at present (D. M. Van Doornik, unpublished data).

For both the Chinook Salmon and steelhead, we used probabilities of 0.8 or greater to assign individual fish to a genetic stock. Comparisons of our genetic estimates with the known origins (from internal tags) of 536 Chinook Salmon and 62 steelhead indicated that this restriction resulted in correct assignments for 85.3% of the Chinook Salmon and 97.8% of the steelhead (the tags were assumed to be correct; L. Weitkamp, unpublished data).

We also used information available for fish tagged with CWTs or PIT tags to determine stock, production type (hatchery or wild; see the next section), and Chinook Salmon age. For these fish, the tags were extracted, the codes were read, and release information was downloaded from the appropriate online database: the Regional Mark Information System for CWTs (Pacific States Marine Fisheries Commission [www.rmpc.org]) or the PIT Tag Information System for PIT tags (Pacific States Marine Fisheries Commission [www.ptagis.org]). Release information included the hatchery, stock, release location, run timing, production type, release size, and release date. This information was used to assign tagged fish to a genetic stock using the criteria provided by Fisher et al. (2014).

Size-Based Age Designations

Juvenile Chinook Salmon enter marine waters as either subyearling (age-0) or yearling (age-1) smolts. These two age-classes are associated with specific life history types (Healey 1983, 1991). In the Columbia River basin, spring runs of Chinook Salmon typically have yearling smolts, fall runs have subyearling smolts, and summer runs can have subyearling or yearling smolts (Waples et al. 2004; the terms "spring," "summer," and "fall" refer to the season in which adults return to freshwater). However, hatchery practices have allowed the production of smolts outside the typical smolt age (e.g., fall Chinook Salmon released as yearlings).

Juvenile Chinook Salmon were assigned to age categories based on their length. In the estuary the cutoff length ranged from 115 mm in April to 140 mm on July 1, while the cutoff used in ocean collections ranged from 120 mm in May to 250 mm in September. These cutoffs were derived from (1) seasonally adjusted length frequency histograms, (2) known

ages based on scale analysis, and (3) known ages determined from PIT tags or CWTs (Pearcy and Fisher 1990; Fisher and Pearcy 1995; Weitkamp, unpublished data; J. Fisher, Oregon State University, unpublished data).

The validity of this age assignment was confirmed by stock-specific plots of fish size versus capture date, which typically indicated a clear separation between yearling and subyearling individuals in both estuarine and marine environments. However, several small unclipped (presumed wild) mid and upper Columbia River and Snake River spring Chinook Salmon caught in the estuary were reassigned as yearlings due to a continuous size distribution (i.e., no obvious size break between small and large individuals) and timing identical to that for larger individuals (i.e., much earlier than that of individuals identified as subyearlings from these stocks). In contrast, several yearling-sized hatchery Spring Creek Group fall Chinook Salmon were known to be subyearlings (from CWTs) and formed a continuous size distribution within this stock. Accordingly, all Spring Creek Group fish caught in the estuary were assigned as subyearlings, as were those less than 300 mm caught in marine waters. These changes involved age reassignments of only small numbers of Chinook Salmon (<3%).

It should be noted that Columbia River steelhead also exhibit variation in smolt age, with hatchery fish typically being released as yearlings (Fish Passage Center Web site) and wild steelhead smolting after 1–3 years in freshwater (Busby et al. 1996). Unlike with Chinook Salmon, however, this age difference is not associated with other life history traits (Busby et al. 1996), so the steelhead in our study were not segregated by age.

Production Types

We determined production type—hatchery or presumed wild—based on both external marks and information from internal tags. Mass-marking programs in the Columbia River externally mark most (>75%) hatchery fish by clipping their adipose fins (Regional Mark Processing Center; Table A.1.1 in the appendix to this article). However, because large numbers of unmarked hatchery fish are released each year, fish with unclipped adipose fins may be either unmarked hatchery fish or wild fish. We were able to determine production type (largely hatchery) for unclipped fish with internal tags (PIT or CWT) from release information, although most unclipped juvenile salmon were not tagged. Consequently, our analysis relies on comparisons between two production types of fish: known hatchery and a combination of wild and unclipped hatchery fish (hereafter referred to as "unclipped" fish).

Data Analysis

Size and capture date in the Columbia River estuary.—For our primary analysis, we compared the size (length and weight) and date at capture in the estuary for yearling and subyearling Chinook Salmon and steelhead by stock, production type, and year. Comparisons of body shape (condition factor) between stocks and production types did not show consistent differences and are not provided here (Weitkamp, unpublished data). Due to changes in sampling effort during the summer months, comparisons of subyearling Chinook Salmon size and timing were restricted to the years 2009–2011.

We modeled differences in length and weight by year, stock, and production type using Bayesian multilevel models (Gelman and Hill 2006). This provides ANOVA-like results while allowing for nonconstant variance and random effects. When modeled as a random effect, a category's batch of parameters (e.g., means for each stock) are allowed to have different values but are assumed to come from a common normal distribution (with estimated mean and standard deviation). This tends to shrink the individual estimates toward the overall mean, with more shrinkage being applied to individual estimates further from the overall mean and/or with higher uncertainty. This is referred to as partial pooling and provides a compromise between complete pooling across a category (e.g., assuming that groups are equal and combining all stocks into a single group) and making independent estimates for each group.

The models for each species and age-class included all main effects (year, stock, and production type) and all two- and three-way interactions. Within-group variance was assumed to be equal across groups except for subyearling Chinook Salmon length and weight, where the variance varied visibly by stock and was therefore allowed to vary by stock. The group means for production type were modeled independently (i.e., no partial pooling) since there were only two categories (hatchery and unclipped).

The fish sampled in this study do not represent random samples from the populations of interest due to the constraints of this type of sampling. However, the regular timing of the sampling events likely produced samples that were approximately representative of the larger population. For the lengths and weights, the variability within sampling events was large relative to the differences between adjacent sampling events. Therefore, ignoring the discrete-sampling-event structure (a violation of the assumption of independence) is unlikely to have had a large effect on the results.

However, in the timing data all of the arrival times are exactly the same within each event. This introduces a strong violation of the assumption of independence. In addition, any changes in the spacing of the sampling events (as occurred later in the season) or truncation of the season (e.g., for yearling Chinook Salmon) could introduce substantial bias. Migration timing for groups of fish was therefore modeled by summing the total number of fish within the group for the individual biweekly sampling events and modeling these counts using a smooth function that increased and then decreased as a function of the day of the year. Specifically, we used the normal density function in which the date of predicted highest

abundance (the mean) was modeled like length, with all main effects and two- and three-way interactions being included. The standard deviation (which controls the width of the normal density function) was modeled in the same way as the mean.

Individual models for length, weight, and timing were fit to each species and age-class (yearling and subyearling Chinook Salmon and steelhead). As an overall model summary, we plotted the variance components associated with the different main effects and interactions. This is comparable to conducting a classic analysis of variance (ANOVA) while taking advantage of partial pooling (Gelman 2005; Gelman and Hill 2006). We also plotted the main effects along with their 95% credible intervals (similar to confidence intervals; Gelman et al. 2013). The group-specific 95% credible intervals are also superimposed on plots of the data. We focus on the overall and relative magnitudes of the differences and their biological relevance as opposed to significance tests, as it is unlikely that any two groups would be exactly equal. For specific comparisons we report the mean difference and the 95% credible interval. For other comparisons, nonoverlapping credible intervals serve as an approximate (conservative) test of the difference between two groups.

A complete description of the models, the method of fitting them, and the approaches used to assess model fit are described in Appendix A.2.

Ocean growth rates.—We estimated growth rates during the first weeks or months in marine waters by comparing the mean size and timing of fish from each stock caught in the estuary with those caught in marine waters during our ocean surveys. This analysis was restricted to hatchery fish due to the small numbers of unclipped fish caught in both environments. Ocean growth rates were estimated both across all years (to provide robust sample sizes) and by year for years in which at least five fish from a stock were caught in both estuarine and marine waters. As the two growth rates were highly correlated (Spearman correlation: $r > 0.9$, $P < 0.05$), annual growth rates averaged across years are provided (the standard deviations reflect the interannual variation). Too few steelhead of known origin (from tags) were caught in marine waters to estimate growth rates.

Early ocean growth rates estimated from changes in length (G_L [mm/d]) were calculated as

$$G_L = (L_o - L_e)/(t_o - t_e),$$

where L_o is the mean fork length in the ocean at mean recovery time t_o and L_e is the mean length in the estuary at recovery time t_e. Because changes in weight typically assume an exponential rather than a linear form (Ricker 1975), instantaneous changes in weight (G_W [g $\cdot$ g^{-1} $\cdot$ d^{-1}]) were calculated as

$$G_W = [\ln(W_o) - \ln(W_e)]/(t_o - t_e),$$

where W_o is the mean weight in the ocean at mean recovery time t_o and W_e is the mean weight in the estuary at recovery time t_e. We expressed G_w as percent body weight per day (% BW/d) by taking the antilog and multiplying it by 100.

Many juvenile salmon were still being caught in the estuary during the ocean surveys. To minimize the influence of possible size differences between early- and late-migrating fish within a stock, we restricted estimates of L_e and W_e to fish caught in the estuary before the 20th of each month in the estuary surveys because the mean start dates for the ocean surveys were May 22 (range, May 19–24), June 21 (range, June 20–23), and September 22 (range, September 20–23), respectively.

RESULTS

Available Data

We had information on the time and length at capture for 724 yearling and 1,289 subyearling Chinook Salmon and 641 steelhead of known origin that were collected in the lower Columbia River estuary during 2007–2011. These fish belonged to 5 different stocks of steelhead and 11 stocks of Chinook Salmon (Table 1). The totals represent stock information provided by 604 CWTs and 58 PIT tags. Chinook Salmon stocks that were abundant in the estuary were then compared with the same groups collected in marine waters, specifically 1,668 yearling and 664 subyearling Chinook Salmon.

Stock-Specific Size and Timing in the Estuary

We observed considerable variation in the size and timing of juvenile salmon in the Columbia River estuary. These differences were generally largest among stocks, with occasional large differences between production types and in almost all cases much smaller differences among years (Figure 2). The variances associated with the length and weight residuals were higher than those of any factor, indicating high within-stock variation (Figure 2). Yearling Chinook Salmon and steelhead originating from lower-river locations generally had earlier timing than those from upper-river locations. In contrast, the timing of subyearling Chinook Salmon was related to hatchery release timing and was more variable between hatchery and unclipped fish (Figures 3–5). The length and weight data showed comparable patterns (see Appendix A.2 for the weight data).

Steelhead.—The estimated peak arrival date of juvenile steelhead in the estuary varied primarily by stock, with the two stocks originating closest to the river's mouth (the lower Columbia and Willamette River stocks) arriving several days earlier than other stocks (Figure 3). There was some evidence of slightly earlier arrival times for hatchery fish, but this varied by stock and year. Length varied primarily by stock and origin (Figure 2), with the lower Columbia River steelhead being

more than 10 mm shorter than the other stocks (Figure 3) and hatchery fish an estimated 25 mm longer (95% CI = 18–31 mm) than unclipped fish.

Yearling Chinook Salmon.—The timing of yearling Chinook Salmon varied primarily by stock and to a lesser degree by year (Figure 2), with the estimated peak arrival date for the Willamette River spring stock being 30 d earlier (95% CI = 19–41 d) than that for the upper Columbia River summer–fall stock (Figure 4). Length varied most by stock and production type, with the Snake River spring and mid and upper Columbia River spring stocks being 10–20 mm shorter than the other stocks. The estimated average length of hatchery fish was 11 mm longer (95% CI = 5–16 mm) than that of unclipped fish. However, this result was driven primarily by the mid and upper Columbia River spring stock, since the other stocks had very few unclipped fish (Figure 4).

Subyearling Chinook Salmon.—The timing of subyearling Chinook Salmon varied by stock and to a lesser extent by production type (Figure 2). The estimated peak arrival date for the earliest-arriving stock (the Spring Creek Group fall stock) was 89 d earlier (95% CI = 64–126 d) than that for the latest stock (the west Cascade fall stock; Figure 5). The estimated peak arrival date for hatchery fish was on average 22 d earlier than that for wild fish (95% CI = 8–37 d). Length varied by stock, following the same pattern as peak arrival date (i.e., stocks with smaller fish arrived earlier). The Spring Creek Group fall stock was estimated to be 30 mm shorter (95% CI = 22–38 mm) than the west Cascade fall stock (Figure 5).

Early Ocean Growth Rates

Juvenile salmon from stocks that had been in the ocean more than a few weeks had higher growth rates than those that had just arrived in marine waters. This pattern was consistent across the yearling and subyearling Chinook Salmon age-classes.

Yearling Chinook Salmon.—Willamette River and west Cascade spring Chinook Salmon had the earliest capture dates in the estuary (Figure 4) and the highest estimated growth rates (≥1.6 mm/d, >3.6% BW/d) when captured in the ocean in late May, 3–4 weeks after ocean entry (Figure 6). In contrast, the growth rates estimated for later-migrating stocks were much lower (≤1.1 mm/d, ≤2.9% BW/d) by the May ocean surveys, consistent with their recent arrival (~2.5 weeks) in marine waters. By the June surveys, however, the growth rates of both early- and late-migrating stocks were similar (Figure 6), reflecting an extended residence (≥4 weeks) in marine waters for all stocks.

Subyearling Chinook Salmon.—The relationship between time of ocean entry and growth rate for subyearling Chinook Salmon followed the same pattern as for yearling Chinook Salmon (i.e., at the time of ocean sampling, earlier migrants had grown more than later migrants). The early-migrating Spring Creek Group fall and Snake River fall Chinook Salmon stocks were the only subyearling stocks caught in sufficient

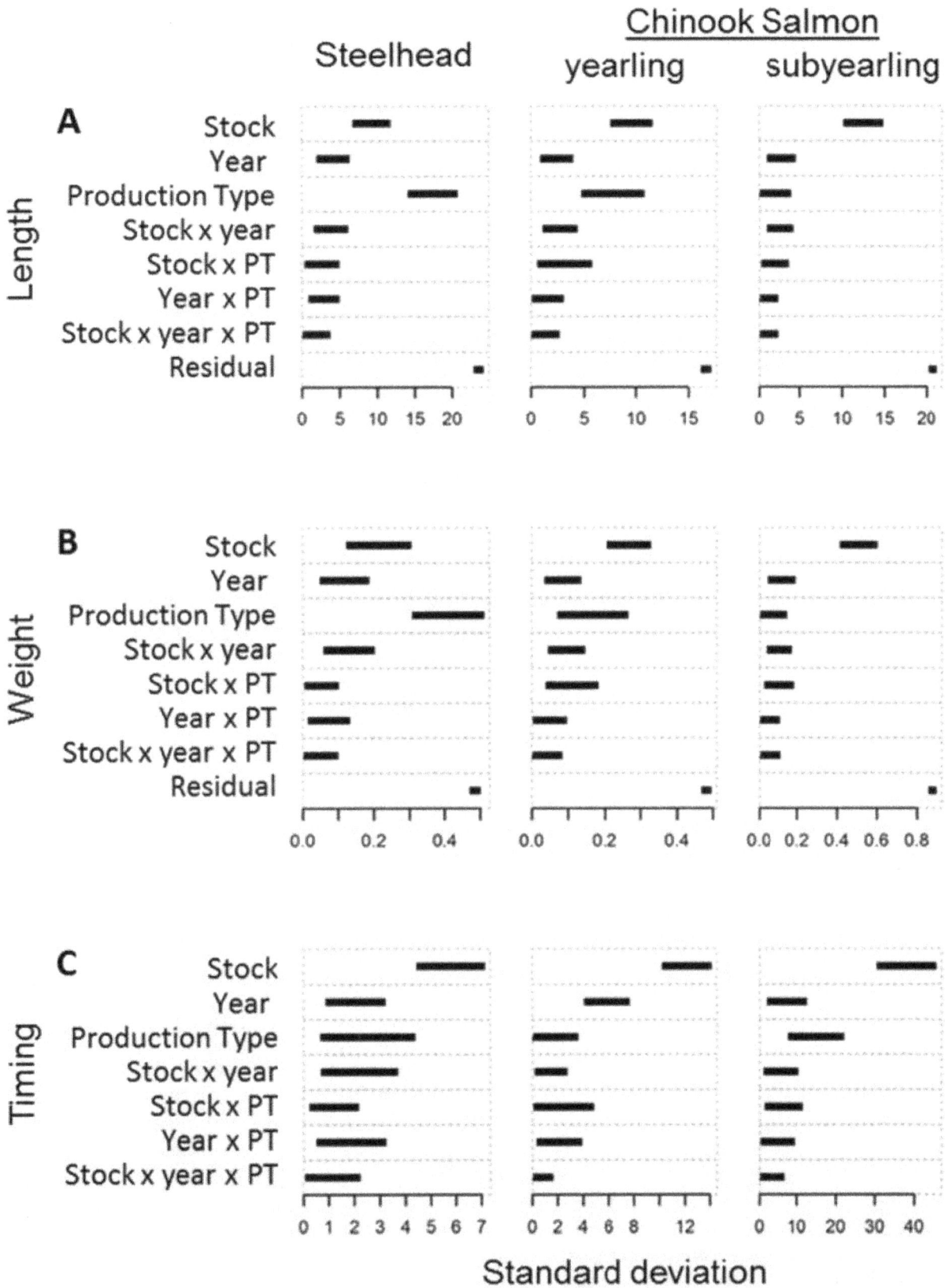

FIGURE 2.   Summary of the variance components for each model (rows) for the three species and/or age groups of juvenile salmonids (columns) from analyses of **(A)** length, **(B)** weight, and **(C)** timing of ocean entry using Bayesian multilevel models. The bars represent 80% credible intervals for the estimated standard deviations between groups for the main effects (stock, year, and production type [PT, i.e., hatchery or unclipped {presumed wild} fish]) and their interactions. Due to the highly unbalanced data, this variation does not indicate the degree of variability explained.

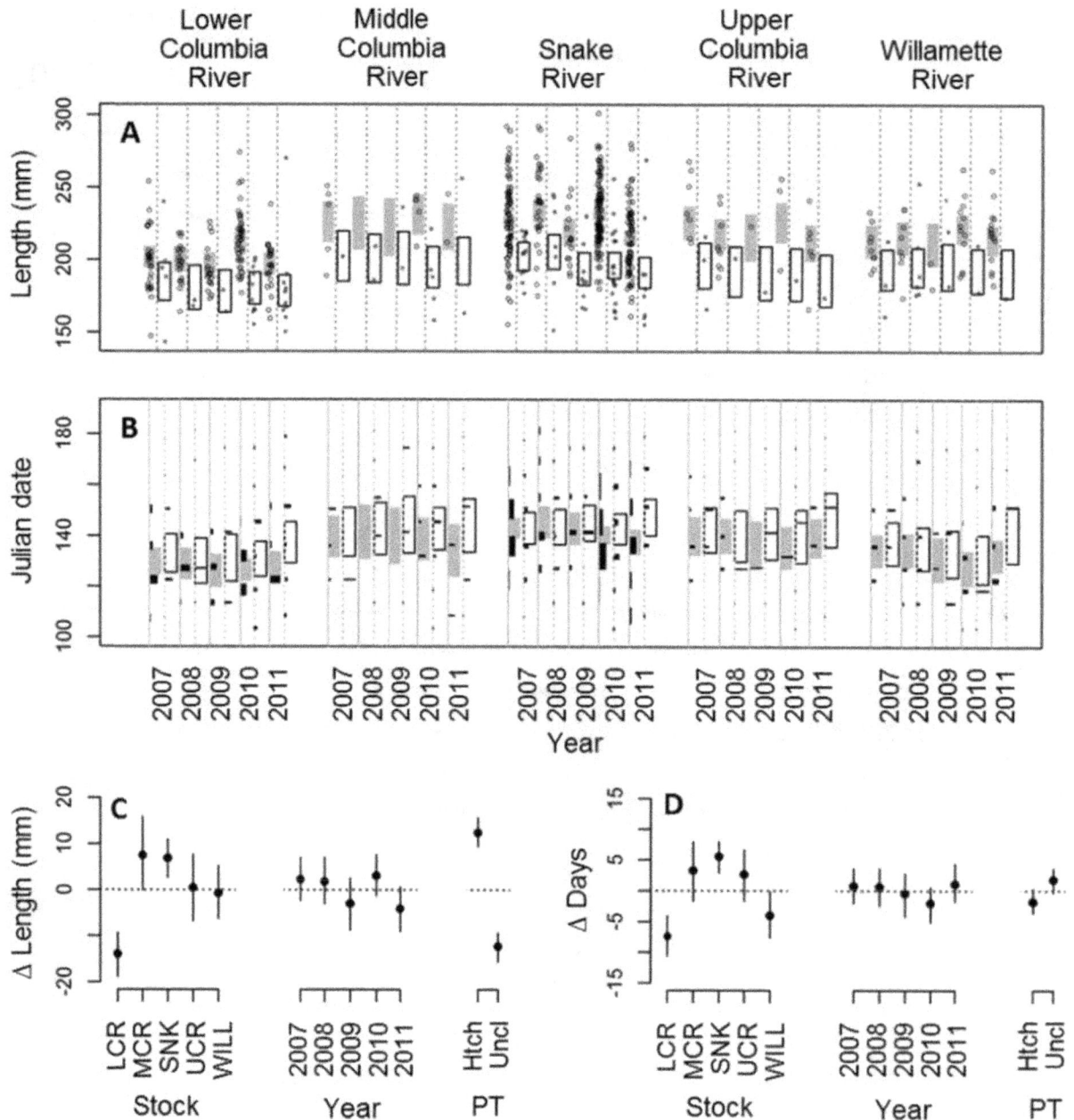

FIGURE 3.   Comparisons of mean length and date at capture in the estuary for juvenile steelhead by stock, production type, and year. Panels (A) and (B) show plots of mean length and peak migration date, with model-based 95% credible intervals. The dots indicate individual fish. Gray bars and gray dots represent hatchery fish, open bars and black dots represent unclipped fish. Panels (C) and (D) pertain to the main effects for length and timing. The points are the estimated deviations from the mean for each level of the main effect, and the vertical lines are the corresponding 95% credible intervals. The stock abbreviations are as follows: LCR = lower Columbia River, MCR = mid Columbia River, SNK = Snake River, UCR = upper Columbia River, and WILL = Willamette River. The production type (PT) abbreviations are as follows: Htch = hatchery and Uncl = unclipped.

numbers during the June ocean survey to estimate growth, at which time Spring Creek Group fish had higher growth rates (1.0 mm/d, 2.6% BW/d) than Snake River fall Chinook Salmon (0.1 mm/d, 0.3% BW/d; Figure 7), consistent with the former's approximately 2-week-earlier timing (Figure 5). By the September ocean surveys, however, juveniles from all subyearling stocks had been in marine waters for at least 1 month; at that time the estimated growth rates were fairly similar (0.8–

1.1 mm/d, 2.1–2.6% BW/d) among stocks (Figure 7) despite the fact that the average dates of ocean entry varied by nearly 90 d (Figure 5).

DISCUSSION

We have demonstrated that there are stock-specific differences in the size at and timing of ocean entry for juvenile

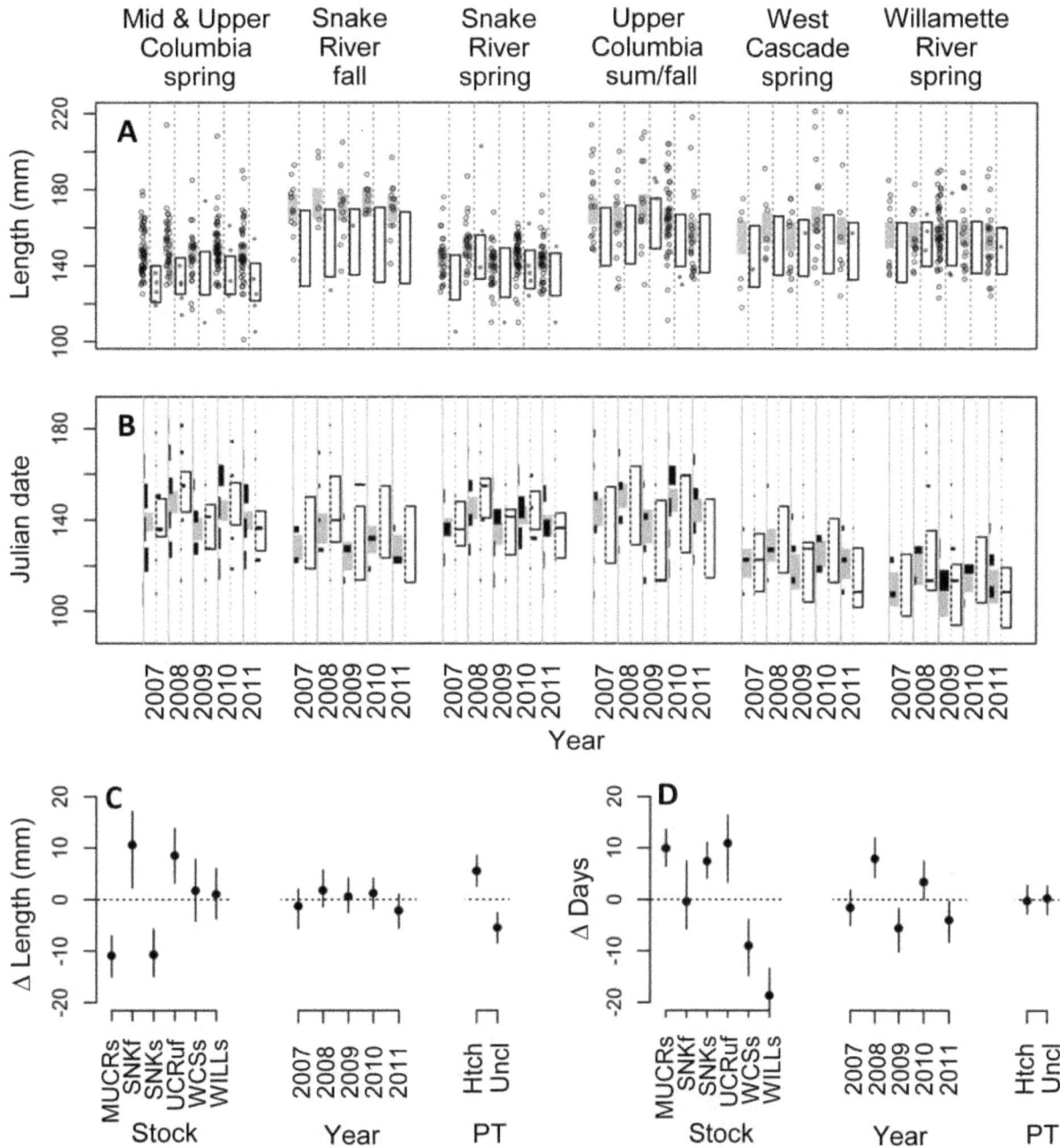

FIGURE 4. Comparisons of mean date and length at capture in the estuary for yearling Chinook Salmon by stock, production type, and year. The stock abbreviations are as follows: MUCRs = mid and upper Columbia River, spring; SNKf = Snake River, fall; SNKs = Snake River, spring; UCRuf = upper Columbia River, summer–fall; WCSs = west Cascade, spring; and WILLs = Willamette River, spring. See Figure 3 for additional details.

Columbia River Chinook Salmon and steelhead that correspond to differences in initial growth opportunities in marine waters. These size and timing differences likely interact with a suite of other factors, including prey availability and predator abundances, to influence survival in estuarine and marine waters. For example, stock-specific variation in the consumption of juvenile salmon by avian predators in the Columbia River estuary has been attributed to stock differences in size,

timing, and behavior (Collis et al. 2001; Ryan et al. 2003; Sebring et al. 2013). While there may be conditions under which early timing results in beneficial growth opportunities or survival (Scheuerell et al. 2009; Satterthwaite et al. 2014), other conditions may select against early timing, leading to survival advantages for later timing (Ryan et al. 2003; Beamish et al. 2013). Similar advantages and disadvantages also likely occur for variation in fish size (e.g., Willette et al.

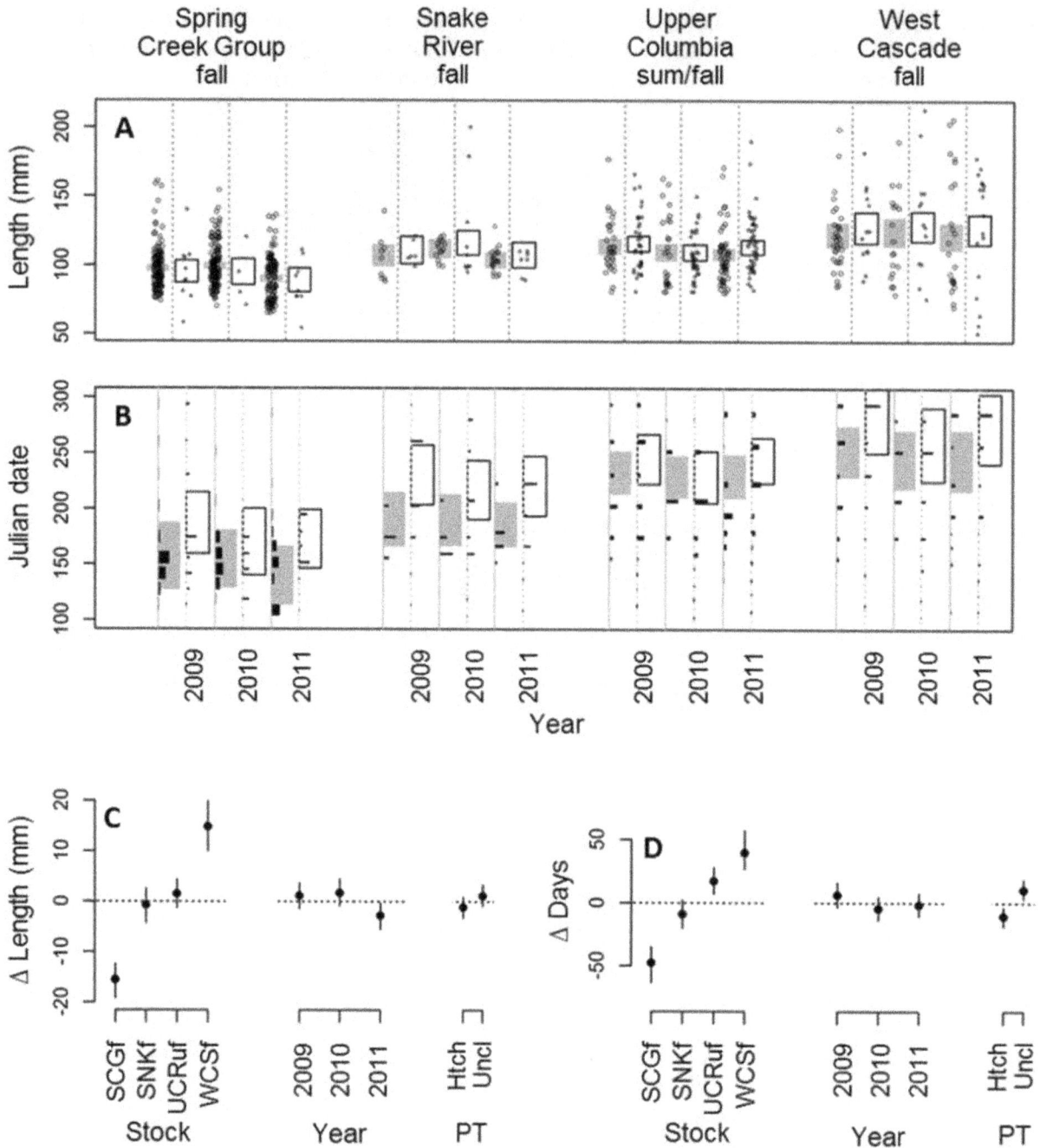

FIGURE 5. Comparisons of mean date and length at capture in the estuary for subyearling Chinook Salmon by stock, production type, and year. The stock abbreviations are as follows: SCGf = Spring Creek Group, fall; SNKf = Snake River, fall; UCRuf = upper Columbia River, summer–fall; and WCSf = west Cascade, fall. See Figure 3 for additional details.

2001). This idea is supported by studies which demonstrate that early timing or large size has survival advantages in some years but not others (Fisher and Pearcy 1988; Tomaro et al. 2012; Miller et al. 2014).

Although we observed stock-specific differences in timing, high within-stock variability resulted in many stocks being present in the estuary for a month or more. Consequently, multiple stocks of salmon from throughout the Columbia River basin were present in the estuary at the same time, including hatchery and unclipped individuals (Figures 3–5). This high temporal and spatial overlap, which has also been observed in shallow habitats in the Columbia River estuary (Teel et al. 2014), suggests high potential for competitive interactions between hatchery and presumed wild fish if resources are limited. The larger size of hatchery individuals may also give them a competitive advantage over small wild fish if larger

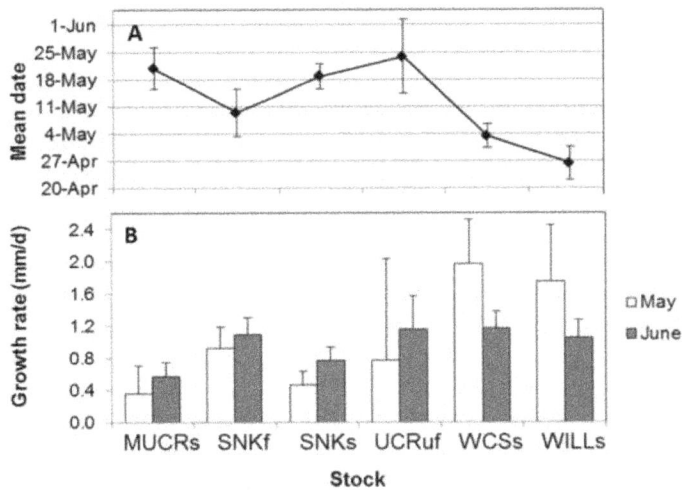

FIGURE 6. (A) Mean dates found in the Columbia River estuary and (B) growth rates as estimated by changes in length for yearling Chinook Salmon, by stock. The stock abbreviations are as follows: MUCRs = mid and upper Columbia River, spring; SNKf = Snake River, fall; SNKs = Snake River, spring; UCRuf = upper Columbia River, summer–fall; WCSs = west Cascade, spring; and WILLs = Willamette River, spring. The error bars represent standard deviations (estimated between years).

size is beneficial in such interactions (Tatara and Berejikian 2012).

High temporal overlap among stocks in the estuary may force wild fish to interact with abundant hatchery fish from other basins, even if hatchery production in their "home" basin is deliberately limited to minimize ecological or behavioral

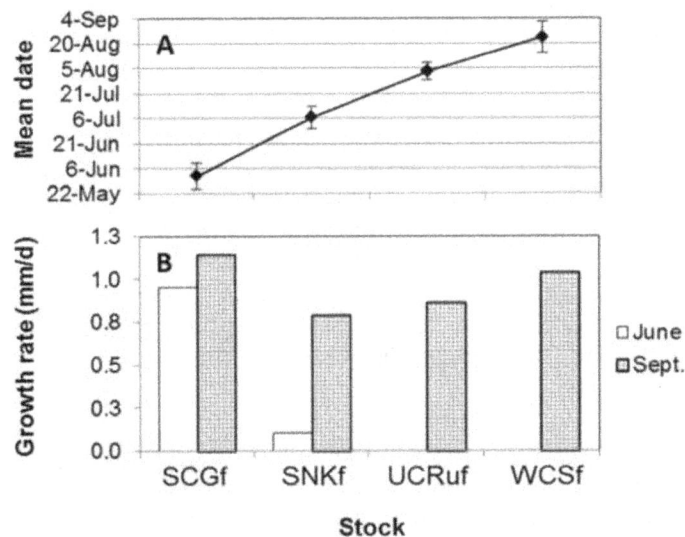

FIGURE 7. (A) Mean dates found in the Columbia River estuary and (B) growth rates as estimated by changes in length for subyearling Chinook Salmon, by stock. The stock abbreviations are as follows: SCGf = Spring Creek Group, fall; SNKf = Snake River, fall; UCRuf = upper Columbia River, summer–fall; WCSf = west Cascade, fall. The error bars represent standard deviation (estimated between years).

interactions (Paquet et al. 2011). Whether wild fish are negatively affected by interactions with hatchery fish in the estuary has not been evaluated, but our findings, paired with those of Teel et al. (2014) for shallow habitats, indicate that the opportunity for such interactions clearly exists in the Columbia River estuary.

Comparisons of size and timing between hatchery and presumed wild salmon in the estuary also indicate the extent to which the patterns documented for abundant hatchery fish may serve as proxies for those for scarce wild fish. The timing of hatchery yearling Chinook Salmon and steelhead stocks was generally similar to that of wild stocks, suggesting that the timing of hatchery fish can be used to represent that of wild fish in the absence of data on the latter. By contrast, the size of yearling Chinook Salmon and steelhead and the size and timing of subyearling Chinook Salmon differed markedly between hatchery and wild fish. For these groups, using data from hatchery fish for wild fish would clearly misrepresent the timing and/or size of wild fish. While there may be situations in which the complete absence of data for wild populations requires the use of hatchery-based data, our results provide insight into when it might (and might not) be appropriate.

Our results also emphasize the direct connection between freshwater and marine habitats for juvenile salmon, because freshwater conditions that affect ocean entry timing also influence initial marine growth opportunities. For example, the large size difference between Willamette River (202 mm, 100 g) and Snake River (149 mm, 35 g) spring Chinook Salmon caught in marine waters in May might be explained by differential use of marine habitats of differing quality (e.g., Tucker et al. 2009). However, when one considers that Willamette River spring Chinook Salmon have occupied marine waters for a month longer than Snake River spring Chinook Salmon (Figure 4), it is apparent that the size difference largely reflects time spent in productive marine waters rather than location (and therefore habitat quality) within those waters. A similar conclusion was reached by a recent study of juvenile Sockeye Salmon O. nerka in marine waters, which found that stock-specific size at recovery was related to smolt size and ocean entry timing in addition to marine growth rates (Beacham et al. 2014). Furthermore, if the ocean entry timing of hatchery fish largely reflects hatchery practices (see the next section), hatchery practices can directly influence initial marine growth opportunities.

Influence of Hatchery Practices on Observed Patterns

Our results indicate that the size at and timing of the ocean entry of hatchery fish largely results from hatchery practices, including release timing, distance to the river's mouth, and size at release. Our prediction that stocks with early migration timing would originate closer to the Columbia River mouth than those with later timing was strongly supported by our results for yearling Chinook Salmon and steelhead (Figures 3, 4).

Data from tagged individuals also indicate that this earlier timing is influenced by migration distance rather than migration speed because lower-river stocks migrated at slower rates than those farther upriver (Table A.1.2), a phenomenon also observed by Dawley et al. (1986). However, ocean entry timing is also affected by hatchery release timing, as illustrated by the late release date and late timing of west Cascade fall subyearling Chinook Salmon despite their location near the river's mouth (Table A.1.1; Figure 5). The extremely early timing of Willamette and west Cascade spring Chinook Salmon likely results from their location low in the basin and their early hatchery release times (Table A.1.1).

The size of yearling hatchery Chinook Salmon and steelhead in the estuary can be explained by hatchery release size. The unusually large size of Snake River steelhead in the estuary (Figure 3) is consistent with their large size at release (97 g) relative to other stocks (<90 g; Table A.1.1). Yearling Chinook Salmon stocks that were exceptionally large (Snake River, spring) or small (mid and upper Columbia River, spring) when captured in the estuary (Figure 3) were also large (63 g) or small (28 g) at their release from hatcheries. Overall, the size at recovery in the estuary of individuals tagged with CWTs was positively related to the size at hatchery release for both yearling Chinook Salmon and steelhead (linear regression: $r^2 \geq 0.23$, $P < 0.01$; $n = 322$ yearling Chinook Salmon, 57 steelhead).

In contrast, the relationship between the size of hatchery subyearling Chinook Salmon in the estuary and their size at release was much weaker (linear regression: $r^2 = 0.04$, $P < 0.01$; $n = 197$). This is likely because some subyearling Chinook Salmon stocks have extended residency in the estuary before entering marine waters (Reimers and Loeffel 1967; Sebring et al. 2013) and continue to grow throughout the summer (Campbell 2010). This is particularly true for west Cascade fall Chinook Salmon, which are released at approximately the same size (8.5–8.8 g) as other subyearling stocks (Table A.1.1) but which are recovered at the mouth much later and at larger sizes than other stocks (Figure 5).

Differences between Hatchery and Wild Fish

We predicted that hatchery fish would be consistently larger than unclipped fish, and our expectations were confirmed for yearling Chinook Salmon and steelhead. This size difference was most pronounced for upper Columbia River steelhead, with hatchery fish being up to 15% longer and weighing 42% more than unclipped fish from the same stock, even though hatchery fish were 1–2 years younger than wild fish (Busby et al. 1996). In contrast, hatchery subyearling Chinook Salmon were not larger than wild fish from the same stocks, but their timing was up to a month earlier. The size of subyearling Chinook Salmon increases throughout the summer (Roegner et al. 2012; Weitkamp, unpublished data), so that if the timing of hatchery and wild subyearling Chinook Salmon from the same

stock were identical hatchery fish would be larger than wild fish, as we predicted.

We did not expect that the timing of hatchery and presumed wild yearling Chinook Salmon and steelhead would be as similar as they were (Figures 3, 4), given that the release timing of hatchery fish is artificially controlled. This suggests that once fish leave a hatchery they initiate active migration by responding to the same environmental cues that wild fish use (e.g., temperature, flow, and photoperiod; Whalen et al. 1999; Beckman et al. 2000; Achord et al. 2007; Sykes et al. 2009). Evidence that fish released from hatcheries do not begin migrating immediately comes from comparisons of the migration rates to the river's mouth estimated for juvenile salmon tagged with CWTs and released from hatcheries with the "active" migration rates for salmon tagged with PIT tags and detected at intermediate dams (Table A.1.2). The active migration rates that we estimated (61–79 km/d) and that others have reported (>50 km/d; Welch et al. 2008; Harnish et al. 2012) were much higher than the CWT-based migration rates that we estimated (3–35 km/d; Table A.1.2) or those reported by Dawley et al. (1986) (3–36 km/d), suggesting that fish do not immediately initiate downstream migration when released from hatcheries.

Subyearling Chinook Salmon were unique because the timing of hatchery fish caught in the estuary was up to a month earlier than that of unclipped fish from the same stock (Figure 5). This large difference may be explained by differential habitat use by hatchery and wild fish. Wild subyearling Chinook Salmon in the Columbia River estuary make greater use of shallow-water habitats and have broader temporal distributions than subyearling hatchery fish, which are abundant in deep-channel habitats (Dawley et al. 1986; Roegner et al. 2012; Weitkamp et al. 2012; Teel et al. 2014). Consequently, the environmental conditions experienced by wild fish are likely different from those experienced by hatchery fish, which results in differences in timing cues and migration timing. Although the range of ocean entry size and timing that we observed for subyearling Chinook Salmon may be less than that reported historically (Rich 1920; Burke 2004), our findings suggest that substantial diversity still exists in the use of estuarine habitats by both hatchery and wild subyearling Chinook Salmon, resulting in ocean entry times spanning the period from May through October (Figure 5).

Validity of Ocean Growth Assumptions

We estimated initial marine growth rates by sampling the same stocks of salmon in estuarine and marine environments. This analysis was based on the assumption that the fish caught in the estuary and ocean were representative of their respective stocks and that the changes in size were due to growth rather than to other factors, such as emigration/immigration in either habitat, interannual variation in size or timing among stocks, and size-selective mortality. Several lines of evidence indicate that this assumption is reasonable, as is our primary finding: stocks that have only recently arrived in marine waters have lower growth rates than those that have spent more than a few weeks in productive marine habitats.

An important feature of our analysis of growth rates is that our sampling was conducted in the primary freshwater emigration and early–ocean rearing habitats for the stocks we examined. Research over the last four decades has consistently demonstrated that yearling and subyearling Chinook Salmon are abundant in the deep waters of the Columbia River estuary (Dawley et al. 1986; Bottom and Jones 1990; Harnish et al. 2012; Weitkamp et al. 2012) and that the adjacent marine region offers major rearing and migration habitats for these same stocks (Fisher and Pearcy 1995; Fisher et al. 2014; Teel et al. 2015). We also captured over 1,000 Chinook Salmon from 197 CWT release groups (i.e., individuals within a group have the same tag code) in both estuarine and marine habitats. These recoveries confirm that we sampled the same groups of fish in the two habitats, and the growth rates estimated from this subset of fish showed the same pattern of slow growth for stocks that had just arrived in marine waters (Weitkamp, unpublished data).

Although annual variability is an important consideration in evaluating growth, we estimated growth rates across years to maximize our sample sizes for comparisons among stocks. Our analysis likely benefited from consistent hatchery releases (in terms of size, timing, and abundance), river flow, and ocean conditions during our study years (Fish Passage Center Web site; Peterson et al. 2014), factors which could alter stock-specific growth patterns.

Whether our reported growth rates are influenced by size-specific mortality is more difficult to determine—and a source of potential bias for all of the studies that use this method (e.g., Fisher and Pearcy 1988; Beamish et al. 2008; MacFarlane 2010). We recalculated the May growth rates for the two stocks with the earliest ocean entry timing (Willamette and west Cascade spring Chinook Salmon) using ocean recovery sizes that were 10% smaller than those we observed (to mimic size-selective predation), and the resulting growth rates (0.9 mm/d, 2.2% BW/d) were still generally higher than our estimated growth rates for stocks with later timing (Figure 6). Numerous predators are known to prey on juvenile salmon in estuarine and marine habitats (Emmett 1997; Collis et al. 2001; Emmett et al. 2006; Zamon et al. 2013), although the extent to which this predation is size selective is not known. However, it is difficult to imagine a predation scenario that would result in the patterns that we observed in marine waters in May and June: high predation (to produce high growth rates) on some stocks but not on others within the same general area.

Studies designed to evaluate size-selective mortality have either failed to find evidence of it or have found mixed results (occurring in some but not all years) for Chinook Salmon, including studies focused on Columbia River salmon

(Claiborne et al. 2011, 2014; Tomaro et al. 2012; Miller et al. 2013, 2014). Early ocean growth rates estimated from otoliths for three of the stocks included here (and some of the same individuals) were very close to our estimated growth rates (Tomaro et al. 2012; Claiborne et al. 2014; Miller et al. 2014), suggesting that our results are reasonably representative of the true growth rates.

We may have overestimated growth rates in two cases: those of Willamette River spring and west Cascade spring Chinook Salmon caught in marine waters in late May, for which our rates are extremely high ($\geq$1.75 mm/d, >3.7% BW/d; Figure 6). We suspect that these high rates are due, in part, to poor estimates of ocean entry timing. The abundance of the Willamette River spring stock in the estuary was highest when sampling commenced in mid-April, while that of the west Cascade spring stock peaked approximately 1 week later (Figure 4), so that our sampling schedule might miss early-migrating individuals. Juvenile Willamette River spring Chinook Salmon are caught off the west coast of Vancouver Island (300–550 km from the mouth of the Columbia River) in April, and both Willamette River and west Cascade spring Chinook Salmon are caught in Southeast Alaska (over 1,300 km from the river) in June (Tucker et al. 2011; Fisher et al. 2014), which is consistent with extremely early ocean entry. Consequently, the high growth rates that we report for these two stocks may exceed the actual growth rates if we missed early migrants in the estuary and therefore estimated a later date of ocean entry than actually occurred. For example, the growth rates for Willamette River and west Cascade spring Chinook Salmon calculated using ocean entry dates 3 weeks earlier (April 4 and 11, respectively; 0.9 mm/d, 2% BW/d) are similar to those of yearling stocks with later ocean entry timing that were caught in June. Our estimates of ocean entry time for other stocks are consistent with back-calculated ocean entry dates determined by otolith chemical and structural analyses (Tomaro et al. 2012; Claiborne et al. 2014; Miller et al. 2014).

Variation in Life History Traits

It has long been recognized that different stocks of salmon—including those originating from the same river—often differ in life history traits that are easily measured while the fish are in freshwater (e.g., Rich 1920; Groot and Margolis 1991; Quinn 2005). Our study adds to an increasing body of literature demonstrating that stock-specific variation in marine life history traits rivals that of freshwater traits. This variation begins with stock-specific size and timing at ocean entry (Roegner et al. 2012; our study), continues as stock-specific migration rates, routes, and behaviors during the first summer or two of ocean residence (e.g., Trudel et al. 2009; Tucker et al. 2011; Burke et al. 2013; Fisher et al. 2014; Teel et al. 2015), includes a poorly-understood winter period when salmon may occupy mid-ocean habitats and are logistically difficult to sample

(Groot and Margolis 1991; Myers et al. 1996; Larson et al. 2013), and concludes with stock-specific differences in marine distributions as adult salmon return to their natal streams (Milne 1957; Wright 1968; Weitkamp 2010; Sharma and Quinn 2012)—differences which have long been exploited by managers to structure fisheries (e.g., Killick 1955).

While our ability to detect such differences in marine life history has only recently been made possible by advances in genetic technology and extensive tagging programs, we should not be surprised that they exist given the life history variation that salmon exhibit in freshwater. Furthermore, this stock-specific variation likely contributes to the overall resilience of salmon populations, allowing species to persist despite unpredictable environmental variation that may favor some strategies over others in given time periods (e.g., Thorpe 1999; Schindler et al. 2010; Bottom et al. 2011). Clearly, stock-specific traits are critical to successfully transitioning from one life stage to another, yet they are often overlooked when different stocks originate from a common river basin.

ACKNOWLEDGMENTS

This research was conducted under Oregon scientific research permits OR2007-3920, OR2008-3265, 14480, 15374, and 16308 and NOAA Fisheries Service ESA permits 1290-6M and 1290-7R. The work was only made possible by an exceptional field crew, including C. Johnson, T. Sandel, P. Peterson, M. Litz, A. Claiborne, A. Claxton, and S. Sebring, and boat operators C. Taylor, B. Kelly, and R. Nelson. D. Kuligowski collected the genetic data used in the study. The study was funded by the Northwest Fisheries Science Center and the Bonneville Power Administration. The original manuscript was greatly improved by constructive comments by J. Myers, B. Burke, M. Trudel, and one anonymous reviewer.

REFERENCES

Achord, S., R. W. Zabel, and B. P. Sandford. 2007. Migration timing, growth, and estimated parr-to-smolt survival rates of wild Snake River spring–summer Chinook Salmon from the Salmon River basin, Idaho, to the lower Snake River. Transactions of the American Fisheries Society 136:142–154.

Beacham, T. D., R. J. Beamish, J. R. Candy, C. Wallace, S. Tucker, J. Moss, and M. Trudel. 2014. Stock-specific size of juvenile Sockeye Salmon in British Columbia waters and the Gulf of Alaska. Transactions of the American Fisheries Society 143:876–889.

Beamish, R. J., C. Mahnken, and C. M. Neville. 2004. Evidence that reduced early marine growth is associated with lower marine survival of Coho Salmon. Transactions of the American Fisheries Society 133:26–33.

Beamish, R. J., R. M. Sweeting, K. L. Lange, and C. M. Neville. 2008. Changes in the population ecology of hatchery and wild Coho Salmon in the Strait of Georgia. Transactions of the American Fisheries Society 137:503–520.

Beamish, R. J., R. M. Sweeting, and C. M. Neville. 2013. Late ocean entry timing provides resilience to populations of Chinook and Sockeye Salmon in the Fraser River. North Pacific Anadromous Fish Commission Technical Report 9:38–44.

Beckman, B. R., D. A. Larsen, C. Sharpe, B. Lee-Pawlak, C. B. Schreck, and W. W. Dickhoff. 2000. Physiological status of naturally reared juvenile spring Chinook Salmon in the Yakima River: seasonal dynamics and changes associated with smolting. Transactions of the American Fisheries Society 129:727–753.

Blankenship, S. M., M. R. Campbell, J. E. Hess, M. A. Hess, T. K. Kassler, C. C. Kozfkay, A. P. Matala, S. R. Narum, M. M. Paquin, M. P. Small, J. J. Stephenson, and K. I. Warheit. 2011. Major lineages and metapopulations in Columbia River Oncorhynchus mykiss are structured by dynamic landscape features and environments. Transactions of the American Fisheries Society 140:665–684.

Bottom, D. L., and K. K. Jones. 1990. Species composition, distribution, and invertebrate prey of fish assemblages in the Columbia River estuary. Progress in Oceanography 25:243–270.

Bottom, D. L., K. K. Jones, C. A. Simenstad, and C. L. Smith. 2011. Reconnecting social and ecological resilience in salmon ecosystems. Pages 3–38 in D. L. Bottom, K. K. Jones, C. A. Simenstad, C. L. Smith, and R. Cooper, editors. Pathways to resilience. Oregon State University Press, Corvallis.

Burke, B. J., M. C. Liermann, D. J. Teel, and J. J. Anderson. 2013. Environmental and geospatial factors drive juvenile Chinook Salmon distribution during early ocean migration. Canadian Journal of Fisheries and Aquatic Sciences 70:1167–1177.

Burke, J. L. 2004. Life histories of juvenile Chinook Salmon in the Columbia River estuary, 1916 to the present. Master's thesis. Oregon State University, Corvallis.

Busby, P. J., T. C. Wainwright, A. J. Bryson, L. J. Lierheimer, R. S. Waples, F. W. Waknitz, and I. V. Lagomarsino. 1996. Status review of West Coast steelhead from Washington, Idaho, Oregon, and California. NOAA Technical Memorandum NMFS-NWFSC-27.

Campbell, L. A. 2010. Life histories of juvenile Chinook Salmon (Oncorhynchus tshawytscha) in the Columbia River estuary as inferred from scale and otolith microchemistry. Master's thesis. Oregon State University, Corvallis.

Chittenden, C. M., J. L. A. Jensen, D. Ewart, S. Anderson, S. Balfry, E. Downey, A. Eaves, S. Saksida, B. Smith, S. Vincent, D. Welch, and R. S. McKinley. 2010. Recent salmon declines: a result of lost feeding opportunities due to bad timing? PLoS (Public Library of Science) ONE [online serial] 5(8):e12423.

Claiborne, A. M., J. P. Fisher, S. A. Hayes, and R. L. Emmett. 2011. Size at release, size-selective mortality, and age of maturity of Willamette River hatchery yearling Chinook Salmon. Transactions of the American Fisheries Society 140:1135–1144.

Claiborne, A. M., J. A. Miller, L. A. Weitkamp, D. J. Teel, and R. L. Emmett. 2014. Evidence for selective mortality in marine environments: the role of fish migration size, timing, and production type. Marine Ecology Progress Series 515:187–202.

Collis, K., D. D. Roby, D. P. Craig, B. A. Ryan, and R. D. Ledgerwood. 2001. Colonial waterbird predation on juvenile salmonids tagged with passive integrated transponders in the Columbia River estuary: vulnerability of different salmonid species, stocks, and rearing types. Transactions of the American Fisheries Society 130:385–396.

Dawley, E. M., R. Ledgerwood, T. H. Blahm, C. W. Sims, J. T. Durkin, R. A. Kirn, A. E. Rankis, G. E. Monan, and F. J. Ossiander. 1986. Migrational characteristics, biological observations, and relative survival of juvenile salmonids entering the Columbia River estuary, 1966–1983. Final Report to the Bonneville Power Administration, Project 81-102, Portland, Oregon.

Emmett, R. L. 1997. Estuarine survival of salmonids: the importance of interspecific and intraspecific predation and competition. NOAA Technical Memorandum NMFS-NWFSC-29:147–158.

Emmett, R. L., G. K. Krutzikowsky, and P. Bentley. 2006. Abundance and distribution of pelagic piscivorous fishes in the Columbia River plume during spring/early summer 1998–2003: relationship to oceanographic conditions, forage fishes, and juvenile salmonids. Progress in Oceanography 68:1–26.

Fisher, J. P., and W. G. Pearcy. 1988. Growth of juvenile Coho Salmon (Oncorhynchus kisutch) off Oregon and Washington, USA, in years of differing coastal upwelling. Canadian Journal of Fisheries and Aquatic Sciences 45:1036–1044.

Fisher, J. P., and W. G. Pearcy. 1995. Distribution, migration, and growth of juvenile Chinook Salmon, *Oncorhynchus tshawytscha*, off Oregon and Washington. U.S. National Marine Fisheries Service Fishery Bulletin 93:274–289.

Fisher, J. P., L. A. Weitkamp, D. J. Teel, S. A. Hinton, M. Trudel, J. F. T. Morris, M. E. Thiess, R. M. Sweeting, J. A. Orsi, and E. V. Farley Jr. 2014. Early ocean dispersal patterns of Columbia River Chinook and Coho Salmon. Transactions of the American Fisheries Society 143:252–272.

Ford, M. J., editor. 2011. Status review update for Pacific salmon and steelhead listed under the Endangered Species Act: Pacific Northwest. NOAA Technical Memorandum NMFS-NWFSC-113.

Gelman, A. 2005. Analysis of variance: why it is more important than ever. Annals of Statistics 33:1–53.

Gelman, A. 2006. Prior distributions for variance parameters in hierarchical models (comment on article by Browne and Draper). Bayesian Analysis 1:515–534.

Gelman, A., J. B. Carlin, H. S. Stern, D. B. Dunson, A. Vehtari, and D. B. Rubin. 2013. Bayesian data analysis, 3rd edition. CRC Press, Boca Raton, Florida.

Gelman, A., and J. Hill. 2006. Data analysis using regression and multilevel/hierarchical models. Cambridge University Press, Cambridge, UK.

Groot, C., and L. Margolis, editors. 1991. Pacific salmon life histories. University of British Columbia Press, Vancouver.

Harnish, R. A., G. E. Johnson, G. A. McMichael, M. S. Hughes, and B. D. Ebberts. 2012. Effect of migration pathway on travel time and survival of acoustic-tagged juvenile salmonids in the Columbia River estuary. Transactions of the American Fisheries Society 141:507–519.

Healey, M. C. 1983. Coastwide distribution and ocean migration patterns of stream- and ocean-type Chinook Salmon, *Oncorhynchus tshawytscha*. Canadian Field-Naturalist 97:427–433.

Healey, M. C. 1991. Life history of Chinook Salmon (*Oncorhynchus tshawytscha*). Pages 311–393 *in* C. Groot and L. Margolis, editors. Pacific salmon life histories. University of British Columbia Press, Vancouver.

Holtby, L. B., B. C. Andersen, and R. K. Kadowaki. 1990. Importance of smolt size and early ocean growth to interannual variability in marine survival of Coho Salmon (*Oncorhynchus kisutch*). Canadian Journal of Fisheries and Aquatic Sciences 47:2181–2194.

Ivlev, V. S. 1961. Experimental ecology of the feeding of fishes. Yale University Press, New Haven, Connecticut.

Jonsson, N., B. Jonsson, and L. P. Hansen. 2003. The marine survival and growth of wild and hatchery-reared Atlantic Salmon. Journal of Applied Ecology 40:900–911.

Kalinowski, S. T., K. R. Manlove, and M. L. Taper. 2007. ONCOR: a computer program for genetic stock identification. Montana State University, Bozeman. Available: http: www.montana.edu/kalinowski/Software/ONCOR. htm. (April 2013).

Killick, S. R. 1955. The chronological order of Fraser River Sockeye Salmon during migration, spawning, and death. International Pacific Salmon Fisheries Commission Bulletin 7:1–95.

Larson, W. A., F. M. Utter, K. W. Myers, W. D. Templin, J. E. Seeb, C. M. Guthrie III, A. V. Bugaev, and L. W. Seeb. 2013. Single-nucleotide polymorphisms reveal distribution and migration of Chinook Salmon (*Oncorhynchus tshawytscha*) in the Bering Sea and North Pacific Ocean. Canadian Journal of Fisheries and Aquatic Sciences 70:128–141.

MacFarlane, R. B. 2010. Energy dynamics and growth of Chinook Salmon (*Oncorhynchus tshawytscha*) from the Central Valley of California during the estuarine phase and first ocean year. Canadian Journal of Fisheries and Aquatic Sciences 67:1549–1565.

Miller, J. A., D. J. Teel, A. Baptista, and C. Morgan. 2013. Disentangling bottom-up and top-down effects during a critical period in the life history of an anadromous fish. Canadian Journal of Fisheries and Aquatic Sciences 70:617–629.

Miller, J. A., D. J. Teel, W. T. Peterson, and A. Baptista. 2014. Assessing the relative importance of local and regional processes on the survival of a threatened salmon population. PLoS (Public Library of Science) ONE [online serial] 9(6):e99814.

Milne, D. J. 1957. Recent British Columbia spring and Coho Salmon tagging experiments, and a comparison with those conducted from 1925 to 1930. Fisheries Research Board of Canada Bulletin 113:1–56.

Mittelbach, G. G., and L. Persson. 1998. The ontogeny of piscivory and its ecological consequences. Canadian Journal of Fisheries and Aquatic Sciences 55:1454–1465.

Moss, J. H., D. A. Beauchamp, A. D. Cross, K. W. Myers, E. V. Farley, J. M. Murphy, and J. H. Helle. 2005. Evidence for size-selective mortality after the first summer of ocean growth by Pink Salmon. Transactions of the American Fisheries Society 134:1313–1322.

Myers, K. W., K. Y. Aydin, and R. V. Walker. 1996. Known ocean ranges of stocks of Pacific salmon and steelhead as shown by tagging experiments, 1956–1995. University of Washington, Fisheries Research Institute Report FRI-UW-9614, Seattle.

Myers, J. M., R. G. Kope, G. J. Bryant, D. Teel, L. J. Lierheimer, T. C. Wainwright, W. S. Grant, F. W. Waknitz, K. Neely, S. T. Lindley, and R. S. Waples. 1998. Status review of Chinook Salmon from Washington, Idaho, Oregon, and California. NOAA Technical Memorandum NMFS-NWFSC-35.

Naish, K. A., J. E. Taylor, P. S. Levin, T. P. Quinn, J. R. Winton, D. Huppert, and R. Hilborn. 2008. An evaluation of the effects of conservation and fishery enhancement hatcheries on wild populations of salmon. Advances in Marine Biology 53:61–194.

NRC (National Research Council). 1996. Upstream: salmon and society in the Pacific Northwest. National Academy Press, Washington, D.C.

Paquet, P., T. Flagg, A. Appleby, J. Barr, L. Blankenship, D. Campton, M. Delarm, T. Evelyn, D. Fast, J. Gislason, P. Kline, D. Maynard, L. Mobrand, G. Nandor, P. Seidel, and S. Smith. 2011. Hatcheries, conservation, and sustainable fisheries: achieving multiple goals:—results of the Hatchery Scientific Review Group's Columbia River basin review. Fisheries 36: 542–561.

Pearcy, W. G. 1992. Ocean ecology of North Pacific salmonids. University of Washington Press, Seattle.

Pearcy, W. G., and J. P. Fisher. 1990. Distribution and abundance of juvenile salmonids off Oregon and Washington, 1981–1984. NOAA Technical Report NMFS-93.

Pearcy, W. G., and S. M. McKinnell. 2007. The ocean ecology of salmon in the northeast Pacific Ocean: an abridged history. Pages 7–30 *in* C. B. Grimes, R. D. Brodeur, L. J. Haldorson, and S. M. McKinnell, editors. The ecology of juvenile salmon in the northeast Pacific Ocean: regional comparisons. American Fisheries Society, Symposium 57, Bethesda, Maryland.

Peterson, W. T., J. L. Fisher, J. O. Peterson, C. A. Morgan, B. J. Burke, and K. L. Fresh. 2014. Applied fisheries oceanography: ecosystem indicators of ocean conditions inform fisheries management in the California Current. Oceanography 27:80–89.

Plummer, M. 2003. JAGS: a program for analysis of Bayesian graphical models using Gibbs sampling. Available: http://citeseer.ist.psu.edu/plummer 03jags.html. (August 2014).

Quinn, T. P. 2005. The behavior and ecology of Pacific salmon and trout. University of Washington Press, Seattle.

R Core Team. 2014. R: a language and environment for statistical computing. R Foundation for Statistical Computing, Vienna.

Rand, P. S., B. Berejikian, A. Bidlack, D. Bottom, J. Gardner, M. Kaeriyama, R. Lincoln, M. Nagata, T. Pearsons, M. Schmidt, W. Smoker, L. Weitkamp, and L. A. Zhivotovsky. 2012. Ecological interactions between wild and hatchery salmon and key recommendations for research and management actions in selected regions of the North Pacific. Environmental Biology of Fishes 94:1–6.

Rannala, B., and J. L. Mountain. 1997. Detecting immigration by using multilocus genotypes. Proceedings of the National Academy of Sciences of the USA 94:9197–9201.

Reimers, P. E., and R. E. Loeffel. 1967. The length of residence of juvenile fall Chinook Salmon in selected Columbia River tributaries. Fish Commission of Oregon Research Briefs 13:5–19.

Rich, W. H. 1920. Early history and seaward migration of Chinook Salmon in the Columbia and Sacramento rivers. U.S. Bureau of Fisheries Bulletin 37:1–74.

Ricker, W. E. 1975. Computation and interpretation of biological statistics of fish populations. Fisheries Research Board of Canada Bulletin 191.

Roegner, G. C., R. A. McNatt, D. J. Teel, and D. L. Bottom. 2012. Distribution, size, and origin of juvenile Chinook Salmon in shallow-water habitats of the lower Columbia River and estuary, 2002–2007. Marine and Coastal Fisheries: Dynamics, Management, and Ecosystem Science [online serial] 4:450–472.

Ryan, B. A., S. G. Smith, J. M. Butzerin, and J. W. Ferguson. 2003. Relative vulnerability to avian predation of juvenile salmonids tagged with passive integrated transponders in the Columbia River estuary, 1998–2000. Transactions of the American Fisheries Society 132:275–288.

Satterthwaite, W. H., S. M. Carlson, S. D. Allen-Moran, S. Vincenzi, S. J. Bograd, and B. K. Wells. 2014. Match–mismatch dynamics and the relationship between ocean-entry timing and relative ocean recoveries of Central Valley fall run Chinook Salmon. Marine Ecology Progress Series 511:237–248.

Scheuerell, M. D., R. W. Zabel, and B. P. Sandford. 2009. Relating juvenile migration timing and survival to adulthood in two species of threatened Pacific salmon (Oncorhynchus spp.). Journal of Applied Ecology 46:983–990.

Schindler, D. E., R. Hilborn, B. Chasco, C. P. Boatright, T. P. Quinn, L. A. Rogers, and M. S. Webster. 2010. Population diversity and the portfolio effect in an exploited species. Nature 465:609–612.

Sebring, S. H., M. C. Carper, R. D. Ledgerwood, B. P. Sandford, G. M. Matthews, and A. F. Evans. 2013. Relative vulnerability of PIT-tagged subyearling fall Chinook Salmon to predation by Caspian terns and double-crested cormorants in the Columbia River estuary. Transactions of the American Fisheries Society 142:1321–1334.

Seeb, L. W., A. Antonovich, M. A. Banks, T. D. Beacham, M. R. Bellinger, S. M. Blankenship, M. Campbell, N. A. Decovich, J. C. Garza, C. M. Guthrie III, T. A. Lundrigan, P. Moran, S. R. Narum, J. J. Stephenson, K. J. Supernault, D. J. Teel, W. D. Templin, J. K. Wenburg, S. F. Young, and C. T. Smith. 2007. Development of a standardized DNA database for Chinook Salmon. Fisheries 32:540–552.

Sharma, R., and T. P. Quinn. 2012. Linkages between life history type and migration pathways in freshwater and marine environments for Chinook Salmon, Oncorhynchus tshawytscha. Acta Oecologica 41:1–13.

Sogard, S. M. 1997. Size-selective mortality in the juvenile stage of teleost fishes: a review. Bulletin of Marine Science 60:1129–1157.

Spence, B. C., and J. D. Hall. 2010. Spatiotemporal patterns in migration timing of Coho Salmon (Oncorhynchus kisutch) smolts in North America. Canadian Journal of Fisheries and Aquatic Sciences 67:1316–1334.

Stephenson, J. J., M. R. Campbell, J. E. Hess, C. Kozfkay, A. P. Matala, M. V. McPhee, P. Moran, S. R. Narum, M. M. Paquin, O. Schlei, M. P. Small, D. M. Van Doornik, and J. K. Wenburg. 2009. A centralized model for creating shared, standardized, microsatellite data that simplifies interlaboratory collaboration. Conservation Genetics 10:1145–1149.

Sykes, G. E., C. J. Johnson, and J. M. Shrimpton. 2009. Temperature and flow effects on migration timing of Chinook Salmon smolts. Transactions of the American Fisheries Society 138:1252–1265.

Tatara, C. P., and B. A. Berejikian. 2012. Mechanisms influencing competition between hatchery and wild juvenile anadromous Pacific salmonids in freshwater and their relative competitive abilities. Environmental Biology of Fishes 94:7–19.

Teel, D. J., D. L. Bottom, S. A. Hinton, D. R. Kuligowski, G. T. McCabe, R. McNatt, G. C. Roegner, L. A. Stamatiou, and C. A. Simenstad. 2014. Genetic identification of Chinook Salmon in the Columbia River estuary: stock-specific distributions of juveniles in shallow tidal freshwater habitats. North American Journal of Fisheries Management 34:621–641.

Teel, D. J., B. J. Burke, D. R. Kuligowski, C. A. Morgan, and D. M. Van Doornik. 2015. Genetic identification of Chinook Salmon: stock-specific distributions of juveniles along the Washington and Oregon coasts. Marine and Coastal Fisheries: Dynamics, Management, and Ecosystem Science [online serial] 7:274–300.

Thorpe, J. E. 1999. Flexible life history strategies: a context for understanding migration in salmonids. Bulletin of the Tohoku National Fisheries Research Institute 62:151–164.

Tomaro, L. M., D. J. Teel, W. T. Peterson, and J. A. Miller. 2012. When is bigger better? Early marine residence of middle and upper Columbia River spring Chinook Salmon. Marine Ecology Progress Series 452:237–252.

Trudel, M., J. Fisher, J. A. Orsi, J. F. T. Morris, M. E. Thiess, R. M. Sweeting, S. Hinton, E. A. Fergusson, and D. W. Welch. 2009. Distribution and migration of juvenile Chinook Salmon derived from coded wire tag recoveries along the continental shelf of western North America. Transactions of the American Fisheries Society 138:1369–1391.

Tucker, S., M. Trudel, D. W. Welch, J. R. Candy, J. F. T. Morris, M. E. Thiess, C. Wallace, and T. D. Beacham. 2011. Life history and seasonal stock-specific ocean migration of juvenile Chinook Salmon. Transactions of the American Fisheries Society 140:1101–1119.

Tucker, S., M. Trudel, D. W. Welch, J. R. Candy, J. F. T. Morris, M. E. Thiess, C. Wallace, D. J. Teel, W. Crawford, E. V. Farley Jr., and T. D. Beacham. 2009. Seasonal stock-specific migrations of juvenile Sockeye Salmon along the West Coast of North America: implications for growth. Transactions of the American Fisheries Society 138:1458–1480.

Van Doornik, D. M., D. J. Teel, D. R. Kuligowski, C. A. Morgan, and E. Casillas. 2007. Genetic analyses provide insight into the early ocean stock distribution and survival of juvenile Coho Salmon (Oncorhynchus kisutch) off the coasts of Washington and Oregon. North American Journal of Fisheries Management 27:220–237.

Waples, R. S. 1991. Pacific salmon, Oncorhynchus spp., and the definition of "species" under the Endangered Species Act. Marine Fisheries Review 53(3):11–12.

Waples, R. S., D. J. Teel, J. M. Myers, and A. R. Marshall. 2004. Life history divergence in Chinook Salmon: historic contingency and parallel evolution. Evolution 58:386–403.

Weitkamp, L. A. 2010. Marine distribution of Chinook Salmon (Oncorhynchus tshawytscha) from the West Coast of North America determined by coded wire tag recoveries. Transactions of the American Fisheries Society 139:147–170.

Weitkamp, L. A., P. B. Bentley, and M. N. C. Litz. 2012. Seasonal and interannual variation in juvenile salmonids and associated fish assemblage in open waters of the lower Columbia River estuary, U.S.A. U.S. National Marine Fisheries Service Fishery Bulletin 110:426–450.

Welch, D. W., E. L. Rechisky, M. C. Melnychuk, A. D. Porter, C. J. Walters, S. Clements, B. J. Clemens, R. S. McKinley, and C. Schreck. 2008. Survival of migrating salmon smolts in large rivers with and without dams. PLoS (Public Library of Science) Biology [online serial] 6:2101–2108.

Whalen, K. G., D. L. Parrish, and S. D. McCormick. 1999. Migration timing of Atlantic Salmon smolts relative to environmental and physiological factors. Transactions of the American Fisheries Society 128:289–301.

Willette, T. M., R. T. Cooney, V. Patrick, D. M. Mason, G. L. Thomas, and D. Scheel. 2001. Ecological processes influencing mortality of juvenile Pink Salmon (Oncorhynchus gorbuscha) in Prince William Sound, Alaska. Fisheries Oceanography 10(Supplement 1):14–41.

Williams, R. N., editor. 2006. Return to the river: restoring salmon to the Columbia River. Elsevier Press, Oxford, UK.

Wright, S. G. 1968. Origin and migration of Washington's Chinook and Coho Salmon. Washington Department of Fisheries, Information Booklet 1, Olympia.

Zabel, R. W., and S. Achord. 2004. Relating size of juveniles to survival within and among populations of Chinook Salmon. Ecology 85:795–806.

Zamon, J. E., E. M. Phillips, and T. J. Guy. 2013. Marine bird aggregation at tidally driven plume fronts of the Columbia River. Deep Sea Research II: Topical Studies in Oceanography 107:85–95.

Appendix A.1: Additional Data

TABLE A.1.1.   Mean size at and time of release of juvenile salmon from Columbia River hatcheries, 2007–2011, by stock. Data are from the Regional Mark Processing Center (www.rmpc.org) and Fish Passage Center (www.fpc.org) databases.

Stock	Number released (millions)	% Fin clipped	Release date	Release weight (g)
Steelhead				
Lower Columbia River	3.0	86.5	Apr 19	75.1
Mid Columbia River	1.2	54.3	Apr 22	88.6
Snake River	8.3	86.8	Apr 14	97.3
Upper Columbia River	1.1	71.1	Apr 26	70.5
Willamette River	1.0	99.6	Apr 15	89.8
All steelhead	14.6	83.7		
Yearling Chinook Salmon				
Mid Columbia River, spring	6.3	87.0	Apr 12	27.9
Snake River, fall	0.9	56.9	Apr 9	47.4
Snake River, spring	9.8	92.3	Apr 20	62.7
Snake River, summer	2.3	93.9	Apr 9	23.0
Upper Columbia River, spring	3.1	67.6	Apr 24	28.7
Willamette River, spring	4.8	97.2	Mar 11	42.1
Upper Columbia River, summer	2.3	98.8	Apr 21	42.0
Upper Columbia River, fall	0.4	37.7	Mar 22	39.8
West Cascade, spring	3.8	92.7	Mar 11	46.2
All yearling Chinook Salmon	33.7	88.7		
Subyearling Chinook Salmon				
Spring Creek Group, fall	18.8	96.3	Apr 23	5.0
Upper Columbia River, fall	19.9	52.7	May 15	8.8
West Cascade, fall	23.8	88.0	Jun 17	8.7
Snake River, fall	4.4	42.6	May 27	8.5
Upper Columbia River, summer	1.4	65.3	May 27	15.1
All subyearling Chinook Salmon	72.6	75.7		

TABLE A.1.2. Downstream migration rates for juvenile salmon tagged with coded wire tags (CWTs) or passive integrated transponder (PIT) tags and collected in the Columbia River estuary. Only groups for which at least five coded-wire-tagged or three PIT-tagged individuals were recovered are included.

Stock		CWT fish[a]			PIT fish[b]	
	n	Days at large	Distance (km)	Rate (km/d)	n	Rate (km/d)
Steelhead						
Snake River	61	37.7	1,033	27.4	14	79.5
Upper Columbia River	12	33.3	841	25.2		
Yearling Chinook Salmon						
Mid Columbia River, spring	53	44.1	386	11.2		
Snake River, fall	52	30.2	697	24.6	3	61.4
Snake River, spring	58	51.1	853	18.1	11	67.1
Snake River, summer	10	63.3	1,100	19.0		
Upper Columbia River, fall	10	24.1	588	31.3		
Upper Columbia River, spring	65	42.1	830	21.6		
Upper Columbia River, summer	80	37.6	790	24.1		
Willamette River, spring	18	49.1	326	10.9		
Lower Columbia River, spring	14	56.0	143	2.6		
Subyearling Chinook Salmon						
Spring Creek, fall	77	37.8	180	7.9		
Snake River, fall	54	27.2	814	35.2	7	64.8
Upper Columbia River, fall	55	30.0	369	17.2		
Upper Columbia River, summer	6	35.5	813	24.1		
West Cascade, fall	24	44.7	181	5.1		

[a]Estimated from release at the hatchery to recovery in the estuary. Release locations and dates were determined from the Regional Mark Processing Center database (www.rmpc.org).

[b]Estimated from detection at an intermediate dam to recovery in the estuary. Dates and intermediate detection locations were determined from the PTAGIS database (www.ptagis.org).

Appendix A.2: Model Details

Models

Length and weight models.—The application of traditional ANOVA models to the length and weight data was not possible due to the extreme lack of balance and differences in variance between groups. Thus we used a Bayesian multilevel model in which group means within a category (i.e., different stocks) were assumed to come from a normal distribution for which the mean and standard deviation were estimated. This is comparable to a random effect in a mixed-effects model. We used the same models for length and weight. Here we present the model in terms of length.

Length was assumed to have a normal distribution,

$$L_i \sim N(\text{mean}_i, s_{\text{resid}}),$$

with the mean

$$\alpha + \beta_{\text{year}} + \gamma_{\text{stock}} + \delta_{\text{year, stock}} + \eta_{\text{hat}} + \theta_{\text{year, hat}} + \zeta_{\text{stock, hat}}$$
$$+ \iota_{\text{year, stock, hat}},$$

where "hat" indicates the ratio of hatchery-origin fish to wild ones and the other subscripts are self-explanatory. The year, stock, and interaction specific constants have normal distributions:

$$\beta_{\text{year}} \sim N(0, s_{\text{year}})$$
$$\gamma_{\text{stock}} \sim N(0, s_{\text{stock}})$$
$$\delta_{\text{year, stock}} \sim N(0, s_{\text{year, stock}})$$
$$\theta_{\text{year, hat}} \sim N(0, s_{\text{year, hat}})$$
$$\zeta_{\text{stock, hat}} \sim N(0, s_{\text{stock, hat}})$$
$$\iota_{\text{year, stock, hat}} \sim N(0, s_{\text{year, stock, hat}})$$

The constant term α (the intercept) and groups in categories with less than three groups (i.e., hat) were assigned diffuse normal priors, namely, $N(0, 1,000)$. The residual standard deviation (s_{resid}) was assumed to follow an inverse gamma distribution $InvGamma(0.01, 0.01)$ unless there were obvious differences in the variance between groups within a category (e.g., stocks). In that case the group residual standard deviations were assumed to follow a normal distribution in the same way as the main effects and interactions above. The standard deviation parameters describing between-group variability (e.g., variability between years, s_{year}) were assumed to follow a uniform distribution, *uniform*(0, 100), (e.g., Gelman 2006).

Migration timing model.—The number of fish observed during a particular sampling event is assumed to follow the negative binomial distribution,

$$\text{count}_i \sim negative\ binomial(\text{mean}_i, \text{dispersion}_i),$$

where the mean is a function of time (a normal density function) that first increases and then decreases over the period of migration as function of the day of the year (Figure A.2.1):

$$\text{mean}_i = \text{scale}_i \times N(\text{day of the year}_i, \text{center}_i, \text{spread}_i).$$

The center parameter represents the date with highest expected count and is modeled in the same way as length above, namely,

$$\text{center}_i = \alpha + \beta_{\text{year}} + \gamma_{\text{stock}} + \delta_{\text{year, stock}} + \eta_{\text{hat}}$$
$$+ \theta_{\text{year, hat}} + \zeta_{\text{stock, hat}} + \iota_{\text{year, stock, hat}}.$$

The spread parameter is the standard deviation of the normal density function, which describes how the counts are spread across time. It is modeled in the same way as center. The scale parameter accounts for the total number of fish counted over the migration period (by adjusting the height of the normal density curve) and is allowed to vary by group:

$$\text{scale}_i = exp(\nu + \mu_{\text{year,stock,hat}}).$$

In this equation the constant ν (the intercept) is assumed to follow a diffuse normal distribution, i.e., $N(0, 1,000)$.

The dispersion parameter for the negative binomial distribution describes the relationship between the mean and the variance, where decreasing values equate to a larger variance for a given mean and the distribution collapses to a Poisson distribution as the dispersion gets very large. We assigned a

diffuse normal distribution to the log of the dispersion parameter.

For all size and timing models, we transformed the main effects and interactions to satisfy the constraint that they sum to zero. That is, all of the main effects sum to zero and the margins of the interactions are zero. For example,

$$\sum_{\text{stock}} \sum_{\text{year}} (\mu_{\text{year, stock, hat}}) = \begin{bmatrix} 0 \\ 0 \end{bmatrix}.$$

This is equivalent to treating the categories as fixed effects or the groups as a finite population (Gelman 2006), where one is interested in making comparisons between specific levels (e.g., between the lower and upper Columbia River stocks). This is independent of whether or not there is partial pooling for the category.

Assessing Model Fit

We assessed the fit of the model to the data by means of plots. The plots of the data for length and weight (length: Figures 2–5 in the main text; weight: Figures A.2.2–A.2.4)[1] include 95% credible intervals for the group-specific means. This allowed us to look for poor fits for the individual groups, unequal variances across groups, and severe violations of the normal-distribution assumption. Because the group means were assumed to come from a common distribution across the category (e.g., stock-specific means), they will be shrunk slightly toward the grand mean for that category. Therefore, groups with few observations may have 95% credible regions that are not precisely centered over the data. In addition, it is possible to estimate the group mean for groups without data. The observed residuals from the length and weight models were compared with the posterior predictive distributions using quantiles.

For the timing model we again used the plots of the data, along with estimates of peak arrival time, to look for any obvious lack of fit. To assess the fit of the normal-distribution curve to the timing distributions, we examined the data along with several fitted curves from the posterior distribution for each group. We examined the fit of the negative binomial distribution by comparing the observed and expected quantiles.

Markov Chain–Monte Carlo Convergence Diagnostics

We assessed the convergence of the Markov chain–Monte Carlo algorithm by means of trace plots of individual chains and the difference between the estimates and

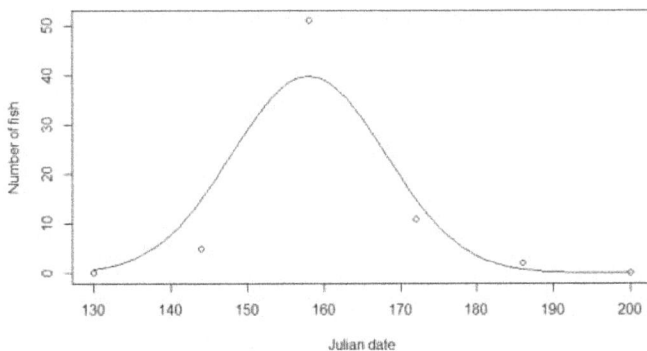

FIGURE A.2.1. Example data and the normal density function used to describe the temporal pattern. The values on the y-axis are the number of fish from a particular group caught during a given sampling event. The points are the counts, and the curve represents the normal density function fit to the data (i.e., the model of migration timing for the group).

[1]Because the length and weight data were very similar and there were more missing weight data, we only presented the length data in the main text. Here we provide comparable plots for the weights of the steelhead, yearling Chinook Salmon, and subyearling Chinook Salmon in our samples.

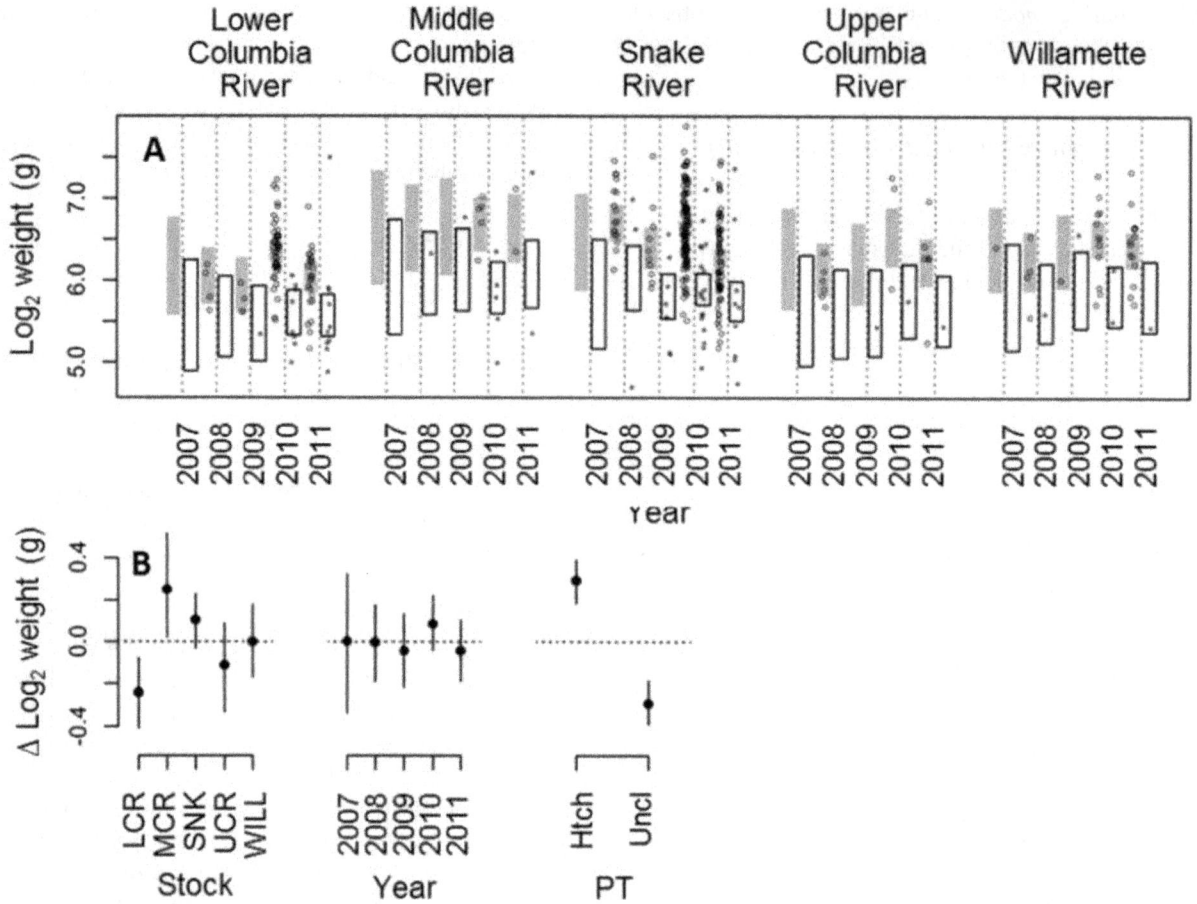

FIGURE A.2.2. Comparisons of log$_2$ transformed mean weight at capture in the estuary for juvenile steelhead by stock, production type, and year. Panel **(A)** shows plots of weight, with model-based 95% credible intervals. The dots indicate individual fish. Gray bars and gray dots represent hatchery fish, open bars and black dots represent unclipped fish. Panel **(B)** pertains to the main effects in the weight model. The points are the estimated deviations from the mean for each level of the main effect, and the vertical lines are the corresponding 95% credible intervals. The stock abbreviations are as follows: LCR = lower Columbia River, MCR = mid Columbia River, SNK = Snake River, UCR = upper Columbia River, and WILL = Willamette River. The production type (PT) abbreviations are as follows: Htch = hatchery and Uncl = unclipped.

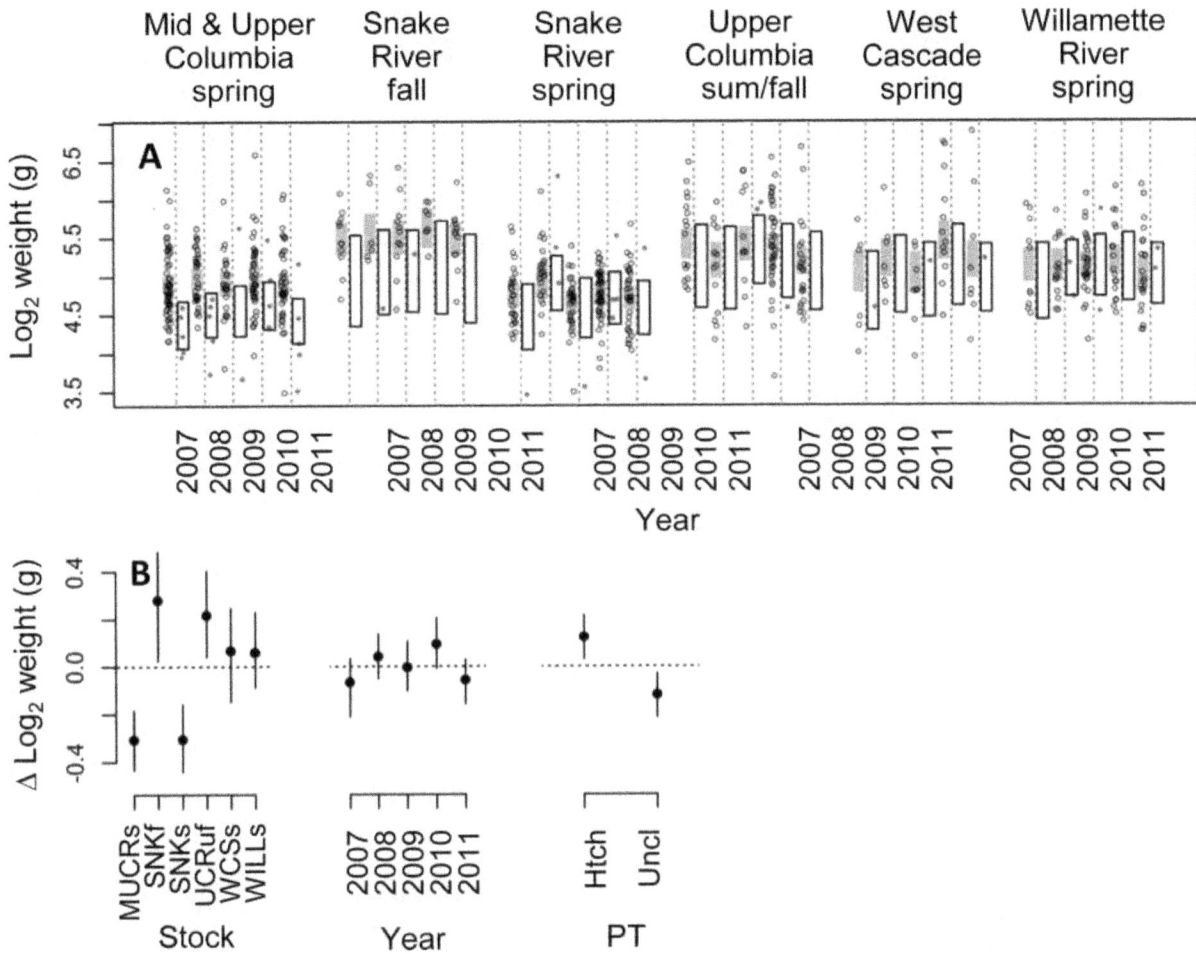

FIGURE A.2.3.   Comparisons of log$_2$ transformed mean weight at capture in the estuary for yearling Chinook Salmon by stock, production type, and year. The stock abbreviations are as follows: MUCRs = mid and upper Columbia River, spring; SNKf = Snake River, fall; SNKs = Snake River, spring; UCRuf = upper Columbia River, summer–fall; WCSs = west Cascade, spring; and WILLs = Willamette River, spring. See Figure A.2.2 for additional details.

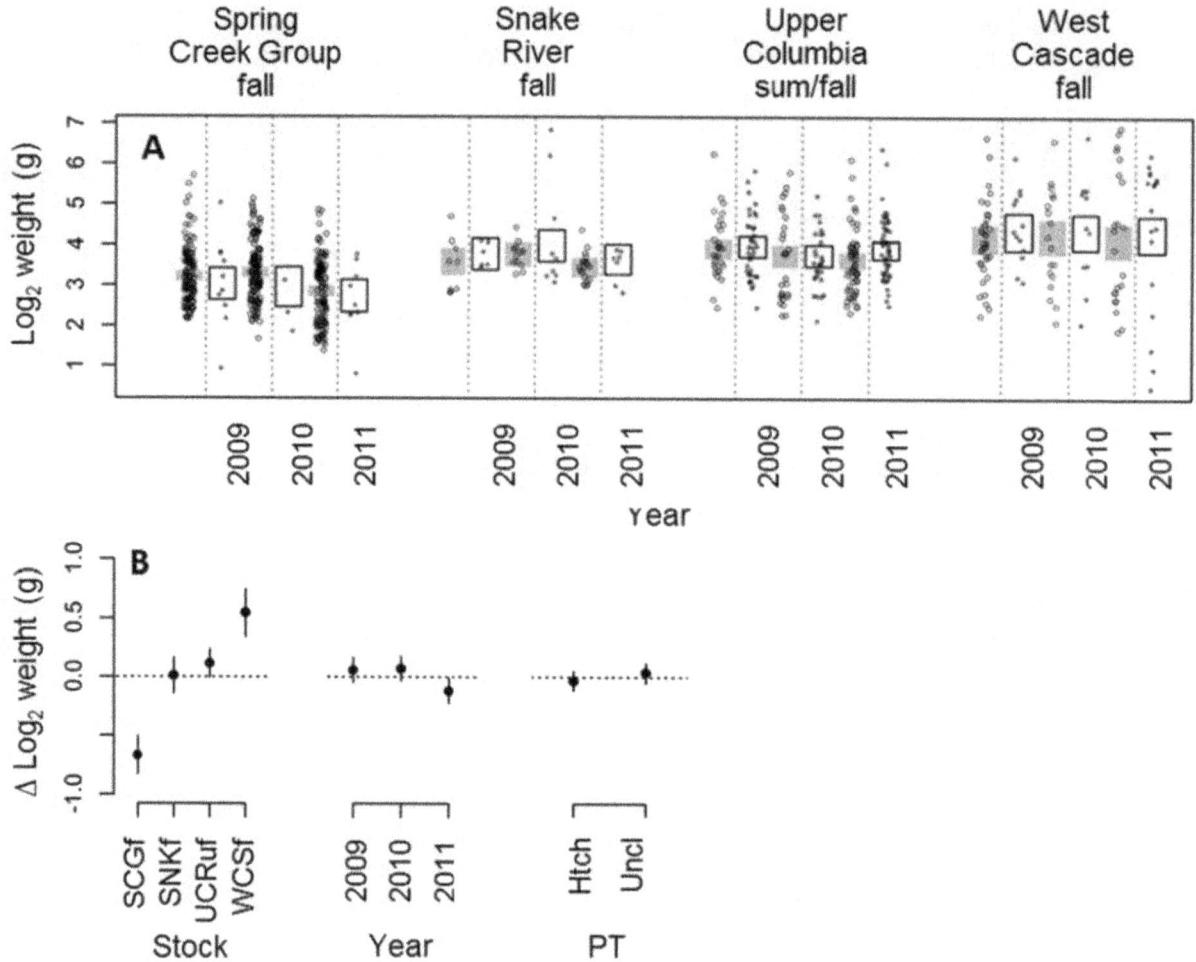

FIGURE A.2.4. Comparisons of $\log_2$ transformed mean weight at capture in the estuary for subyearling Chinook Salmon by stock, production type, and year. The stock abbreviations are as follows: SCGf = Spring Creek Group, fall; SNKf = Snake River, fall; UCRuf = upper Columbia River, summer–fall; and WCSf = west Cascade, fall. See Figure A.2.2 for additional details.

the credible intervals resulting from multiple chains. Chains were run until the thinned trace plots indicated good mixing and the estimates and credible intervals from separate chains converged.

Software

The models were fit using Markov chain–Monte Carlo sampling as implemented in JAGS software (Plummer 2003). The R language (R Core Team 2014) was used for data manipulation and plotting.

APPENDIX REFERENCES

Gelman, A. 2006. Prior distributions for variance parameters in hierarchical models (comment on article by Browne and Draper). Bayesian Analysis 1:515–534.

Plummer, M. 2003. JAGS: a program for analysis of Bayesian graphical models using Gibbs sampling. Available: http://citeseer.ist.psu.edu/plummer03jags.html. (August 2014).

R Core Team. 2014. R: a language and environment for statistical computing. R Foundation for Statistical Computing, Vienna.

11

Quantification of Habitat and Community Relationships among Nearshore Temperate Fishes Through Analysis of Drop Camera Video

Ryan R. Easton*

College of Earth, Ocean, and Atmospheric Sciences, Oregon State University,
104 CEOAS Administration Building, 101 Southwest 26th Street, Corvallis, Oregon 97331, USA;
and Oregon Department of Fish and Wildlife, Marine Resources Program,
2040 Southeast Marine Science Drive, Newport, Oregon 97365, USA

Selina S. Heppell

Department of Fisheries and Wildlife, Oregon State University, 104 Nash Hall, Corvallis,
Oregon 97331, USA

Robert W. Hannah

Oregon Department of Fish and Wildlife, Marine Resources Program,
2040 Southeast Marine Science Drive, Newport, Oregon 97365, USA

Abstract

Temperate nearshore reefs along the Pacific coast of North America are highly valuable to commercial and recreational fisheries yet comprise a small fraction of the seabed. Monitoring fisheries resources in this region is difficult; high-relief structural complexity and adverse sea conditions have led to a paucity of information on temperate reef species assemblage patterns. Reliable, inexpensive tools and methods for monitoring are needed, as many traditional tools are both logistically complicated and expensive, limiting the frequency of their implementation over a large scale. Video drop cameras of varying designs have previously been employed to estimate fish abundance and distribution. We surveyed a nearshore rocky reef off the northern Oregon coast with a video lander (a video camera mounted on a landing platform so it can be dropped to the seafloor) over the spring and winter of 2011. We designed a 272-point systematic grid to document the species assemblage and the distribution and habitat associations of the reef species, including two overfished rockfishes: Canary Rockfish *Sebastes pinniger* and Yelloweye Rockfish *Sebastes ruberrimus*. Species assemblages differed significantly across the reef by depth and by season for the outer part of the reef. Well-defined habitat associations existed for many species; Canary Rockfish were associated with complex moderate-relief habitat types such as large boulders and small boulders, while Yelloweye Rockfish were associated with high-relief habitats like vertical walls. Species associations were evaluated pairwise to identify nearshore complexes. We compared our site with five exploratory reef sites off the central Oregon coast and found that nearshore reefs differed from our site, while offshore reefs were more similar. Video landers provide a solution to the need for increased sampling of temperate reef systems that are subject to difficult conditions and can contribute to habitat mapping, fish abundance indices, and fish assemblage information for monitoring and management of fisheries resources.

Subject editor: Donald Noakes, Vancouver Island University, Nanaimo, British Columbia

*Corresponding author: ryan.r.easton@gmail.com

Temperate nearshore reefs along the Pacific coast of North America represent areas of valuable economic resources for commercial and recreational fisherman (Williams and Ralston 2002; Fox et al. 2004; Gunderson et al. 2008). In Oregon's territorial sea (out 5.6 km), nearshore rocky reefs make up only a small fraction of the total area (~7%), with the remaining region comprised predominately of sand and unconsolidated sediments (D. Fox, Oregon Department of Fish and Wildlife, personal communication). However, these reefs constitute much of the Essential Fish Habitat designated for many pelagic and demersal fishes, which currently inhabit the nearshore region (PFMC 2005; PFMC 2011).

Monitoring fisheries resources and habitat in temperate reefs is difficult due to their high-relief structural complexity, adverse sea conditions that are common in temperate regions, and depths that often exceed those safe for visual surveys by scuba (Adams et al. 1995; Williams et al. 2010). Therefore, there is a paucity of information on species assemblage patterns on temperate reefs at different times of the year and along depth gradients. More information is needed to determine how the reefs function as critical habitat and to refine the fine-scale habitat associations of the fish that utilize those habitats (Gunderson et al. 2008). Annual trawl surveys conducted by the National Marine Fisheries Service currently cover the continental shelf from Cape Flattery, Washington, to the USA–Mexico border, but this survey does not come close to shore and trawls are unable to adequately catch fish that typically reside among rock escarpments and boulders (Zimmermann 2003; Cordue 2007). There have been repeated calls for more comprehensive sampling, particularly for reef-associated species that are considered to be below or near overfishing thresholds (Yoklavich et al. 2007; Williams et al. 2010).

Visual survey tools, such as remotely operated vehicles (ROVs) and human-occupied vehicles, are useful to survey untrawlable rocky habitats. These methods collect valuable information regarding the distribution, relative abundance, and species–habitat associations of various fish species, further aiding Essential Fish Habitat designation (Stein et al. 1992; Krieger 1993; Adams et al. 1995; Johnson et al. 2003; Yoklavich et al. 2007). However, the expense and expertise required for these methods, as well as confounding depth and sea conditions, can make them prohibitive to employ over a large scale within shallow (<70 m), highly productive nearshore waters. Recently, in the nearshore waters of the U.S. West Coast, there have been efforts to establish Marine Protected Areas and Marine Reserves, as well as to expand comprehensive multibeam seafloor mapping. Reliable, inexpensive tools and methods to monitor the effects of protected areas on local fish stocks are needed, as traditional tools (ROVs, autonomous underwater vehicles, and human-occupied vehicles) are both logistically complicated and expensive, generally limiting the frequency of their implementation. Video lander systems (a video camera mounted on a landing platform so it can be dropped to the seafloor) may be the solution to the increased sampling needed to ground-truth habitat maps and determine the relative abundance and distribution of nearshore fisheries resources over broad areas and in winter months.

In northeastern Pacific Ocean waters, highly variable and changing conditions represent obstacles to those researchers who are looking to obtain nonextractive visual data on nearshore Pacific rocky-reef fishes. Video drop cameras of varying designs have previously been employed as noninvasive tools for estimating fish abundance and distribution. Baited underwater video stations and baited remote underwater video stations have been shown to be effective in estimating the relative abundance of many fish that are mobile or solitary and that have low population sizes or avoid other visual survey methods (Ellis and DeMartini 1995; Priede and Merrett 1996; Willis and Babcock 2000; Cappo et al. 2004; Harvey et al. 2007; Stobart et al. 2007; Hannah and Blume 2012; Wakefield et al. 2013). Nonbaited underwater photo and video lander platforms have been effective in capturing accurate and repeatable fish and habitat data throughout a wide range of depths and habitat types without artificially attracting fish with bait, which would bias potential species–habitat associations (Gledhill et al. 1996; Roberts et al. 2005; Hannah and Blume 2012).

The objectives of this research were multifaceted, with a primary aim to determine the ability of a low-cost drop camera system to comprehensively survey a temperate nearshore rocky reef. We documented the species assemblage and the distribution and habitat associations of nearshore Pacific rocky-reef fishes in spring and winter to describe the distribution of key fished species, including two overfished rockfishes that are under intensive "stock rebuilding plans" by the Pacific Marine Fisheries Council. The video lander was evaluated as a survey tool for monitoring protected areas in nearshore temperate reef complexes, while concurrently assessing the ability to be used as a comprehensive ground-truthing tool to identify habitat types that are currently used by multibeam sonar surveys. Finally, we analyzed fish community composition with a recently developed quantitative method that measures pairwise species co-occurrence (Stone and Roberts 1990; Ulrich and Gotelli 2010; Groundfish Management Team 2013).

METHODS

The video lander we used is an autonomous underwater video system designed and built for use in high-relief rocky habitat by the Oregon Department of Fish and Wildlife's Marine Resources Program (Hannah and Blume 2012). The video lander is composed of an aluminum tube frame, a Deep Sea Power and Light (DSPL) Multi-SeaCam 2060, dual DSPL LED Ritelites (850 lm, 3,000 K), and an aluminum pressure housing containing two 13.2V rechargeable NiMH battery packs, a controller board, a Sony TRV-11 digital camcorder (recording video received from the DSPL Multi-SeaCam 2060) recording onto 60-min Mini DVC cassette tapes, and either a depth activated pressure switch used in waters deeper than 18 m or a push activated switch for shallower depths (Figure 1). The sacrificial base is designed so that, if stuck in rocky habitat, the lander can release from the base and rotate around multiple attachment points to maximize the retrieval probability in high-relief rocky habitat. The digital video footage from each lander drop was transferred from the original Sony DVC 60-min cassettes into digital format on a personal desktop computer with Adobe Premiere Pro through a firewire-connected Sony GV-HD700 portable video recorder deck. The video lander was deployed unbaited to avoid drawing in fish from other habitat types near the sampling point.

We selected the Three Arch Rocks rocky-reef complex located off Oceanside, Oregon, approximately 11 km south of the entrance to Tillamook Bay, for its broad depth range and known species diversity (Figure 2). The structure of the reef is a horseshoe pattern, running approximately 5 km east–west and 2 km north–south (Figure 3). Three Arch Rocks reef has a broad depth range, from surface to approximately 75 m as it runs east to west, and is known to support a high diversity of marine species (E. Schindler, Oregon Department of Fish and Wildlife, personal communication).

The video lander was initially deployed on the Three Arch Rocks reef on 12 separate days between April 17 and June 28,

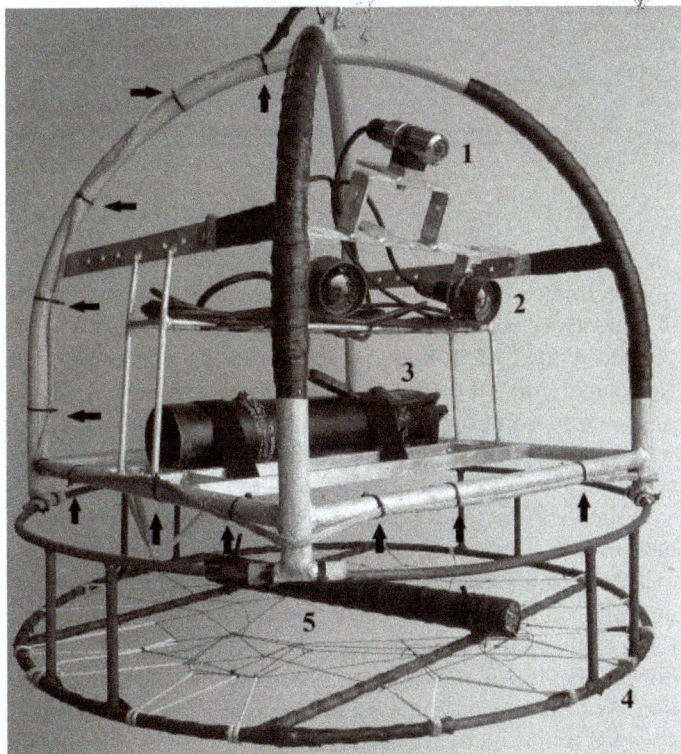

FIGURE 1. The video lander platform utilized at Three Arch Rocks reef. Displayed in the photo are the Deep Sea Power and Light (DSPL) Multi-Sea-Cam 2060 (1), dual DSPL LED Mini-Sealites (2), pressure tube containing batteries and Sony TRV-11 digital camcorder (3), sacrificial (breakaway) base (4), and steel-rod weight bar (5), with arrows showing the break-away connection points.

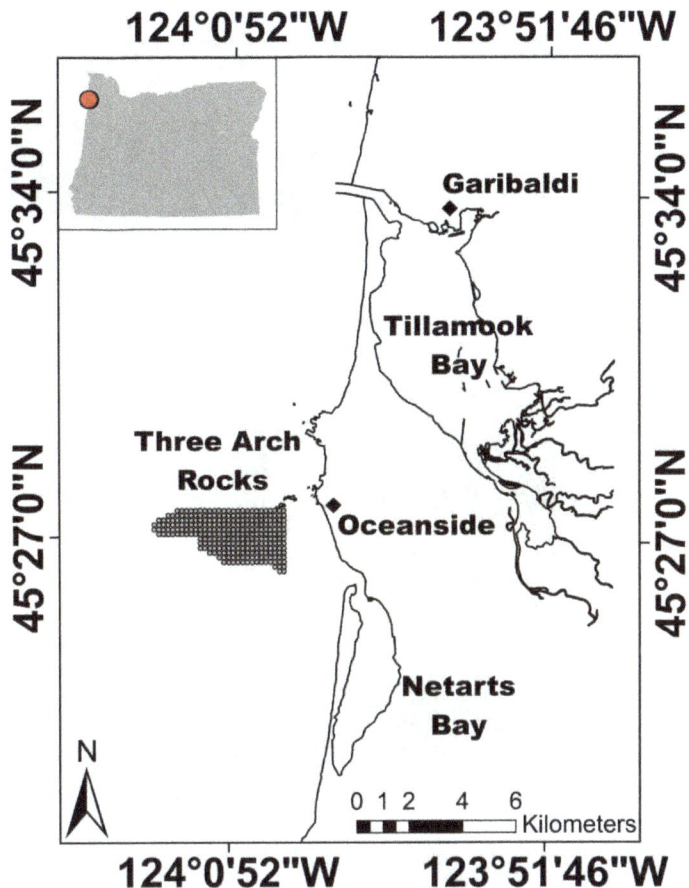

FIGURE 2. Three Arch Rocks rocky-reef study area off of Oceanside, Oregon, located approximately 11 km south of the entrance to Tillamook Bay. The study area covers approximately 15 km².

2011. A systematic grid consisting of 272 individual drop points, spaced 175 m apart, was designed to maximize coverage of the reef structure while capturing all possible habitat types throughout the reef's entire depth range (Figure 3). The grid spacing was designed to maintain independence, while reducing the possibility for double counting individual fish during any given sampling day (Matthews 1990a, 1990b, 1992; Pacunski and Palsson 2002). The entire grid was blocked into seven regions prior to the survey, with the goal of completing at least one blocked section each day. The video lander platform used for the spring survey was again employed for winter sampling in December 2011, with the addition of a set of 10-cm paired scaling lasers to better quantify substrate grain size. Two separate attempts were made to complete the grid between December 1 and December 9, 2011; however, due to poor underwater visibility conditions we were only able to successfully survey a contiguous block of 70 drops comprising the outer quarter of the grid.

The video lander was deployed for a fixed duration during daylight hours at each sampling location, following the protocols in Hannah and Blume (2012). Each video sample consisted of 5 min of recorded bottom time, beginning at

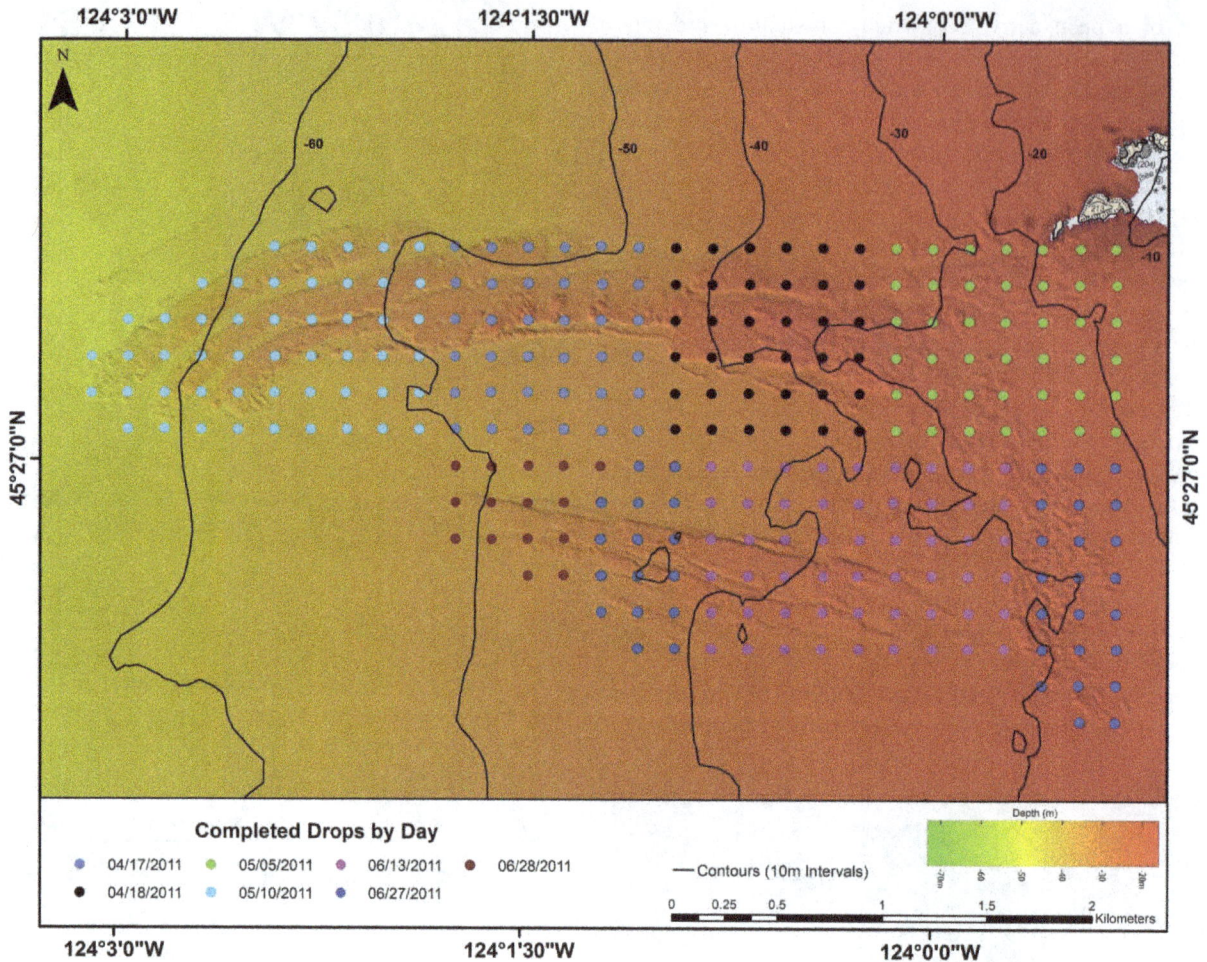

FIGURE 3.   The video lander systematic survey grid of the Three Arch Rocks reef completed in the spring (April–June) of 2011. Each point represents an individual drop site within the grid by calendar day (175-m spacing).

the estimated time of the lander reaching the seafloor and ending when retrieval began. Five minutes of recorded bottom time allowed for sufficient sediment settling, as well as capturing a maximum count of species present. Completed video tapes were reviewed aboard the vessel to determine underwater visibility and if any drops needed to be repeated based on low water clarity, camera orientation, or visual obstruction.

During the spring survey, 415 individual drops were completed over 12 boat-days, providing over 48.5 h of video footage. Of the 12 total sampling days, 7 d showed good bottom visibility, 1 d showed moderate bottom visibility, and 4 d showed little to no underwater visibility on the bottom, where turbidity obscured the view to the point that neither habitat nor fish were discernable, thus requiring resampling of these sites. A total of 143 drops had to be repeated due to poor underwater visibility or an obstructed view. The final sample size for analysis was 272 drops, a single drop for each of the 272 sampling locations. We used the drop with the highest score for visibility and view in cases when an individual site required resampling because habitat type or species were indiscernible due to

high turbidity or marine snow. Acceptable weather for the winter survey occurred between December 6 and 9, 2011, and yielded a total of 108 usable drops across the Three Arch Rocks reef grid. Of these 108 usable drops, a continuous block of 70 drops comprising the outer reef section were used for species composition comparison with the spring survey results.

Following the field deployments, videos were reviewed in the laboratory by the primary author. Video review consisted of two separate components. The initial review was used to describe camera visibility and view, as well as topographic relief (Table 1). Topographic relief was defined as flat, low, or high depending on the observed habitat type at each drop location. We did not have a way to accurately measure distance sampled by the camera but roughly classified visibility conditions on a scale from 0 to 2 (poor, medium, good) following the criteria in Table 1 (Hannah and Blume 2012). Primary habitat (dominant habitat type in the camera's view) and secondary habitat (second most abundant habitat feature in view) were classified for each drop based on the habitat criteria shown in Table 2 (Hannah and Blume 2012). Drops that had

TABLE 1. Criteria used to classify relief, underwater visibility, and view when reviewing video lander footage from Three Arch Rocks reef. Video footage had to receive at least a 1 in the Visibility or View categories to be used in further analysis.

Category	Class	Description
Relief	0	Flat (sand, flat bedrock, gravel or pebble, hash)
	1	Low (cobble, small boulder, bedrock)
	2	High (large boulder, vertical wall, crevice)
Visibility	0	Poor = view of surrounding substrate completely obscured by turbidity or marine snow
	1	Medium = view of surrounding substrate is not obscured but viewing distance is limited by variable turbidity or marine snow or both
	2	Good = view of surrounding substrate is clear to the limit of the lighted area
View	0	Completely obscured by habitat very close to the camera (includes lander tipped on side, looking down or up)
	1	Partially restricted by habitat very close to the camera
	2	Unrestricted view

view and visibility scores of 0 were excluded from analysis. The second review was used to identify fish observed on the video, which were identified to the lowest taxonomic level possible, usually to species. The maximum count of individuals from each species in any single frame (MaxN) was recorded for each drop to eliminate the potential for double counting of individuals (Harvey et al. 2007). Fish observed as the video lander was being retrieved were not included in the maximum count. Habitat observations made from the review of video lander footage were compared with a habitat classification map of the Three Arch Rocks region developed and provided by the Active Tectonics and Seafloor Mapping Lab (ATSML) at Oregon State University. The ATSML habitat maps are developed from multibeam sonar scans that collect data on a 4-m × 4-m (16 m^2) grid pixel size. To develop a habitat map, these grid boxes are smoothed into 10-m × 10-m (100 m^2) mapping units for which the dominant habitat type is displayed in a habitat box.

We investigated the species–habitat associations of the 9 most abundant fish species observed on the Three Arch

Rocks reef during the spring survey and the 10 most abundant fish species observed during the winter survey. These included the following: Black Rockfish *Sebastes melanops*, Blue Rockfish *Sebastes mystinus*, Canary Rockfish *Sebastes pinniger*, Quillback Rockfish *Sebastes maliger*, Yelloweye Rockfish *Sebastes ruberrimus*, Yellowtail Rockfish *Sebastes flavidus*, Lingcod *Ophiodon elongatus*, Kelp Greenling *Hexagrammos decagrammus*, Spotted Ratfish *Hydrolagus colliei* (winter only), and Pile Perch *Damalichthys vacca*. The number of individual fish species was totaled for each drop location to determine species richness. A Fisher's exact test was used to compare presence–absence data for each of the most abundant fish species on the reef to both of the primary and secondary habitat types identified. Unidentified adult fish were excluded from all analyses; however, unidentified juvenile rockfish *Sebastes* spp. were included in the analyses and treated as their own category.

Differences in species assemblages and relative abundance (maximum count per drop in a single frame; MaxN) between sections of the reef were analyzed with a pairwise one-way

TABLE 2. Habitat criteria used to classify the primary and secondary habitat types observed at the Three Arch Rocks reef complex from video lander survey footage (Hannah and Blume 2012).

Abbreviation	Substrate interpretation	Description
FLB	Flat bedrock	Rock with little to no relief
BR	Bedrock outcrop	Solid rock with some relief extending across the view
LB	Large boulder	Boulders approximately 1–3 m in diameter (includes angular blocks broken off from bedrock)
SB	Small boulder	Boulders approximately 0.25–1.00 m in diameter
CO	Cobble	Cobble approximately 6–25 cm in diameter
GP	Gravel pebble	Gravel or pebble approximately 2–60 mm in diameter
SA	Sand	Sand or mud with grain size 0.06–2.00 mm in diameter
CR	Crevice	Crevices in rock up to 1 m high by 1–3 m wide
VW	Vertical wall	Rock wall higher than 2 m and greater than 80° to the horizontal
HA	Hash	Small broken bits of shells

analysis of similarity (ANOSIM), adjusted for multiple comparisons with a step-down sequential Bonferroni correction, using the software package Paleontological Statistics 2.15. The degree of difference in the species composition was then measured using the Bray–Curtis dissimilarity index (BCDI) in the SIMPER routine (Bray and Curtis 1957; Hammer et al. 2001; Hannah and Blume 2012). Every drop, including those with no observations, was included in both analyses. The relative abundance of each species observed among sections of the study site was compared using a nonparametric Wilcoxon test, as has been previously used with video lander data (Hannah and Blume 2012). Additionally, the species assemblage at Three Arch Rocks reef was compared with the species assemblages of five exploratory sites surveyed by a video lander and presented in Hannah and Blume (2012). This assemblage comparison was again performed using pairwise one-way ANOSIMs, adjusted for multiple comparisons using a step-down sequential Bonferroni correction in Paleontological Statistics 2.15.

Pairwise species co-occurrence was evaluated using the checkerboard score (C-score) metric to provide a single score of co-occurrence of a pair of species using presence–absence data (Stone and Roberts 1990; Ulrich and Gotelli 2010; Groundfish Management Team 2013):

$$C_{ij} = \frac{(K_i - S_{ij}) \times (K_j - S_{ij})}{K_i \times K_j},$$

where K_i = the number of occurrences of species i, K_j = the number of occurrences of species j, and S_{ij} = the number of co-occurrences of species i and j. The C-score analysis provides a normalized value of 0–1, where 1 indicates perfect segregation between the two species and 0 indicates complete overlap. The Groundfish Management Team considers C-scores above 0.70 as a strong indication that the two species are segregated and scores of 0.30 and below as the two species exhibiting a high degree of overlap.

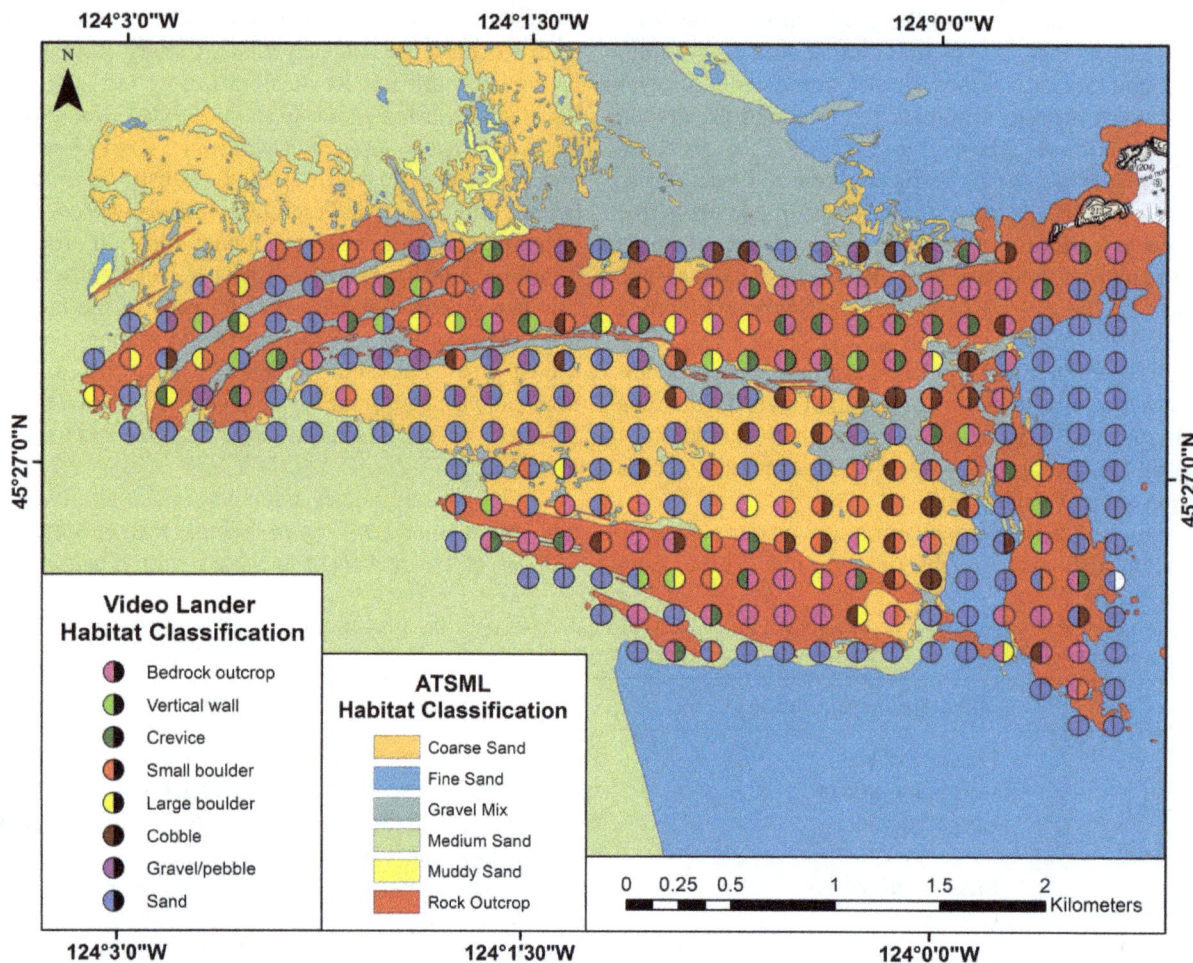

FIGURE 4. Total distribution of habitat types (see Table 2) across the Three Arch Rocks reef as observed by the video lander (April through June 2011) overlaid on a habitat classification map (≥ 100-m² mapping unit patch size) developed by Oregon State University's Active Tectonics and Seafloor Mapping Lab (ATSML) for the state waters mapping project. Video-lander-observed habitat classifications are shown as a divided circle with primary habitat type on the left and secondary habitat type on the right; if only one habitat type was observed the circle is shown as a contiguous color. (For simplification, the video lander habitat classification legend shows only the reference color of the primary habitat type observed).

RESULTS

Habitat characterization over the entire survey grid revealed that sand was the most abundant habitat type for both primary (41.2%) and secondary (35.3%) habitat. Bedrock outcrop was the second most frequent primary habitat type (21.7%), with the high-relief habitat types (large boulder [4.0%], vertical wall [5.9%], and crevice [3.7%]) and moderate-to-low-relief habitat types (cobble [9.2%], gravel–pebble [5.9%], and small boulder [8.4%]) registering lower in overall frequency (Figure 4). We compared the habitat classifications characterized from the review of video lander footage with habitat maps developed by Oregon State University's ATSML for the state waters mapping project. We found an 80.1% agreement of primary habitat type between our habitat classifications and the ATSML classifications at the lowest resolution of the two classifications. Differences in agreement between the two methodologies is primarily due to scale; the video lander provides a view over a relatively small area, while habitat maps generated from multibeam sonar and backscatter data generally blend multiple similar habitat types into more uniform classifications. This habitat smoothing inherently leads to a loss of fine-scale habitat resolution in multibeam sonar habitat mapping, which is information the video lander is able to provide. Discrepancies along the edges between habitat types may indicate important areas for resampling.

Species Abundance and Distribution

While many (46% in spring, 36% in winter) of the drops performed over the course of the survey did not have any fish observed, the majority of the sites surveyed by the video lander had one or more fish species present. Over the course of the spring survey, the nine most abundant species observed by the video lander showed distinct habitat and depth associations,

TABLE 3. Habitat associations (based on *P*-values obtained from Fisher's exact test) of the 10 most abundant species observed by the video lander during both the spring (April–June 2011) and winter (December 2011) surveys of the Three Arch Rocks reef complex. Kelp Greenling is the only fish with clear sexual dimorphism that could be observed in the videos. Significant *P*-values are color-coded by survey period: yellow = spring significant positive association, blue = winter significant positive association, green = spring and winter significant positive association, and red = significant negative association. If a significant *P*-value was identified during both survey periods, the "less significant" of the two seasons is presented. The abbreviations for habitat type are as follows: bedrock outcrop (BR), large boulder (LB), small boulder (SB), crevice (CR), vertical wall (VW), cobble (CO), gravel–pebble (GP), and sand (SA).

Common and scientific name	Primary habitat type							
	BR	LB	SB	CR	VW	CO	GP	SA
Black Rockfish *Sebastes melanops*	0.0068	0.0357	>0.1000	0.0004	>0.1000	>0.1000	>0.1000	<0.0001
Blue Rockfish *Sebastes mystinus*	>0.1000	0.0335	>0.1000	0.0255	0.0024	>0.1000	>0.1000	0.0005
Canary Rockfish *Sebastes pinniger*	>0.1000	0.0081	0.0372	>0.1000	0.0743	0.0492	>0.1000	<0.0001
Quillback Rockfish *Sebastes maliger*	>0.1000	0.0016	0.0535	>0.1000	0.0791	0.0701	>0.1000	<0.0001
Yelloweye Rockfish *Sebastes ruberrimus*	0.0552	0.0073	>0.1000	0.0004	0.0362	>0.1000	>0.1000	<0.0001
Yellowtail Rockfish *Sebastes flavidus*	>0.1000	0.0157	>0.1000	0.0156	0.0017	>0.1000	>0.1000	0.0003
Kelp Greenling *Hexagrammos decagrammus*	0.0162	0.0008	0.0110	>0.1000	>0.1000	>0.1000	>0.1000	<0.0001
Male	0.0033	0.0070	0.0432	>0.1000	>0.1000	>0.1000	>0.1000	<0.0001
Female	>0.1000	>0.1000	>0.1000	>0.1000	>0.1000	0.0421	>0.1000	0.0051
Lingcod *Ophiodon elongatus*	0.0845	>0.1000	>0.1000	0.0794	>0.1000	>0.1000	>0.1000	<0.0001
Pile Perch *Damalichthys vacca*	>0.1000	>0.1000	>0.1000	0.0874	>0.1000	>0.1000	>0.1000	0.0005
Spotted Ratfish *Hydrolagus colliei*	>0.1000	>0.1000	>0.1000	>0.1000	>0.1000	0.0475	>0.1000	>0.1000

(extended on next page)

while three other identified fishes (Copper Rockfish *Sebastes caurinus*, Cabezon *Scorpaenichthys marmoratus*, and unidentified juvenile rockfishes) were not observed with enough regularity to identify any significant associations (Table 3).

In total, 745 individual rockfish were observed, among 939 total individual fish of all species, as well as 17 Dungeness crab *Metacarcinus magister* (Table 4). Over the course of the spring survey, Black Rockfish was the most abundant species overall (34 stations, 277 individuals), with Canary Rockfish being the second most abundant (41 stations, 225 individuals) (Table 4). The most abundant and frequently observed demersal rockfish species was Yelloweye Rockfish (22 stations, 27 individuals) (Table 4). Kelp Greenlings were the most frequently observed fish other than rockfish on the reef (67 stations, 90 individuals), while Lingcod were second (48 stations, 61 individuals) (Table 4). These two hexagrammid fishes

were also the species with the broadest habitat associations and depth distributions.

During the winter survey, pelagic schooling rockfish (Black Rockfish, Blue Rockfish, and Yellowtail Rockfish) were the most abundant group of fishes observed on the reef, with Yellowtail Rockfish (14 stations, 92 individuals) being the single most abundant and frequently observed of the three (Table 4). Canary Rockfish (19 stations, 40 individuals) was the second most abundant species overall, excluding Northern Anchovy *Engraulis mordax*, and were observed in the greatest frequency (Table 4). The most abundant and frequently observed demersal rockfish was Yelloweye Rockfish (10 stations, 15 individuals), followed by Quillback Rockfish (4 stations, 5 individuals) (Table 4).

In winter, Kelp Greenlings (16 stations, 17 individuals) exhibited a broad distribution across depth and habitat, while

TABLE 3.   Extended.

Common and scientific name	Secondary habitat type							
	BR	LB	SB	CR	VW	CO	GP	SA
Black Rockfish *Sebastes melanops*	0.0087	0.0351	>0.1000	0.0024	>0.1000	>0.1000	0.0532	<0.0001
Blue Rockfish *Sebastes mystinus*	>0.1000	0.0072	0.0009	<0.0288	>0.1000	>0.1000	>0.1000	0.0003
Canary Rockfish *Sebastes pinniger*	>0.1000	0.0085	0.0095	>0.1000	>0.1000	>0.1000	>0.1000	<0.0001
Quillback Rockfish *Sebastes maliger*	>0.1000	0.0560	0.0143	0.0269	>0.1000	>0.1000	>0.1000	0.0045
Yelloweye Rockfish *Sebastes ruberrimus*	>0.1000	0.0003	0.0626	>0.1000	>0.1000	>0.1000	>0.1000	0.0343
Yellowtail Rockfish *Sebastes flavidus*	>0.1000	>0.1000	0.0067	<0.0001	>0.1000	>0.1000	>0.1000	0.0130
Kelp Greenling *Hexagrammos decagrammus*	0.0105	0.0487	>0.1000	0.0177	>0.1000	>0.1000	>0.1000	<0.0001
Male	0.0053	0.0224	>0.1000	0.0030	>0.1000	0.0868	0.0529	<0.0001
Female	>0.1000	>0.1000	>0.1000	0.0955	>0.1000	>0.1000	>0.1000	0.0019
Lingcod *Ophiodon elongatus*	0.0431	>0.1000	0.0010	0.0332	0.0814	>0.1000	>0.1000	0.0206
Pile Perch *Damalichthys vacca*	>0.1000	>0.1000	0.0569	>0.1000	>0.1000	>0.1000	>0.1000	0.0225
Spotted Ratfish *Hydrolagus colliei*	>0.1000	>0.1000	0.0008	>0.1000	>0.1000	>0.1000	>0.1000	>0.1000

TABLE 4. Maximum count (n), number of drop sites each species was observed at (Drops), relative abundance (%; n/total number of fish observed), and rank abundance of observed fish taxa on the Three Arch Rocks reef over both the spring (April–June 2011) and winter (December 2011) surveys; NA denotes lack of observation.

Common and scientific name and totals	Total		Spring (272 drops)				Winter (108 drops)				Spring and winter comparison	
	n	Rank	n	Drops	%	Rank	n	Drops	%	Rank	Spring (70 drops) n	Winter (70 drops) n
Black Rockfish *Sebastes melanops*	303	1	277	34	29.5	1	26	6	9.5	3	6	0
Canary Rockfish *Sebastes pinniger*	265	2	225	41	24.0	2	40	19	14.7	2	154	28
Yellowtail Rockfish *Sebastes flavidus*	147	3	55	55	5.9	6	92	14	33.7	1	35	84
Blue Rockfish *Sebastes mystinus*	139	4	124	19	13.2	3	15	9	5.5	6	11	6
Kelp Greenling *Hexagrammos decagrammus*	107	5	90	67	9.6	4	17	16	6.2	5	27	9
Lingcod *Ophiodon elongatus*	67	6	61	48	6.5	5	6	5	2.2	10	16	6
Yelloweye Rockfish *Sebastes ruberrimus*	42	7	27	27	2.9	7	15	10	5.5	6	21	14
Quillback Rockfish *Sebastes maliger*	26	8	21	18	2.2	9	5	4	1.8	11	12	5
Pile Perch *Damalichthys vacca*	24	9	24	14	2.6	8	NA	0	NA	NA	9	0
Spotted Ratfish *Hydrolagus colliei*	20	10	NA	NA	NA	NA	20	11	7.3	4	0	18
Unidentified fish	20	10	5	4	0.5	12	15	2	5.5	6	2	15
Unidentified rockfish *Sebastes* spp.	19	12	5	5	0.5	12	14	5	5.1	9	1	12
Unidentified flatfish, order Pleuronectiformes	10	13	9	6	1.0	10	1	1	0.4	13	2	1
Unidentified juvenile rockfish *Sebastes* spp.	9	14	4	2	0.4	14	5	3	1.8	11	1	5
Copper Rockfish *Sebastes caurinus*	6	15	6	6	0.6	11	NA	NA	NA	NA	1	0
Cabezon *Scorpaenichthys marmoratus*	3	16	2	2	0.2	15	1	1	0.4	13	2	0
Unidentified surfperch, family Embiotocidae	2	17	2	2	0.2	15	NA	NA	NA	NA	0	0
Tiger Rockfish *Sebastes nigrocinctus*	1	18	NA	NA	NA	NA	1	1	0.4	13	0	1
Wolf-eel *Anarrhichthys ocellatus*	1	18	1	1	0.1	16	NA	NA	NA	NA	1	0
China Rockfish *Sebastes nebulosus*	1	18	1	1	0.1	16	NA	NA	NA	NA	0	0
Northern Anchovy *Engraulis mordax*	>1,500	NA	NA	NA	NA	NA	>1,500	16	NA	NA	0	>1,500
Dungeness crab *Metacarcinus magister*	18	NA	17	4	NA	NA	1	1	NA	NA	0	1
Total number of rockfish	958		745		79.3		213		78.0		242	155
Total number of fish (excluding Northern Anchovy)	1,212		939				273				301	205

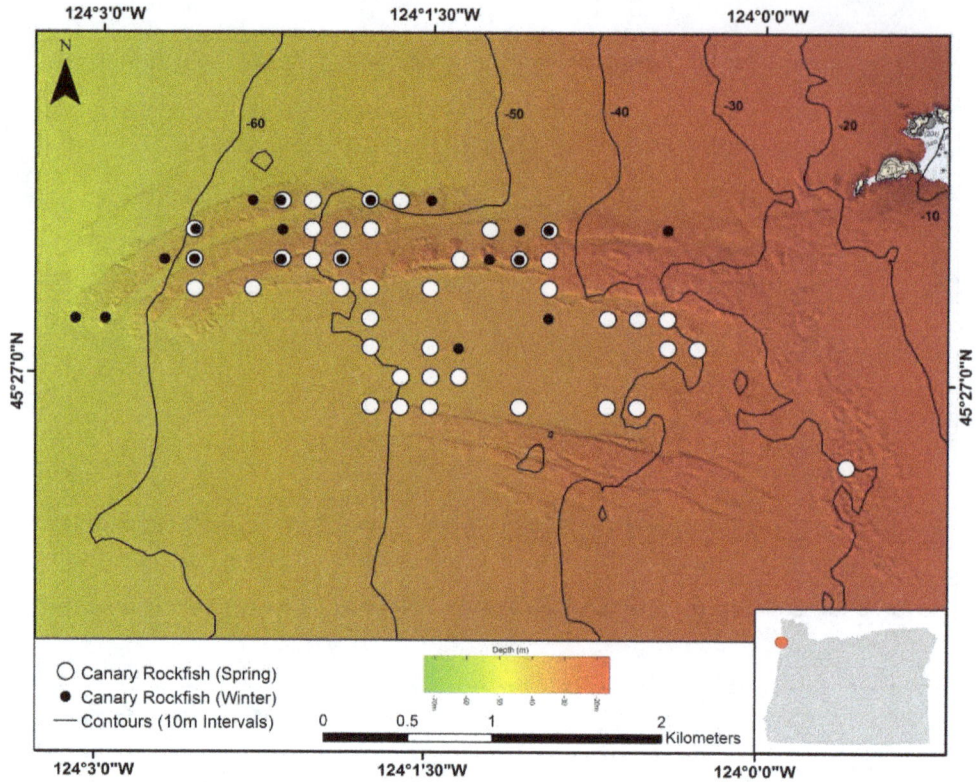

(a)

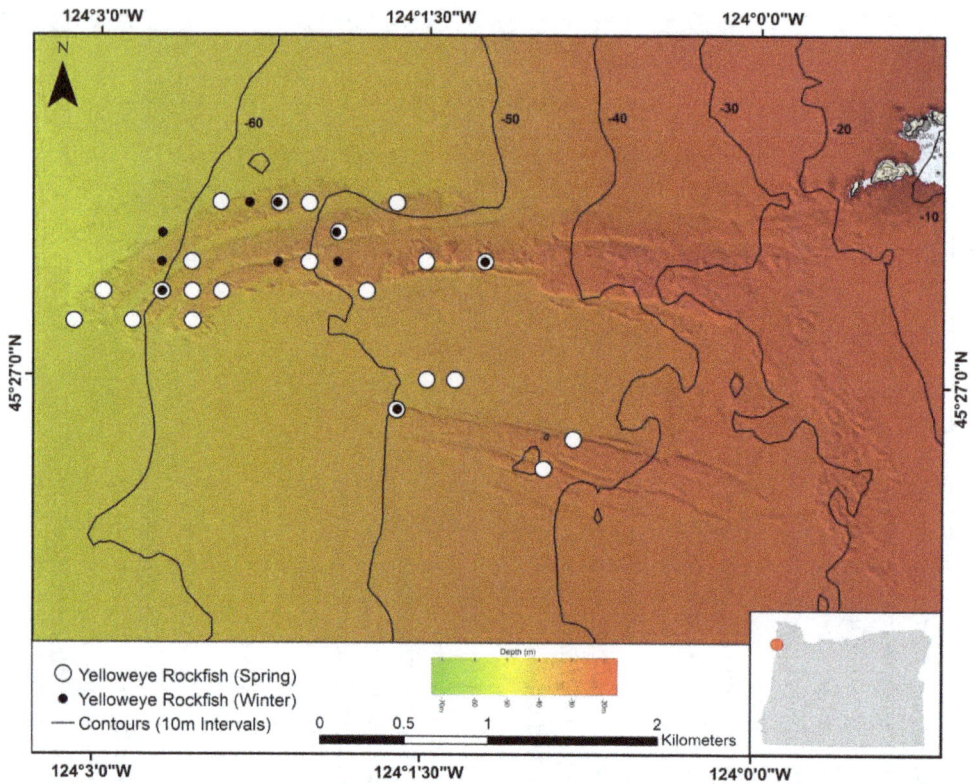

(b)

FIGURE 5. Distribution of (A) Canary Rockfish and (B) Yelloweye Rockfish across the Three Arch Rocks reef as observed by the video lander over both the spring (April–June 2011) and winter (December 2011) surveys combined. Slight differences in position location between spring and winter are due to the current shifting the final recorded drop location between seasons.

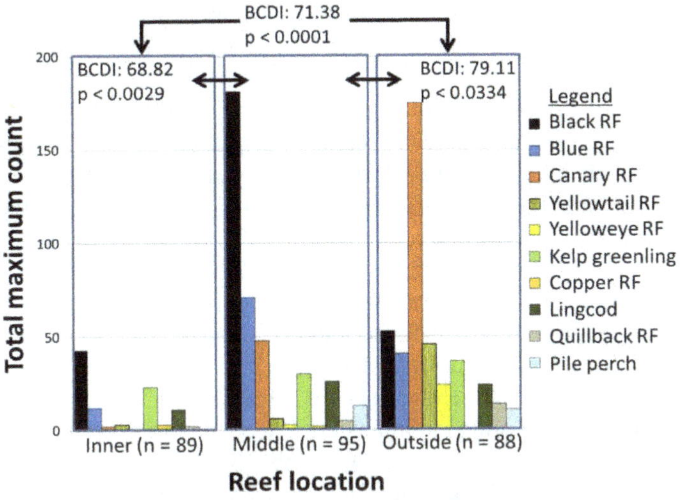

FIGURE 6. Spring (April–June 2011) species composition of Three Arch Rocks reef divided into three depth categories; inner (89 drops), middle (95 drops), and outside (88 drops) across the reef based on video lander video analysis (arrows represent the reef sections compared for species composition and abundance with corresponding Bray–Curtis dissimilarity index [BCDI] values and *P*-values from pairwise one-way ANOSIMs); RF = rockfish.

Lingcod appeared noticeably absent (5 stations, 6 individuals) (Table 4). This absence of Lingcod may be due to spawning migration, as Lingcod are known to nest in shallow-water habitats in winter months (Matthews 1992; O'Connell 1993; Martell et al. 2000). Spotted Ratfish, which were not observed during the spring survey, were observed at 11 stations during December over multiple habitat types. Additionally, schools of varying size of Northern Anchovy, also not observed during the spring survey, were observed at 16 drop stations across the reef in December (Table 4). In total, 213 identified rockfish were observed, among 273 total fish (excluding Northern Anchovy) and one Dungeness crab (Table 4). Some of the species showed consistency in sighting locations between spring and winter surveys. For instance, of the 10 sites where

Yelloweye Rockfish were observed in December, 5 were the same as the spring survey, while Kelp Greenlings were observed again at 8 of 16 total sites. It is impossible to know if these were the same individual fish, but consistency does provide strong evidence for habitat association.

Habitat associations were often significant but varied for some species between winter and spring sampling (Table 3). Nearly all of the species we observed showed negative correlations with sand habitat. Schooling pelagic rockfish (Black Rockfish, Blue Rockfish, and Yellowtail Rockfish) all showed a qualitative relationship with the reef structure, exhibiting a distribution pattern that mirrored the shape of the reef. Yelloweye Rockfish and Canary Rockfish, both currently managed as overfished stocks on the West Coast, exhibited significant associations with a variety of habitat types across the Three Arch Rocks reef (Table 3). During each survey period, Canary Rockfish had the broadest distribution across depths and habitat types of any of the observed rockfish species (Figure 5A). Yelloweye Rockfish on the other hand, while exhibiting significant relationships with high-vertical-relief habitat types (Table 4), showed a more restricted distribution than Canary Rockfish, with the vast majority of observations occurring on the outer third portion of the reef (Figure 5B).

Reef Fish Community Composition

We conducted an analysis of the species composition across the reef to investigate how species composition changed with depth and location. The survey grid was divided into three segments moving east to west across the reef: inside (89 drops), middle (95 drops), and outside (88 drops) (Figure 6). Comparison of the habitat composition across the three reef sections (inside, middle, outside) showed that only the inner and outer sections differed significantly (BCDI = 75.88; ANOSIM: *P* < 0.0427). This result was primarily driven by differences in

TABLE 5. Normalized *C*-scores (pairwise associations) of the 10 most abundant species observed by the video lander during the spring survey. A *C*-score of 0 indicates the total overlap of two species, while a *C*-score of 1 would indicate that the two species are never found together. In general, *C*-scores less than 0.3 (dark gray) are thought to signify species pairs that commonly associate, scores between 0.3 and 0.7 (light gray) indicate moderate association, while scores over 0.7 indicate very low association (unshaded).

Fish species	Blue Rockfish	Canary Rockfish	Yellowtail Rockfish	Yelloweye Rockfish	Kelp Greenling	Copper Rockfish	Lingcod	Quillback Rockfish	Pile Perch
Black Rockfish	0.15	0.70	0.59	0.79	0.47	0.46	0.56	0.84	0.40
Blue Rockfish		0.71	0.36	0.65	0.51	0.79	0.54	0.89	0.66
Canary Rockfish			0.40	0.37	0.49	0.46	0.32	0.51	0.41
Yellowtail Rockfish				0.32	0.57	0.58	0.69	0.58	0.87
Yelloweye Rockfish					0.61	0.80	0.58	0.48	0.78
Kelp Greenling						0.65	0.43	0.55	0.52
Copper Rockfish							0.47	0.79	0.77
Lingcod								0.58	0.29
Quillback Rockfish									0.88

sand, bedrock, and small boulder (secondary habitat) between these two regions of the reef. The species assemblages of the three sections all differed significantly in composition (Figure 6). The inside section, dominated primarily by Black Rockfish and Kelp Greenling, had the lowest overall species abundance and richness. This section also differed significantly in species composition from the middle (BCDI = 68.82; ANOSIM: $P < 0.0029$) and the outside (BCDI = 71.38; ANOSIM: $P < 0.0001$) sections of the reef. The middle section was dominated by Black Rockfish, followed by Blue Rockfish and Canary Rockfish. The middle section also differed significantly from the outside section (BCDI = 79.11; ANOSIM: $P < 0.0334$), which was dominated by Canary Rockfish, but also had the highest abundance of Yellowtail Rockfish, Yelloweye Rockfish, and Quillback Rockfish (Figure 6).

A comparison of the species composition between the spring and winter surveys, restricted to the contiguous outermost 70 stations of the Three Arch Rocks reef survey that were sampled in both seasons, revealed significant differences in species composition (BCDI = 76.41; ANOSIM: $P < 0.0155$). This difference was driven by the winter presence of Spotted Ratfish at the reef and the overall lower winter abundance of Canary Rockfish (Wilcoxon test: $P < 0.0168$), Kelp Greenling ($P < 0.0241$), and Lingcod ($P < 0.0456$).

We investigated species correlations using the technique utilized by the Northwest Fisheries Science Center to identify species that commonly associate with others (C-score; Groundfish Management Team 2013) (Table 5). Black Rockfish and Blue Rockfish were commonly associated, with C-scores of 0.15 and 0.07 in the spring and winter, respectively. During the winter survey, strong pairwise associations were observed for Yelloweye Rockfish and Quillback Rockfish, Blue Rockfish and Yellowtail Rockfish, and Yelloweye Rockfish and Canary Rockfish. Other species pairs exhibited intermediate scores, likely due to low sample sizes (Table 5).

Nearshore Species Assemblage Variation on the Oregon Coast

The presence or absence and relative abundance of nearshore temperate fishes is likely to be quite variable from reef to reef due to differences in environmental conditions and fishing pressure, as well as year-to-year or seasonal movements and recruitment events (Gunderson et al. 2008). Three Arch Rocks reef is unique in its structure in that it runs east to west, spanning a wide range of depths, giving it a very diverse assemblage of temperate Pacific reef fishes. We compared our results from the spring survey with those from five exploratory sites, four nearshore (Cape Perpetua, East Siletz, West Siletz, and Seal Rocks) and one offshore (Stonewall Bank), off the central Oregon coast (presented in Hannah and Blume 2012) and found some interesting differences (Table A.1 in the appendix). The species assemblage at the Three Arch Rocks

reef was significantly different from that of Cape Perpetua, approximately 150 km south (BCDI = 86.02; ANOSIM: $P < 0.0011$). The Cape Perpetua reef, which runs north to south, showed greater relative abundances of Canary Rockfish and juvenile rockfish but lower relative abundances of Kelp Greenling and Black Rockfish during the July survey (Table A.1 in the appendix). The differences were less conclusive for the February survey at Cape Perpetua (BCDI = 81.07; ANOSIM: $P < 0.064$), but this was likely due to the smaller numbers of fish observed in winter months. The Eastern Siletz reef showed the greatest disparity in species assemblage in comparison with the Three Arch Rocks reef (BCDI = 88.88; ANOSIM: $P < 0.0001$) due to the high relative abundance of Blue Rockfish and Kelp Greenling at the Eastern Siletz site (Table A.1 in the appendix). Similar results were observed when comparing with the Seal Rocks site (BCDI = 85.75; ANOSIM: $P < 0.0002$) due to the greater relative abundance of Black Rockfish, Canary Rockfish, Kelp Greenling, and juvenile rockfish observed at Seal Rocks (Table A.1 in the appendix). In contrast, the Western Siletz reef site was more similar to Three Arch Rocks (BCDI = 78.62; ANOSIM: $P < 0.0721$) (Table A.1 in the appendix). Similar results were observed when comparing the species assemblage to Stonewall Bank (BCDI = 70.63; ANOSIM: $P < 0.0574$), due primarily to the absence of Black Rockfish at Stonewall Bank and the relative greater abundance of Canary Rockfish and juvenile rockfish there but also the greater relative abundance of Lingcod at Three Arch Rocks (Table A.1 in the appendix).

DISCUSSION

Assessing and monitoring the distribution and abundance of fishes inhabiting rocky-reef structures along the West Coast of North America is a significant challenge. Typically, scientists must take a dynamic approach, utilizing a myriad of sources of data and a variety of data collection techniques to identify the relative abundance of species and their associated habitat types across geographic regions (Francis 1986; Parker et al. 2000). In order for resource surveys to be effective, a sizeable amount of planning, funding, and personnel from multiple agencies is often needed. However, even with substantial time and effort, the results of these surveys can still be highly variable, with poor representation of some habitat types in areas that are difficult to sample (Krieger 1993; Jagielo et al. 2003). Our study provides a comprehensive look at how video drop camera survey data can be analyzed to contribute to our understanding of nearshore fishes and their habitat associations.

The limitations of the most commonly used survey gear types (i.e., bottom trawl) are well documented and understood both by those who employ the gear and those who utilize the data in stock assessments (Adams et al. 1995; Williams and Ralston 2002). The question, however, is how to integrate different sources of data and promote the use of novel sampling methods that can overcome the shortcomings of currently used sampling methods for nearshore rocky-reef species. Previous work has demonstrated that surveys utilizing various direct video observation techniques, such as video landers, baited underwater video stations, and scuba, are effective at surveying nearshore, shallow-water fish populations (Watson et al. 2005; Langlois et al. 2010; Hannah and Blume 2012). Additionally, video data analyzed from direct observations has been shown to provide relative abundance estimates of a variety of fish species through nonextractive, fishery-independent means (Gledhill et al. 2006; Yoklavich et al. 2007; Coleman et al. 2011; Merritt et al. 2011). Much of this work, however, has been performed in areas such as Hawaii, southern California, and the Gulf of Mexico— regions which are not inhibited to the same extent by the survey challenges present in the highly productive and often turbulent waters of the northeastern Pacific Ocean. Our research demonstrates that a video lander platform can comprehensively survey a temperate nearshore reef under difficult conditions, collecting broad-scale fish assemblage, relative abundance, and habitat data. This illustrates the opportunity for video landers to be developed into a more widely utilized, cost-effective survey tool to monitor nearshore rocky reefs along the Pacific coast of North America and other poorly surveyed systems.

Visual surveys, however, are not without their own limitations. Visibility is the single most important factor when it comes to a successful video survey, be it with the video lander, scuba divers, an ROV, or a submersible. The video lander has an advantage over other visual methods in its simplicity and ease of use and deployment, allowing minimal time and effort for data collection relative to conventional ROVs or submersibles (Stein et al. 1992; Krieger and Ito 1999; Johnson et al. 2003; Yoklavich et al. 2007; Pacunski et al. 2008). The relative simplicity, small size, and ease of use allows for rapid deployment of the video lander when weather conditions become favorable, as well as minimal cost to abort a survey when conditions prove unsuitable for sampling.

The data from this study provide information on habitat associations for overfished and rarely surveyed species in both spring and winter seasons that can further contribute to stock assessment and spatial management. The species–habitat associations of many demersal species determined from the spring survey were reinforced by the winter survey (e.g., Yelloweye Rockfish), while others were expanded upon (e.g., Quillback Rockfish) (Table 3). However, due to the low sample size of these species, more drops will be required to refine and substantiate these results. Additionally, the ability of the video lander to accurately identify species–habitat associations for pelagic schooling rockfish (Black Rockfish, Blue Rockfish, and Yellowtail Rockfish) may be limited and requires further investigation. Furthermore, these species have reduced interactions with the benthic substrate compared with demersal species (i.e., Yelloweye Rockfish, Quillback Rockfish), with much of their time spent schooling in the water column, potentially decreasing the capacity of the video lander to survey these species.

As with previous studies, the video lander exhibited limitations in its ability to identify and count flatfishes and very cryptic species like Cabezon due to the oblique angle of the camera and the standard definition resolution of the camera system (Hannah and Blume 2012). Additionally, we were unable to identify juvenile rockfish and other small fish to species due to the low resolution of standard-definition video, as well as some fish being too distant from the camera, an issue that will be improved with advances in high-definition and stereo video camera systems (Hannah and Blume 2014). The extent to which the video lander attracts or repels fish, as well as the limited and highly variable size of the area viewed, represent potential sampling biases which are currently not quantified. Despite this, extensive research has shown maximum counts from video surveys to be an accurate, although conservative, index of relative abundance (Willis and Babcock 2000; Stoner et al. 2008; Merritt et al. 2011). Other research has suggested using a mean count of fish observed in single snapshots over the entire course of the video in place of a single maximum count to improve the estimate of true abundance (Schobernd et al. 2014). Continued analysis of these methods of estimating relative abundance will be needed to further evaluate the video lander as an effective survey tool.

The C-score analysis of the video lander datasets offers another dimension with which to examine the distribution of nearshore species across the reef. Previous C-score analysis focused on deepwater slope rockfish and "other" roundfish found in trawl samples enumerated by the West Coast Groundfish Observer Program and the Alaska Fisheries Science Center (Groundfish Management Team 2013). These are robust datasets, both spatially and temporally, but are lacking in nearshore species like those observed by the video lander at Three Arch Rocks. Because C-scores are calculated from presence–absence data, the video lander provides ideal high-resolution data amenable to C-score analysis, enabling it to contribute to an additional assessment need. The video lander provides data on both overlap and segregation of species pairs, allowing for potential targeted management actions at the species, or species pair, level. Further surveys are required to expand on the results as many of the observed species had low total counts; however, preliminary analysis shows the versatility and robustness of video lander data for this type of analysis. Our analysis of C-scores for key groundfish species compliments the analyses performed by the Groundfish Management Team by providing results for nearshore, high-relief habitats that are not sampled in existing surveys.

Our comparison of observations at Three Arch Rocks reef to those of other experimental survey sites off the central Oregon coast displays the uniqueness and complexity of the species assemblage at different sites (Table A.1 in the appendix). While these locations are within 150 km of Three Arch Rocks reef, the east-to-west distribution of Three Arch Rocks reef creates a larger depth range and distance from shore, with a reef structure that is home to both shallow (Blue Rockfish and

Black Rockfish) and deeper-water (Yelloweye Rockfish and Canary Rockfish) species. This combination of depths and species diversity at Three Arch Rocks reef highlights the ecological diversity of this area and the need for additional surveys to characterize nearshore species assemblages and habitats in Oregon.

The ability to survey nearshore reefs off the Pacific coast is sporadic, and the weather windows are generally short. The video lander therefore is an ideal tool to use in all seasons because of its rapid, intensive survey capability combined with a short preparation and implementation schedule. The winter survey yielded interesting results, including the first video observations of Spotted Ratfish at Three Arch Rocks reef. While visibility issues plagued the winter survey, it provided new insight into this nearshore rocky-reef environment in winter. Given acceptable ocean conditions, the video lander has the capability to perform at the same level in the winter as it did in the spring.

Finally, the video lander may serve as a critical tool for evaluations of Marine Protected Areas and Marine Reserves, which have recently been employed as a conservation tool in Oregon and elsewhere in the United States. Nonextractive monitoring methods are needed, as many of these areas are closed to fishing, even for scientific study. The high spatial coverage, low operational and logistical cost, and nonextractive nature of video lander surveys make them an ideal tool for Marine Protected Area or Marine Reserve assessment, monitoring, and habitat ground-truthing. Further refinement of video quality, processing, and distance-area sampling will make video landers essential components of nearshore reef monitoring.

ACKNOWLEDGMENTS

Steve and Ray Dana provided the commercial passenger fishing vessel *Blue Water Too* as a sampling platform for the spring survey. Al Pazar provided the research vessel *Pacific Surveyor* as a sampling platform for the winter survey. Chris Goldfinger of the ATSML at Oregon State University provided funding for the winter survey, while Chris Romsos of ATSML provided habitat maps and technical support. Matthew T.O. Blume from the Marine Resources Program in the Oregon Department of Fish and Wildlife assisted with field sampling during the spring survey and provided technical support throughout. This manuscript was improved through the input of W. Waldo Wakefield at the National Oceanic and Atmospheric Administration's Northwest Fisheries Science Center in Newport, Oregon, Kevin Thompson and Scarlett Arbuckle at Oregon State University, and two anonymous reviewers.

REFERENCES

Adams, P. B., J. L. Butler, C. H. Baxter, T. E. Laidig, K. A. Dahlin, and W. Wakefield. 1995. Population estimates of Pacific coast groundfishes from video transects and swept-area trawls. U.S. National Marine Fisheries Service Fishery Bulletin 93:446–455.

Bray, J. R., and J. T. Curtis. 1957. An ordination of the upland forest communities of southern Wisconsin. Ecological Monographs 27:325–349.

Cappo, M., P. Speare, and G. De'ath. 2004. Comparison of baited remote underwater video stations (BRUVS) and prawn (shrimp) trawls for assessments of fish biodiversity in inter-reefal areas of the Great Barrier Reef Marine Park. Journal of Experimental Marine Biology and Ecology 302:123–152.

Coleman, F. C., K. M. Scanlon, and C. C. Koenig. 2011. Groupers on the edge: shelf edge spawning habitat in and around marine reserves of the northeastern Gulf of Mexico. Professional Geographer 63:456–474.

Cordue, P. L. 2007. A note on non-random error structure in trawl survey abundance indices. ICES Journal of Marine Science 64:1333–1337.

Ellis, D. M., and E. E. DeMartini. 1995. Evaluation of a video camera technique for indexing abundances of juvenile Pink Snapper *Pristipomoides filamentosus*, and other Hawaiian insular shelf fishes. U.S. National Marine Fisheries Service Fishery Bulletin 93:67–77.

Fox, D., A. Merems, M. Amend, H. Weeks, C. Romsos, and M. Appy. 2004. Comparative characterization of two nearshore rocky reef areas: a high-use recreational fishing reef vs. an unfished reef. Oregon Department of Fish and Wildlife, Newport.

Francis, R. C. 1986. Two fisheries biology problems in West Coast groundfish management. North American Journal of Fisheries Management 6:453–462.

Gledhill, C. T., G. W. Ingram Jr, K. R. Rademacher, P. Felts, B. Trigg, and M. S. Pascagoula. 2006. NOAA Fisheries reef fish video surveys: yearly indices of abundance for Gag (*Mycteroperca microlepis*), SEDAR10-DW-12. Southeast Data Assessment and Review, North Charleston, South Carolina.

Gledhill, C. T., J. Lyczkowski-Shultz, K. Rademacher, E. Kargard, G. Crist, and M. A. Grace. 1996. Evaluation of video and acoustic index methods for assessing reef-fish populations. ICES Journal of Marine Science 53:483–485.

Groundfish Management Team. 2013. Groundfish management team report on methods and results that may be used to evaluate alternatives for stock complex reorganization. Pacific Fishery Management Council, June 2013 Briefing Book, Agenda Item F.8.b, Portland.

Gunderson, D. R., A. M. Parma, R. Hilborn, J. M. Cope, D. L. Fluharty, M. L. Miller, R. D. Vetter, S. S. Heppell, and H. G. Greene. 2008. The challenge of managing nearshore rocky reef resources. Fisheries 33:172–179.

Hammer, O., D. A. T. Harper, and P. D. Ryan. 2001. PAST: Paleontological statistics package for education and data analysis. Paleontologia Electrobica [online serial] 4:article 4.

Hannah, R. W., and M. T. O. Blume. 2012. Tests of an experimental unbaited video lander as a marine fish survey tool for high-relief deepwater rocky reefs. Journal of Experimental Marine Biology and Ecology 430–431:1–9.

Hannah, R. W., and M. T. O. Blume. 2014. The influence of bait and stereo video on the performance of a video lander as a survey tool for marine demersal reef fishes in Oregon waters. Marine and Coastal Fisheries: Dynamics, Management, and Ecosystem Science [online serial] 6:181–189.

Harvey, E. S., M. Cappo, J. J. Butler, N. Hall, and G. A. Kendrick. 2007. Bait attraction affects the performance of remote underwater video stations in assessment of demersal fish community structure. Marine Ecology Progress Series 350:245–254.

Jagielo, T., A. Hoffmann, J. Tagart, and M. Zimmermann. 2003. Demersal groundfish densities in trawlable and untrawlable habitats off Washington: implications for the estimation of habitat bias in trawl surveys. U.S. National Marine Fisheries Service Fishery Bulletin 101:545–565.

Johnson, S. W., M. L. Murphy, and D. J. Csepp. 2003. Distribution, habitat, and behavior of rockfishes, *Sebastes* spp., in nearshore waters of southeastern Alaska: observations from a remotely operated vehicle. Environmental Biology of Fishes 66:259–270.

Krieger, K. J. 1993. Distribution and abundance of rockfish determined from a submersible and by bottom trawling. U.S. National Marine Fisheries Service Fishery Bulletin 91:87–96.

Krieger, K. J., and D. H. Ito. 1999. Distribution and abundance of Shortraker Rockfish, *Sebastes borealis*, and Rougheye Rockfish, *S. aleutianus*, determined from a manned submersible. U.S. National Marine Fisheries Service Fishery Bulletin 97:264–272.

Langlois, T. J., E. S. Harvey, B. Fitzpatrick, J. J. Meeuwig, G. Shedrawi, and D. L. Watson. 2010. Cost-efficient sampling of fish assemblages: comparison of baited video stations and diver video transects. Aquatic Biology 9:155–168.

Martell, S. J. D., C. J. Walters, and S. S. Wallace. 2000. The use of marine protected areas for conservation of Lingcod (*Ophiodon elongatus*). Bulletin of Marine Science 66:729–743.

Matthews, K. R. 1990a. An experimental study of the habitat preferences and movement patterns of Copper, Quillback, and Brown rockfishes (*Sebastes* spp.). Environmental Biology of Fishes 29:161–178.

Matthews, K. R. 1990b. A telemetric study of the home ranges and homing routes of Copper and Quillback rockfishes on shallow rocky reefs. Canadian Journal of Zoology 68:2243–2250.

Matthews, K. R. 1992. A telemetric study of the home ranges and homing routes of Lingcod *Ophiodon elongatus* on shallow rocky reefs off Vancouver Island, British Columbia. U.S. National Marine Fisheries Service Fishery Bulletin 90:784–790.

Merritt, D., M. K. Donovan, C. Kelley, L. Waterhouse, M. Parke, K. Wong, and J. C. Drazen. 2011. BotCam: a baited camera system for nonextractive monitoring of bottomfish species. U.S. National Marine Fisheries Service Fishery Bulletin 109:56–67.

O'Connell, V. M. 1993. Submersible observations of Lingcod, *Ophiodon elongatus*, nesting below 30 m off Sitka, Alaska. Marine Fisheries Review 55:19–24.

Pacunski, R. E., and W. A. Palsson. 2002. Macro- and micro-habitat relationships of adult and sub-adult rockfish, Lingcod, and Kelp Greenling in Puget Sound. Washington Department of Fish and Wildlife, Olympia.

Pacunski, R. E., W. A. Palsson, H. G. Greene, and D. Gunderson. 2008. Conducting visual surveys with a small ROV in shallow water. Pages 109–128 *in* J. R. Reynolds and H. G. Greene, editors. Marine habitat mapping technology for Alaska. Alaska Sea Grant for North Pacific Research Board, Fairbanks.

Parker, S. J., S. A. Berkeley, J. T. Golden, D. R. Gunderson, J. Heifetz, M. A. Hixon, R. Larson, B. M. Leaman, M. S. Love, J. A. Musick, V. M. O'Connell, S. Ralston, H. J. Weeks, and M. M. Yoklavich. 2000. Management of Pacific rockfish. Fisheries 25(3):22–30.

PFMC (Pacific Fishery Management Council). 2005. Pacific coast groundfish management plan for the California, Oregon, and Washington groundfish fishery. Appendix B, part 1. Assessment methodology for groundfish essential fish habitat. PFMC, Portland, Oregon.

PFMC (Pacific Fishery Management Council). 2011. Pacific coast groundfish fishery management plan for the California, Oregon, and Washington groundfish fishery. PFMC, Portland, Oregon.

Priede, I. G. and N. R. Merrett. 1996. Estimation of abundance of abyssal demersal fishes; a comparison of data from trawls and baited cameras. Journal of Fish Biology 49:207–216.

Roberts, J. M., O. C. Peppe, L. A. Dodds, D. J. Mercer, W. T. Thomson, J. D. Gage, D. T. Meldrum, A. Freiwald, and J. M. Roberts. 2005. Monitoring environmental variability around cold-water coral reefs: the use of a benthic photolander and the potential of seafloor observatories. Pages 483–502 *in* A. Freiwald, editor. Cold-water corals and ecosystems. Springer Berlin-Heidelberg, Berlin.

Schobernd, Z. H., N. M. Bacheler, and P. B. Conn. 2014. Examining the utility of alternative video monitoring metrics for indexing reef fish abundance. Canadian Journal of Fisheries and Aquatic Sciences 71:464–471.

Stein, D. L., B. N. Tissot, M. A. Hixon, and W. H. Barss. 1992. Fish-habitat associations on a deep reef at the edge of the Oregon continental shelf. U.S. National Marine Fisheries Service Fishery Bulletin 90:540–551.

Stobart, B., J. A. García-Charton, C. Espejo, E. Rochel, R. Goñi, O. Reñones, A. Herrero, R. Crec'hriou, S. Polti, C. Marcos, S. Planes, and A. Pérez-Ruzafa. 2007. A baited underwater video technique to assess shallow-water Mediterranean fish assemblages: methodological evaluation. Journal of Experimental Marine Biology and Ecology 345:158–174.

Stone, L., and A. Roberts. 1990. The checkerboard score and species distributions. Oecologia 85:74–79.

Stoner, A. W., B. J. Laurel, and T. P. Hurst. 2008. Using a baited camera to assess relative abundance of juvenile Pacific Cod: field and laboratory trials. Journal of Experimental Marine Biology and Ecology 354:202–211.

Ulrich, W., and N. J. Gotelli. 2010. Null model analysis of species associations using abundance data. Ecology 91:3384–3397.

Wakefield, C. B., P. D. Lewis, T. B. Coutts, D. V. Fairclough, and T. J. Langlois. 2013. Fish assemblages associated with natural and anthropogenically-modified habitats in a marine embayment: comparison of baited videos and opera-house traps. PLoS (Public Library of Science) ONE [online serial] 8:e59959.

Watson, D. L., E. S. Harvey, M. J. Anderson, and G. A. Kendrick. 2005. A comparison of temperate reef fish assemblages recorded by three underwater stereo-video techniques. Marine Biology 148:415–425.

Williams, E. H., and S. Ralston. 2002. Distribution and co-occurrence of rockfishes (family: Sebastidae) over trawlable shelf and slope habitats of California and southern Oregon. U.S. National Marine Fisheries Service Fishery Bulletin 100:836–855.

Williams, K., C. N. Rooper, and R. Towler. 2010. Use of stereo camera systems for assessment of rockfish abundance in untrawlable areas and for recording Pollock behavior during midwater trawls. U.S. National Marine Fisheries Service Fishery Bulletin 108:352–362.

Willis, T. J., and R. C. Babcock. 2000. A baited underwater video system for the determination of relative density of carnivorous reef fish. Marine and Freshwater Research 51:755–763.

Yoklavich, M. M., M. S. Love, and K. A. Forney. 2007. A fishery-independent assessment of an overfished rockfish stock, Cowcod (*Sebastes levis*), using direct observations from an occupied submersible. Canadian Journal of Fisheries and Aquatic Sciences 64:1795–1804.

Zimmermann, M. 2003. Calculation of untrawlable areas within the boundaries of a bottom trawl survey. Canadian Journal of Fisheries and Aquatic Sciences 60:657–669.

APPENDIX: Nearshore Species Assemblage Variation on the Oregon Coast

Table A.1.   Comparison of video lander survey data, including the total numbers of fish observed (summed maximum counts across stations of the 25 most abundant species between the five study locations), by species or group and survey area (*n* denotes the number of stations sampled). All data not from the Three Arch Rocks reef was obtained from Hannah and Blume 2012.

Fish species, totals, and depth	Three Arch Rocks (Apr–Jun, $n = 272$)	Three Arch Rocks (Dec, $n = 108$)	Cape Perpetua (Feb, $n = 30$)	Cape Perpetua (Jul, $n = 30$)	Seal Rocks ($n = 43$)	East Siletz ($n = 36$)	West Siletz ($n = 30$)	Stonewall Bank ($n = 173$)
Black Rockfish	277	26	62	72	182	15	11	0
Blue Rockfish	124	15	2	1	18	187	47	54
Brown Rockfish *Sebastes auriculatus*	0	0	3	1	0	0	0	0
Canary Rockfish	225	40	173	156	47	18	74	202
China Rockfish	1	0	0	0	0	0	0	0
Copper Rockfish	6	0	6	2	5	0	0	0
Unidentified juvenile rockfish	4	5	7	47	56	3	0	281
Quillback Rockfish	21	5	9	8	6	2	3	1
Rosethorn Rockfish *Sebastes helvomaculatus*	0	0	0	0	0	0	0	18
Tiger Rockfish	0	1	0	0	0	0	0	0
Unidentified rockfish	5	14	0	0	0	0	0	0
Yellowtail Rockfish	55	92	44	14	18	2	5	95
Yelloweye Rockfish	27	15	3	0	1	6	3	22
Cabezon	2	1	2	3	0	0	0	0
Kelp Greenling	90	17	14	18	23	16	8	13
Lingcod	61	6	8	7	15	23	12	13
Northern Anchovy	0	>1,500	0	0	0	0	0	0
Pacific Halibut *Hippoglossus stenolepis*	0	0	0	0	0	0	0	8
Pile Perch	24	0	68	11	6	0	2	0
Spotted Ratfish	0	20	0	0	0	0	0	0
Unidentified fish	5	15	0	0	0	0	0	0
Unidentified flatfish	9	1	0	0	0	0	0	0
Unidentified sculpin, family Cottidae	0	0	1	1	0	2	0	4
Unidentified surfperch	2	0	0	0	0	0	0	0
Wolf-eel	1	0	0	0	0	0	0	0
Total number of rockfish	745	213	309	301	333	233	143	673
Total number of fish (excluding Northern Anchovy)	939	273	402	341	377	274	165	711
Mean station depth (m)	41.0	52.8	51.3	51.3	30.7	33.6	40.4	54.3

Vertical Distribution of Age-0 Walleye Pollock during Late Summer: Environment or Ontogeny?

Sandra L. Parker-Stetter,[*][1] **John K. Horne, and Samuel S. Urmy**[2]
School of Aquatic and Fishery Sciences, University of Washington, Box 355020, Seattle, Washington 98195-5020, USA

Ron A. Heintz
National Oceanic and Atmospheric Administration, National Marine Fisheries Service, Alaska Fisheries Science Center, Auke Bay Laboratories, 17109 Point Lena Loop Road, Juneau, Alaska 99801, USA

Lisa B. Eisner
National Oceanic and Atmospheric Administration, National Marine Fisheries Service, Alaska Fisheries Science Center, 7600 Sand Point Way NE, Seattle, Washington 98115, USA

Edward V. Farley
National Oceanic and Atmospheric Administration, National Marine Fisheries Service, Alaska Fisheries Science Center, Auke Bay Laboratories, 17109 Point Lena Loop Road, Juneau, Alaska 99801, USA

Abstract

Variability in the late-summer vertical distribution of age-0 Walleye Pollock *Gadus chalcogrammus* in the southeastern Bering Sea has been attributed to a range of physical and biological factors. Using acoustic data (38 and 120 kHz) collected during the 2010 Bering Aleutian Salmon International Survey (BASIS) and dedicated high-resolution surveys (HR1 and HR2), we evaluated whether late-summer distributions could be explained by water column properties (environment) or whether sampling was likely occurring during the ontogenetic shift of age-0 Walleye Pollock from near-surface habitat to demersal habitat (ontogeny). Neither water column attributes (temperature, relative temperature, salinity, dissolved oxygen, and density gradient) nor the acoustic density of zooplankton prey strongly predicted the acoustic estimates of age-0 Walleye Pollock vertical presence or density. At 6 of 10 paired BASIS–HR1 stations, age-0 Walleye Pollock shifted deeper in the water column between BASIS sampling and the HR1 sampling conducted 8–34 d later. There were no consistent differences in FL ($P > 0.05$ for 2 of 4 station pairs) or energy density ($P > 0.05$ for 3 station pairs) between age-0 Walleye Pollock caught in near-surface trawls and those caught in midwater trawls. Our data suggest that the observation of both near-surface and midwater age-0 Walleye Pollock during late summer is likely due to an ontogenetic habitat shift; however, the causative factor was not clear given the limited sample sizes and explanatory variables. The timing of the ontogenetic shift, which appears to have begun before August 18, 2010, can ultimately affect survey strategies, and knowledge of this timing can provide additional insight into factors affecting the overwinter survival of age-0 Walleye Pollock.

Subject editor: Kenneth Rose, Louisiana State University, Baton Rouge

*Corresponding author: sandy.parker-stetter@noaa.gov
[1]Present address: National Oceanic and Atmospheric Administration, National Marine Fisheries Service, Northwest Fisheries Science Center, Fisheries Resource Assessment and Monitoring Division, 2725 Montlake Boulevard East, Seattle, Washington 98112, USA.
[2]Present address: School of Marine and Atmospheric Sciences, Stony Brook University, 239 Montauk Highway, Southampton, New York 11968, USA.

In the eastern Bering Sea (EBS), Walleye Pollock *Gadus chalcogrammus* have been identified in different portions of the water column during their first year of life. In the spring and early summer, age-0 Walleye Pollock are typically found near the thermocline (~20-m depth) and/or in the upper portion of the water column ($\leq$50-m depth; Smart et al. 2013) during the day. By early summer of the following year (i.e., at age 1), they have completed their ontogenetic transition into deeper water and are found in midwater and semi-demersal habitats (Honkalehto et al. 2010; Lauth 2010). The location of age-0 Walleye Pollock within the water column between those two time periods has not been well documented. Sampling with a surface trawl during summer through early fall has caught age-0 Walleye Pollock near and above the thermocline (Moss et al. 2009), but acoustics and midwater trawling have also observed and caught age-0 Walleye Pollock deeper in the water column (>75-m depth) during that same time of year (Bailey 1989; Parker-Stetter et al. 2013). The observed variability in late-summer vertical distribution can ultimately influence the design and accuracy of surveys that are used to estimate the abundance and distribution of this commercially valuable species.

Studies in the EBS have suggested a variety of physical and biological explanations for the vertical distribution of age-0 Walleye Pollock. Environmental factors have included thermal preference (Tang et al. 1996; Swartzman et al. 1999), water column stratification (Francis and Bailey 1983), light levels (Olla and Davis 1990), and high potential growth at a given location (Ciannelli et al. 2002). Conversely, fish length (Bailey 1989; Miyake et al. 1996; Swartzman et al. 2002), food availability and prey size (Olla and Davis 1990; Schabetsberger et al. 2003), and cannibalism (Bailey 1989) have been suggested as biological explanations for the observed differences in age-0 Walleye Pollock vertical distribution. Combining physical and biological explanations, laboratory experiments have suggested that age-0 Walleye Pollock migrate to temperatures that benefit their energetic status, with food-deprived juveniles being found in colder water than fish with higher rations (Sogard and Olla 1996).

An alternative explanation for the distribution of age-0 Walleye Pollock in both near-surface and midwater regions is that late-summer surveys are sampling during the age-0 fish's ontogenetic transition from near-surface to semi-demersal habitats. In the Gulf of Alaska, Brodeur and Wilson (1996) observed that the mean depth of capture for age-0 Walleye Pollock generally increased from July–August to October. Although it is known that age-0 Walleye Pollock shift from near-surface to demersal habitats within their first year in the EBS, the timing and causative factors are unclear.

We evaluated factors potentially influencing the late-summer vertical distribution of age-0 Walleye Pollock from two complementary perspectives: environmental conditions and ontogenetic transition. We predicted that if late-summer surveys are characterizing the distribution and abundance of age-0 Walleye Pollock before they undergo the ontogenetic transition to deeper water, then the variability in age-0 vertical distribution will be related to water column properties and/or the vertical distribution of potential zooplankton prey. We also predicted that if late-summer observations of age-0 Walleye Pollock coincide with an ontogenetic transition to deeper water, then (1) age-0 vertical distributions and aggregation attributes (school size and relative density) will change between observations at the same location separated in time and (2) surface- and midwater-caught fish will differ in FL, energy density, or both.

METHODS

Survey Area and Design

The Bering Aleutian Salmon International Survey (BASIS; see Farley et al. 2009) was conducted in the EBS on the National Oceanic and Atmospheric Administration (NOAA) FSV *Oscar Dyson* between August 18 and September 16, 2010 (Parker-Stetter et al. 2013). Immediately after BASIS, high-resolution (HR) surveys (HR1 and HR2) were conducted during September 17–25, 2010. Operations occurred during daytime from 1 h after sunrise to 1 h before sunset.

The HR study area was selected to encompass a region where age-0 Walleye Pollock were observed at a range of locations within the water column during the preceding BASIS, with the goal of resurveying acoustic transects and reoccupying a subset of BASIS stations in that area. The study area was bounded by 54°45′N in the south, 57°00′N in the north, 167°00′W in the west, and 166°00′W in the east (Figure 1). Bottom depth contours in the surveyed area ranged between 75 and 350 m.

For the HR surveys, north–south acoustic transects were spaced 00°30′ apart (29.6–31.5 km at survey latitudes), with a shorter transect occurring on 166°30′W due to weather (Figure 1). The HR transects were surveyed between September 17 and September 24 (HR1), and the 167°00′W transect was acoustically surveyed a second time on September 25, 2010 (HR2; Table 1; Figure 1). Sections of the 167°00′W and 166°00′W transects were surveyed during BASIS between August 18 and September 9, 2010 (Table 1).

Oceanographic data were collected at all HR1 stations at a spacing of 00°15′. At reoccupied BASIS stations, a surface trawl was always performed, and a pycnocline or midwater trawl was also conducted if targets were observed below the surface trawl fishing depth. At all other stations, surface and midwater trawls were both conducted if targets were observed on the echosounder (Figure 1).

Data Collection and Processing

Conductivity–temperature–depth unit deployment and processing.—Oceanographic data were collected using a Sea-Bird Model 911plus conductivity–temperature–depth (CTD)

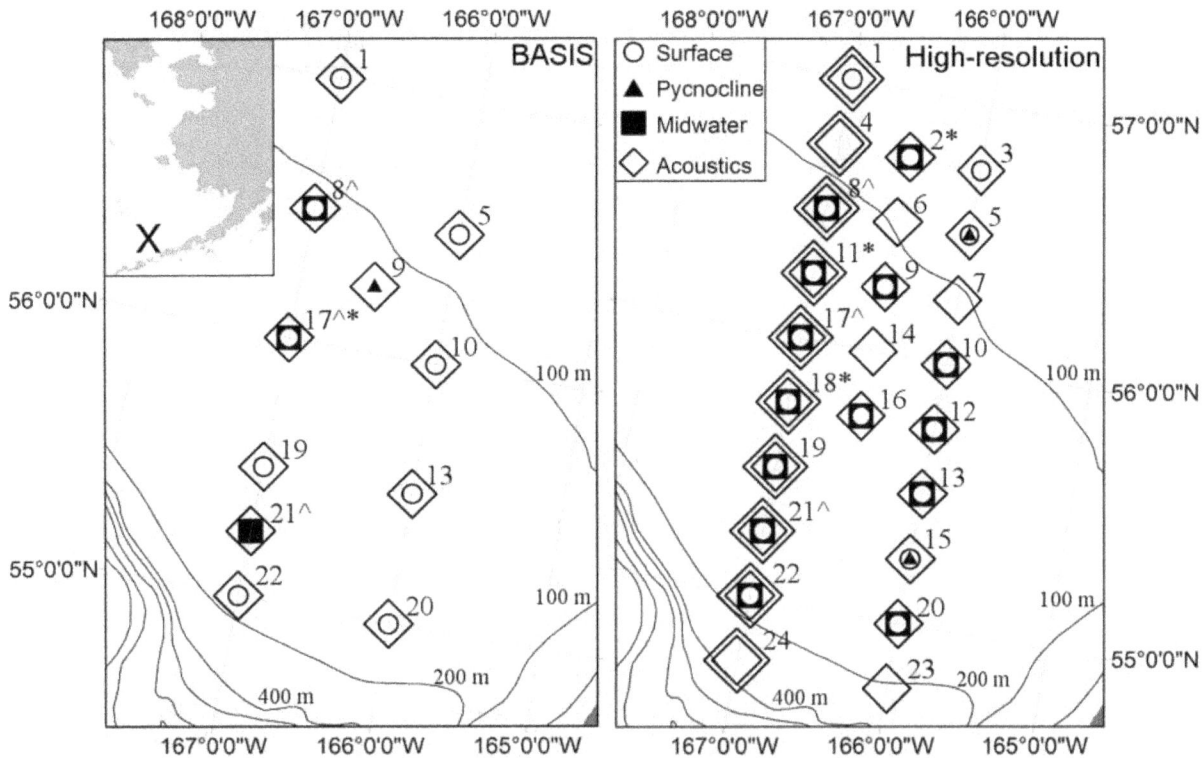

FIGURE 1. Acoustic data and trawl (surface, pycnocline, and midwater) locations used during the 2010 Bering Aleutian Salmon International Survey (BASIS; left, sampled August 18–September 9) and the 2010 high-resolution (HR1 and HR2) surveys (right, sampled September 17–25). Station numbers are referred to throughout the text. Inset shows the study area's location in the Bering Sea. The double-diamond symbol indicates sampling during HR1 and HR2; an asterisk denotes a surface–midwater comparison; and the "^" symbol denotes a BASIS–HR comparison with sufficient samples to be included in analyses.

unit (Sea-Bird Electronics, Bellevue, Washington). Casts were made from the surface to 5–10 m from the bottom. Downcast data were error-checked and averaged into 1-m vertical bins.

The Rossby radius of deformation (L_R) was used to estimate the horizontal distance from each CTD cast location at which water properties were assumed homogeneous (cf. Alenius et al. 2003). The two-layer L_R (Rossby 1938; Gill 1982) was calculated for each CTD cast station as

$$L_R = \sqrt{g'D}/f, \qquad (1)$$

where D is the depth (m) of the upper layer, f is the Coriolis parameter (per second), and g' is the reduced gravity, given by

$$g' = \frac{g\,(\rho_2 - \rho_1)}{\bar{\rho}}, \qquad (2)$$

where g is the acceleration due to gravity (m/s^2), ρ_1 and ρ_2 are the mean densities (kg/m^3) in the upper and lower layers, and $\bar{\rho}$ is the mean density (kg/m^3) over the whole water column.

Trawling.—A Cantrawl 400/601 rope trawl (25–30-m vertical opening; 1.2-cm-mesh liner in the cod end) equipped with 5-m alloy trawl doors (NET Systems, Bainbridge Island,

Washington) was used to identify targets observed on the echosounder and to collect specimens. For surface trawls, the headrope was equipped with floats, and the net's position at the surface was maintained by minor adjustments in the length of wire out. In limited instances when high backscatter from age-0 Walleye Pollock was observed near or immediately below the footrope of the surface trawl, the net was deployed without floats to headrope depths of 12–30 m in a pycnocline set (Parker-Stetter et al. 2013). For midwater trawls, the floats were removed and a Simrad FS-70 trawl sonar (Kongsberg Maritime, Horten, Norway) was used to monitor net depth and fish catch in real time (Parker-Stetter et al. 2013).

At all BASIS stations and at the reoccupied BASIS stations during HR sampling, surface trawl sets lasted 30 min. At other surface trawl stations during HR sampling, the sets lasted 15 or 30 min depending on the apparent target density indicated by the acoustics. Trawls within the pycnocline lasted 7–15 min depending on the target density shown by the FS-70 sonar. Surface and pycnocline samples were kept separate throughout our analyses. Midwater trawls (headrope depths = 50–95 m) lasted for 4–19 min depending on the target density observed using the echosounder and the FS-70 sensor during the set. For comparison, all catches were standardized to catch per 30 min of effort (number of fish/30-min trawl).

TABLE 1. Station numbers (see Figures 1 and 2), sampling dates (number of days since previous sampling is shown in parentheses), and sample type (A = acoustic; T = trawl) for the Bering Aleutian Salmon International Survey (BASIS) and the high-resolution surveys (HR1 and HR2) conducted in 2010.

Station	Bottom depth (m)	BASIS			HR1			HR2		
		Date	A	T	Date	A	T	Date	A	T
1	75	Sep 9	×	×	Sep 17 (8 d)	×	×	Sep 25 (8 d)	×	
2	80				Sep 23	×	×			
3	80				Sep 23	×	×			
4	90				Sep 17	×		Sep 25 (8 d)	×	
5	90	Sep 1	×	×	Sep 23 (22 d)	×	×			
6	100				Sep 22	×				
7	100				Sep 22	×				
8	105	Sep 9	×	×	Sep 17 (8 d)	×	×	Sep 25 (8 d)	×	
9	110	Sep 1	×	×	Sep 24 (23 d)	×	×			
10	115	Aug 19	×	×	Sep 22 (34 d)	×	×			
11	120				Sep 18	×	×	Sep 25 (7 d)	×	
12	120				Sep 22	×	×			
13	120	Aug 18	×	×	Sep 21 (34 d)	×	×			
14	125				Sep 24	×				
15	125				Sep 21	×	×			
16	130				Sep 24	×	×			
17	135	Sep 1	×	×	Sep 18 (17 d)	×	×	Sep 25 (7 d)	×	
18	135				Sep 18	×	×	Sep 25 (7 d)	×	
19	135	Sep 8	×	×	Sep 19 (11 d)	×	×	Sep 25 (6 d)	×	
20	135	Aug 18	×	×	Sep 21 (34 d)	×	×			
21	145	Sep 8	×	×	Sep 19 (11 d)	×	×	Sep 25 (6 d)	×	
22	160	Sep 8	×	×	Sep 20 (12 d)	×	×	Sep 25 (5 d)	×	
23	195				Sep 20	×				
24	350				Sep 20	×		Sep 25 (5 d)	×	

Trawl catches were sorted to species, weighed, and processed for biological information. Catches less than 1 metric ton were processed in their entirety. For catches greater than 1 metric ton, the entire catch was weighed, and a subsample of approximately 1 metric ton was randomly selected by splitting the catch. Measurements of FL (nearest mm) were taken on at least 50 individuals from each species. Up to 20 age-0 Walleye Pollock from a trawl were individually wrapped and flash-frozen for analysis of energy density (kg/J wet weight) via the method of Heintz et al. (2013).

Acoustic data collection.—Acoustic data were collected at five frequencies (18, 38, 70, 120, and 200 kHz) using a Simrad EK-60 echosounder. The 38-kHz (ES38-B; 2,000 W) and 120-kHz (ES120-7C; 500 W) data were used in this analysis. Transducers were mounted on a centerboard that was lowered to 9.15 m below the surface during the survey. Ship speed was 5.1–6.2 m/s (10–12 knots), and data were collected using a 1.024-ms pulse duration at a maximum rate of 1 pulse/s. All frequencies were calibrated prior to the survey using standard-sphere methods (Foote et al. 1987).

Acoustic data processing.—On-transect acoustic data were processed in Echoview version 5.10 (Echoview Pty Ltd, Hobart,

Australia). A sound speed of 1,470 m/s and absorption coefficients of 0.00998 dB/m (at 38 kHz) and 0.029057 dB/m (at 120 kHz) were used during processing. Noise spikes and dropped pings were removed from all files. Vessel noise was removed via linear subtraction (Watkins and Brierley 1996; Korneliussen 2000), and a 10-dB signal-to-noise ratio filter was used during processing. The bottom was detected in Echoview, inspected, and manually corrected as needed. To account for transducer depth (9.15 m) and twice the near-field range of the 38-kHz unit, data from within 15 m of the surface were excluded from the analysis. Data from within 0.5 m of the bottom were also excluded.

Age-0 Walleye Pollock aggregation acoustics.—As a preprocessing step for aggregation detection, a 3-sample × 3-ping Gaussian blur convolution was calculated using the 38-kHz native resolution (0.18 m vertical × 1 ping horizontal) data (Reid and Simmonds 1993). Aggregations were detected on the blurred data by using the Echoview SHAPES algorithm (Barange 1994) with aggregation detection parameter settings as follows: a minimum total/candidate school length of 0.2 m; a minimum total/candidate school height of 0.2 m; a maximum vertical linking distance of 2.0 m; a maximum horizontal linking distance of 15.0 m; and a minimum volume backscattering strength (S_v) threshold of -60.9 dB re 1 m^{-1}. The 15.0-m maximum horizontal linking distance was selected based on the maximum distance covered by the vessel during a single pulse transmission. The -60.9 dB minimum S_v threshold represents the lower 1 SD of verified S_v sample data for age-0 Walleye Pollock (Parker-Stetter et al. 2013) and was used to conservatively delineate the outer edges of all aggregations. Aggregations were then exported from the 38-kHz native resolution data by using a minimum S_v threshold of -67 dB (Parker-Stetter et al. 2013). Aggregations that had an edge within 0.5 m of the bottom or surface exclusion lines were removed from the analysis, as their geometry may have been incorrect due to overlap with the surface or bottom exclusion zones.

Three attributes of aggregations were exported from the acoustic data files: aggregation length (horizontal; m), aggregation thickness (vertical; m), and mean S_v (hereafter, "S_v acoustic density"; a logarithmic measure in dB). Uncorrected descriptors of aggregation geometry were used in analyses because many aggregations were too small relative to the beam width for accurate geometric correction (cf. Diner 2001). To account for potential overestimation or underestimation of aggregation length due to beam width, we calculated a minimum aggregation length (length $- 0.5 \cdot$[3-dB beam width]) and a maximum aggregation length (length $+ 0.5 \cdot$[3-dB beam width]).

Along-transect age-0 Walleye Pollock aggregation data were integrated within 1-m vertical × 6-ping horizontal bins. Vertical bin size was selected to match the CTD data's 1-m vertical resolution. The 6-ping horizontal bin size corresponded to the 90th percentile for the number of pulses within an aggregation; this was selected in an effort to maximize the number of aggregations that would be contained within a single bin rather than be split across bins. Areas around aggregations were assigned 999 dB (a value ≈ 0 in the linear domain) for integration.

Zooplankton acoustics.—Because specific algorithms to categorize EBS copepods are lacking, zooplankton (i.e., copepods and euphausiids) were identified following the methods of Murase et al. (2009). The 38- and 120-kHz data were resampled into 1-m vertical × 3-ping horizontal bins; any sample (0.18 m vertical × 1 ping horizontal) containing an age-0 Walleye Pollock aggregation was excluded from the resample calculation as "no data." Using a -90 dB S_v threshold, a frequency difference (120 kHz–38 kHz) value was generated for each resampled bin (Murase et al. 2009). Bins containing zooplankton were identified using a frequency difference range of 7.1–22.0 dB (Murase et al. 2009). This frequency difference range also encompasses the frequency difference (120 kHz–38 kHz) values for euphausiids (10.9–16.7 dB; De Robertis et al. 2010). Methods used to determine the frequency difference range in the Murase et al. (2009) study were based on the copepod species *Neocalanus*

cristatus, which is larger than the *Calanus* spp. copepods that are typically consumed by age-0 Walleye Pollock during cold years (Coyle et al. 2011) in the Bering Sea. As our analysis was based on the relative vertical distribution of zooplankton rather than absolute abundance, any potential underestimates of *Calanus* spp. caused by using the frequency difference range from Murase et al. (2009) should be consistent within the water column and should not affect our results.

Along-transect zooplankton S_v acoustic density was exported from the 120-kHz data in 1-m vertical × 6-ping horizontal bins by using a -90 dB data threshold (Murase et al. 2009). In empty bins (i.e., those that had been entirely filled with an age-0 Walleye Pollock aggregation), we assumed that zooplankton backscatter in surrounding bins was representative of what would be found in the empty 1-m × 6-ping bin, and we filled those cells with the mean zooplankton S_v acoustic density from the surrounding eight cells (i.e., an 18-ping × 3-m window).

Environment: Water Column Properties and Vertical Distribution of Age-0 Walleye Pollock

The HR1 data set was used to evaluate whether the vertical distribution of age-0 Walleye Pollock was related to environmental factors as it was more synoptically collected (sampling area ≈ 7,500 km^2; time = 8 d) than the EBS-wide BASIS data (Parker-Stetter et al. 2013). We used CTD cast data and zooplankton acoustic data to represent the vertical distribution of factors that might influence age-0 Walleye Pollock use of the water column as habitat.

Acoustic integration data for age-0 Walleye Pollock and zooplankton (1 m vertical × 6 ping horizontal) within the L_R of an HR1 CTD cast were associated with that cast for the analysis. All 24 HR1 stations were used in this analysis. Because most environmental variables were only available as one-dimensional vertical profiles from CTD casts, acoustic age-0 Walleye Pollock and zooplankton integration bins within each L_R were averaged into 1-m depth intervals, thereby reducing the two-dimensional acoustic echo integration data to a one-dimensional profile with the same vertical resolution as the CTD cast. Each depth interval at each station was then treated as a data point in the analysis.

Age-0 Walleye Pollock backscatter was patchy and included many empty cells, even after averaging into one-dimensional profiles. Because these data were zero-inflated, we represented the probability of Walleye Pollock occurrence and the expected Walleye Pollock acoustic density (using logarithmic S_v acoustic density, conditional on occurrence) as separate processes in a two-part model (Stefánsson 1996; Hollowed et al. 2012). The probability of occurrence was modeled using a generalized linear model (GLM) with a logit link function and a binomial error structure; acoustic density was modeled using an identity-link GLM with a Gaussian error structure (i.e., ordinary linear regression).

We used an information-theoretic approach (Burnham and Anderson 2002) to compare the ability of seven environmental variables (hypotheses) to explain the vertical distribution of age-0 Walleye Pollock. Variables were selected to reflect common ideas in the literature as well as our own hypotheses; the variables included depth from the surface (Depth; m), vertical density gradient (Stratification; kg/m^3), water temperature (Temp; °C), relative water temperature (RelTemp; unitless), dissolved oxygen (Oxygen; mg/L), salinity (Salinity; unitless), and zooplankton S_v acoustic density (ZoopS$_v$; dB) within the 1-m vertical bins. Relative temperature at a given depth was calculated by remapping the measured temperatures on a scale of zero to 1.0 (minimum to maximum). A second set of models was run using the same variables but with the inclusion of an autoregressive component, as all previous model runs had autocorrelated residuals. This autoregressive component (AdjPollock), used as a measure of age-0 Walleye Pollock aggregating behavior, was calculated for each depth interval as the average of Walleye Pollock S_v acoustic density in adjacent bins above and below each depth interval. This approach is equivalent to a first-order conditional autoregressive spatial model (Cressie and Wikle 2011).

Support for each variable or hypothesis was evaluated based on Akaike weight (w_i), an estimate of the probability that the hypothesis would be selected as the best if the analysis was repeated on a new data set (Whittingham et al. 2006). Both occurrence (binary: 0, 1) and density (numeric: S_v acoustic density) models were fitted for each hypothesis by using the glm function in R software (R Development Core Team 2013). To assess goodness of fit, we also calculated an R^2 value for each model fit. For the binomial occurrence models, R^2 values based on the residual sum of squares could not be calculated, so we calculated pseudo-R^2 values (McFadden 1974) based on the likelihood ratio of the fitted and null models (Table 2, analysis A).

We also conducted a post hoc analysis of all possible combinations of the seven variables, again by using w_i to identify the combinations of variables that were best supported by the data. The exhaustive models were run for occurrence and density, both excluding and including the autoregressive component AdjPollock. Hypotheses within the 90% confidence set, which represents the probability that the best-approximating model occurred within that set of models (Johnson and Omland 2004), are presented (Table 2, analysis B).

Ontogeny: Changes in Vertical Distribution and Aggregations over Time

In evaluating whether age-0 Walleye Pollock vertical distribution was related to ontogeny, we considered both space (i.e., sampling location) and time (i.e., sampling date) because our analyses could not decouple the two. Given the bathymetric gradient at the site, bottom depth was used to represent station

location. To represent time, sampling dates were converted to day of year.

Trawl distribution.—Trawling was conducted at 18 HR1 stations, with 18 surface trawls, 2 pycnocline trawls, and 14 midwater trawls (Figure 1; Table 1). Eleven of the HR1 stations were BASIS stations that were reoccupied between 8 and 34 d later. During BASIS, nine surface trawls, one pycnocline trawl, and three midwater trawls were conducted.

Trawl efficiency for the Cantrawl is unknown (Parker-Stetter et al. 2013); therefore, catches of age-0 Walleye Pollock were categorized in order-of-magnitude bins for comparison: ≤3 (0–3 fish), ≤30 (4–30 fish), ≤300 (31–300 fish), ≤3,000 (301–3,000 fish), ≤30,000 (3,001–30,000 fish), and ≤300,000 fish/30-min trawl. These bins were used to compare trawls (surface and/or midwater) conducted at the same location during both BASIS and the HR1 survey (Table 2, analysis C).

Aggregation attributes.—Mean aggregation length, thickness, and S_v acoustic density were used to describe potential changes in the size or density of age-0 Walleye Pollock aggregations between BASIS, HR1, and/or HR2. Attributes were summarized using all aggregations within the L_R of each station. The SDs were calculated for aggregation thickness and S_v acoustic density. Given the potential effects of acoustic beam width on estimates of aggregation length, we present mean length with ±0.5·beam width instead of SD. In subsequent statistical analysis, minimum aggregation length is used rather than mean length to avoid potential beam width biases.

We evaluated whether aggregation attributes for a given station during BASIS, HR1, and/or HR2 were different at subsequent reoccupations of that station by using either the Dunnett-modified Tukey–Kramer test (DTK test; Dunnett 1980) or the Welch two-sample t-test (Welch test; Welch 1947). The DTK test, which corrects for multiple comparisons, was used for stations with data available from BASIS, HR1, and HR2. The Welch test was used for stations with only HR1 and BASIS data or those with only HR1 and HR2 data (Table 2, analysis D). Neither test assumes equal sample sizes or equal variances (Dunnett 1980; Ruxton 2006). For the DTK test and the Welch test, we evaluated whether the 95% confidence interval included zero as a measure of potential differences at $P < 0.05$. We acknowledge that biological significance must be evaluated using additional background information, such as theory and plausibility (Goodman 1999; Lew 2012); therefore, we use the phrase "potential significance" in reporting our results. The DTK test and Welch test methods are referenced in additional comparisons described below.

Metrics of age-0 Walleye Pollock vertical distribution.—Summary metrics were used to describe the vertical distribution of age-0 Walleye Pollock aggregations (Woillez et al. 2007; Urmy et al. 2012) during BASIS, HR1, and/or HR2. The center of acoustic mass (a measure of weighted mean depth) and the S_v acoustic density (a measure of water column density) were calculated from the 6-ping × 1-m integration data following Urmy et al. (2012). We calculated the mean

TABLE 2. Summary of analyses used to evaluate the potential effects of environment or ontogeny on the vertical distribution of age-0 Walleye Pollock (BASIS = 2010 Bering Aleutian Salmon International Survey; HR1 and HR2 = 2010 high-resolution surveys; Depth = depth from the surface, m; Stratification = vertical density gradient, kg/m^3; Temp = water temperature, °C; RelTemp = relative water temperature, unitless; Oxygen = dissolved oxygen concentration, mg/L; Salinity = salinity, unitless; ZoopS$_v$ = zooplankton S$_v$ acoustic density; dB; AdjPollock = autoregressive component [see Methods]; GLM = generalized linear model).

Analysis	Data set(s)	Data type	Key characteristics	y-variable(s)	x-variable(s)	Test	Figure or table
Environment: was age-0 Walleye Pollock vertical distribution related to environmental conditions?							
A	HR1	S$_v$ acoustic density	24 stations	1. Occurrence 2. Density	Depth, Stratification, Temp, RelTemp, Oxygen, Salinity, and ZoopS$_v$ (with and without AdjPollock)	GLM, Akaike weights, R^2, or pseudo-R^2	Table 2
B	HR1	As above	As above	As above	As above	Post hoc all-subsets and Akaike weights	Table 3
Ontogeny: did age-0 Walleye Pollock trawl catch change over time?							
C	BASIS, HR1	Trawl data	9 BASIS–HR1 surface pairs; 3 BASIS–HR1 midwater pairs	Fish/30-min trawl		None	Figure 2
Ontogeny: did age-0 Walleye Pollock aggregation attributes change over time?							
D	BASIS, HR1, HR2	Acoustic attributes of aggregations	10 BASIS–HR1 station pairs; 10 HR1–HR2 pairs	1. Mean aggregation length 2. Mean aggregation thickness 3. S$_v$ acoustic density		Dunnett-modified Tukey–Kramer test (DTK test) or Welch's two-sample t-test (Welch test)	Figure 4, Table 4
Ontogeny: did age-0 Walleye Pollock vertical distribution change over time?							
E	BASIS, HR1, HR2	Acoustic metrics of vertical distribution	10 BASIS–HR1 station pairs; 10 HR1–HR2 pairs	1. Center of acoustic mass 2. S$_v$ acoustic density		DTK test or Welch test	Figure 5

(Continued on next page)

TABLE 2. Continued.

Analysis	Data set(s)	Data type	Key characteristics	y-variable(s)	x-variable(s)	Test	Figure or table
Ontogeny: were there spatial differences in age-0 Walleye Pollock fork length or energy density?							
F	HR1	Trawl data	8 surface; 14 midwater	FL	Bottom depth	GLM	Figure 6
G	HR1	Trawl data	4 surface; 11 midwater	Energy density	Bottom depth	GLM	Figure 6
Ontogeny: was age-0 Walleye Pollock fork length or energy density different in surface versus midwater trawls?							
H	BASIS, HR1	Trawl data	1 BASIS surface–midwater pair; 3 HR1 surface–midwater pairs	FL		Welch test	Figure 6
I	BASIS, HR1	Trawl data	1 BASIS surface–midwater pair; 2 HR1 surface–midwater pairs	Energy density		Welch test	Figure 6
Ontogeny: did age-0 Walleye Pollock fork length or energy density change over time?							
J	BASIS, HR1	Trawl data	3 BASIS–HR1 midwater pairs	1. FL 2. Energy density		Welch test	Figure 6

and SD of the center of mass and S_v acoustic density for bins within the L_R of each station. Comparisons of metric values for a given station sampled during BASIS, HR1, and/or HR2 were made using either the DTK test or the Welch test (Table 2, analysis E).

Ontogeny: Fork Length and Energy Density of Age-0 Walleye Pollock

Surface versus midwater.—The mean and SD of age-0 Walleye Pollock FL and energy density were calculated for all surface, pycnocline, and midwater trawls conducted during BASIS and the HR1 survey. For these analyses, data are presented for individual stations, arranged by increasing bottom depth.

To examine potential spatial differences in Walleye Pollock FL and energy density within the HR1 data, GLMs were used to evaluate relationships between fish FL or energy density at the surface and/or midwater and bottom depth for HR1 (Table 2, analyses F and G).

At stations where age-0 Walleye Pollock were caught both at the surface and midwater, we evaluated whether the FL and energy density of fish caught in surface trawls differed from those of fish caught in midwater trawls. Four stations (one from BASIS; three from HR1) were available for evaluating FL, and three stations (one from BASIS; two from HR1) were available for evaluating energy density, as surface and midwater trawls both caught over 10 fish. A Welch test was conducted to evaluate whether mean FL or energy density differed between surface and midwater trawl samples obtained at the same station (Table 2, analyses H and I).

Changes in fork length or energy density over time.—Data for this analysis came from three stations at which midwater trawls were performed during both BASIS and HR1 surveys. The limited number of comparisons resulted from the high number of HR1 surface trawls with a catch of 0 fish/30-min trawl. Mean FL and energy density of age-0 Walleye Pollock were calculated for trawls at all paired BASIS–HR1 stations and were plotted against day of year. For the three stations at which midwater trawls were conducted during both BASIS and the HR1 survey, we performed a Welch test to compare the FL or energy density of midwater-caught age-0 Walleye Pollock between the surveys (Table 2, analysis J).

RESULTS

General Results

Trawling.—Trawling was conducted at 11 previously sampled BASIS stations that were reoccupied during the HR surveys (Figure 1; Table 1). The BASIS surface trawl catches ($n = 9$ trawls; Figure 2) ranged from 0 to 6,558 fish/30-min trawl, with a mean of 1,256 fish/30-min trawl. Age-0 Walleye Pollock (40–94 mm FL) were caught in seven of the nine

BASIS surface trawls, constituting between 0% and 79% (mean = 27%) of the fish catch by number. Age-0 Pacific Cod *Gadus macrocephalus* (42–88 mm FL) were frequently caught; Prowfish *Zaprora silenus* (51–117 mm FL), Bering Wolffish *Anarhichas orientalis* (161–202 mm FL), and age-0 rockfishes *Sebastes* spp. (36–56 mm FL) were also caught in three or more trawls. The single trawl within the pycnocline had a catch of 2,493 fish/30-min trawl, and 94% of the catch consisted of age-0 Walleye Pollock (43–83 mm FL). Age-0 Pacific Cod (55–93 mm FL) were also caught in the pycnocline trawl. Midwater trawl catches during BASIS contained 590–19,844 fish/30-min trawl (mean = 7,118 fish/30-min trawl; $n = 3$ trawls; headrope = 80 m; Figure 2). Age-0 Walleye Pollock (44–91 mm FL) comprised 97, 72, and 100% of the catch in the three midwater trawls, but age-0 Pacific Cod (56–91 mm FL) and Prowfish (45–91 mm FL) were also caught.

The HR survey completed trawling at 24 stations along the three acoustic transects (Figure 1; Table 1). Surface trawl catches during the HR survey were consistently low (mean = 796 fish/30-min trawl; minimum = 0 fish/30-min trawl; maximum = 6,408 fish/30-min trawl; $n = 18$ trawls) and typically contained age-0 Walleye Pollock (56–108 mm FL), age-0 Pacific Cod (57–88 mm FL), Prowfish (57–143 mm FL), and age-0 rockfishes (27–63 mm FL). Threespine Sticklebacks *Gasterosteus aculeatus* (52–82 mm FL) and juvenile Atka Mackerel *Pleurogrammus monopterygius* (150–216 mm FL) were also caught in at least three trawls. Age-0 Walleye Pollock constituted a mean of 25% (minimum = 0%; maximum = 100%) of the surface trawl catch by number and was the dominant constituent in all trawls with catches greater than 300 fish/30-min trawl. The two trawls within the pycnocline yielded catches of 24 and 1,624 fish/30-min trawl; age-0 Walleye Pollock (62–98 mm FL) constituted 25% and 96% of the catch in these trawls. Age-0 Pacific Cod (64–94 mm FL), Capelin *Mallotus villosus* (79–116 mm FL), age-0 rockfishes (35–52 mm FL), and Pacific Herring *Clupea pallasii* (323 mm FL) were also present in pycnocline catches. The HR midwater catches (mean = 27,816 fish/30-min trawl; minimum = 60 fish/30-min trawl; maximum = 213,721 fish/30-min trawl; $n = 14$ trawls; headrope = 50–95 m) were at least an order of magnitude higher than surface trawl catches. Midwater trawl catches consistently contained few species other than age-0 Walleye Pollock (55–104 mm FL), but age-0 Pacific Cod (58–100 mm FL), Prowfish (56–139 mm FL), and age-0 rockfishes (46–62 mm FL) were also caught in low numbers (≤300 individuals) during three or more trawls. Age-0 Walleye Pollock made up a mean of 90% (minimum = 65%; maximum = 100%) of the midwater trawl catches and constituted over 95% of the fish in all catches greater than 300 fish/30-min trawl.

Acoustics.—During BASIS, acoustic data were collected along approximately 275 km of transect within the HR study area. Overall, 3,540 aggregations occurred within the L_R (3.6–5.5 km) of BASIS CTD stations (Figure 3). In water column analyses, these aggregations occupied 791 six-ping horizontal

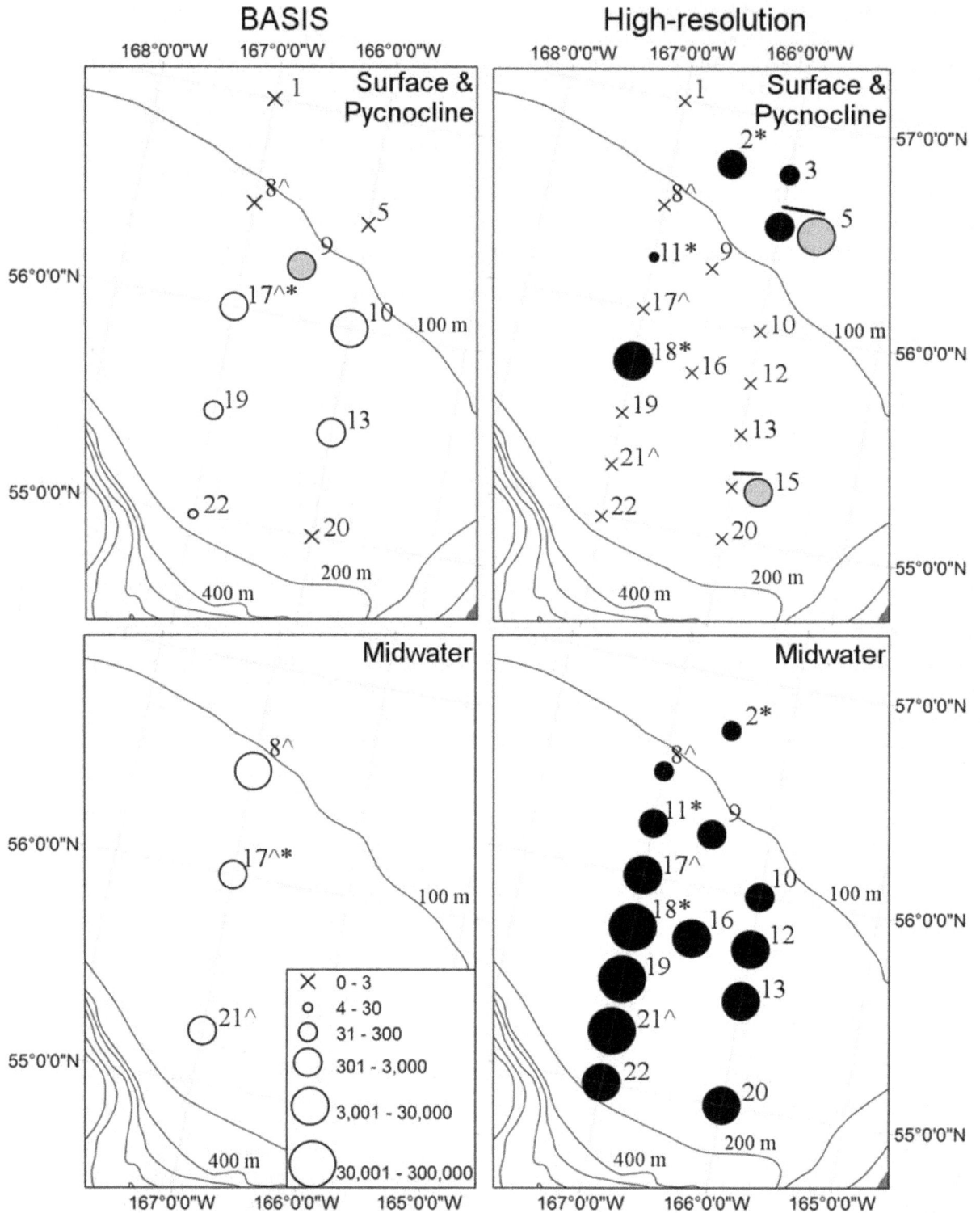

FIGURE 2.   Trawl catch of age-0 Walleye Pollock during the 2010 Bering Aleutian Salmon International Survey (BASIS; left panels, open symbols) and the high-resolution (HR1) survey (right panels, filled symbols) for surface or pycnocline trawling (upper panels; gray symbols = pycnocline trawls) and midwater trawling (lower panels). Surface and pycnocline trawls at the same location are indicated with a black line. Scale (number of fish/30-min trawl) is common among plots. Station numbers are repeated from Figure 1.

bins. We removed 10 surface aggregations and 15 bottom aggregations from this analysis due to their proximity to surface or bottom exclusion lines.

Acoustic data were collected along approximately 600 km of HR transects. In total, 6,034 aggregations were identified within the L_R (3.4–5.5 km) of HR CTD stations: 4,522 aggregations were identified during HR1, and 1,512 aggregations were identified during the second pass (HR2) on 167°W (Figure 3). These aggregations were contained within 4,568 sixping horizontal bins (3,479 bins during HR1; 1,089 bins during

HR2). Overall, 49 surface aggregations and 260 bottom aggregations were removed from the HR1 or HR2 data due to the aggregations' proximity to surface or bottom exclusion lines.

Environment: Water Column Properties and Vertical Distribution of Age-0 Walleye Pollock

Explanatory water column or zooplankton variables had significant but weak effects on the vertical distribution of age-0 Walleye Pollock aggregations during the HR1 survey

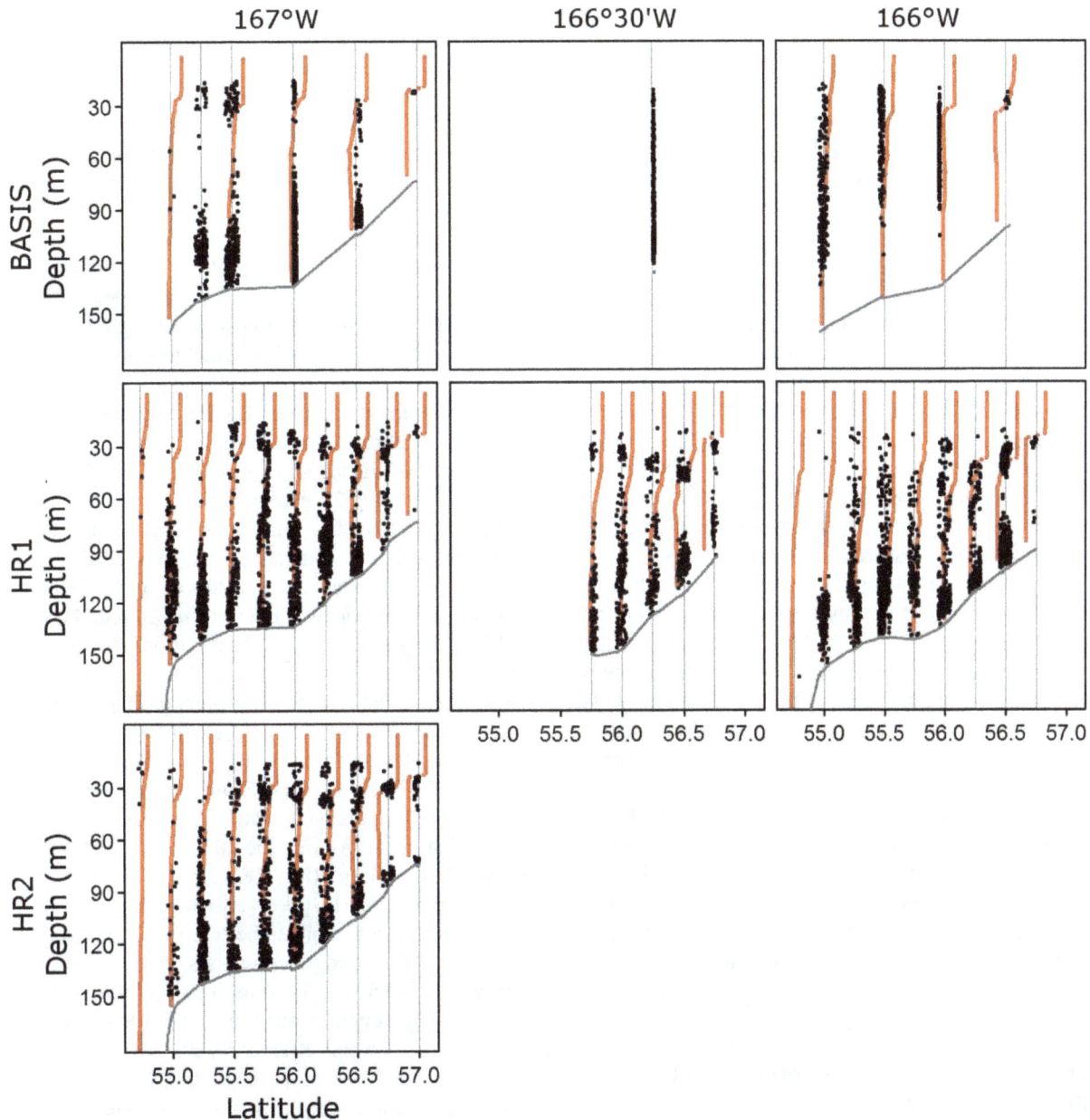

FIGURE 3.   Vertical location (midpoint) of age-0 Walleye Pollock aggregations within the Rossby radius (λ_R) of each survey station on 167°W, 166°30′W, and 166°W (BASIS = 2010 Bering Aleutian Salmon International Survey; HR1 and HR2 = high-resolution surveys). Results of conductivity–temperature–depth (CTD) measurements at each station are provided for reference; gray vertical lines denote the 5°C temperature line for the CTD cast (red): values to the left are below 5°C, and values to the right are above 5°C. Bottom depth along the transect is shown (dark gray line).

TABLE 3. Hypothesis model results for the vertical distribution of age-0 Walleye Pollock occurrence and density during the high-resolution (HR1) survey in 2010. Hypothesis/variable, difference in Akaike's information criterion corrected for small sample sizes between the given model and the best-performing model in the set [AIC_c for the best model in parentheses], Akaike weight (w_i), and R^2 statistic are provided. Bold italics indicate results for the best hypothesis. The R^2 value determined after the autoregressive component was added (R^2 AR) is also given.

Hypothesis/variable	Occurrence				Density			
	ΔAIC_c	w_i	R^2	R^2 AR	ΔAIC_c	w_i	R^2	R^2 AR
Depth	24,752	0	0.03	0.31	2,049	0	<0.01	0.64
Temperature	18,584	0	0.03	0.30	446	0	<0.01	0.66
Relative temperature	6,520	0	0.15	0.35	405	0	0.02	0.66
Dissolved oxygen	*0 (89,249)*	*1*	*0.21*	0.34	269	0	0.09	0.67
Salinity	6,452	0	0.15	0.29	*0 (11,305)*	*1*	*0.21*	0.67
Stratification	20,678	0	<0.01	0.25	364	0	<0.01	0.66
Zooplankton S_v	24,538	0	0.02	0.28	1,559	0	<0.01	0.63

(Table 2, analysis A). For the vertical occurrence of age-0 Walleye Pollock, the Oxygen hypothesis had the most support ($w_i = 1.0$; Table 3). Oxygen had a negative coefficient, suggesting that the probability of occurrence increased as Oxygen decreased within the water column. The Oxygen hypothesis had an R^2 value of only 0.21, but the addition of AdjPollock increased the R^2 to 0.34. For the vertical density of age-0 Walleye Pollock, Salinity was the best hypothesis ($w_i = 1.0$; Table 3). Salinity had a positive coefficient, suggesting that age-0 Walleye Pollock density increased as Salinity increased within the water column. The R^2 value for the Salinity hypothesis was initially 0.21, and the value increased to 0.67 when AdjPollock was added to the model.

Models with most or all of the available variables had the greatest support in the exhaustive all-subsets model comparison for HR1 (Table 2, analysis B; Table 4). For both occurrence and density of age-0 Walleye Pollock, there were only one or two models in the 90% confidence set. Although these models were supported by the Akaike's information criterion (AIC) values, they explained little of the variance ($R^2 = 0.27$ for occurrence and 0.23 for density). When AdjPollock was added, it was included in all of the top models for occurrence and density (Table 4). For occurrence, there was a single superior model that included all eight variables and AdjPollock ($R^2 = 0.41$). For density model selection with AdjPollock, 12 models were within the 90% confidence set, and they contained between four and six of the eight explanatory variables. The model with the most support had an R^2 value of 0.66, but the AIC value of that model was within 3 units of six other models.

Ontogeny: Changes in Vertical Distribution and Aggregation Attributes over Time

In Figures 4, 5, and 6, means of the respective measure are presented with 1 SD (except for aggregation length, which is $\pm 0.5 \cdot$ beam width) to provide insight into the distribution of data (Cumming et al. 2007).

Trawl distributions of age-0 Walleye Pollock.—Surface trawl catches of age-0 Walleye Pollock remained the same or decreased at eight of nine paired BASIS–HR1 stations (Table 2, analysis C). At five surface trawl locations, age-0 Walleye Pollock catch decreased by one to four orders of magnitude (i.e., to ≤3 fish/30-min trawl) between BASIS sampling and the HR1 sampling that occurred 11–34 d later (Figure 2). At three of the paired surface trawl locations (sampled 8–34 d apart), the catch of age-0 Walleye Pollock remained at ≤3 fish/30-min trawl between BASIS and the HR1 survey. At only one location did the surface trawl catch increase (from ≤3 to <3,000 fish/30-min trawl) between BASIS sampling and HR1 sampling 22 d later.

Paired BASIS–HR1 midwater trawls occurred at only three locations (Table 2, analysis C). At two paired stations (sampled 11 or 17 d apart), the catch of age-0 Walleye Pollock increased by one or two orders of magnitude between BASIS and HR1 (Figure 2). At the third station, where HR1 sampling occurred 8 d after BASIS, the midwater catch of age-0 Walleye Pollock decreased by two orders of magnitude.

Attributes of age-0 Walleye Pollock aggregations—The aggregation length (mean length $\pm$ [0.5·beam angle]) for age-0 Walleye Pollock varied among stations but typically increased with increasing station bottom depth (Table 2, analysis D; Figure 4). Among the 20 repeated observations (10 for BASIS–HR1; 10 for HR1–HR2), there were potential differences ($P < 0.05$) for 14 observations, suggesting that minimum aggregation length changed between sampling dates (Table 5). For those 14 paired observations, there was no consistent increase or decrease in minimum aggregation length (see Methods, Aggregation attributes) between BASIS and HR1 sampling or between HR1 and HR2 sampling (Table 5). For the HR2 survey (day of the year 268), eight stations had minimum aggregation lengths within a smaller range of values (13–24 m; Figure 4) than was observed during either BASIS or the HR1 survey.

The thickness of age-0 Walleye Pollock aggregations typically increased and became more variable as the station

TABLE 4. The 90% confidence set for exhaustive selection of models describing the vertical distribution of age-0 Walleye Pollock occurrence and density during the high-resolution (HR1) survey, both with and without the autoregressive (AR) component AdjPollock (ΔAIC_c = difference in Akaike's information criterion corrected for small sample sizes, as calculated between the given model and the best-performing model in the set [AIC_c for the best model in parentheses]; w_i = Akaike weight; R^2 AR = the R^2 value determined after the AR component was added). Bold italics indicate results for the best hypothesis. Abbreviations for the variables are defined in Table 2.

Hypothesis	ΔAIC_c	w_i	R^2	R^2 AR
Occurrence				
~ Depth + Temp + RelTemp + ZoopS$_v$ + Stratification + Oxygen + Salinity	*0 (80,627)*	0.90	0.27	
Occurrence with AR				
~ Depth + AdjPollock + Temp + RelTemp + ZoopS$_v$ + Stratification + Oxygen + Salinity	*0 (51,800)*	1.00		0.41
Density				
~ Temp + RelTemp + ZoopS$_v$ + Stratification + Oxygen + Salinity	*0 (10,712)*	0.56	0.23	
~ Depth + Temp + RelTemp + ZoopS$_v$ + Stratification + Oxygen + Salinity	0.8	0.38	0.23	
Density with AR				
~ AdjPollock + ZoopS$_v$ + Stratification + Salinity	*0 (8,002)*	0.22		0.66
~ AdjPollock + RelTemp + ZoopS$_v$ + Stratification + Salinity	1.2	0.12		0.66
~ AdjPollock + Temp + ZoopS$_v$ + Stratification + Salinity	1.6	0.10		0.66
~ Depth + AdjPollock + ZoopS$_v$ + Stratification + Salinity	1.7	0.10		0.66
~ AdjPollock + ZoopS$_v$ + Stratification + Oxygen + Salinity	1.9	0.09		0.66
~ AdjPollock + Temp + RelTemp + ZoopS$_v$ + Stratification + Salinity	2.6	0.06		0.66
~ AdjPollock + RelTemp + ZoopS$_v$+ Stratification + Oxygen + Salinity	3.0	0.05		0.66
~ Depth + AdjPollock + RelTemp + ZoopS$_v$+ Stratification + Salinity	3.2	0.04		0.66
~ Depth + AdjPollock + Temp + ZoopS$_v$+ Stratification + Salinity	3.6	0.04		0.66
~ AdjPollock + Temp + ZoopS$_v$ + Stratification + Oxygen + Salinity	3.6	0.04		0.66
~ Depth + AdjPollock + ZoopS$_v$ + Stratification + Oxygen + Salinity	3.7	0.03		0.66
~ AdjPollock + ZoopS$_v$ + Stratification + Salinity	3.7	0.03		0.66

bottom depth increased (Table 2, analysis D; Figure 4). Among the 20 repeated observations (10 for BASIS–HR1; 10 for HR1–HR2), 13 stations had aggregation thicknesses that were potentially different ($P < 0.05$; Table 5). During HR2, nine stations had mean aggregation thicknesses within a range of values (0.8–1.9 m; Figure 4) that was smaller than the range for BASIS or HR1.

The S_v acoustic density of age-0 Walleye Pollock aggregations was fairly consistent as station bottom depth increased, but it was higher and more variable at bottom depths of 125 m or greater (Table 2, analysis D; Figure 4). The S_v acoustic density was potentially different ($P < 0.05$) for 13 of the 20 repeated observations, but there was no consistent pattern of increase or decrease between sampling dates (Table 5). During HR2, S_v acoustic density values were within a range (−57.5 to −51 dB) similar to that observed during BASIS and the HR1 survey.

Metrics of age-0 Walleye Pollock vertical distribution.—As station bottom depth increased, the acoustic center of mass for age-0 Walleye Pollock aggregations moved deeper in the water column (Table 2, analysis E; Figure 5). At stations 1–3, the center of mass was above the 1°C isotherm and typically in water temperatures less than 2°C. At stations 4 and 5, the center of mass was in 2°C water. At station 6, the center of mass was in 1°C water. At stations 7–24, water temperatures exceeded 3°C throughout the water column, and the center of

mass was typically deeper. Among the 20 repeated observations, 11 involved potential differences ($P < 0.05$) in the center of mass between BASIS and HR1 sampling or between HR1 and HR2 sampling, with the center of mass located deeper in the water column for nine observations (Table 5). During the HR2 survey, the acoustic center of mass was shallow (between 23 and 42 m) over the coldest water at stations 1 and 4 and at the deepest survey station (station 24, with 350-m bottom depth). All other stations had center-of-mass values between 74 and 125 m during HR2.

Water column S_v acoustic density of age-0 Walleye Pollock varied across stations but tended to be higher and more variable at stations with bottom depths of 120 m or greater (Table 2, analysis E; Figure 5). There were potential differences ($P < 0.05$) in 14 of the 20 repeated observations, but S_v acoustic density did not consistently increase or decrease between samples (Table 5). The range of S_v acoustic density values during the HR2 survey (65 to 54.5 dB) was smaller than that observed during the HR1 survey or BASIS.

Ontogeny: Fork Length and Energy Density of Age-0 Walleye Pollock

Surface versus midwater.—For age-0 Walleye Pollock sampled during the HR1 survey or BASIS, no overall pattern in FL was apparent across station bottom depths (Table 2, analysis

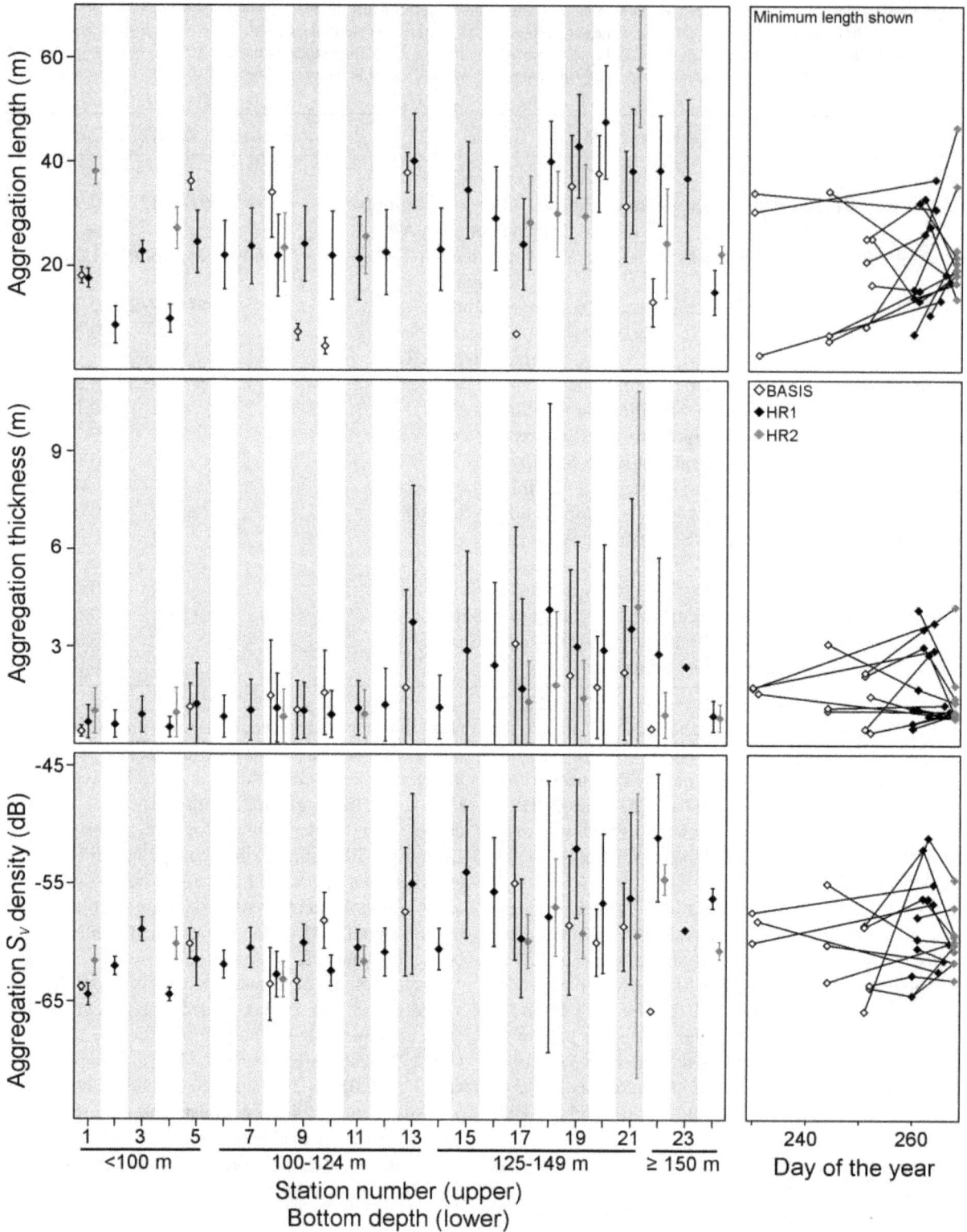

FIGURE 4.   Length (upper panels), thickness (middle panels), and S_v acoustic density (lower panels) of age-0 Walleye Pollock aggregations during the 2010 Bering Aleutian Salmon International Survey (BASIS) and the high-resolution (HR1 and HR2) surveys. Means ($\pm$[0.5·beam width] for aggregation length; $\pm$SD for thickness and S_v acoustic density) for each station and bottom depth are presented (left panels). Minimum aggregation length, mean thickness, and mean S_v acoustic density within a station by day of the year are also shown (right panels). Symbol descriptions apply to all panels.

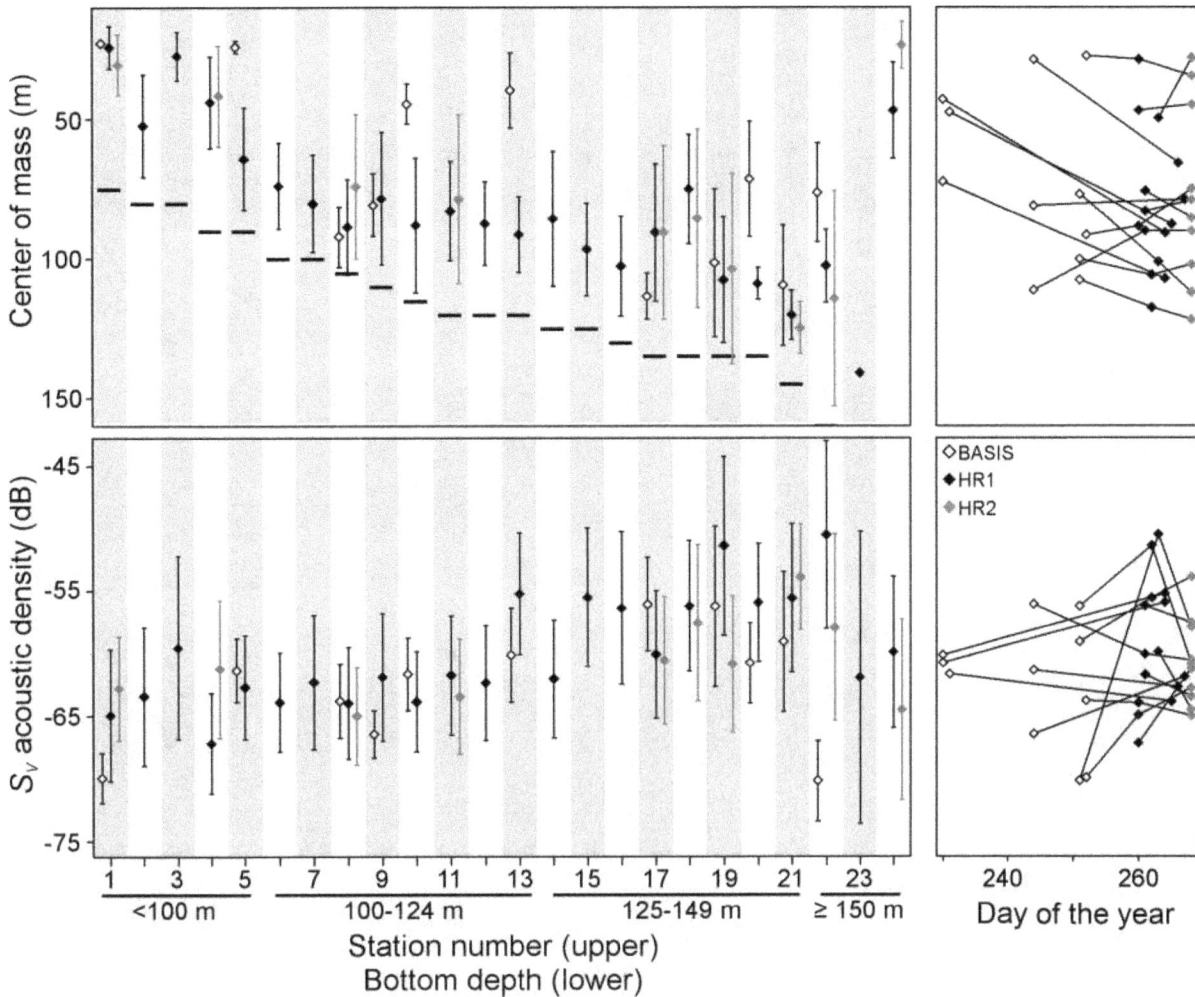

FIGURE 5.   Acoustic center of mass (upper panels) and water column S_v acoustic density (lower panels) of age-0 Walleye Pollock vertical distributions during the 2010 Bering Aleutian Salmon International Survey (BASIS) and the high-resolution (HR1 and HR2) surveys. Means (±SD) for each station and bottom depth (left panels) and means within a station by day of the year (right panels) are shown. On the upper left panel, horizontal bars indicate bottom depth. Symbol descriptions apply to all panels.

F; Figure 6). A GLM evaluating spatial differences in age-0 Walleye Pollock FL with bottom depth during HR1 indicated potential differences ($P < 0.001$) for surface-caught fish (df = 404; $n = 8$ stations; bottom depth = 74–135 m) and had a coefficient of -0.22, suggesting that as bottom depth increased the mean FL of age-0 Walleye Pollock caught in surface trawls decreased. The P-value for the surface GLM did not change when stations with low catches were removed or when pycnocline-caught fish were included in the analysis. The GLM for midwater-caught fish had a P-value of 0.15 (df = 1,296; $n = 14$ stations; bottom depth = 82–161 m), suggesting that Walleye Pollock FL did not differ among midwater station samples.

There was no consistent pattern between age-0 Walleye Pollock energy density and bottom depth across stations (Table 2, analysis G), but the GLM evaluating energy density of surface-caught fish versus station bottom depth during HR1 had a P-value of 0.008 (df = 32; $n = 4$ stations). The

coefficient for the surface GLM was -0.02, suggesting that as bottom depth increased the energy density of age-0 Walleye Pollock caught at the surface decreased. The GLM for energy density of midwater-sampled fish had a P-value of 0.40 (df = 89; $n = 11$ stations), suggesting that the energy density of fish caught in midwater samples did not differ with bottom depth.

There was also no consistent difference in FL between age-0 Walleye Pollock that were caught at the surface and those caught midwater at the one BASIS station and three HR1 stations available for this comparison (Table 2, analysis H; Figure 6). At two of the three HR1 stations, the FLs of surface and midwater fish were potentially different ($P < 0.05$), with surface fish being larger at station 2 and midwater fish being larger at station 11. At the two remaining stations (BASIS station 17; HR1 station 18), the FLs of surface- and midwater-caught fish were not different ($P > 0.05$). The maximum within-station difference in mean FL was 4.01 mm.

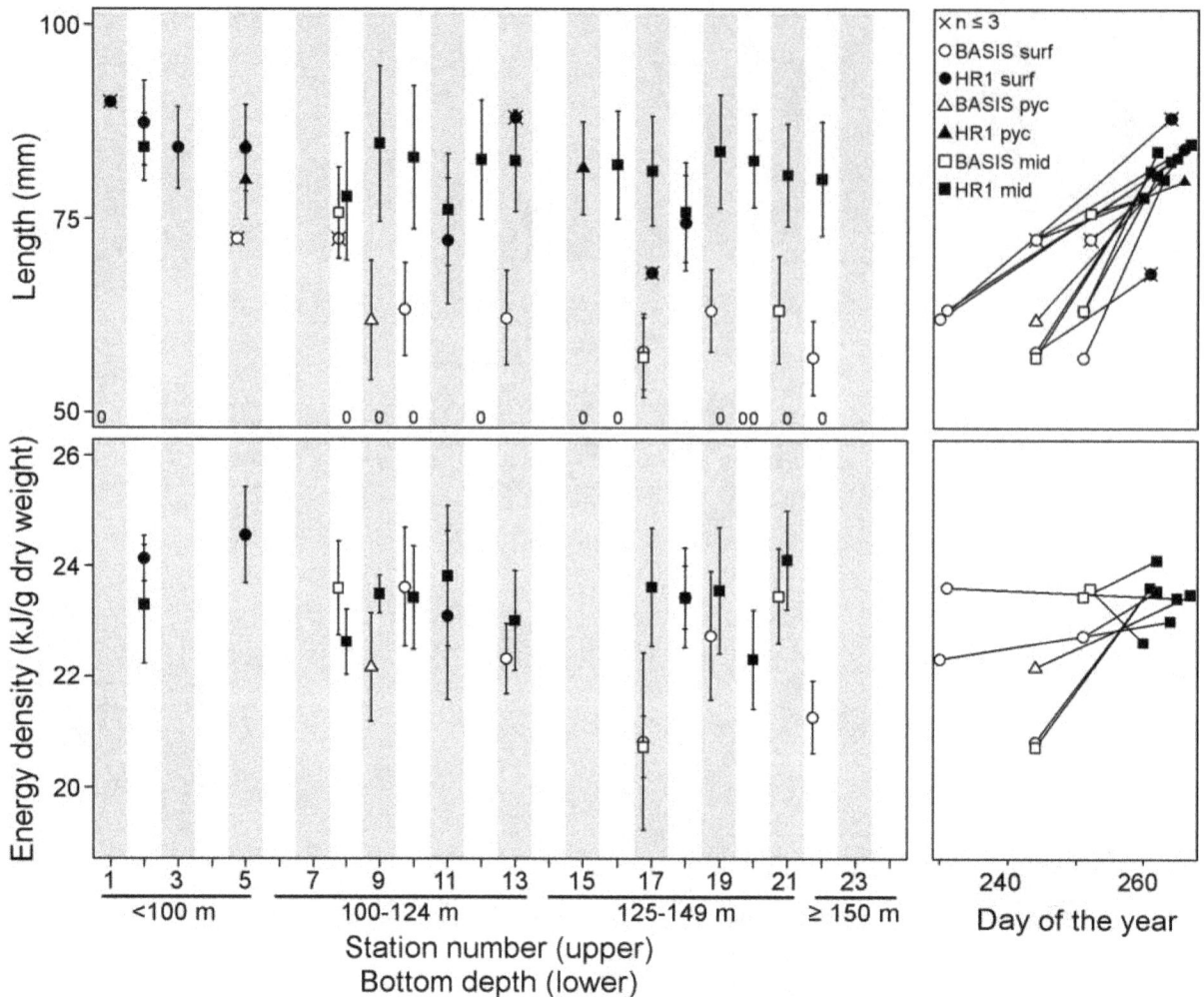

FIGURE 6. Fork length (upper panels) and energy density (lower panels) of age-0 Walleye Pollock during the 2010 Bering Aleutian Salmon International Survey (BASIS) and the high-resolution (HR1) survey. Means (±SD) for each station and bottom depth (left panels) and means within a station by day of the year (right panels) are shown. An "×" behind a symbol indicates that the sample contained three or fewer fish; a zero along the x-axis of the upper left panel indicates that the surface trawl catch was zero. Symbol descriptions apply to all panels (surf = surface trawl; pyc = pycnocline trawl; mid = midwater trawl).

Energy density comparisons between surface- and midwater-caught age-0 Walleye Pollock were possible for BASIS station 17 and HR1 stations 2 and 18 (Table 2, analysis I; Figure 6). The comparisons indicated that surface and midwater fish did not differ in energy density ($P > 0.05$). The maximum within-station difference in mean energy density was 0.8 J/g dry mass.

Changes in fork length or energy density over time.—As expected, mean FLs of age-0 Walleye Pollock were higher during the HR1 survey than during BASIS (Table 2, analysis J; Figure 6). Among the three stations where midwater trawls were performed during both BASIS and HR1, the mean FL of age-0 Walleye Pollock potentially increased between sampling periods at two stations (stations 17 and 21; $P < 0.05$) but not at the third (station 8; $P > 0.05$; Figure 6). Contrary to our expectation, the FLs of age-0 Walleye Pollock sampled during BASIS were not more similar to the FLs of HR1-sampled fish

when the number of days between sampling events decreased (Figure 6). During HR1 sampling, the mean FL of fish—regardless of whether they were caught near the surface or midwater—ranged from 72.1 to 87.3 mm at stations with more than three fish in the catch.

Energy densities of age-0 Walleye Pollock were similar within most stations regardless of whether the fish were sampled during BASIS or the HR1 survey (Table 2, analysis J; Figure 6). Among the three paired BASIS–HR1 midwater trawls, two stations had energy densities that were potentially different ($P < 0.05$) between sampling periods; energy density increased between BASIS and the HR1 survey at station 17, whereas it decreased between surveys at station 8. The third station (station 21) had a P-value greater than 0.05 for the comparison of energy density between BASIS and HR1 samples. Mean energy density during the HR1 survey was between 22.3 and 24.4 J/g dry mass.

TABLE 5. Summary of comparisons for age-0 Walleye Pollock aggregation attributes (length [minimum], thickness, and S_v acoustic density) and metrics of vertical distribution (center of mass and S_v acoustic density) between surveys at each station (B–H1 = comparison between the 2010 Bering Aleutian Salmon International Survey [BASIS] and high-resolution survey HR1; H1–H2 = comparison between high-resolution surveys HR1 and HR2). Days between surveys are given for each station ("∧" symbol indicates that the descriptor increased between observations at $P < 0.05$; "∨" symbol indicates that the descriptor decreased between observations at $P < 0.05$; "ns" indicates no significant change, $P \geq 0.05$).

| | Days between | | Aggregation attributes | | | | | | Metrics of vertical distribution | | | |
| | | | Minimum length | | Thickness | | S_v acoustic density | | Center of mass | | S_v acoustic density | |
Station	B–H1	H1–H2	B–H1	H1–H2	B–H1	H1–H2	B–H1	H1–H2	B–H1	H1–H2	B–H1	H1–H2
1	8	8	ns	∧	ns	ns	ns	ns	ns	∧	ns	ns
4		8		∧		∧		∧		ns		∧
5	22		∨		ns		ns		∧		∨	
8	8	8	∨	∧	∨	∨	ns	ns	ns	∨	ns	ns
9	23		∧		ns		∧		ns		∧	
10	34		∧		∨		∨		∧		∨	
11		7		∧		∨		∨		ns		∨
13	34		ns		∧		∧		∧		∧	
17	17	7	∧	∧	∨	ns	∨	ns	∨	ns	∨	ns
18		7		∨		∨		ns		∧		∨
19	11	6	ns	∨	∧	∨	∧	∨	∧	ns	∧	∨
20	34		ns		∧		∧		∧		∧	
21	11	6	ns	∧	∧	ns	∧	∨	∧	∧	∧	∧
22	12	5		∨		∨		∨	ns			∨
24		5		ns		ns		∨		ns		ns

DISCUSSION

Our study examined whether environmental factors (e.g., water temperature or location of zooplankton prey) or ontogeny affected the vertical distribution of age-0 Walleye Pollock. Based on previous studies and common hypotheses, we assumed that the vertical distribution of age-0 Walleye Pollock represented a response to environmental factors, but an HR survey that sampled as many environmental and biological factors as possible was needed to refine our understanding. We also recognized that vertical distribution could be related to ontogeny, particularly since the timing of the transition from near-surface to demersal waters by EBS age-0 Walleye Pollock is not well understood (Duffy-Anderson et al. 2015).

Contrary to our initial prediction, water column properties alone did not have a strong effect on the vertical distribution of age-0 Walleye Pollock. High autocorrelation—attributed to aggregating behavior—played a larger role in explaining vertical distribution than did environmental variables. This result was not surprising given that schooling among juvenile Walleye Pollock increases their foraging success (Baird et al. 1991; Ryer and Olla 1992) and is generally thought to reduce predation risk (Pitcher and Parrish 1993). Previous studies have primarily evaluated age-0 Walleye Pollock distributions relative to environmental factors, including the thermocline (e.g., Swartzman et al. 1999, 2002), water column stability (Francis and Bailey 1983; Coyle et al. 2008), and zooplankton prey (Olla and Davis 1990; Ciannelli et al. 2004; Coyle et al.

2008). In our analysis, these variables were not significant predictors of the occurrence or density of age-0 Walleye Pollock. Instead, we found that salinity was the best predictor of age-0 occurrence and that oxygen was the best predictor of age-0 density. Most of the age-0 Walleye Pollock in this analysis occurred in the midwater of the outer domain, which is characterized by low variability in salinity and oxygen (Coachman 1986); thus, the inclusion of these variables in final model selections was not surprising.

We characterized the potential zooplankton prey field for age-0 Walleye Pollock by using acoustics rather than net tows. Acoustics provided an estimate of relative zooplankton biomass at locations within the water column, whereas net tows produce vertically integrated water column estimates (e.g., Coyle et al. 2011). Although the adopted methods from Murase et al. (2009) may have resulted in underestimation of Calanus spp. abundances, our analysis was based on the relative distribution of zooplankton (i.e., copepods and euphausiids) in the water column rather than absolute abundance. Additional acoustic research effort is needed to refine and expand EBS-specific acoustic classification methods for zooplankton (e.g., De Robertis et al. 2010). Knowledge of the vertical location of zooplankton relative to age-0 Walleye Pollock during the day is important, as an analysis of age-0 Walleye Pollock diets from 27 surface, midwater, crepuscular, and night trawls during the HR1 survey suggested that they fed on Calanus spp. and euphausiids throughout the day and were not restricted to feeding during a diurnal cycle (N. Kuznetsova, Pacific Fisheries Research Center [TINRO], Vladivostok, Russia, personal communication).

Despite the comprehensive set of variables used in our analyses, two potential factors that have been suggested to influence Walleye Pollock vertical distributions were not included in this study. Olla and Davis (1990) and Ryer and Olla (1992) found that light levels affected the vertical distribution and shoaling of age-0 Walleye Pollock under laboratory conditions. Unfortunately, light levels at depth were not measured during our surveys due to equipment difficulties. Given that Kotwicki et al. (2013) found that bottom trawl catches of age-1 and older (age-1+) Walleye Pollock decreased with increasing near-bottom light levels, this factor should be evaluated in future studies of age-0 vertical distribution. Ryer and Olla (1992) also demonstrated that predator presence affected the degree of age-0 Walleye Pollock shoaling in laboratory experiments, but they did not evaluate the effect of predator presence on vertical distribution. Primary predators of age-0 Walleye Pollock during late summer include Arrowtooth Flounder *Atheresthes stomias* and age-1+ Walleye Pollock (Lang et al. 2000), with both species occupying demersal or semi-demersal habitats during daytime. During our surveys, we captured no Arrowtooth Flounder or age-1+ Walleye Pollock in 27 surface trawls and 17 midwater trawls. An acoustic trawl survey conducted in June–August 2010 suggested that age-1+ Walleye Pollock were primarily located over 150 km to the northwest of our study area (Honkalehto et al. 2012). Combining (1) the evidence that predator density was low in our study area with (2) our primary interest in evaluating potential factors influencing the movement of age-0 Walleye Pollock from near-surface waters to the midwater zone, we felt that the omission of predator variables from our modeling was justified.

From an ontogenetic perspective, data on trawl catch and acoustic center of mass suggested that the vertical distribution of age-0 Walleye Pollock shifted deeper in the water column between BASIS and the HR surveys. This pattern matches observations in the Gulf of Alaska by Brodeur and Wilson (1996), who reported that the depth of age-0 Walleye Pollock capture increased through the season. The presence of age-0 Walleye Pollock in the midwater zone during the 2010 BASIS (Parker-Stetter et al. 2013) suggests that movement of age-0 Walleye Pollock out of near-surface waters and into deeper waters began before August 18. Based on our single year of data, we cannot state whether the timing of the descent is consistent among years. Contrary to the concern stated by Hollowed et al. (2012) that surface trawl distributions are only representative of smaller fish (per Swartzman et al. 2002), we found no difference in FL between surface- and midwater-caught age-0 Walleye Pollock. Instead, our data suggest that the age-0 Walleye Pollock length distributions derived from the surface trawl data may represent only the proportion of the population that had not moved deeper in the water column by the time of sampling.

We predicted that aggregations of age-0 Walleye Pollock would become longer, thicker, and denser in the late summer and early fall. This pattern has been observed for other species, such as Pacific Herring (Sigler and Csepp 2007). The only consistent pattern across sampling dates was that the mean aggregation length and thickness tended to cluster around a general aggregation size (13–24 m for length; 0.8–1.9 m for thickness) by the date of the HR2 survey. This aggregation size may reflect a balance between predation risk and foraging success (cf. Pitcher and Parrish 1993). Aggregation length, thickness, and S_v acoustic density tended to increase, but characteristics became more variable with increasing bottom depth. In a study of age-1–3 Walleye Pollock, Stienessen and Wilson (2008) found that aggregation thickness increased with depth in the water column but that the density of fish (approximately comparable to our S_v acoustic density) decreased with depth. The aggregation lengths and thicknesses observed in our study were considerably smaller than those of age-0 Walleye Pollock surveyed off Japan, where maximum aggregation length was about 500 m and mean thickness was approximately 18 m during the day (Kang et al. 2006). Observation conditions for the Kang et al. (2006) study differed from ours in that sampling took place earlier in the year (June–July) and over shallower bottom depths (~75 m) than our study, so survey timing and bottom depth may have affected aggregation characteristics.

Our approach also evaluated differences in age-0 Walleye Pollock vertical distribution using growth characteristics of fish FL and energy density. Larger fish have been predicted to occur deeper than smaller fish (Miyake et al. 1996; Swartzman et al. 2002). A parallel prediction could be made for the energy densities of age-0 Walleye Pollock if midwater temperatures and food rations are bioenergetically favorable (Sogard and Olla 1996). Our limited data suggested that although station bottom depth may explain differences in FLs of surface-caught age-0 Walleye Pollock across space, midwater-caught age-0 Walleye Pollock were not consistently longer and did not have higher energy densities than surface-caught fish. When stations were resampled over a range of temporal lags, the slopes of lines connecting samples at a given station differed, but FLs and energy densities converged to a small range of values independent of the capture depth in the water column. This observed convergence of FLs and energy densities may reflect the switch between "grow longer" and "grow fatter" responses to seasonal changes, as has been proposed for EBS age-0 Walleye Pollock (Siddon et al. 2013a). We do not believe that the observed convergence of FLs was due to net selectivity, since the Cantrawl is used to catch salmon of up to 475 mm (cf. Farley et al. 2011). The range of energy densities observed during the HR surveys is similar to or higher than means reported for the 2008 and 2009 cohorts (Siddon et al. 2013b) and is similar to the mean energy densities of age-0 Walleye Pollock sampled during 2007–2012 (based on Heintz et al. 2013). Future work should increase the number of paired surface trawl and midwater trawl samples and should involve sampling earlier in the year, when the probability of catching both surface and midwater fish is higher.

There is currently no standard methodology to determine age in juvenile Walleye Pollock; therefore, we were unable to estimate growth rates or to evaluate the influence of daily age on our results. Previous efforts to estimate the daily ages of age-0 Walleye Pollock have been inconsistent or unable to identify daily growth rings (reviewed by Duffy-Anderson et al. 2015). It is possible that differences in FL among locations could be due to different cohorts being advected away from spawning locations (cf. Smart et al. 2012). It is also conceivable that the initiation of age-0 Walleye Pollock movement from near-surface waters to the midwater zone is a function of daily age, but we observed no clear patterns in paired samples with either FL or energy density. Davis and Ottmar (2009) suggested that annual age (0+ and 1+) had a minor effect on the vertical distribution of Pacific Cod over a range of environmental conditions in the laboratory. Additional work on techniques to determine the daily age of late-juvenile Walleye Pollock (e.g., Bailey 1989) would provide insight into the condition and timing of movement from near-surface waters to deeper waters.

Two major assumptions were implicit in our processing of the acoustic data. First, we assumed that the -60 dB threshold in combination with school detection parameters would exclude jellyfish from the backscatter categorized as age-0 Walleye Pollock. De Robertis and Taylor (2014) verified that jellyfish had low backscatter contributions relative to small pelagic fishes such as age-0 Walleye Pollock, so this assumption was likely met. The second assumption was that detected aggregations of acoustic backscatter were age-0 Walleye Pollock. Age-0 Pacific Cod were present in near-surface waters of the EBS during BASIS in 2010 but at lower densities than age-0 Walleye Pollock (Parker-Stetter et al. 2013). Because age-0 Walleye Pollock were typically caught in low numbers at trawl stations with few aggregations, we felt that this assumption was valid. If aggregations contained age-0 Pacific Cod, then the location of near-surface aggregations may reflect a mixture of species rather than exclusively age-0 Walleye Pollock.

The modeling approach used here was chosen to evaluate *why* vertical distributions differed from station to station, not *whether* they differed. The use of a traditional ANOVA with "station" as a factor or the use of GLMs with separate intercepts for each station would have resulted in significant differences between or among stations. Fitting separate intercepts for each station would have overfitted the regression models since they would tell us nothing about the factors that made age-0 Walleye Pollock density higher at station x than at station y. Our intent was to infer the processes influencing Walleye Pollock vertical distribution rather than to quantify the distribution patterns. To examine whether the

1-m vertical data resolution introduced distributional artifacts, we re-ran the models with acoustic data averaged in 2-m and 5-m vertical bins. Hypotheses with the most support did not change in occurrence or density models, and R^2 remained at values less than 0.20 in the absence of the autoregressive component.

Our finding that ontogeny rather than biological or physical water column properties was likely responsible for differences in age-0 Walleye Pollock vertical distribution within and between surveys should be considered in the design of population abundance estimate surveys. If surface trawling occurs while age-0 Walleye Pollock are transitioning from near-surface to midwater depths, then abundance will be underestimated and the distribution will represent fish in the near-surface water rather than the population. Sampling earlier in the season would likely allow age-0 Walleye Pollock to be surveyed while they are in near-surface waters above or within the pycnocline (Parker-Stetter et al. 2013). Although age-0 Walleye Pollock would be mixed with jellyfish and other fish species at that time, consistent vertical distribution might increase the catches of age-0 Walleye Pollock and minimize bias. Alternatively, sampling later in the year could provide vertical separation between age-0 Walleye Pollock and other species, enabling more accurate acoustic-based estimates of population abundance and distribution.

ACKNOWLEDGMENTS

We thank the scientific staff, captain, and crew of the NOAA FSV *Oscar Dyson* for assistance during the 2010 BASIS and HR surveys; Natalia Kuznetsova (TINRO, Russia) for processing diet samples from BASIS and HR1; and the Alaska Fisheries Science Center's (AFSC) Midwater Assessment and Conservation Engineering Program for the use of equipment and software during the surveys and providing calibration data for scientific echosounders. The AFSC's Fish, Energy, Diet, Zooplankton Laboratory is gratefully acknowledged for their assistance in species identification. We also thank David Barbee and Jennifer Nomura (University of Washington [UW]) for assistance with acoustic data collection, processing, and analysis; David McGowan (UW) and John Pohl (Northwest Fisheries Science Center) for reviewing earlier versions of the manuscript; and Kenneth Rose (Louisiana State University) and two anonymous reviewers for suggestions that improved the final product. The findings and conclusions in this paper are those of the authors and do not necessarily represent the views of the NOAA National Marine Fisheries Service. Reference to trade names does not imply endorsement by the NOAA National Marine Fisheries Service. This paper is Publication Number 542 of the North Pacific Research Board and is Publication Number 161 of the Bering Ecosystem Study (BEST)–Bering Sea Integrated Ecosystem Research Program (BSIERP) Bering Sea Project.

REFERENCES

Alenius, P., A. Nekrasov, and K. Myrberg. 2003. Variability of the baroclinic Rossby radius in the Gulf of Finland. Continental Shelf Research 23:563–573.

Bailey, K. M. 1989. Interaction between the vertical distribution of juvenile Walleye Pollock *Theragra chalcogramma* in the eastern Bering Sea, and cannibalism. Marine Ecology Progress Series 53:205–213.

Baird, T. A., C. H. Ryer, and B. L. Olla. 1991. Social enhancement of foraging on an ephemeral food source in juvenile Walleye Pollock, *Theragra chalcogramma*. Environmental Biology of Fishes 31:307–311.

Barange, M. 1994. Acoustic identification, classification and structure of biological patchiness on the edge of the Agulhas Bank and its relation to frontal features. South African Journal of Marine Science 14:333–347.

Brodeur, R. D., and M. T. Wilson. 1996. A review of the distribution, ecology and population dynamics of age-0 Walleye Pollock in the Gulf of Alaska. Fisheries Oceanography 5:148–166.

Burnham, K. P., and D. R. Anderson. 2002. Model selection and multi-model inference: a practical information-theoretic approach. Springer, New York.

Ciannelli, L., R. D. Brodeur, G. L. Swartzman, and S. Salo. 2002. Physical and biological factors influencing the spatial distribution of age-0 Walleye Pollock (*Theragra chalcogramma*) around the Pribilof Islands, Bering Sea. Deep Sea Research Part II: Topic Studies in Oceanography 49:6109–6126.

Ciannelli, L., B. W. Robson, R. C. Francis, K. Aydin, and R. D. Brodeur. 2004. Boundaries of open marine ecosystems: an application to the Pribilof Archipelago, southeast Bering Sea. Ecological Applications 14:942–953.

Coachman, L. K. 1986. Circulation, water masses, and fluxes on the southeastern Bering Sea shelf. Continental Shelf Research 5:23–108.

Coyle, K. O., L. B. Eisner, F. J. Mueter, A. I. Pinchuk, M. A. Janout, K. D. Cieciel, E. V. Farley, and A. G. Andrews. 2011. Climate change in the southeastern Bering Sea: impacts on Pollock stocks and implications for the oscillating control hypothesis. Fisheries Oceanography 20:139–156.

Coyle, K. O., A. I. Pinchuk, L. B. Eisner, and J. M. Napp. 2008. Zooplankton species composition, abundance and biomass on the eastern Bering Sea shelf during summer: the potential role of water-column stability and nutrients in structuring the zooplankton community. Deep Sea Research Part II: Topic Studies in Oceanography 55:1775–1791.

Cressie, N., and C. K. Wikle. 2011. Statistics for spatio-temporal data. Wiley, Hoboken, New Jersey.

Cumming, G., F. Fidler, and D. L. Vaux. 2007. Error bars in experimental biology. Journal of Cell Biology 177:7–11.

Davis, M., and M. Ottmar. 2009. Vertical distribution of juvenile Pacific Cod *Gadus macrocephalus*: potential role of light, temperature, food, and age. Aquatic Biology 8:29–37.

De Robertis, A., D. R. McKelvey, and P. H. Ressler. 2010. Development and application of an empirical multifrequency method for backscatter classification. Canadian Journal of Fisheries and Aquatic Sciences 67:1459–1474.

De Robertis, A., and K. Taylor. 2014. In situ target strength measurements of the scyphomedusa *Chrysaora melanaster*. Fisheries Research 153:18–23.

Diner, N. 2001. Correction on school geometry and density: approach based on acoustic image simulation. Aquatic Living Resources 14:211–222.

Duffy-Anderson, J. T., S. Barbeaux, E. Farley, R. Heintz, J. Horne, S. Parker-Stetter, C. Petrik, E. C. Siddon, and T. I. Smart. 2015. State of knowledge review and synthesis of the first year of life of Walleye Pollock (*Gadus chalcogrammus*) in the eastern Bering Sea, with comments on implications for recruitment. Deep Sea Research Part II: Topical Studies in Oceanography. DOI: 10.1016/j.dsr2.2015.02.001.

Dunnett, C. W. 1980. Pairwise multiple comparisons in the unequal variance case. Journal of the American Statistical Association 75:796–800.

Farley, E. V. Jr., J. Murphy, J. Moss, A. Feldmann, and L. Eisner. 2009. Marine ecology of western Alaska juvenile salmon. Pages 307–329 in C. C. Krueger and C. E. Zimmerman, editors. Pacific salmon: ecology and management of western Alaska's populations. American Fisheries Society, Symposium 70, Bethesda, Maryland.

Farley, E. V., A. Starovoytov, S. Naydenko, R. Heintz, M. Trudel, C. Guthrie, L. Eisner, and J. R. Guyon. 2011. Implications of a warming eastern Bering Sea for Bristol Bay Sockeye Salmon. ICES Journal of Marine Science 68:1138–1146.

Foote, K. G., H. P. Knudsen, G. Vestnes, D. N. MacLennan, and E. J. Simmonds. 1987. Calibration of acoustic instruments for fish density estimation. ICES (International Council for the Exploration of the Sea) Cooperative Research Report 44.

Francis, R. C., and K. M. Bailey. 1983. Factors affecting recruitment of selected gadoids in the northeast Pacific and east Bering Sea. Pages 35–60 in W. S. Wooster, editor. From year to year: interannual variability of the environment and fisheries of the Gulf of Alaska and the Eastern Bering Sea. Washington State Sea Grant, Seattle.

Gill, A. E. 1982. Atmosphere-ocean dynamics. Academic Press, New York.

Goodman, S. N. 1999. Toward evidence-based medical statistics I: the *P*-value fallacy. Annals of Internal Medicine 130:995–1004.

Heintz, R. A., E. C. Siddon, E. V. Farley, and J. M. Napp. 2013. Correlation between recruitment and fall condition of age-0 Pollock (*Theragra chalcogramma*) from the eastern Bering Sea under varying climate conditions. Deep Sea Research Part II: Topical Studies in Oceanography 94:150–156.

Hollowed, A. B., S. J. Barbeaux, E. D. Cokelet, E. Farley, S. Kotwicki, P. H. Ressler, C. Spital, and C. D. Wilson. 2012. Effects of climate variations on pelagic ocean habitats and their role in structuring forage fish distributions in the Bering Sea. Deep Sea Research Part II: Topical Studies in Oceanography 65–70:230–250.

Honkalehto, T., A. McCarthy, P. Ressler, S. Stienessen, and D. Jones. 2010. Results of the acoustic-trawl survey of Walleye Pollock (*Theragra chalcogramma*) on the U.S. and Russian Bering Sea shelf in June–August 2009 (DY0909). National Marine Fisheries Service, Alaska Fisheries Science Center, Processed Report 2010-03, Seattle.

Honkalehto, T., A. McCarthy, P. Ressler, K. Williams, and D. Jones. 2012. Results of the acoustic-trawl survey of Walleye Pollock (*Theragra chalcogramma*) on the U.S. and Russian Bering Sea shelf in June–August 2010 (DY1006). National Marine Fisheries Service, Alaska Fisheries Science Center, Processed Report 2012-01, Seattle.

Johnson, J. B., and K. S. Omland. 2004. Model selection in ecology and evolution. Trends in Ecology and Evolution 19:101–108.

Kang, M., S. Honda, and T. Oshima. 2006. Age characteristics of Walleye Pollock school echoes. ICES Journal of Marine Science 63:1465–76.

Korneliussen, R. 2000. Measurement and removal of echo integration noise. ICES Journal of Marine Science 57:1204–1217.

Kotwicki, S., A. De Robertis, J. N. Ianelli, A. E. Punt, J. K. Horne, and J. M. Jech. 2013. Combining bottom trawl and acoustic data to model acoustic dead zone correction and bottom trawl efficiency parameters for semipelagic species. Canadian Journal of Fisheries and Aquatic Sciences 70:208–219.

Lang, G., R. D. Brodeur, J. M. Napp, and R. Schabetsberger. 2000. Variation in groundfish predation on juvenile Walleye Pollock relative to hydrographic structure near the Pribilof Islands, Alaska. ICES Journal of Marine Science 57:265–271.

Lauth, R. R. 2010. Results of the 2009 eastern Bering Sea continental shelf bottom trawl survey of groundfish and invertebrate resources. NOAA Technical Memorandum NMFS-AFSC-204.

Lew, M. J. 2012. Bad statistical practice in pharmacology (and other basic biomedical disciplines): you probably don't know *P*: statistical inference using *P*-values. British Journal of Pharmacology 166:1559–1567.

McFadden, D. 1974. Conditional logit analysis of qualitative choice behavior. Pages 105–142 in P. Zarembka, editor. Frontiers in econometrics. Academic Press, New York.

Miyake, H., H. Yoshida, and Y. Ueda. 1996. Distribution and abundance of age-0 juvenile Walleye Pollock, *Theragra chalcogramma*, along the Pacific coast of southeastern Hokkaido, Japan. NOAA Technical Report NMFS-126.

Moss, J. H., E. V. Farley, A. M. Feldmann, and J. N. Ianelli. 2009. Spatial distribution, energetic status, and food habits of eastern Bering Sea age-0 Walleye Pollock. Transactions of the American Fisheries Society 138:497–505.

Murase, H., M. Ichihara, H. Yasuma, H. Watanabe, S. Yonezaki, H. Nagashima, S. Kawahara, and K. Miyashita. 2009. Acoustic characterization of biological backscatterings in the Kuroshio–Oyashio inter-frontal zone and subarctic waters of the western North Pacific in spring. Fisheries Oceanography 18:386–401.

Olla, B. L., and M. W. Davis. 1990. Behavioral responses of juvenile Walleye Pollock *Theragra chalcogramma* Pallas to light, thermoclines and food: possible role in vertical distribution. Journal of Experimental Marine Biological and Ecology 135:59–68.

Parker-Stetter, S. L., J. K. Horne, E. V. Farley, D. H. Barbee, A. G. Andrews, L. B. Eisner, and J. M. Nomura. 2013. Summer distributions of forage fish in the eastern Bering Sea. Deep Sea Research Part II: Topical Studies in Oceanography 94:211–230.

Pitcher, T. J., and J. K. Parrish. 1993. The functions of shoaling behaviour. Pages 363–439 *in* T. J. Pitcher, editor. The behaviour of teleost fishes, 2nd edition. Chapman and Hall, London.

R Development Core Team. 2013. R: a language and environment for statistical computing. R Foundation for Statistical Computing, Vienna. Available: www.R-project.org. (July 2015).

Reid, D. G., and E. J. Simmonds. 1993. Image analysis techniques for the study of fish school structure from acoustic survey data. Canadian Journal of Fisheries and Aquatic Sciences 50:886–893.

Rossby, C. G. 1938. On the mutual adjustment of pressure and velocity distributions in certain simple current systems, parts I, II. Journal of Marine Research 2:239–263.

Ruxton, G. D. 2006. The unequal variance *t*-test is an underused alternative to Student's *t*-test and the Mann–Whitney *U*-test. Behavioral Ecology 17:688–690.

Ryer, C. H., and B. L. Olla. 1992. Social mechanisms facilitating exploitation of spatially variable ephemeral food patches in a pelagic marine fish. Animal Behaviour 44:69–74.

Schabetsberger, R., M. Sztatecsny, G. Drozdowski, R. D. Brodeur, G. L. Swartzman, M. T. Wilson, A. G. Winter, and J. M. Napp. 2003. Size-dependent, spatial, and temporal variability of juvenile Walleye Pollock (*Theragra chalcogramma*) feeding at a structural front in the southeast Bering Sea. Marine Ecology 24:141–164.

Siddon, E. C., R. A. Heintz, and F. J. Mueter. 2013a. Conceptual model of energy allocation in Walleye Pollock (*Theragra chalcogramma*) from age-0 to age-1 in the southeastern Bering Sea. Deep Sea Research Part II: Topical Studies in Oceanography 94:140–149.

Siddon, E. C., T. Kristiansen, F. J. Mueter, K. K. Holsman, R. A. Heintz, and E. V. Farley. 2013b. Spatial match–mismatch between juvenile fish and prey provides a mechanism for recruitment variability across contrasting climate conditions in the eastern Bering Sea. PLoS (Public Library of Science) ONE [online serial] 8:e84526.

Sigler, M. F., and D. J. Csepp. 2007. Seasonal abundance of two important forage species in the North Pacific Ocean, Pacific Herring and Walleye Pollock. Fisheries Research 83:319–331.

Smart, T. I., J. T. Duffy-Anderson, J. K. Horne, E. V. Farley, C. D. Wilson, and J. M. Napp. 2012. Influence of environment on Walleye Pollock eggs, larvae, and juveniles in the southeastern Bering Sea. Deep Sea Research Part II: Topical Studies in Oceanography 65–70:196–207.

Smart, T. I., E. C. Siddon, and J. T. Duffy-Anderson. 2013. Vertical distributions of the early life stages of a common gadid in the eastern Bering Sea (*Theragra chalcogramma*, Walleye Pollock). Deep Sea Research Part II: Topical Studies in Oceanography 94:201–210.

Sogard, S. M., and B. L. Olla. 1996. Food deprivation affects vertical distribution and activity of a marine fish in a thermal gradient: potential energy-conserving mechanisms. Marine Ecology Progress Series 133:43–55.

Stefánsson, G. 1996. Analysis of groundfish survey abundance data: combining the GLM and delta approaches. ICES Journal of Marine Science 53:577–588.

Stienessen, S. C., and C. D. Wilson. 2008. Juvenile Walleye Pollock aggregation structure in the Gulf of Alaska. Pages 271–287 *in* G. H. Kruse, K. Drinkwater, J. N. Ianelli, J. S. Link, D. L. Stram, V. Wespestad, and D. Woodby, editors. Resiliency of gadid stocks to fishing and climate change. Alaska Sea Grant College Program, Fairbanks.

Swartzman, G., R. Brodeur, J. Napp, G. Hunt, D. Demer, and R. Hewitt. 1999. Spatial proximity of age-0 Walleye Pollock (*Theragra chalcogramma*) to zooplankton near the Pribilof Islands, Bering Sea, Alaska. ICES Journal of Marine Science 56:545–560.

Swartzman, G., J. Napp, R. Brodeur, A. Winter, and L. Ciannelli. 2002. Spatial patterns of Pollock and zooplankton distribution in the Pribilof Islands, Alaska, nursery area and their relationship to Pollock recruitment. ICES Journal of Marine Science 59:1167–1186.

Tang, Q., X. Jin, F. Li, J. Chen, W. Wang, Y. Chen, X. Zhao, and F. Dai. 1996. Summer distribution and abundance of age-0 Walleye Pollock, *Theragra chalcogramma*, in the Aleutian Basin. Ecology of juvenile Walleye Pollock, *Theragra chalcogramma*. NOAA Technical Report NMFS-126.

Urmy, S. S., J. K. Horne, and D. H. Barbee. 2012. Measuring the vertical distributional variability of pelagic fauna in Monterey Bay. ICES Journal of Marine Science 69:184–196.

Watkins, J. L., and A. S. Brierley. 1996. A post-processing technique to remove background noise from echo integration data. ICES Journal of Marine Science 53:339–344.

Welch, B. L. 1947. The generalization of "Student's" problem when several different population variances are involved. Biometrika 34:28–35.

Whittingham, M. J., P. A. Stephens, R. B. Bradbury, and R. P. Freckleton. 2006. Why do we still use stepwise modelling in ecology and behaviour? Journal of Animal Ecology 75:1182–1189.

Woillez, M., J.-C. Poulard, J. Rivoirard, P. Petitgas, and N. Bez. 2007. Indices for capturing spatial patterns and their evolution in time, with application to European Hake (*Merluccius merluccius*) in the Bay of Biscay. ICES Journal of Marine Science 64:537–550.

13

Understanding Social Resilience in the Maine Lobster Industry

Anna M. Henry* and **Teresa R. Johnson**

School of Marine Sciences, University of Maine, 200 Libby Hall, Orono, Maine 04469, USA

Abstract

The Maine lobster *Homarus americanus* fishery is considered one of the most successful fisheries in the world due in part to its unique comanagement system, the conservation ethic of the harvesters, and the ability of the industry to respond to crises and solve collective-action problems. However, recent threats raise the question whether the industry will be able to respond to future threats as successfully as it has to ones in the past or whether it is now less resilient and can no longer adequately respond to threats. Through ethnographic research and oral histories with fishermen, we examined the current level of social resilience in the lobster fishery. We concentrated on recent threats to the industry and the ways in which it has responded to them, focusing on three situations: a price drop beginning in 2008, a recovery in 2010–2011, and a second collapse of prices in 2012. In addition, we considered other environmental and regulatory concerns identified by fishermen. We found that the industry is not responding effectively to recent threats and identified factors that might explain the level of social resilience in the fishery.

The Maine lobster *Homarus americanus* fishery is heralded for its cultural status, the participatory nature of its regulatory scheme, the conservation ethic of its harvesters, and more recently, its seemingly infinite increase in landings and value (Acheson 2003; Acheson and Gardner 2010). Despite these characteristics, during the summer of 2012, this iconic fishery experienced the lowest prices in 30 years. The media reported examples of the way industry members described the 2012 season: "I don't see any winners in this, this year." (Seelye 2012); "It's down to a point now where it's not worth it to go out. It's ridiculous." (Wickenheiser 2012); and "There ain't no money right now to be made" (Lobstermen tying up their boats 2012).

This may seem to be a familiar narrative, as the historical booms and busts of the lobster fishery have been well documented (Acheson and Steneck 1997; Acheson 2003; Acheson and Gardner 2010). However, over the past three decades the lobster fishery has experienced much more boom than bust (Acheson 2003), as exemplified by a steady increase in landed weight and value since the mid-1990s (Figure 1). While this positive trend has been attributed to a combination of external factors, including the reduced abundance of predators and favorable environmental conditions (Acheson and Steneck 1997; Boudreau and Worm 2010; Steneck et al. 2011), the industry's success is often also attributed to its unique comanagement system and its ability to respond to crises and solve collective-action problems (Acheson 2003; Acheson and Gardner 2010; Wilson et al. 2013).

The 2012 crisis raises the question whether the industry will be able to respond to future threats as successfully as it has to ones in the past or whether it is less resilient now and thus no longer able to adequately respond to threats. We examined social resilience in the Maine lobster fishery in terms of the

Subject editor: Syma Ebbin, University of Connecticut, Avery Point Campus

*Corresponding author: anna.henry@maine.edu

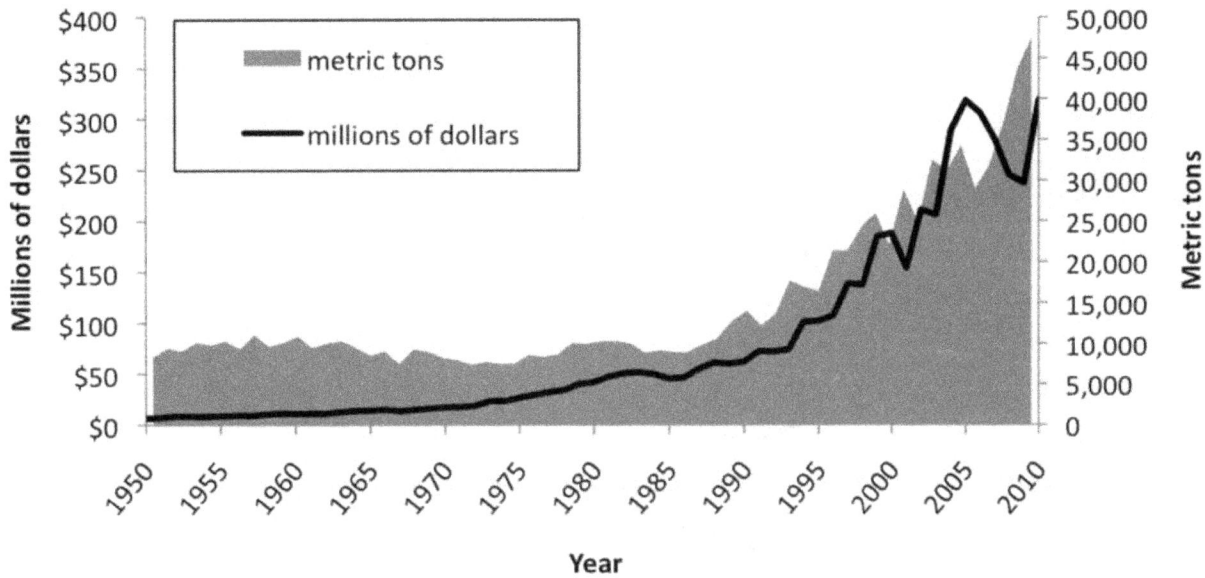

FIGURE 1. Maine lobster landings and value (Maine DMR 2012b).

specific threats facing fishermen and their ability to respond to these threats.

THEORETICAL PERSPECTIVE

Although the concept of resilience is pervasive in the ecological literature, we focus on social resilience or "the ability of groups or communities to cope with external stresses and disturbances as a result of social, political, and environmental change" (Adger 2000:347). It is difficult to discuss resilience without also addressing vulnerability, which we define as "the state of susceptibility to harm from exposure to stresses associated with environmental and social change and from the absence of capacity to adapt" (Adger 2006:281). Communities and individuals are more vulnerable if they are not able to adapt or if they are less resilient. Following a "people ecology" perspective, we sought to understand resilience by focusing on the differential threats faced by individuals and groups and their ability to respond to these threats (Vayda and McCay 1975; McCay 1978).

Threats can vary by frequency, intensity, and duration (Cutter 1996). Responses also vary in terms of the time that has to be invested and the magnitude of the adjustment necessary. We classify responses as *coping* (smaller, short-term reactions that can easily be reversed and modified as threats change) and *adaptations* (longer-term adjustments that require more investment and organization and are more difficult to alter in the future). The level of response is often determined by the condition of the threat. The theory of the "economics of flexibility" suggests that the likelihood and timing of these different response types relate to the depth of the threat (McCay 2002), i.e., that responses that require smaller investments (coping) will occur first, reserving some "flexibility" with which to respond to potential future threats or an intensification of the current threat. In this way responders ration their capacity for resilience, reserving *adaptation* responses for threats that are of larger magnitude or for use after lower-level *coping* responses have failed (McCay 2002:357).

METHODS

This paper is one component of a larger study assessing vulnerability and resilience in Maine fishing communities. As part of this project we conducted 18 semistructured (Bernard 2002) and 26 oral history interviews (Ritchie 2003) with fishermen, other community members, and government officials in four fishery-dependent communities in Maine from October 2010 to December 2011. These interviews lasted from 1 to 2 h and focused on the threats that fishermen have faced and the ways in which they have responded to these threats. We began our study with key-informant interviews and then relied on snowball sampling to broaden the representation until theoretical saturation (Bernard 2002). We selected respondents to represent the diverse marine fishery–related occupations in each community. The fishermen interviewed ranged in age from 34 to 80, with an average age of 54. While not all respondents were lobster fishermen, the majority of those interviewed held lobster licenses, representing zones A and D from the Canadian border to the midcoast region of Maine. All interviews were recorded, and all of the oral histories and a majority of the semistructured interviews were transcribed. Detailed notes were taken from other semistructured interviews. Three focus groups were held in June 2012 to gather more insight from fishermen and community members ($n = 13$) and to ground-truth some of the findings from the interviews; these sessions were recorded and detailed notes were taken from the audio files. We used QSR NVivo 9 to analyze data following a modified grounded theory approach that involved multiple

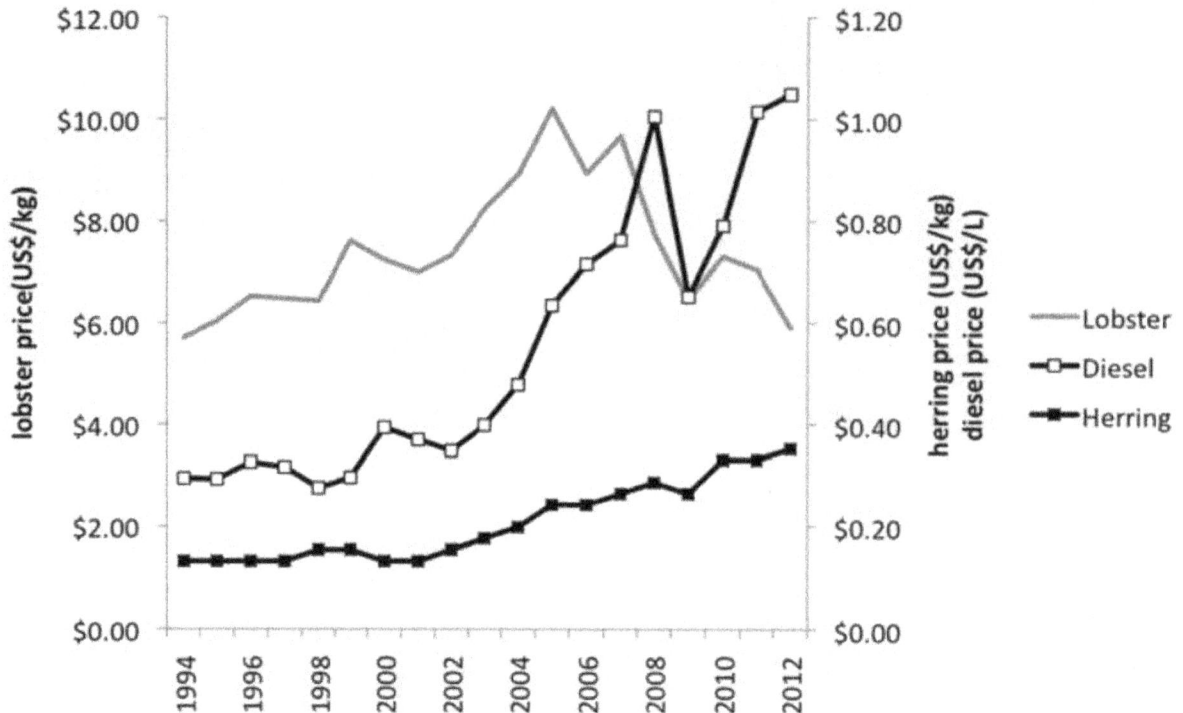

FIGURE 2.   Lobster and herring prices are exvessel value (Maine DMR 2012b). While this accurately reflects the average price a lobster harvester receives at the boat, the price that he/she would pay for herring as bait would be much higher than the exvessel value and may increase at a more rapid rate. For example, it has been reported that a barrel of herring bait (approximately 91 kg) cost $25 in 2000 and $150 in 2010 (Acheson and Acheson 2010). The price of diesel fuel is taken from the U.S. Energy Information Administration and may not reflect the actual price paid by harvesters. Additionally, all of these prices may vary based on the quantity purchased or sold. However, the trend of increasing expenses and decreasing lobster prices is clear.

iterations of coding (Glaser and Strauss 1967; Strauss and Corbin 1998). In this paper we draw on themes that emerged from this analysis along with news articles and notes from public meetings that occurred after completion of the interviews.

In our analysis, we concentrated on recent threats to the industry and the ways in which it has responded to them, focusing on three situations: a drop in lobster prices beginning in 2008, a recovery of prices in 2010–2011, and a second collapse of prices in 2012. We also considered other environmental and regulatory concerns identified by fishermen. We discuss the current level of vulnerability in the lobster fishery and what we have learned about its resilience based on responses to recent threats. We also note factors that may explain the level of social resilience in this fishery and why it has not been able to respond effectively to recent threats.

MANAGEMENT OF THE MAINE LOBSTER FISHERY

Lobster management in Maine relies on a combination of informal and formal institutions (Acheson 2003). Historically, access to the lobster fishery has been restricted informally through "harbor gangs," small groups of fishermen who "maintain a fishing territory for the use of its members" (Acheson 2003:24). Membership in a harbor gang is restricted, and territory is defended from incursion by adjacent gangs through

harassments ranging from verbal threats and abuse to the destruction of gear. Territoriality persists today, and reports of gear molestation from trap cutting to suspicious boat sinkings occur annually.

Formal management of lobster fishing has relied on effort controls (limited entry and trap limits). Enacted by the Maine state legislature in 1995, the Zone Management Law established a statewide trap limit, an apprenticeship program for new entrants, and a trap tag program that links traps to their owners. Additionally, the law created a formal comanagement system in the form of lobster management zones. Councils of fishermen elected by other license holders in the zone allow members to modify existing rules and propose new rules regarding trap limits and limited entry with a two-thirds majority vote. These proposals, if approved by the commissioner, are then transformed into state regulations by the Department of Marine Resources (DMR). As a result of this process, all zones have limits of 600–800 traps and six of the seven zones have restricted entry with varying entry-to-exit ratios.[1] This

---

[1]To remove effort from the fishery, most zones have created entry-to-exit ratios; a system where the number of new licenses issued in a year is dependent upon the number of licenses (more specifically trap tags) retired in the previous year. Licenses in the zone without entry-to-exit ratios have only increased by 2% from 1997 to 2011, while across all zones licenses have decreased 12% (Dayton and Sun 2012).

has helped to slow entry into the fishery and continue the sense of resource stewardship and empowerment of license holders. Lobster management also includes minimum and maximum size restrictions and a prohibition on the harvest of reproductive females. These measures, known as the double-gauge law and V-notch program, evolved through state legislation as a result of lobbying from the industry and are an example of successful collective action by this industry. Although these regulations were enacted by the legislature, they have long been supported by the industry because they follow fishermen's conservation logic, namely, to preserve lobsters during their most vulnerable stages of life (Acheson 2003:218).

2008 GLOBAL ECONOMIC CRISIS

In 2008, lobster fishermen faced a crisis with many of the same characteristics as the predicament faced in the summer of 2012. While the threats were similar—low prices, lack of market, and excess product—the mechanisms behind them were very different. In 2008, the dismal global economic climate created a chain of events resulting in a loss of market for Maine lobsters, which carried into much of the 2009 season.

As much as 70% of the lobsters landed in Maine are shipped to Canada for processing, reshipment, or export back to the United States (Steneck et al. 2011). Many of these processing plants were funded by Icelandic banks until October 2008, when the Icelandic banking system collapsed, pulling funding from the processors and forcing a majority to cease operations. The resulting reduction in demand forced Maine lobster prices down to record lows; as one lobsterman described the situation, "I made more money when I was 15 years old fishing out of a skiff" (Richardson 2010). Another described it as "an economic disaster the size of Katrina" (Lobster solutions hard to come by 2008).

Exacerbating the effect of these market pressures and low prices on lobster harvesters was the fact that bait and fuel prices were increasing simultaneously (Figure 2). One fisherman whom we interviewed put these price changes into historical perspective as follows:

> In 1994 ... we got $2.60 a pound [$5.73/kg] for our product, fuel was 70 cents a gallon [$0.18/L], and our bait was $8 a bushel [$0.12/kg]. Last year [2009] we got $2.60 a pound [$5.73/kg], the bait was $21 a bushel [$0.32/kg], and fuel was $2.89 [$0.76/L].[2]

The cumulative impact of these threats led the lobster industry to adjust their strategies in creative ways to keep their businesses afloat.

The most common responses by lobster harvesters were changes in fishing behavior to increase profits. Some increased fishing effort to increase their landings and compensate for the

low prices. Others limited the number of traps they hauled, hoping to reduce the glut in the market and drive the price up, or stopped hauling altogether because the prices they received were not enough to cover their expenses. These opposing strategies had little influence on overall landings and price, prompting some lobstermen to call for an industrywide tie-up to reduce production until prices recovered. However, due to a 1958 consent decree lobstermen are prohibited from refraining to harvest lobsters until a minimum price is reached and from compelling others to refrain from harvesting. Because Maine lobstermen are owner-operators, this type of organized tie-up violates U.S. antitrust laws related to collusion and price fixing. While an industry-led tie-up was prohibited, many fishermen hoped the state would intervene; however, the state has the authority to shut down the fishery only when it is required to protect the resource, not in response to economic conditions.

Unable to impact prices at the dock, many fishermen adopted strategies to reduce their expenses, such as dropping their sternmen and fishing alone. One fisherman describes his strategy to cut costs during 2008 as follows:

> We still kept fishing, but ... you probably didn't haul as often, you know, and you didn't run your boat so hard. You didn't burn as much fuel. You know, you just kept going. You just tried to ride it out and stay with it.

According to our interviews, such responses are customary among fishermen who are having a particularly bad year or experience unexpected circumstances that increase their costs (such as a large repair or maintenance issue or other unforeseen personal expenses). Many fishermen refer to "belt tightening," i.e., reducing their expenses and living off as little as possible as their strategy to get through tough times. In 2008, the belt tightening was industrywide.

Due to the widespread nature of the crisis in 2008, there were additional responses that were practiced industrywide or that required organization from multiple facets of the industry. The Maine Lobster Promotion Council worked with local grocery chains to run lobster promotions in order to increase demand as well as publicizing the crisis with an ad campaign geared toward local consumers intended to increase demand in the fall when the tourist season is over. The public sector also offered support. U.S. Senator Olympia Snowe organized meetings between industry members and federal agencies to develop processing plants in Maine. The U.S. Department of Agriculture Trade Adjustment Assistance program was created to provide training and financial assistance to farmers or fishermen who have been negatively impacted by foreign imports[3] and is assisting over 4,000 fishermen in New England

---

[2]Unattributed quotes are from interviews or focus groups conducted for this project.

[3]While the specific threat to the lobster industry in 2008 was related to a lack of export markets, the TAA program was developed to increase domestic production of seafood overall, which suffers from a $9 billion deficit (TAA 2010).

(Northeast lobstermen begin to realize benefits from the USDA's Trade Adjustment Assistance (TAA) program 2012). In October 2008, Maine Governor Jon Baldacci signed an executive order creating a Task Force on the Economic Sustainability of Maine's Lobster Industry to examine possible long-term solutions to buffer the lobster fishery from global economic conditions. The task force drafted a strategic plan that included recommendations to increase markets within Maine, improve product quality and adjust the timing of landings to maximize price, and create promotional opportunities for Maine lobster (Mosely Group 2009). While the report identified strategies to promote Maine lobster, implementation of the plan required $7.50–8.25 million of funding per year, a majority of which would be provided by industry assessments. The Maine Lobstermen's Association voted against such assessments, and due to the lack of industry support no legislation was drafted to implement the recommendations of the strategic plan.

RECOVERY IN 2010–2011

Despite the "crisis of 2008," interviews conducted in 2010 and 2011 indicate that most fishermen characterized the Maine lobster fishery as resilient. Market demand had increased, and while prices had not rebounded to their pre-2008 levels, they were rising (Figure 2). The opinion that the lobster fishery was financially viable was almost unanimous, as described by one interview respondent: "Last year [2010] them guys made a fortune off that lobster." Fishermen without lobster licenses who were interviewed during this period commonly expressed regret that they had not gotten licenses when they were available or had let their licenses lapse in favor of participating in other fisheries. These attitudes signify the perceived health of the industry and the importance of lobster as part of a diversified, resilient harvesting strategy. As one fisherman put it, "the only thing I'm missing is my lobster license, which is a big one. I'd like to have that card in my deck." Similarly, another fisherman stated, "it seems to me like in the last few years they've done pretty good. I mean they must be doing alright because they're all trading in their boats every year for bigger, newer boats."

While the overall perception was that the resource was doing well, anxiety about the future status of the fishery remained. Concerns during this period focused on both potential stock declines and market volatility. One fisherman explained his concern as follows: "In my opinion, lobster fishing has never been better than it has been in the last 15–20 years. How long is it gonna last? That's the big question. Nobody knows the answer." Even absent any reductions in stock abundance, the instability in prices gave rise to market concerns for the future:

> The stocks have been pretty much increasing. There were a couple lull years, oh, I would say around 2008, 2009 were down years a

TABLE 1. Comments illustrating lobster harvesters' perceptions of market-related problems in the industry.

"Lobster price has rebound some, but if that goes to the shit again at $2 a pound, which it can, you're screwed."

"We need to do a better job marketing our product."

"Our financial business plan is based on a $4 boat price. If we get below $4, it doesn't work." (Lobstering in 2010, 2010)

". . . if they'll do that and start processing all the Maine lobsters, don't let Canada have our lobsters for nothing, keep the lobsters in the United States, the price would go up."

". . . another thing that we should be doing is instead of lugging our lobsters to here and there, we should be buying them and processing them here. Take the meats. I mean we can now dissect our lobsters and take the meat and freeze it and can it, do this and that with it. It's something we could be doing in this community right here."

little bit, but since then the stocks have been increasing again. The stocks look really, really good for the future. . . . Now if the market doesn't adjust to it, it doesn't mean it's a good thing. I'd rather catch 1,000 lb [454 kg] at $5.00 a pound [$11/kg] than catch 3,000 lb [1,360 kg] at $2.00 a pound [$4.41/kg].

This uncertainty, combined with increased dependence on a single species, is frequently cited as a source of vulnerability for the industry's future (Steneck et al. 2011; Wahle et al. 2013).

There are approximately 7,000 license holders on the coast of Maine, many living in towns with limited alternative economic opportunities. One fisherman described the dependence on lobster in his town this way:

> So your lobster stocks collapse . . . this town's screwed because we're not diverse enough to handle something like that and probably in the '90s when it was diverse, it was scallopers, draggers, lobstermen, all of the above, and everybody made a living doing a little bit of everything, but now it's basically all their eggs are in lobstering except for a scattering few.

It is generally established that diversity leads to increased resilience in social–ecological systems (Folke et al. 2002). This increase in resilience occurs both during disturbances (when diversity reduces the impact of threats by spreading the risk) and after disturbances (because a diverse system has a greater capacity to respond to change) (Folke et al. 2002; Turner et al. 2003). The strategy of diversification is also utilized specifically by fishermen. Much like maintaining a stock portfolio with high- and low-risk investments, fishermen who harvest multiple species can buffer the effects of a price collapse or reduced catch of a particular species by shifting to other fisheries. This strategy is particularly important in an environment with high uncertainty, such as fishing, as it is difficult to predict future hazards (Berkes 2007). However, given increasingly restrictive regulatory environments, a diverse harvesting strategy is more difficult to employ, reducing resilience in many fisheries (Tuler et al. 2008; Murray et al. 2010). Historically, many fishermen in Maine utilized this strategy and targeted a diverse mix of species, harvesting what was abundant and in season. As many species such as groundfish began to decline and regulatory restrictions were strengthened, many fishermen transitioned to harvesting lobster or intensified their lobster harvesting efforts. This dependence on lobster has led some scholars to describe the Gulf of Maine as a lobster monoculture (Wilson et al. 2007; Steneck et al. 2011). Such a shift may have been a successful response to threats at the time, but the increased dependence on a single fishery has also reduced the current level of resilience in the fishing industry.

Despite the responses and adjustments made by the industry in 2008, many fishermen continued to refer to the vulnerabilities that had been identified that year and spoke to the need to increase the economic sustainability of the fishery (Table 1). Looking back, it appears that their concerns about market stability and lobster prices were extremely prescient, as these vulnerabilities were exposed again just 2 years later.

2012 ENVIRONMENTAL CRISIS

The summer lobster season of 2012 materialized in many ways as a déjà vu of 2008, with lobster prices falling dramatically. While the external price shock experienced by the fishermen was much the same, the mechanism creating the threat was very different, however. Unseasonably warm water temperatures caused a glut of soft-shell lobsters in the late spring and early summer—too early for the Maine tourist season and its associated markets and overlapping the period when Canadian processors are at full capacity processing domestic lobster. Because soft-shell lobsters do not package well, they cannot be shipped live, further reducing the available market. Additionally, soft-shell lobsters fetch lower prices, as they contain less meat than comparable hard-shell ones. These factors combined in the summer of 2012 to produce the lowest lobster prices in 30 years, down to an exvessel value of $1.35 per pound ($2.98/kg) in some ports (Seelye 2012).

Confronting threats similar to those faced in 2008, the industry unsurprisingly reacted with almost identical responses. Many lobstermen started to increase their fishing effort to make up for the lower prices. One fisherman stated, "I didn't used to need to come in with a huge haul to make a living.... Now I do" (Seelye 2012). Other lobstermen took the opposite approach and began tying up their boats, hoping that a reduction in supply would increase prices. Again, there were calls for an organized tie-up but the commissioner of the DMR said that the state could not shut down the fishery for economic reasons and would not tolerate any type of peer pressure, including cutting traps to encourage other boats to tie up (Lobstermen tying up their boats 2012).

While the DMR did not have the authority to intervene, the state government reiterated its support for the industry, creating a committee of Lobster Advisory Council members, processors, dealers, exporters, and industry representatives to "consider whether there are changes that could be made in the lobster fishery to improve the quality of the product landed and the profitability of the industry" (Maine DMR 2012a). The governor also announced that the state would investigate ways to encourage additional processing capacity in Maine. One new Maine processing plant shipped its first load of lobsters in August 2012 and is anticipating processing 4.8 million pounds (2.2 kg) per year when running at full capacity (Hall 2012). Additional plants opened in 2013, increasing Maine's overall processing capacity to 10–12 million pounds; approximately 60–70% of Maine lobster landings are still shipped to Canada for processing, however (Canfield 2013).

Although this new processing capacity will undoubtedly help reduce the dependence on Canadian processors, some say that it is putting the cart before the horse: "If you don't focus on marketing, [having] more processors in Maine is just going to force the price down" (Trotter 2012). Echoing the sentiment expressed in 2008 as well as the conclusions of the governor's task force of that year, many lobstermen feel that the focus needs to be on increased marketing and branding of Maine lobster as a product: "What we have that (Canadians) don't have is a great brand.... We just need to be innovative. In the U.S., the Maine brand is strong, there are huge untapped markets right here in this country" (Trotter 2012). The recent increase in processing capacity in Maine may facilitate this branding, as more product will be available that can be labeled "Maine made" because it will not have traveled to Canada for processing. The federal government has supported this marketing strategy, and Maine congresswoman Chellie Pingree was able to negotiate with two major cruise lines to commit to purchasing 8,800 lb (3,991 kg) of Maine lobster to be served to passengers on ships that visit Maine (McCracken 2012). While this is an important step that will help increase domestic markets and create demand for the product, it may be no more than a symbolic effort, as the order accounts for less than one one-hundredth of a percent of total lobster landings in Maine.

OTHER THREATS FACING FISHERMEN

In addition to the drop in prices in 2008 and 2012, our research identified other factors that threaten the resilience of the lobster fishery in Maine. First, the limited-entry system has made it difficult for many young people to obtain lobster licenses. One fisherman explained the situation as follows: "You gotta jump through hoops and breathe fire to get in the fishery now ... you have to apprentice and log days and hours to get on a waitlist and will be dead and gone before (you) ever gets into lobstering." This has many fishermen concerned about the future of the lobster industry as current fishermen begin to age and retire without a matching influx of young fishermen:

> I think we'll see a drastic dip in the number of people in the fishing business, because we've limited entry drastically. You take a town like this one that has 60 or 80 fishermen in it, half of those guys probably won't be in the business 20 years from now, and I don't see 20 or 30 [new entrants into the fishery] coming along.

Members of the industry are not the only ones concerned about the rising age structure of harvesters and the restrictions of the current limited-entry system. During the 125th legislative session, the Maine legislature directed the DMR to commission an independent evaluation of the costs and benefits of the limited-entry licensing system. The resulting report suggested that one deficiency of the system is the long waiting period for receiving a new license and that the current average tenure on the waiting list is 6 years (Dayton and Sun 2012).[4]

Additionally, fishermen are concerned about future environmental conditions that may threaten the industry. The sea surface temperatures of the northeastern continental shelf were higher in the first 6 months of 2012 than they have been in the last 70 years, and preliminary data suggest that this has affected temperatures throughout the water column, including bottom temperatures (Dawicki 2012). This has many fishermen worried that the Gulf of Maine could experience stock declines similar to those seen in southern New England in the late 1990s. One fisherman explained his concern about future environmental conditions this way:

> One of the things that I worry about more than anything else is environmental conditions because we've seen in Long Island Sound and places south of Cape Cod where the fishery can be wiped out almost overnight because of environmental factors—pollution, warm water.... I think Maine has always been protected because of its cold water.... I worry that if it warms up just a little bit, we're gonna have major problems.

The combination of warm water temperatures and an increased density of lobsters has led to concern that shell disease will be a threat in the future; although the prevalence of the disease has increased since 2010, it is still seen in less than 1% of the lobsters sampled in Maine.

DISCUSSION

The parallels between the 2008 and 2012 crises are difficult to ignore. Both were characterized by external threats that affected lobster prices and markets. While the drivers behind these threats differed—global economic and environmental conditions, respectively—the results were very much the

same. So what does all this mean in terms of the resilience of the Maine lobster fishery?

One would think that having experienced the crisis of 2008, the industry would have been better prepared to respond to the threats faced in 2012, but as Patrice McCarron, executive director of the Maine Lobsterman's Association stated, "Unfortunately, this summer's crisis revealed that little progress has been made since 2008" (MacPherson 2012). The fact that the industry was not more prepared to respond to a similar crisis 4 years later indicates that the lobster industry is not as resilient as we think.

We classify many of the responses to the crises of the past 4 years as *coping* strategies (short-term changes in behavior designed to withstand a perturbation) rather than *adaptation* (longer-term strategies that require larger investments and that are more difficult to reverse) (Tuler et al. 2008). In the past the lobster industry has been remarkably successful in responding to threats, enacting institutional changes from trap limits to V-notching, size gauges, and the zone council system (Acheson 2003). So why has it not adapted similarly to the new market pressures and historically low prices? We offer five possible explanations: (1) adaptation and institutional change take time, and because the current threats are relatively new the industry has not had adequate time to adapt to them; (2) the crisis is not perceived as extreme or imminent enough to require long-term adaptation; (3) the current management scheme is unable to adapt to these types of challenges; (4) the new economic threats are external and on a broader scale, requiring larger market-based responses that are outside the scope of the harvesting operation; and (5) the unpredictable timing and nature of these new threats has led to coping responses.

Adaptation Takes Time

The process followed by Maine lobster fishermen throughout the 20th century to devise the rules and institutions that currently regulate the Maine lobster fishery illustrates that institutional change and adaptation is a slow, complex process (North 1990). First, the lobstermen had to agree that changes were necessary and would have a positive impact on the fishery. Obtaining agreement in a fishery often characterized as fiercely individual can be an arduous task. One fisherman describes the difficulty in organizing lobster fishermen as follows: "You could have 3,000 guys agreeing on doing something and you have one guy saying, 'No, I'm not going. I'm gonna do it my way,' and the 3,000 will rapidly join him." Once consensus was achieved, the lobstermen had to convince the state legislature to enact legislation to create the institutions necessary to implement reform. This has not been a quick, easy process; it has required 70 years of evolving biological, social, and political conditions to create the current lobster management scheme (Acheson 2003).

The recent crisis in the lobster fishery has spanned just 5 years, and with the rebound in prices from 2010 to 2011 the

---

[4]While the report made numerous recommendations to remedy this, none have been implemented as of the publication of this article.

actual period in which the threats were experienced is even shorter. Therefore, it may be premature to expect the lobster industry to have devised new institutions that increase their resilience to these price shocks. Some fishermen are aware of the time required for this process. As one explained, "There is no quick fix to this. We do not need to overreact and act fast by putting in some regulation that just won't work in the long run" (Lobster solutions hard to come by 2008). The historical ability of the industry to devise institutions to adapt to variability in stock abundance generates confidence in its capacity, given adequate time, to adapt to the new threats and to create new institutions that will address the issues.

Level of Crisis

The level of a threat can be determined by its intensity, frequency, and duration (Cutter 1996), which in turn affect the magnitude of the response to the threat (Kasperson et al. 1995; Dow 1999). While the crises of 2008 and 2012 created difficult economic times for many in the lobster industry, the effects have been relatively short-lived. We recognize that the crises were catastrophic for lobstermen who lost their boats or who were forced out of the industry, but for those who were able to continue fishing our interviews showed that during the years 2010 and 2011 many still perceived the fishery to be doing well. Just 2 years after experiencing the crisis of 2008, fishermen spoke of threats to the lobster industry in vague, futuristic terms.

The perception that threats to the lobster fishery are not imminent or of high enough degree to require substantial adaptation is supported by the widespread success of the lobster fishery for the past 30–40 years. With increases in landings and relatively stable markets since the mid-1980s, many in the industry have not experienced significant misfortune. As one fisherman explained,

> The fishermen who are 40 and under have never known struggle. They really haven't.... All they've ever known for the last 20–25 years is ever-increasing catches, ever-increasing wallets, and they may think they have, okay, but they've never known struggle.... To me, struggling is no matter what you have for bills you can't catch enough lobsters to pay for 'em. And they've never known that.

The recent success of the fishery has left younger members of the industry without the "social memory" of strategies that have been successful in responding to threats in the past. Social memory is a key aspect of resilience, as it provides a wealth of information regarding the diversity of responses available to different threats and their likely outcomes (Folke et al. 2005). This lack of social memory reduces the resilience of the lobster industry, as it cannot utilize the "head start" in responding to threats that social memory provides.

This lack of experience with previous threats may also lead the industry to underestimate the level of current threats. Due to the recent positive trend in landings, there is little perception that there is any current threat to the abundance of Maine lobster stocks. One fisherman in 2011 described the status of the resource as follows:

> The ocean is full of lobsters, it's full of them. There's nothing to get 50 lobsters in a pair of traps; it's not keepers, you understand, but lobsters overall.... I've never seen that in any of my lifetime, so things look good in the lobster industry for a while if they all live.

Because of this, many fishermen believe that they will be able to compensate for lower prices by fishing harder and are therefore less likely to make long-term adaptive changes. Some fishermen recognize the futility of this strategy, however:

> Well, see, lobstermen have a bad business plan. When the price drops they go harder to try to make up for price difference, which you start to use more bait; so when you start to use more bait it increases the bait price and, as you go harder, you burn more fuel and then you start catching even more lobsters that even drives the price of lobsters even further down, so it's not a really good business plan.

Intensifying pressure on the resource may not be a good business plan, but the fact that it is seen as a viable strategy when times get tough may be one factor impeding the institutional change necessary for long-term sustainability in the market.

Previous institutional changes in the lobster fishery have followed crises that were perceived as significant, imminent threats. The transition of the industry from one in which harvesters had a "pirate ethic" and violated laws for personal gain to one with a "conservation ethic" that promoted sustainable regulations and compliance has been attributed to the catastrophic stock collapse in the 1930s, which caused 30% of lobstermen to leave the industry (Acheson and Gardner 2010). After years of division, this crisis shocked the industry, catalyzing the transition to more sustainable regulations and industrywide understanding of the importance of complying with those regulations. Perhaps the price drops of 2008 and 2012 have not been catastrophic enough for the industry to realize institutional change.

Noneconomic threats are also perceived as future threats, but they do not appear urgent to a majority of the industry. Environmental changes, such as shell disease and consistent changes in water temperature have yet to substantially impact harvest levels and are therefore deemed to be less urgent. A shock to the system of the same magnitude as the stock collapse of the 1930s may be necessary before real institutional change will occur.

Adaptability of the Current Management Scheme

The level of vulnerability of the lobster industry after the shocks of the last 4 years remains to be seen. Some of this will hinge on future market and environmental conditions that are unknowable, but it will also be determined by the flexibility of the institutions that regulate the industry. At a recent meeting at which lobstermen were asked to list elements of the current management system that are working, many fishermen found it difficult to think of a bright spot. One fisherman summarized it as "the whole thing's broke." The comanagement system of the lobster fishery relies heavily on harvester participation, which has declined in recent years. Fishermen who have attended meetings say that as a whole the industry is "apathetic until after the fact" and that opinions are not voiced until after policy decisions have been made. This may be due to fatigue, as lobstermen have attempted to change regulations in the past only to meet interference from the state. One fisherman described the process as "good intentions go in and garbage comes out."

One of Elinor Ostrom's[5] design principles for long-enduring institutions is that "external authorities do not challenge the rights of appropriators to devise their own institutions" (Ostrom 1990:101). The state has ultimate authority over which industry proposals are adopted in the form of regulations, and while this provision lends organization and authority to the process, in its current form it may not prove responsive enough to industry needs. If the institutions stay rigid and change does not occur, the industry could be at a precipice, where threats that were previously absorbed become catastrophic (Holling 1986). However, if the system remains flexible, the lobster industry may increase its resilience and ability to adapt to recent and future threats.

Perhaps the zone council system is "not adaptive to industry," as some fishermen have stated. This alone does not preclude the ability of the industry to respond to current threats. One aspect of resilient systems is that a disturbance or threat "has the potential to create opportunity for doing new things, for innovation, and for development" (Folke et al. 2005:253). The lobster industry could demonstrate its resilience by devising new ways of responding outside of the zone council or legislative system. This type of response may require additional leadership or political entrepreneurs to initiate reform. Political entrepreneurs are people who "do more than work for the public good; they also offer information, expertise, and public resources" and "are the means by which [the rules] are negotiated" (Acheson 2003:72, 79). They can be the catalyst required to generate new rules and institutions to respond to threats. As the fishing community ages, there seems to be less interest in spending the time required to make the connections and persuade the right people to make regulatory changes. This lack of leadership reduces social resilience and the ability to initiate new responses to change. Without a new generation of political entrepreneurs to take the reins, it may be difficult for any new system of management, or increased flexibility of current management schemes, to come to fruition.

---

[5]Nobel Prize winner and noted researcher of common pool resource institutions.

Broad-Scale Threats Require Response at a Matching Scale

The current threats facing the industry exhibited by the price drops of 2008 and 2012 are large-scale, external threats; therefore, any successful response must be at an appropriately broad scale. When the global economy falls into a recession that affects lobster prices or water temperature changes the temporal distribution of landings, there are few response options available to the individual harvester. As one lobsterman stated, "The only thing you can do is tighten your belt up and keep on fishing." Individual or collective action from harvesters cannot respond adequately to an external threat of this scale.

Collective action in the lobster fishery is akin to a prisoner's dilemma, and although the industry has collectively devised institutions to respond to threats in the past these threats were largely internal and related to taking collective action for conservation (Acheson and Gardner 2010). Threats of this nature are at a scale to which individual fishermen can respond, as their behavior has a direct effect on stock abundance, resulting in an effective, tight feedback loop that links their response to its effect on the threat. This feedback loop is a key factor in stimulating responses (Berkes 2002). As a threat broadens in scale, the feedback loop becomes less coupled, decreasing the motivation for response. Because the price threats in the lobster fishery are at a broad scale, this feedback loop is less tightly related to the harvesters' responses; therefore, a larger-scale response is required, such as one by the state. However, while the state allows for collective action with respect to conservation (e.g., the double-gauge law and V-notching), it prohibits it with respect to economic objectives (i.e., coordinated tie-ups), further impeding the ability of lobster fishermen to respond to economic changes such as those experienced in 2008 and 2012.

Nested scales of management increase resilience in complex social–ecological systems (Ostrom and Janssen 2004). When multiple scales of management exist, there is a variety of responses available to address threats within the system. The current management institutions in the lobster fishery are appropriate to respond to threats to the resource, but in order to respond to broader, market-based threats new, larger-scale institutions are required. There is evidence that toward the end of the 2012 season the industry acknowledged the need for large-scale market-based responses. In the fall of 2012, the industry was exploring license surcharges of $3 million annually that would fund promotional efforts aimed at expanding local, regional, and global markets (Schreiber 2012). This support for a larger-scale response is an encouraging sign for future resilience in the lobster industry.

Unpredictable Threats Lead to Coping Responses

One reason coping responses have been the main ones thus far is the unpredictable nature of these new threats. Collective action and new institutions were successful in responding to past threats because those threats were predictable and consistent. It was much easier to foresee the threats of increased effort in the fishery and the harvesting of short lobsters than it is to anticipate when market prices will drop. The source, extent, and timing of the more recent threats to the lobster industry are unpredictable. Environmental changes may be easier to respond to if there are gradual changes such as consistent, incremental shifts in the species' distribution or a slow, linear increase in the prevalence of shell disease. But if these changes occur in an abrupt, unpredictable manner with no evidence of changing future conditions, the responses will be equally abrupt and unplanned.

There is a general recognition that the good times cannot last forever. This is exemplified by the ways in which lobstermen describe the future: "There's going to be a huge, huge catch of lobsters for a few years now. But that can be reversed real quick." "No one says lobsters have to stay alive; they could die just as quick as they come." "It's good right now, but it's not going to stay that way." While the uncertainty of the future is almost universally acknowledged, there is no general recognition of what the specific threats will be or when they will occur. Because the threats are unpredictable, reliance on coping responses may be the most logical strategy.

The current resilience in the lobster industry may be best described by (1) the attitude among harvesters that they need to be prepared for future threats, whatever they may be, and (2) the coping responses they are implementing to deal with a less predictable future. This attitude is reflected by a reduction in purchases of new boats. As one lobsterman described the situation, "There's not many boat builders making lobster boats right now because … people aren't buying." Another fisherman described the way he is changing his behavior due to this unpredictability as follows: "The way the fishing is now, I can't see myself doing the tricks I used to do, trade trucks every year, I mean. I think I'm gonna have to keep what I've got." While large-scale, institutional responses will help increase the resilience to future market threats, smaller coping strategies appear to be the best option now available to increase the resilience to other unpredictable, external threats in the future. It remains to be seen whether the resilience created by these coping responses will be adequate to withstand future threats.

ACKNOWLEDGMENTS

This research was funded by Maine Sea Grant (R-10-04), NOAA Saltonstall–Kennedy Grant Program (NA10NMF-4270207), and the Maine Agriculture and Forest Experiment Station (Hatch Grant 08005-10). Focus groups were funded in part by the Sustainability Solutions Initiative supported by funding from National Science Foundation award EPS-0904155 to Maine EPSCoR at the University of Maine. This paper was prepared for a symposium called The American Lobster in a Changing Ecosystem: A U.S.–Canada Science Symposium, held in Portland, Maine, in November 2012.

This paper would not have been possible without the help of all of the fishermen who were willing to spend time in interviews and focus groups. The authors also thank Chris Bartlett from Maine Sea Grant and the University of Maine Cooperative Extension for his assistance with interviews and focus groups and the Cobscook Bay Resource Center for assistance with recruiting interview participants. The authors also acknowledge Cameron Thompson, who conducted some of the interviews and assisted with data analysis, and Jessica Jansujwicz, who helped with the focus groups. We thank James Wilson and three anonymous reviewers for valuable comments and suggestions on previous versions of this manuscript.

REFERENCES

Acheson, J. M. 2003. Capturing the commons: devising institutions to manage the Maine lobster industry. University Press of New England, Lebanon, New Hampshire.

Acheson, J. M., and A. W. Acheson. 2010. Factions, models, and resource regulation: prospects for lowering the Maine lobster trap limit. Human Ecology 38:587–598.

Acheson, J. M., and R. Gardner. 2010. The evolution of conservation rules and norms in the Maine lobster industry. Ocean and Coastal Management 53:524–534.

Acheson, J. M., and R. S. Steneck. 1997. Bust and then boom in the Maine lobster industry: perspectives of fishers and biologists. North American Journal of Fisheries Management 17:826–847.

Adger, W. N. 2000. Social and ecological resilience: are they related? Progress in Human Geography 24:347–364.

Adger, W. N. 2006. Vulnerability. Global Environmental Change 16:268–281.

Berkes, F. 2002. Cross-scale institutional linkages: perspectives from the bottom up. Pages 293–321 in E. Ostrom, T. Dietz, N. Dolsak, P. C. Stern, S. Stonich, and E. U. Weber, editors. Drama of the commons. National Academies Press, Washington, D.C.

Berkes, F. 2007. Understanding uncertainty and reducing vulnerability: lessons from resilience thinking. Natural Hazards 41:283–295.

Bernard, H. R. 2002. Research methods in anthropology: qualitative and quantitative approaches, 3rd edition. Altamira Press, Walnut Creek, California.

Boudreau, S. A., and B. Worm. 2010. Top-down control of lobster in the Gulf of Maine: insights from local ecological knowledge and research surveys. Marine Ecology Progress Series 403:181–191.

Canfield, C. 2013. Maine moves to capture lobster-processing market. Associated Press (July 5). Available: http://bigstory.ap.org/article/maine-moves-capture-lobster-processing-market. (January 2014).

Cutter, S. L. 1996. Vulnerability to environmental hazards. Progress in Human Geography 20:529–539.

Dawicki, S. 2012. Sea surface temperatures reach record highs on northeast continental shelf. National Oceanic and Atmospheric Administration Ecosystem Advisory 2. (18 September 2012).

Dayton, A., and J. Sun. 2012. An independent evaluation of the Maine limited-entry licensing system for lobster and crab. Report of Gulf of Maine Research Institute to Maine Department of Marine Resources, Boothbay Harbor.

Dow, K. 1999. The extraordinary and the everyday in explanations of vulnerability to an oil spill. Geographical Review 89:74–93.

Folke, C., J. Colding, and F. Berkes. 2002. Synthesis: building resilience and adaptive capacity in social–ecological systems. Pages 352–387 in F. Berkes, J. Colding, and C. Folke, editors. Navigating social–ecological systems: building resilience for complexity and change. Cambridge University Press, New York.

Folke, C., T. Hahn, P. Olsson, and J. Norberg. 2005. Adaptive governance of social–ecological systems. Annual Review of Environment and Resources 30:441–473.

Glaser, B., and A. Strauss. 1967. The discovery of grounded theory: strategies for qualitative research. Aldine, Chicago.

Hall, J. 2012. Lobster processing: young Mainer to open seafood plant. Portland Press Herald (August 13). Available: http://www.pressherald.com/news/getting-lobsters-from-trap-to-table_2012-08-14.html. (August 2012).

Holling, C. S. 1986. The resilience of terrestrial ecosystems: local surprise and global change. Pages 292–317 in W. Clark and R. Munn, editors. Sustainable development of the biosphere. Cambridge University Press, Cambridge, UK.

Kasperson, R. E., J. X. Kasperson, B. L. I. Turner, K. Dow, and W. B. Meyer. 1995. Critical environmental regions: concepts, distinctions, and issues. Pages 1–41 in J. X. Kasperson, R. E. Kasperson, and B. L. Turner, editors. Regions at risk: comparisons of threatened environments. United Nations University Press, Tokyo.

Lobster solutions hard to come by. 2008. Commercial Fisheries News. Available: http://www.fish-news.com/cfn/editorial/editorial_11_08/Lobster_crisis-Solutions_will_come_hard.html. (February 2015).

Lobstering in 2010. 2010. Free Press (January 7). Available: http://freepressonline.com/main.asp?Section ID=52&SubSectionID=78&ArticleID=4574. (February 2015).

Lobstermen tying up their boats. 2012. WMTW-TV, Portland, Maine (July 12). Available: http://www.wmtw.com/news/money/Lobstermen-tying-up-their-boats/-/8791814/15499618/-/oskfp7z/-/index.html?absolute=true. (August 2012).

MacPherson, I. 2012. A road map for the future. Landings: News and Views from Maine's Lobstering Community (October 15). Available: http://mainelandings.org/tag/lobster-council-of-canada/. (October 2012).

Maine DMR (Maine Department of Marine Resources). 2012a. Lobster Advisory Council's profitability and lobster quality committee meeting. Maine DMR, Augusta.

Maine DMR (Maine Department of Marine Resources). 2012b. Historical Maine landings data. Available: http://www.maine.gov/dmr/commercialfishing/historicaldata.htm. (August 2012).

McCay, B. J. 1978. Systems ecology, people ecology, and the anthropology of fishing communities. Human Ecology 6:397–422.

McCay, B. J. 2002. Emergence of institutions for the commons: contexts, situations, and events. Pages 361–401 in E. Ostrom, T. Dietz, N. Dolsak, P. C. Stern, S. Stonich, and E. U. Weber, editors. Drama of the commons. National Academies Press, Washington, D.C.

McCracken, C. 2012. Cruise ship buys local lobsters on waterfront: congresswoman Pingree thanked for new business opportunity. Bangor Daily News (September 12). Available: http://bangordailynews.com/community/cruise-ship-buys-local-lobsters-on-waterfront-congresswoman-pingree-thanked-for-new-business-opportunity/. (October 2012.)

Mosely Group. 2009. Maine lobster industry strategic plan. Report for Governor's Task Force on the Economic Sustainability of Maine's Lobster Industry, Augusta.

Murray, G., T. Johnson, B. J. McCay, M. Danko, K. St. Martin, and S. Takahashi. 2010. Creeping enclosure, cumulative effects, and the marine commons of New Jersey. International Journal of the Commons 4:367–389.

North, D. C. 1990. Institutions, institutional change, and economic performance. Cambridge University Press, New York.

Northeast lobstermen begin to realize benefits from the USDA's Trade Adjustment Assistance (TAA) program. 2012. Fishermen's Voice 17:11–12.

Ostrom, E. 1990. Governing the commons. Cambridge University Press, New York.

Ostrom, E., and M. A. Janssen. 2004. Multilevel governance and resilience of social–ecological systems. Pages 239–259 in M. Spoor, editor. Globalisation, poverty, and conflict. Kluwer Academic Publishers, Dordrecht, The Netherlands.

Richardson, J. 2010. Mainers gobbling lobster to keep industry afloat. Portland Press Herald (March 13). Available: http://www.pressherald.com/archive/mainers-gobbling-lobster-to-keep-industry-afloat_2008-10-28.html. (October 2012).

Ritchie, D. A. 2003. Doing oral history: a practical guide. Oxford University Press, New York.

Schreiber, L. 2012. Lobster industry aims bigger on marketing. Working Waterfront (November 14). Available: http://www.workingwaterfront.com/articles/Lobster-Industry-Aims-Bigger-on-Marketing/15081. (December 2012).

Seelye, K. Q. 2012. In Maine, more lobsters than they know what to do with. New York Times (July 28). Available: http://www.nytimes.com/2012/07/29/us/in-maine-fishermen-struggle-with-glut-of-lobsters.html?_r=1&hp. (October 2012).

Steneck, R. S., T. P. Hughes, J. E. Cinner, W. N. Adger, S. N. Arnold, F. Berkes, S. A. Boudreau, K. Brown, C. Folke, L. Gunderson, P. Olsson, M. Scheffer, E. Stephenson, B. Walker, J. Wilson, and B. Worm. 2011. Creation of a gilded trap by the high economic value of the Maine lobster fishery. Conservation Biology 25:904–912.

Strauss, A., and J. Corbin. 1998. Basics of qualitative research: techniques and procedures for developing grounded theory. Sage, Thousand Oaks, California.

TAA (Trade Adjustment Assistance for Farmers). 2010. U.S. lobster exports & the U.S.–Canada relationship. Available: http://taatrain.cffm.umn.edu/LobsterMrktOV/Default.aspx?SectionID=194. (October 2012).

Trotter, B. 2012. Maine ponders industry strategy as lobster protests continue in Canada. Bangor Daily News (August 8). Available: http://bangordailynews.com/2012/08/08/business/maine-ponders-industry-strategy-as-lobster-protests-continue-in-canada/. (February 2015).

Tuler, S., J. Agyeman, P. P. da Silva, K. R. LoRusso, and R. Kay. 2008. Assessing vulnerabilities: integrating information about driving forces that affect risks and resilience. Human Ecology Review 15:171–184.

Turner, B. L., R. E. Kasperson, P. A. Matson, J. J. McCarthy, R. W. Corell, L. Christensen, N. Eckley, J. X. Kasperson, A. Luers, M. L. Martello, C. Polsky, A. Pulsipher, and A. Schiller. 2003. A framework for vulnerability analysis in sustainability science. Proceedings of the National Academy of Sciences of the USA 100:8074–8079.

Vayda, A. P., and B. J. McCay. 1975. New directions in ecology and ecological anthropology. Annual Review of Anthropology 4:293–306.

Wahle, R. A., A. Battison, L. Bernatchez, S. Boudreau, K. Castro, J. H. Grabowski, S. J. Greenwood, C. Guenther, R. Rochette, and J. Wilson. 2013. The American lobster in a changing ecosystem: a U.S.–Canada science symposium, 27–30 November 2012, Portland, Maine. Canadian Journal of Fisheries and Aquatic Resources 70:1571–1575.

Wickenheiser, M. 2012. Maine lobstermen reeling from low prices, seeking cooperation from dealers. Bangor Daily News (July 5). Available: http://bangordailynews.com/2012/07/05/business/maine-lobstermen-reeling-from-low-prices-seeking-cooperation-from-dealers/. (December 2012).

Wilson, J. A., J. M. Acheson, and T. R. Johnson. 2013. The cost of useful knowledge and collective action in three fisheries. Ecological Economics 96:165–172.

Wilson, J., L. Yan, and C. Wilson. 2007. The precursors of governance in the Maine lobster fishery. Proceedings of the National Academy of Sciences of the USA 104:15212–15217.

14

To Adapt or Not Adapt: Assessing the Adaptive Capacity of Artisanal Fishers in the Trondheimsfjord (Norway) to Jellyfish (*Periphylla periphylla*) Bloom and Purse Seiners

Rachel Gjelsvik Tiller*

SINTEF, Fisheries and Aquaculture, Strindvegen 4, 7034 Trondheim, Norway; and Department of Sociology and Political Science, Norwegian University of Science and Technology, Høgskoleringen 1, 7491 Trondheim, Norway

Jarle Mork

Department of Biology, Norwegian University of Science and Technology, Høgskoleringen 1, 7491 Trondheim, Norway

Yajie Liu

SINTEF, Fisheries and Aquaculture, Strindvegen 4, 7034 Trondheim, Norway

Åshild L Borgersen

Department of Biology, Norwegian University of Science and Technology, Høgskoleringen 1, 7491 Trondheim, Norway

Russell Richards

University of Queensland, St. Lucia QLD 4072, Brisbane, Australia

Abstract

Worldwide increases of jellyfish has occurred during the last several decades. A dense population of a large scyphozoan jellyfish, *Periphylla periphylla*, has established itself as top predator in the Trondheimsfjord in Norway, impacting traditional fisheries. On this background we discuss the adaptive capacity of artisanal fishers and stakeholder involvement in environmental management. A serendipitous discovery was that fishers report that their capacity to adapt to the presence of jellyfish in fact was sufficient. What they could not adapt to, within the context of jellyfish proliferation, was top–down decisions from the national government allowing purse seiners into the fjord to harvest Sprat *Sprattus sprattus* and Atlantic Herring *Clupea harengus* rest quotas and thereby also large bycatches of the local codfishes. This harvest was perceived more detrimental to their fishery than was the jellyfish invasion. Relative to fisheries management's choice of regulatory mechanisms during times of climatic change, we argue that by involving stakeholders intimately, the resulting policy advice will be experienced bottom–up and, thus, more legitimate and serendipitous results of a critical nature are more likely to surface.

Subject editor: Syma Ebbin, University of Connecticut, Avery Point Campus

*Corresponding author: rachel.tiller@sintef.no

The last decade has seen mass occurrences of jellyfish blooms globally (Brotz et al. 2012), undesirable effects being reported by commercial fishermen in, among others, Japan, the Mediterranean, South America and Norway (Uye and Ueta 2004; Quiñones et al. 2013; Palmieri et al. 2014; Tiller et al. 2014)[1]. Scientists have speculated that the warming of the oceans is a vital component in the success of these jellyfish blooms (Kawahara et al. 2013). The blooms affect artisanal fishers, i.e, fishing, per Johnson (2006), that is "anchored in household and community based social and economic organization." These small-scale artisanal fisheries occur in the inner Trondheimsfjord and are locally anchored with small catches that are both for consumption and small-scale sale. Impacts in local communities include loss of income, increased hazard to safety, and a rising fear of fishing being eradicated in areas where the infestation is particularly great (Quiñones et al. 2013). The increase in jellyfish populations is also having a negative influence on fish stocks because they prey on the spawn of commercially important fish species. This increased pressure on fish stocks brought about by increased jellyfish numbers is troublesome given that fishing efforts are also increasing, driven by the increased global demand for seafood (Pauly et al. 2002) coupled with population growth and the future need for food security from the marine sector (Garcia and Rosenberg 2010). At the same time, global commercial fish stocks are declining after years of overfishing. Estimates are that 24–36% of global wild fish stocks have collapsed and that 68–72% are overexploited or collapsed (Worm et al. 2006; Pauly 2007; Pauly 2008; FAO Fisheries and Aquaculture Department 2010). Additionally, the European Union through its distant water fleet (DWF) has transferred infrastructure subsidies to developing nations to gain access to their exclusive economic zones (EEZ). These subsidies, however, also contribute to overcapacity in the host country by reducing fishing cost and thereby are contributing to overfishing (Barkin and DeSombre 2013; Le Manach et al. 2013). However, there is a political priority in many nations, including Norway, to protect the traditional industries of coastal communities, specifically the cultural heritage of what some consider remnants of former hunting and gathering past, namely artisanal and commercial fishing (Barnard 1983; Barkin and DeSombre 2013; Ministry of Trade Industry and Fisheries 2013).

This coupling of declining fisheries, natural system uncertainty including jellyfish blooms, and increased stress on marine areas means that fisheries managers are regularly faced with making difficult management decisions while weighing social and ecological concerns against each other in a political setting (Bunnefeld et al. 2011; Tiller et al. 2014). Anticipating the effects of these decisions on

the entire socio-ecological system is difficult given that management decisions are introduced into complex contexts with humans, the environment and economy interacting at multiple temporal and spatial scales. The adaptive capacity of stakeholders to respond to changes to the socio-ecological system is sometimes difficult to foresee by managers, and often there are outcomes that management does not anticipate that can have critical effects on stakeholder groups. Within this context, we discuss the management of increasing jellyfish populations globally. We also examine a local artisanal fishery affected by high concentrations of *Periphylla periphylla* and a declining population of Atlantic Cod *Gadus morhua* (hereafter, "cod").

We argue that the coupling of quantitative economic and biological data with qualitative stakeholder data will give a more complete picture of the impacts of stressors, such as jellyfish blooms, from a global perspective down to the impact on local communities (e.g., Trondheimsfjord, Norway), in line with a significant and growing body of literature on stakeholders and co-management (e.g., see www.sciencedirect.com and search words "stakeholder" or "co-management"). The importance of this inclusive approach is to go beyond a top–down approach to include local knowledge and understanding and to incorporate stakeholder adaptive capacity, or social resilience, when assessing their vulnerability to emerging ecological stressors (e.g., jellyfish blooms) and how they might affect fisheries and the cultural heritage of the artisanal fishing in the area. Furthermore, involving stakeholders can also uncover serendipitous discoveries of importance that also affect the stakeholders' ability to adapt to new situations. According to stakeholder theory, on which we ground our findings, this is critical in management approaches because giving management advice is viewed more legitimately from the vantage point of the stakeholders most directly affected when they have been involved in the process. The results of this involvement is therefore that compliance to regulatory changes, as an indicator of the institutional, as opposed to environmental, effectiveness of such measures (Zürn 1998; Kütting 2000; Tiller, R. G. 2010), are more likely to happen (Österblom et al. 2011). In looking at the occurrence of increased *P. periphylla* population and the flow-on effect on cod stocks and the fishermen, we therefore explored the interrelation among cod, jellyfish and fishing based on a time series data of cod and jellyfish in the Trondheimsfjord in concert with fishermen questionnaires—coupled with information from Råfisklaget about actual price changes of cod in the last decades. We also used systems thinking and Bayesian belief networks (BBNs) to map stakeholder perceptions of causality in light of the theory of stakeholder participation. This information in turn was used to determine how stakeholders can be used in research situation, and how their real life information correlates with biological data on the same topic.

Here, we first give the theoretical and methodological backdrop of the topic, followed by background information about

---

[1]For thorough interdisciplinary background information on the case of the fishermen in the Trondheimsfjord specifically, consult Quiñones et al. (2013) and Tiller et al. (2014).

the growth of *P. periphylla* population globally in the context of multiple drivers (e.g., climate change and warming oceans) and a overview of the situation in the Trondheimsfjord, Norway. We then present a time series of biological data for a specific geographical location in Trøndelag, Norway, where the perceptions of local fishermen on the developments of jellyfish in the area, and their adaptive capacity, have been recently explored through participatory workshops (Tiller et al. 2014). This information is coupled with follow-up questionnaires to the most active (fulltime) fishermen in the area where they have provided the data based on their own logbooks and observations with regards to the changes in catch and catch composition in the last decades. This information is then combined with aggregated data from the Directorate of Fisheries (The Norwegian Fishermen's Sales Organization; "Norges Råfisklag") about the changes in landings and prices for cod in the same period and area. The discussion focuses on perceptions of the effect of inclusive governance from the vantage point of the stakeholders in response to these biological changes and how theory expects a bottom–up approach to coastal management to affect the compliance to management decisions by creating legitimacy in the process. We also present the serendipitous finding that the stakeholders perceive their adaptive capacity to jellyfish to be high; however, their adaptive capacity and the cod population have been more detrimentally affected by a top–down decision to allow purse seiners into this ecologically protected fjord system to harvest "rest quotas" (i.e., the allowance of ocean going fishing vessels have to transfer up to 20% of their quota for a given species to another vessel in their company or to a collaborator within the same vessel group, provided the vessel has fished >30% of their quota prior to selling it to another vessel; Lovdata.no 2005). With both jellyfish and purse seiners preying on the fish stocks targeted by these artisanal fishermen, they feel like they are caught between a rock and a hard place, and are unable to adapt.

THEORY

The complexities of fisheries management arguably necessitates socio-ecological integration, which has shown to lead to stakeholder trust and legitimacy in fisheries management decisions, and an improvement in the rate of compliance as well (Österblom et al. 2011). A stakeholder in general has been defined by the literature as "any group or individual who can affect, or is affected by, the achievement of the organization's objectives" (Freeman 2010). This is a broad definition and leaves the concept of having a stake, or invested interests, unequivocally open to include virtually anything, any topic, and the jurisdiction of a given stakeholder open to anyone. We distinguish between the management and the engagement of stakeholders, referring to their involvement in the actual decision making process. In management, we are looking at persuasive strategies, the mapping of groups of importance, and the assignment of importance to those stakeholders that are in need of attention. In engagement, we are referring to a strategy of involving the stakeholders in the decision making process and making them real participants with mutual responsibility in the results, rather than just recipients of attention. It is therefore necessary to look closer at stakeholders to determine who they are and to what degree they are affected by an objective, such as jellyfish invasions in the current case.

Searches for the keyword "stakeholder" or "co-management" through an online database shows greatly increasing results of studies on these topics over the last 2 decades (www.sciencedirect.com). There may be multiple reasons for this trend, but we argue that this increase in academic interest in the topic of stakeholders is due to a lack of stakeholder compliance to top–down decisions (including in the context of marine-based management). In Europe, for example, the failures of the Common Fisheries Policy (CFP) through a lack of profitability and a plenitude of overfishing, has been blamed on a lack of decision-making transparency and legitimacy with stakeholders, which has created a lack of compliance (Österblom et al. 2011). The lack of stakeholder compliance is of great interest from a political science perspective. Governments of democratic nations continuously make laws that influence stakeholder groups by shaping and regulating them. In order for these groups to be able to influence this system, however, they must abide by the rules and regulations established by the political and social setting within which they operate. In other words, the political culture wherein stakeholder groups manoeuvre is a reflection and a reinforcement of the political context thereof.

Norway is considered a state that is open to a variety of different interest groups accessing management, though there is no pretence of all groups having equal access to power. Organizations in this system are the link between their members and the government and actively participate in committees that are set up, whether they are advisory or permanent. Ultimately, however, they may still be overrun by a strong government in the decisive phase, resulting in decisions being made that are still contrary to what the members of the organization may have preferred (Dryzek et al. 2003). Participation by stakeholders in the committee system of the national government is one of the areas of inclusion that can be found at the heart of Norway's structure, especially in the case of the Ministry of Trade, Industry and Fisheries. These committees involve representatives from a variety of interest groups, as well as politicians and administrators. Often the goal is to produce a report on an issue, to be later used in preparation of parliamentary proposals, which would be commenced with having an open hearing. Having a case out on a hearing means that the government would like comments from affected stakeholders to a proposal they are working on at the moment, and the background for it would normally be to map out potential economic and administrative consequences of a given decision (Ministry of Trade Industry and Fisheries 2007), giving the stakeholders a voice and encouraging legitimacy and transparency. The incorporation of stakeholder groups throughout the decision-making process in the fisheries sector is therefore intense in Norway, and many groups are involved at different levels and at different times. This is in line with findings showing that Norway has had some success with regards to stakeholder trust and legitimacy in fisheries management, as demonstrated by an improvement in the rate of compliance (Österblom et al. 2011). Findings also show though a difference between fisheries organizations in Norway, the main fisheries organization, The Norwegian Fisherman's Association ("Fiskarlaget"), having greater power and influence with the government than does the smaller organization of coastal fishers, the Association of Coastal Fishers ("Norges Kystfiskarlag"; Tiller 2008).

METHODS

Given that stakeholder feelings of policy legitimacy and compliance is of critical importance to the sustainable development of fisheries globally, we used an integrated approach of systems thinking and Bayesian Belief Network (BBN) modeling in developing the stakeholder-driven scenarios and gaining critical insight into the adaptive capacity of the stakeholder group. Systems thinking as a method was used to develop shared mental models of the system, as perceived by the stakeholders involved. This step provides a conceptualization of the system based on the given stakeholder group-level beliefs and experiences and helps identify potential drivers and consequences in the context of the study (i.e., the management of increasing jellyfish populations globally, and in the case study, *P. periphylla* concentrations coupled with declining cod populations and the effect this has on commercial fishermen in the area). This systems thinking process also helps in identifying important elements within the system conceptualization that have influence over, or are influenced by, other elements within the same system. A benefit is that it allows exploration of a complex system at the local scale (in this case, the Trondheimsfjord) based on the expertise of the stakeholders themselves. The system conceptualization was also used to identify and select a priority issue that was further explored using BBN modeling and which uncovered other aspects of the issue. This priority issue represented an element or theme that emerged from the system thinking process that the stakeholders believed strongly about (i.e., *P. periphylla* in Trondheimsfjord). Vensim, a software specifically designed for systems modeling and developed by Ventana Inc. (Vensim.com), was used to develop the system conceptualization during each workshop.

There is a strong motive for engaging with stakeholders in order to access the expertise that they possess (i.e., knowledge-based data), which is characteristically strongly qualitative. For example, the fields of climate change adaptation and resource management have strong human dimensions and therefore draw heavily upon this knowledge-base. However, quantifying this narrative-rich knowledge base for the purpose of making management decisions (e.g., adaptive management scenario testing) is difficult. On these grounds, BBN modeling was selected as the methodological framework for further exploration of the priority issue. In addition, it was chosen because it facilitates participatory modeling and is well-suited to representing causal relationships between variables in the context of variability, uncertainty and subjectivity. Furthermore, BBN modeling is a method that is extremely well suited for coalescing knowledge into a single modeling framework, even if the knowledge comes from a variety of sources (e.g., stakeholders) and is of a variety of completeness. It is particularly effective in eliciting stakeholder opinion through participatory engagement for two reasons. First, the visual aspect of developing the causal maps that characterize Bayesian network models are easily understood and readily accomplished (as confirmed in our experience) by the stakeholders. The impact of this should not be understated because this fosters trust during the stakeholder engagement process. Second, the robust mathematical framework of Bayes theory underpins these models. This aspect, while not necessarily obvious to the stakeholders, provides a mathematical basis for incorporating the beliefs of the stakeholders into the model, something that traditional statistical approaches (e.g., null hypothesis testing) does not allow. It has also demonstrated an ability to use subjective expert opinions to both derive the structure of, and variables within, a BBN (Richards et al. 2013).

The methodological process of developing BBNs through stakeholder engagement is outlined in detail elsewhere (Richards et al. 2013; Tiller et al. 2013). Briefly, however, the structure of a BBN is a network of nodes that are connected by arcs. Each node is treated as a variable and therefore must have more than one state (e.g., if car color is the variable, then the states could include white, red, blue, etc). Furthermore, these states must be mutually exclusive (a variable can only have one state at a time), consistent (i.e., the states must relate to the same variable), and exhaustive such that the states cover all possibilities (e.g. for car, the variable color would require all possible colors be assigned as individual states, or alternatively, the states defined in a way that covers all possibilities: white cars and not white cars). Arcs connect variables and show the direction of causality through the direction of the arrow at the end of the arc. This direct connection between variables represents conditional dependence, which is a fundamental tenet of Bayes theory upon which BBNs are based. Feedback pathways are not allowed in Bayesian networks and therefore the entire network must be acyclical (i.e., one direction of causality). The implications for this constraint include the inability to model the influence of reinforcing (positive feedback) or balancing (negative feedback) pathways on the system being modeled. Such feedback pathways are important for understanding the temporal evolution of a system (i.e., how it changes overtime) and how it might respond to perturbations (Sterman 2000). While there are techniques that can enable feedback pathways in BBNs, these can quickly lead to cumbersome models with a large amount nodes, even for very simple feedbacks (Kjrulff and Madsen 2008). If the purpose of a model is to explore the role of feedback pathways in governing temporal dynamics, then other modeling methodologies such as systems dynamics (Sterman 2000) would be more appropriate to use than Bayesian statistical modeling. However, in our modeling, we are interested in using a methodology that allows straightforward integration of multidisciplinary (environmental, social, and economic) variables, accommodates expert opinion as a data source, and allows models to be developed even when data are relatively scarce. Furthermore, in our work we are focused on scenario analysis (i.e., what if?) where changes in conditions (new evidence) may be used to update our prior understanding of an event (e.g., the priority issue in our model) to posterior understandings. These ideals are well matched by the attributes of BBNs.

The other main component of the BBN is the set of conditional probability tables (CPTs) that quantitatively define the conditional dependence between linked nodes. In the workshop setting outlined in this paper, the perceptions of the stakeholders are used to populate these CPTs with probabilities, quantifying their beliefs about the relative importance of different variables within the network. The underlying probabilistic framework (i.e., Bayes theory) provides a mechanism of directly integrating social, economic, and environmental variables within a single model (Kjærulff and Madsen 2008).

During the workshops used in this study and elsewhere (Richards et al. 2013; Tiller et al. 2013), development of the structure of the BBNs is a group-level exercise. That is, it represents the group-level belief about which variables are included and how arcs connect them. Therefore, this process typically requires negotiation between the stakeholders. Conversely, each stakeholder populates the CPTs with their probabilities, providing individual-level parameterisation. The individually parameterised BBNs can then be combined into a single model because they share the same structure but have different CPTs. This is achieved here by using an auxiliary variable (Kjærulff and Madsen 2008), which weights each of the individual stakeholder CPTs so that the beliefs of one stakeholder can be given more weighting in the model than others. For this study the stakeholders were weighted evenly. Finally, the BBN-development process facilitates the capture of further information through the discussions that accompanied the development of these networks, that narrative providing important context to the importance of different variables during the workshops.

In terms of the time series of cod versus jellyfish, we used the aggregated data from the National statistics for Trøndelag area (including Sør- and Nord-Trøndelag) to illustrate the changes in fisheries resources over the last decade. To capture the changes in cod and jellyfish from this time series and their

associated effects on fishermen and their livelihood, a structured questionnaire survey was conducted among the small-scale commercial fishermen in the inner Trondheimsfjord for cod fishing. The questionnaire were divided into several sections, including basic fishing information (e.g., fishing area, gear, and season), economic components (catch and catch composition), price and cost, views on *P. periphylla* effects, and socio-demographic characteristics of fishermen. The respondents were also given the opportunity to provide commentary to questions. The questionnaire was provided by mail to all the fishermen in the inner Trondheimsfjord, and 50% replied, most with relatively complete answers.

JELLYFISH

The fishermen in the Trondheimsfjord have reported increased jellyfish blooms affecting their fisheries. These blooms, however, are not only a local problem in the Trondheimsfjord. They have also become an increasing global problem in the last few decades (Purcell 2007; Brotz et al. 2012). A number of stressors that include natural ecological fluctuations, anthropogenic activity (e.g., eutrophication; Arai 2001; Richardson et al. 2009), overfishing, habitat modification, chemical pollution, and introduction of exotic species in the marine environment (Hay 2006; Purcell 2007; Richardson et al. 2009) are suggested causes of these blooms. Climatic changes that alter temperatures, and nutrient fluxes also favor jellyfish; they therefore often strike in ecosystems that are out of balance (Lynam et al. 2005; Hay 2006; Purcell 2007; Halpern et al. 2008). The most important direct negative consequences of jellyfish blooms are economic losses, which include reduced tourism in affected areas due to stinging danger. It can also reduce fish catches of artisanal and commercial fishers due to damage to net gear, stinging danger and the resultant longer working hours required to clean and fix fishing nets (Quiñones et al. 2013), fish mortality due to stinging, oxygen deprivation in the aquaculture industry, and blocking of water inlets of power plants (Hay et al. 1990; Båmstedt et al. 1998; Hay 2006; Purcell 2007). These negative consequences can lead to large economic losses through reduced profits and increasing costs, especially for fisheries (Graham et al. 2003; Quiñones et al. 2013), as well as to the whole fishing industry at a sector level (Kim et al. 2012; Nastav et al. 2013).

The reason for their immense impact on the fishing industry is that they are gelatinous zooplankton, including both medusa of the phylum Cnidaria and planktonic members of the phylum Ctenophora (Brotz et al. 2012). They are therefore more resilient than fish in a changing world owing to a suite of attributes they possess that enable them to survive and thrive in disturbed marine environment (Richardson et al. 2009). They are furthermore nonvisual predators, seeking prey without using eyesight. This gives them a great advantage over other predators, like fishes, in waters with reduced light penetration such as during increased spring-flood river run-off or general pollution. *P. periphylla* is naturally distributed in all world oceans in waters with a wide range of temperatures, and like other scyphozoans, it can tolerate low oxygen concentrations better than most fishes. Thus, the annual cycles in light, temperature, salinity and oxygen saturation that are typical of Norwegian fjords may occasionally favor jellyfish over the fishes. Especially in periods of reduced abundance of fish, whether due to natural stock fluctuations or overexploitation, scyphozoans like *P. periphylla* can utilize their superior reproduction capacity (spawning throughout the year), longevity (>30 years) and recruitment.

Different jellyfish species will be affected differently by changes in the marine environment, though. Temperate species that come under stress from higher temperatures will increase in abundance, be able to overwinter and have longer reproductive seasons, which combined will result in larger populations. Tropical species, however, will be stressed under higher temperatures and therefore shift their distribution towards cooler waters and have shorter active seasons (Purcell 2005). Many species of jellyfish bloom more frequently than others (Purcell 2007) with some areas of the world experiencing more devastating consequences of jellyfish blooms than other areas. There is one famous example of a jellyfish causing the collapse of a whole ecosystem as well as the entire fishing industry in the area, namely the introduction of *Mnemiopsis leidyi* to the Black Sea in the 1980s. *Mnemiopsis* sp. was introduced to an already unstable ecosystem (mainly due to pollution and overfishing), and with its extreme ingestion and reproductive rate along with the lack of predators, it was not long until it had outcompeted all other species in the ecosystem. Since the 1980s, *Mnemiposis* sp. has been a nuisance in the Mediterranean and plagued large areas of Western Europe (Gershwin 2013:55–75). In Japan and China, blooms of giant jellyfish *Nemopilema nomurai* (which can reach a diameter of 2 m and a weight of 200 kilos) have caused devastating consequences for the fishing industry. *N. nomurai* blooms in spring in the East China Sea and is then transported with the currents into the Sea of Japan where it dies off in winter. Blooms of the *N. nomurai* jellyfish are extensive. In just one day, >3 million jellyfish can pass the Tsushima Strait. Fishing gear in these areas are torn apart and ruined, their poisonous tentacles sting the fishermen, the fish catch is minimal and working hours are longer (Uye 2011). In Qingdao, China, several deaths have also been reported from contact with this jellyfish (Purcell 2007). A third example of jellyfish blooms comes in the shape of the toxic *Pelagia noctiluca*, which has been a nuisance for tourists in the Mediterranean Sea and the Adriatic Sea, where the tourism industry suffer great economic losses due to stinging danger (Purcell 2005).

In India, furthermore, invasions of jellyfish have been a nuisance for the fishing industry as well as popular beaches in Palolem in Canacona, Utorda in Salcete, Miramar in Tiswadi and some other North Goa beaches (Fernandes 2012). Pollution and permanent parking of vessels in the Mandovi river of Panaji is also giving rise to toxic jellyfish blooms, causing problems for the fishing industry (Nagvenkar 2012). The moon jelly *Aurelia aurita* has also caused problems for fisheries, as well as power plants and aquaculture around the world (Mills 2001). In China, increased number of marine construction where polyps settle and decreased currents (retention) in bays has lead to a higher than usual abundance of moon jellies (Dong et al. 2012). Generally speaking, with increasing size of the predatory jellyfish the predation rate increases (Purcell and Arai 2001), so one can imagine that 200-kg jellyfish appearing in millions in a given area can cause severe impacts on the ecosystem and fishing industry.

TRONDHEIMSFJORD: COD VERSUS JELLYFISH

In Norway the financial losses related to jellyfish have been primarily the aquaculture and fishing industries. Aquaculture has had heavy losses due to jellyfish such as *Apolemia uvaria* and lion's mane *Cyanea capillata* clogging the fish cages and stinging the gills of the fish, causing suffocation and mortality (Båmstedt et al. 1998). The traditional fisheries experience, however, has centered around high densities of jellyfish, specifically the helmet jellyfish *Periphylla periphylla*, which has been clogging their nets, stinging fishers, and preyed on both the larval stages of cod and Atlantic Herring *Clupea harengus* as well as the food of these fish, the redfeed, thereby reducing the catches for the artisanal fishers in the area dramatically over the years.

Helmet jellyfish have established and thrived in many Norwegian fjords in recent years. It has gradually become a predominant species in the inner Trondheimsfjord ecosystem for the last decade (Solheim 2012). These jellyfish have caused a series of problems to the ecosystem and marine resources that coastal fishermen in the areas depend on for their livelihood. Trondheimsfjord is the third longest and seventh deepest fjord in Norway. The innermost part of the fjord (our study area) is divided into three main basins, namely Beitstadfjord, Verrasund, and Verrabotn (Figure 1). It used to be a relatively self-sustained and functional ecosystem containing a number of marine species and resources. Currently, the most important marine resources for the coastal fishermen are cod and Saithe *Pollachius virens*. Emerging species like European Pollack *Pollachius pollachius* and crab have also gradually become important to fishermen as an income supplement in light of declining stocks of the former.

Trondheimsfjord supports a local, self-recruiting cod stock, which traditionally has been the keystone species in the ecosystem and has sustained local fishermen for their livelihood for centuries (Dahl 1899; Mork, Reuterwall et al. 1982; Mork et al. 1985). Outside the spawning season cod are dispersed throughout the fjord, but aggregate on the spawning grounds in the innermost parts of the fjord in spring (March–May), when the annual spawning fishery takes place (Dahl 1899). Local artisanal fishermen use 30–35 ft (9.14–10.67 m) coastal fishing

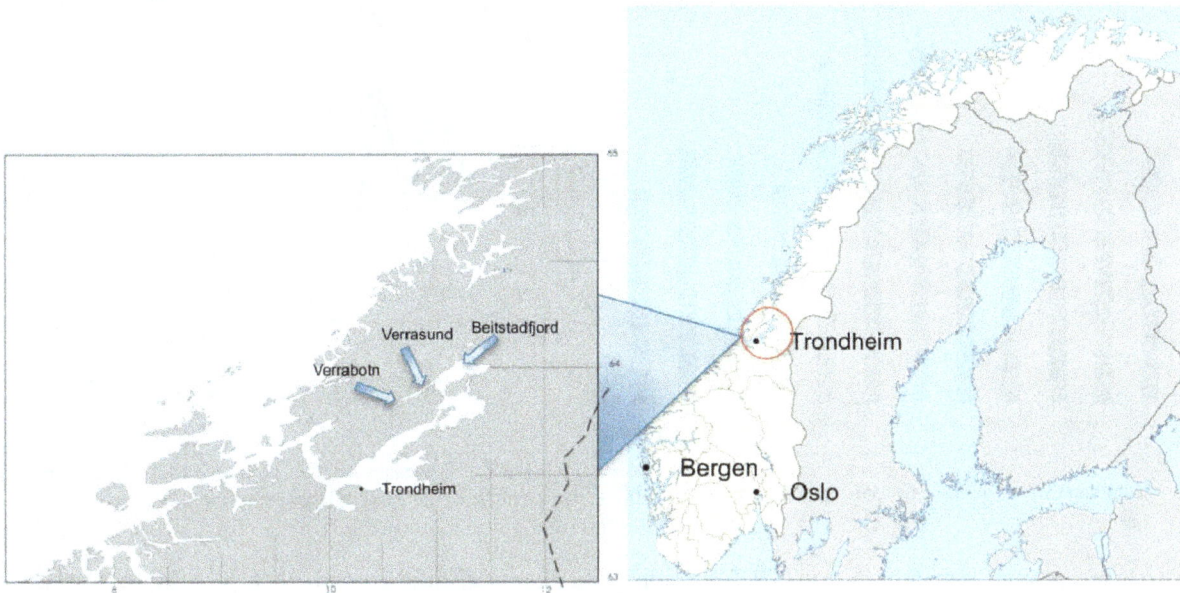

FIGURE 1.   Location of the three fishing areas in the innermost basin of the Trondheimsfjord, in central Norway, where the impacts of jellyfish on stakeholders were examined.

vessels with conventional and low-tech gears such as gill nets to harvest cod. Data from the National Statistics for Trøndelag area (including Sør- and Nord-Trøndelag) illustrate the changes in fisheries resources over the last decade (Figures 2, 3), showing that cod and Saithe are the predominant species in terms of catch and value, although overall they have shown a gradual declining trend in the last few of years. European Hake *Merluccius merluccius* is the most valuable fish species in terms of monetary value, while Saithe gets the market's lowest price. The opposite is true when it comes to quantities caught, as expected. The price gap between cod and Saithe, however, has become smaller, owing to a declining price in cod and increasing price in Saithe (Figure 4). However, cod is still the fishermen's favorite species to catch, according to the responses to the questionnaires in the current study.

The Trondhjem Biological Station (TBS) has collected samples and data from the study area for cod and *P. periphylla* in spring and autumn for the past two decades. The results (Figure 5) clearly reveal that the catch per unit effort (CPUE) of cod has drastically fallen down (solid line) while the CPUE of *P. periphylla* has sharply increased (dotted line) in the same period. This contrasting development suggests that *P. periphylla* possibly may have had negative impacts on cod, and certainly on the fishing patterns of the local fishermen. The visible drop in the spring sample of 2013 is based on a documented mass death of *Periphylla*. The standing crop of *Periphylla* in the fishing areas Verrasundet and Verrabotn has been quite variable as measured by CPUE. Those locations are mainly "fed" with *Periphylla* drifting from the main population in the Beitstadfjord and can probably not themselves sustain large *Periphylla* populations over time. Those two locations are too shallow for

the dial vertical migration requirements of *Periphylla* and too small and meagre with respect to prey for the jellyfish.

Documentation exists on more or less periodic mass deaths of the jellyfish at those two locations. The results of mass deaths have been detected after the winter, possibly pointing to starvation in the cold part of the year when the abundance of prey organisms are reduced in the fjord (Solheim 2012). As measured by CPUE, the mass death in Verrasundet/Verrabotn during winter 2010–2011 wiped out more than 90% of the autumn standing crop there. The mass death was also confirmed on bottom video using a remotely operated vehicle (the ROV "Minerva" of

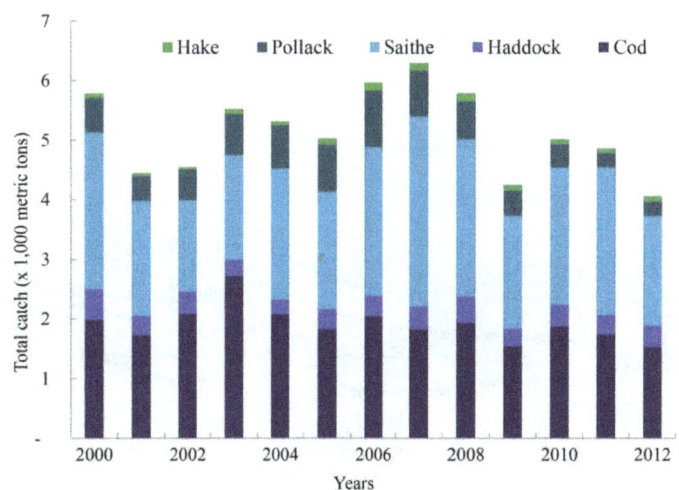

FIGURE 2.   The catches of the major commercial fish species in Trøndelag, Norway.

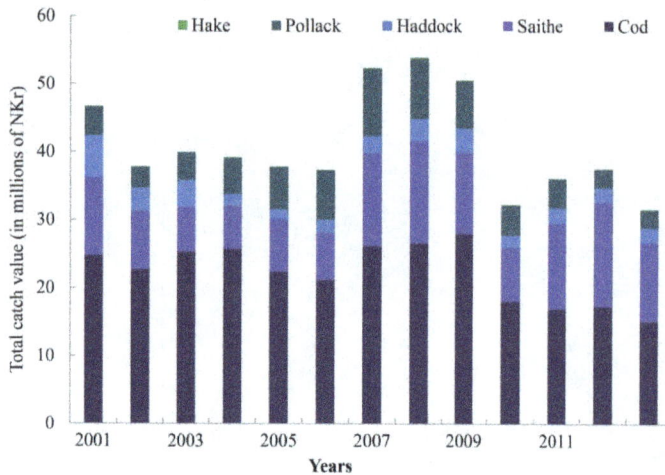

FIGURE 3.   The nominal values of major fish species in Trøndelag, Norway.

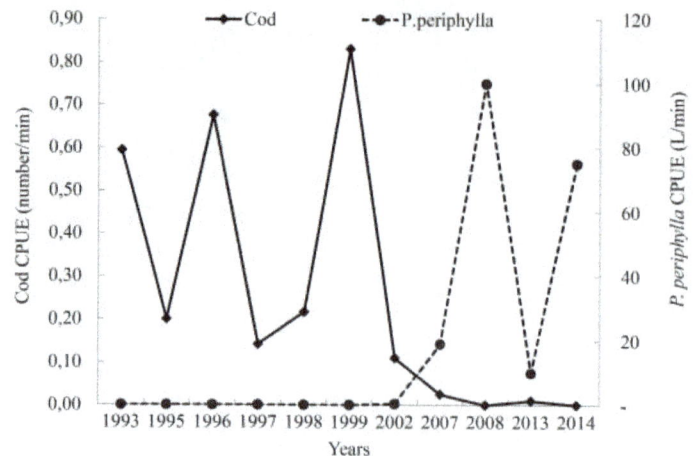

FIGURE 5.   The catch per unit effort (CPUE) of cod versus the jellyfish *Periphylla periphylla* in the Trondheimsfjord, Norway.

NTNU). Also in recent years, a marked drop in CPUE of Periphylla in Verrabotn and Verrasundet after the winters has been observed, although not as severe as in the winter 2010–2011.

RESULTS

The mass deaths aside, the results of the questionnaires sent out to these fishermen were that they had observed *P. periphylla* blooms dating back as far as to the 1998–1999 fishing season. The catch composition in fishermen's catches has changed substantially since then (Figure 6), the catch of cod decreasing from over 60% of the total catch in 2000 to about 30% in 2012 while Saithe increased from 20% to 50%. European Hake and European Pollack have also shown increasing trend since they receive better price in the market due to increasing demand.

In line with the observations of the fishermen, bottom trawl data suggests that *P. periphylla* established itself in the inner

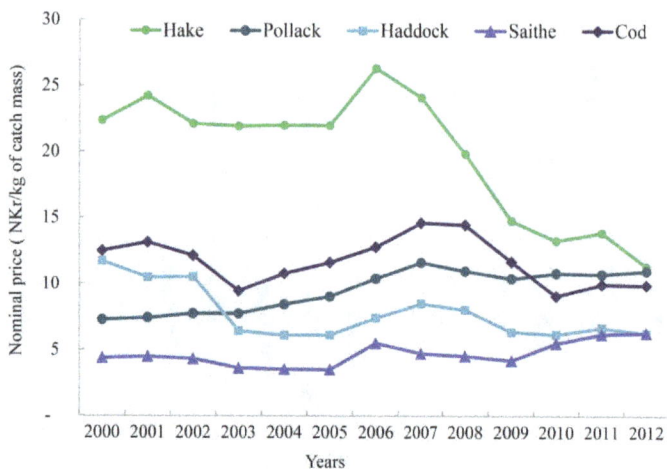

FIGURE 4.   The nominal prices of major commercial fish species in Trøndelag, Norway.

Trondheimsfjord around 1999, probably first in Beitstadfjord, where *P. periphylla* blooms were observed earlier than in other parts of the inner fjord. Some of the fishermen who traditionally used Beitstadfjord as an important fishing ground, reported that they responded to the jelly problem by partly shifting their fishing efforts away from the Beitstadfjord to the more coastward neighbouring location near the Ytterøya Island, where the jelly density was much lower and, hence, less problematic. In general, however, the fishermen indicated that the overall effect of *P. periphylla* on their fishing activities was relatively significant. The questionnaires distributed indicated that the total income from cod fishing had been reduced over the last decade, although *P. periphylla* was not attributed as the primary factor. The main reason was the market price for cod and their increasing cost of fishing in general. However, the their increased fishing cost was partially due to *P. periphylla*, which caused fishermen to go farther out in their fishing zones and spend longer hours in sea and required more time cleaning and repairing nets. The fishermen also pointed out that they would leave the jellyfish-affected fishing areas to go somewhere else to fish, if management would provide alternatives for them, and otherwise would likely have to find alternative income to compensate for the loss (e.g., blue mussel farming). This indicates that fishermen in the area have gradually accepted that they will have to adapt to the situation if management cannot mitigate the jellyfish problem.

Overall, the fishermen did not indicate being worse off because of the income loss from cod fishing, since it had been compensated by the income from other activities like increasing opportunities for emerging species, like crab and pollack or mussel farming. For instance, some fishermen indicated that only half of their current income came from fishing. They further mentioned that they have considered selling their fishing vessels and permits if *P. periphylla* continues to be a problem and alternative options for income become less available, albeit is fishing is their part of their preferred lifestyle, which

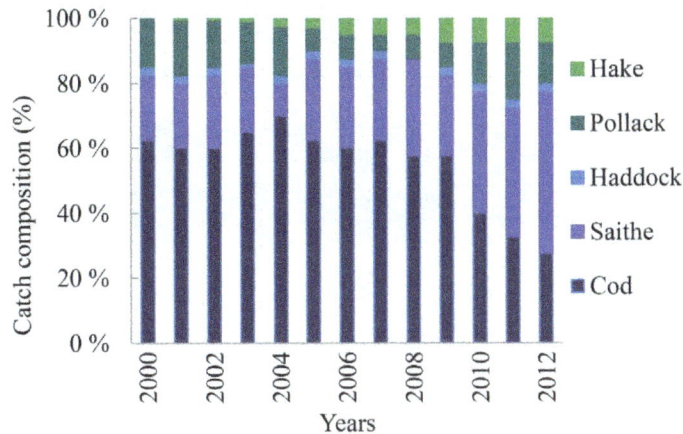

FIGURE 6. Fish species catch composition in the Trondheimsfjord, Norway, from 2001–2012.

they want to continue as long as they can sustain their livelihood. This is in line with ethnographic fieldwork elsewhere in Norway as well, where similar views are expressed (Broch 2013). However, the fishermen generally perceived the future fishing in the Trondheimsfjord as not very promising. They believe that policy and management can help improve their fishing situation, but they did not support new fishing regulations would potentially restrict their fishing activities even further. These different attitudes indicate that the fishermen do not believe policy and management can change the jellyfish situation, but they do need financial support for maintaining their fishing activities, despite the fact that there are fewer fish each passing year. They stressed that if *P. periphylla* can be explored and used for commercial values, however, they would be willing to adapt to this venture, and harvest these instead, if the opportunity arises. This is also in line with the observation of Broch (2013), who quotes a fisher stating, "If some fish stocks disappear there will always be something else to fish."

The questionnaires largely mirrored the group interviews done with the same group of fishermen, where systems thinking and BBN were the tools of investigation. They echoed the issue of *P. periphylla* taking up room in the fishing nets that would otherwise go to commercially desired species like cod. This had caused them to have to travel further to reach enough fish to harvest, increasing fishing cost as well as possible opportunity cost due to the loss of fishing grounds. They also discussed the challenge of having to spend hours cleaning the nets after use, which was both straining and time consuming, as well as hazardous to the fishermen cleaning them. During the last years, one of them had two 3-week medical leaves after getting jelly slime in his eyes. Another effect was that the quality of the fish caught in the net was also reduced due to scarring from jellyfish burns and jelly pigments, which in turn made them less valuable at landing sites. The increased weight of nets filled with jellyfish furthermore heightened the chance of the boat rolling over, which could have dramatic

consequences for the fishermen involved. They furthermore feared that dead and decaying *P. periphylla* that sunk to the bottom would absorb available dissolved oxygen. Overall, the ecological, economic, and social ramifications associated with *P. periphylla* becoming established in the fjord lead the fishermen to believe they might have to find other work, or end up unemployed, which was similar to what they expressed in the later questionnaires.

These causal pathways were reflected in the system conceptualization (Figure 7), and it all centred around jellyfish and the financial ramification of this new player in the local ecosystem.

The BBN modeling process was not only to aid the stakeholders in discussing their adaptive capacity to *P. periphylla*, but also to provide for mitigation options, including political mitigation options (or as presented to the stakeholders, "where something could actually be done"; i.e., the bottom nine variables in Figure 8). This is also the area of the program where the policy maker can make changes and see how these actions play out in projections of other outcomes further up in the network. The starting point for the BBN modeling was the selection of a priority issue from the systems thinking processes (Figure 7), and it was expected that the stakeholders would centre their perceptions around the issue area of the negative implications jellyfish had and would continue to have on their system and how to alleviate this problem. The stakeholders first framed their priority issue of income in their experiences with *P. periphylla* and (under the guidance of the researchers) discretised (the process of categorising the node into discrete states) this priority issue by allocating a desirable and undesirable state to it, which in this case became "liveable" (desirable) and "nonliveable" (undesirable).

Following this, the stakeholders selected a set of primary-level variables that they felt directly influenced their ability of attaining an income that was liveable. They framed their selections around three themes: (1) if the fish biomass was high (fish biomass), (2) if the landing sites were local (fisheries landing sites) and (3) if the commercial harvest of *P. periphylla* was profitable (*harvest Periphylla commercially*) and respectively discretized these with states of high versus low, local versus not local, and profitable versus not profitable. The stakeholders then assigned the secondary-level variables that they believed had direct influence on the primary-level variables using the context of where "something could be done" and assigned two states to each of these; the resulting BBN is presented in Figure 8. We then ran a sensitivity analysis following the results of the conditional probability table, as applied to the BBN model, to formally tests the sensitivity of the BBN (using the priority node as the reference point) to changes in the variable settings.

The priority node (income) was most sensitive to fish biomass, as expected from data on availability of fish stocks declining along with the increase in jellyfish in the area. The importance of fish biomass appears to be a result of these

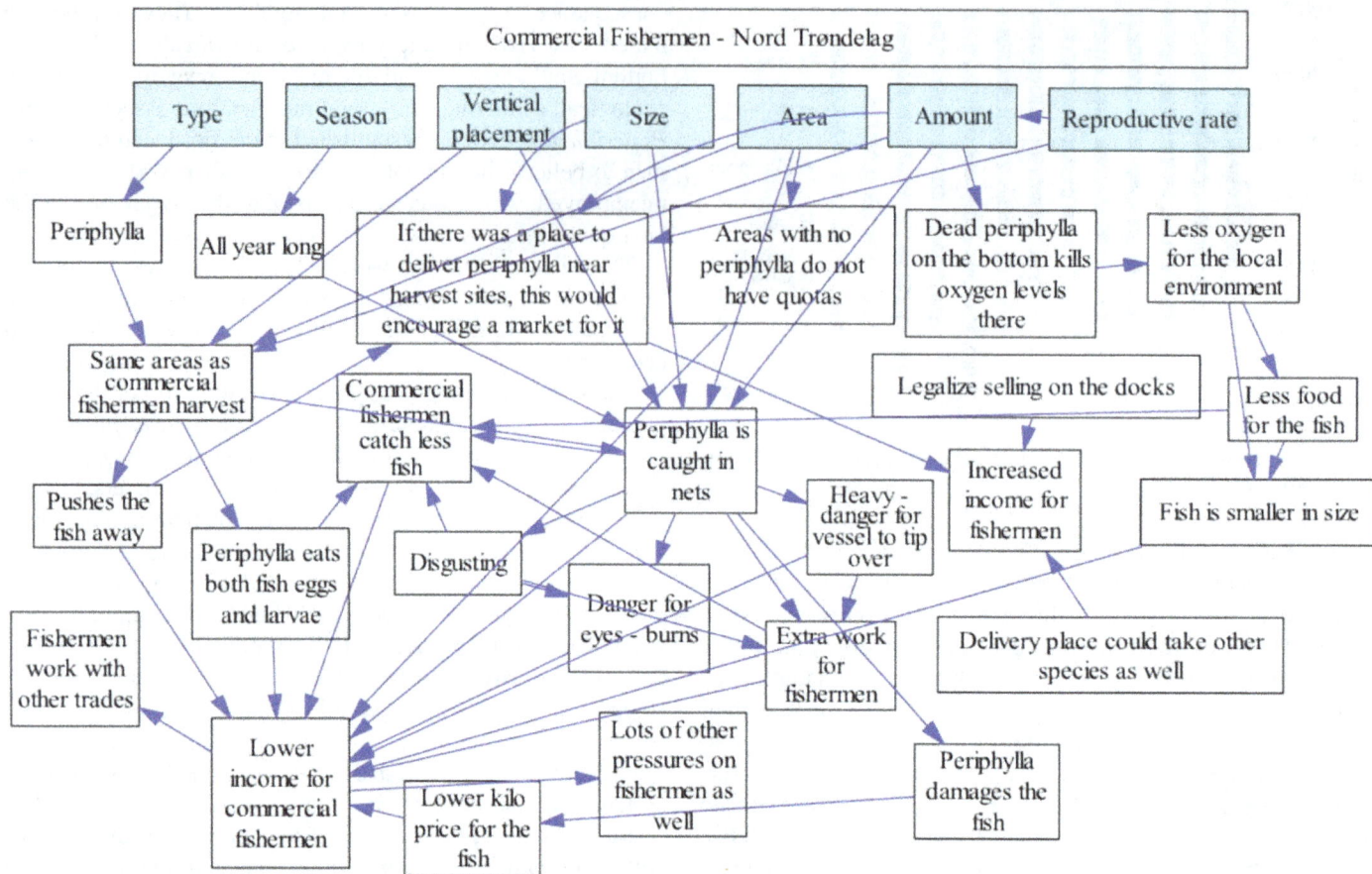

FIGURE 7.   Causal pathways of the systems conceptualization process, as applied to commercial fishermen of Nord Trøndelag.

fishermen stating that they were having difficulty in fulfilling their fish quotas because of *P. periphylla*. However, it was also because of the presence of purse seiners harvesting the food of the cod. In parallel to this narrative, the influence of purse seining for Atlantic Herring and Sprat *Sprattus sprattus* emerged from the sensitivity analysis as the main determinant of whether fish biomass was high or low (these variables shown in red in Figure 8), which was a serendipitous discovery. The group showed clear frustration of what they figuratively perceived to be a "vacuuming" of the waters of the main staple in the cod diet (Atlantic Herring and Sprat), as well as cod, which follow shoals of herring and are therefore caught as bycatch. This vacuuming, they claimed, was by large company purse seiners from other parts of Norway that were allowed to come in to their fjord to harvest Sprat quotas, mirroring an ongoing tension between the coastal and the ocean going fishing fleets. Such statements are based perceptions of threat or fear, and not based on biological or economic data, yet as drivers of stakeholder actions, are important to acknowledge.

The stakeholders then proceeded to rank fish biomass (and hence income) to be much less sensitive to the two other secondary nodes (i.e., *P. periphylla* harvesting scenario and

environmental protection of the fjord), something that was unexpected, since the focus of the workshop had consistently been on how damaging jellyfish are for the fishery. The concept of environmental protection also received much attention during discussion with the stakeholders during the workshop. However, much of this talk too was framed as a contamination issue associated with the purse seiners and the necessity of protecting them (the artisanal fishermen) from these commercial purse seiners. Their frustration with purse seining was very clear and that, while *P. periphylla* had been experienced and was exacerbating the current problem, there appeared greater issue with the purse seiners.

This was also reflected in the questionnaires, where commenting was common in the margins. One of these comments referred to a question as to the yearly catch in 2011–2012; this fishermen answered in this case that "...the fjord was emptied out of sprat by purse seiners [that year]..." On a question on agreement or disagreement to more environmental regulations in the fjord, the response strongly opposed more regulations, including closing off areas in the Trondheimsfjord, e.g., "...there has to be put an end to the sprat and herring fishing in the fjord," again referring to the purse seiners from other areas of Norway.

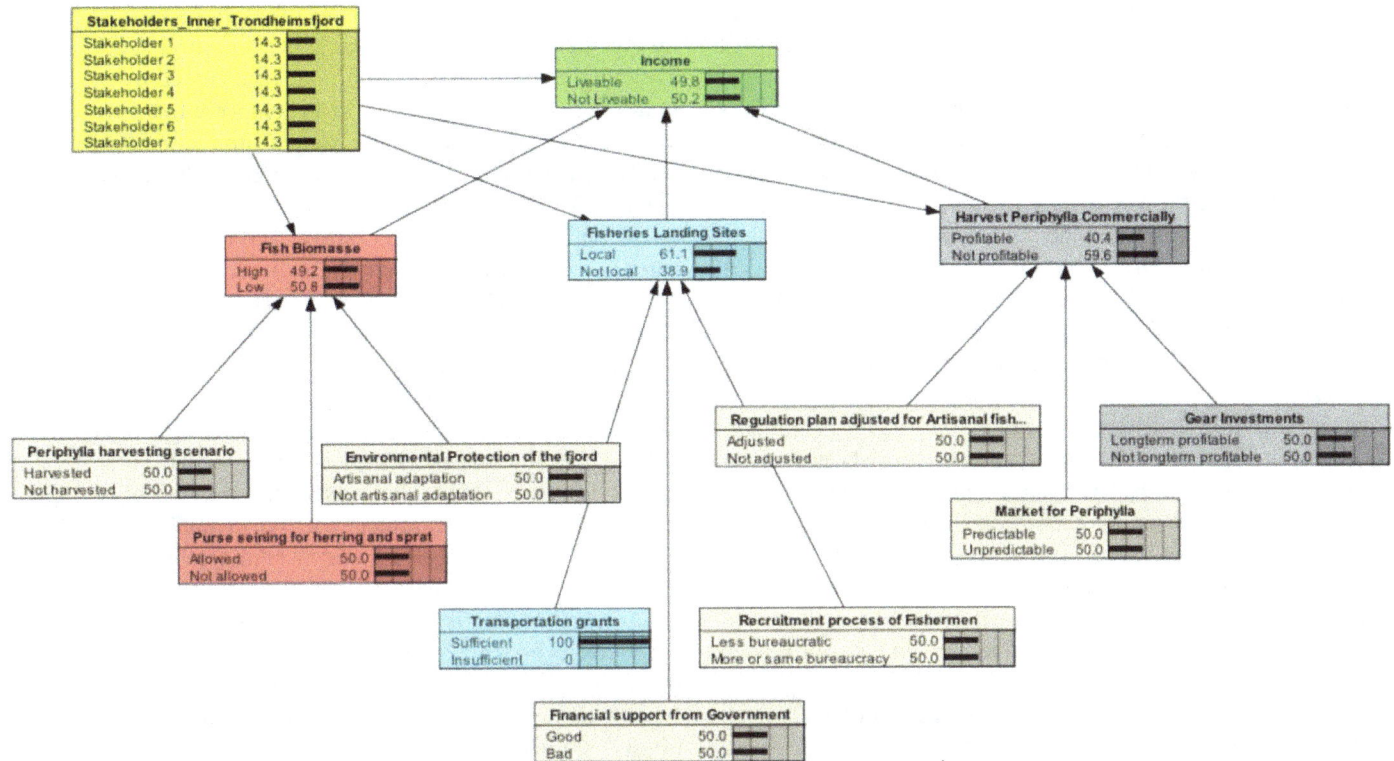

FIGURE 8. Bayesian belief network for artisanal fishermen, showing purse seining for Atlantic Herring and Sprat as having the biggest influence on their priority issue, rather than the jellyfish *Periphylla periphylla*.

DISCUSSION

Since the last decade the ecological structure and the fish abundance of the Trondheimsfjord has changed, not only because of jellyfish, but probably also due to intense exploitation of fisheries, climate change, and market forces. These have combined to generate unwanted effects on fishing behavior and patterns on the local fishers. The fulltime artisanal fishermen depend on harvesting ocean resources for their livelihood. They have rich knowledge about the fjord and resources therein, and have witnessed the changes in the marine resources and fishing conditions over time. Their fishing grounds are in general located in the areas close to their homes in the coastal areas as effected by vessel size and the location of landing sites. In this paper, we have found that these artisanal fishermen are facing increasing complexities within their fishery through the combination of ecological impacts from jellyfish and competition from commercial purse seiners that harvest cod prey as well as the cod as bycatch, elucidating thereby the idiom "Scylla and Charybdis," or being caught between a rock and a hard place. In this case, both the jellyfish and the purse seiners are targeting commercial species fish of utmost interest to the local artisanal fishers. What emerged in this study was that though jellyfish populations were rampant, and there was a distinct correlation between jellyfish increase and cod decrease, yet the focus of the fishers

was still on a human dimension. They consistently referred to competing stakeholder groups—the long-distance purse seiners—as a contributing problem of great scale in their fishing areas, almost equal to that of the jellyfish. The former is an ad hoc political issue, where the Norwegian coastal fishermen often feel discriminated against in national politics and prioritization between the local and long-distance fishing fleet. The fisher's union affiliation of these stakeholders split in the 1990s, demonstrating the underlying feeling of discrimination. In 1987, the Association of Coastal Fishers ("Kystfiskernes Forening"), now called the Association of Norwegian Coastal Fishers ("Norges Kystfiskarlag"), was established. The group of coastal fishermen present at the proposal stage of the association commented upon the increased influence of the Norwegian Fishing Vessel Owners Association ("Fiskebåt"), which has a membership base of around 90% of all Norwegian ocean-going vessels larger than 27.5 m in the Norwegian Fishermans Association ("Norges Fiskarlag"). The Norwegian Fishermans Association is the largest fisheries organization in Norway, with some 5,700 members. It consists of seven local chapters spread along county lines, with the main administration in Trondheim, Norway. Members are both coastal fishermen with small vessels, as well as the crew of ocean-going trawlers, thereby ensuring a broad and encompassing membership base with both employers and employees represented (Norwegian Fishermans Association 2014). Though size clearly is paramount, especially within an interest group set-ting where size and membership numbers matter, there are still voices that may be lost even when the organization is as topic-specific as the Norwegian Fishermans Association. Thus, coastal fishers wanted to create a counterweight to the Norwe-gian Fishing Vessel Owners Association *within* the association to ensure the coastal fisherman a louder voice when decisions were made. However, when the Association of Coastal Fishers applied for group membership in the Norwegian Fishermans Association, similar to Norwegian Fishing Vessel Owners Association in 1990, the application was declined, and the group has since stood in opposition to the Norwegian Fisher-mans Association as a distinct and nonassociated organization. The organization is different from the Norwegian Fishermans Association in that its membership base consists entirely of coastal fishermen; these fishermen believe the rights to and fights for increased amounts of quotas and a livelihood base has, since the 1960s, been consistently taken from them and distributed to trawlers. In 2014 the organization had around 1,000 members and about 600 vessels from the coastal fleet associated with it (Association of Norwegian Coastal Fishers 2014).

The fishers affected by jellyfish in the Trondheimsfjord, believe they were unfairly impacted by top–down decisions allowing for rest quotas being distributed liberally to purse seiners and allowing them to enter the fjord where artisanal fisheries had their livelihoods. This allowance was voiced to be a bigger hindrance of adaptive capacity for the fishermen than the jellyfish, with expressions such as "vacuuming the fjord of fish." We also discovered that, though there is a jelly-fish infestation in this area that could be coupled with a decrease in cod populations, the stakeholders had complemen-tary views on how to achieve sustainable development while still preserving both the cod and the artisanal fisherman in the inner Trondheimsfjord within the framework of their biggest perceived threat, namely the purse seiners. They suggested developing marine protected areas that would ensure that purse seiners were not allowed in to the area at all, while still allowing for artisanal fishermen to harvest the surplus of the fjord area. This protectionist vantage point expressed by the fishermen in the Trondheimsfjord could be a reflec-tion of the discovery of an economic morality in regards to the harvest of a marine surplus (Gezelius 2004). In this study, which looked at compliance to fisheries regulation and the morality of not adhering to said regulations, it was discovered that morality was divided into green and yellow spheres of economic activities at each extreme, and a con-tinuum along which all fishing activities were perceived to land within this area. Large offshore fishing fleets were con-sidered to be more yellow than small inshore fisheries, which were only considered moderately yellow. The fishers in the Canadian case furthermore blamed the groundfish col-lapse, among others, on excessive demands driven by off-shore trawlers (Gezelius 2004).

These are examples of when the bottom–up stakeholder involvement can be formalized. However, purse seining in the Trondheimsfjord is not as common as perceived by the fishers in this area, and the views of the fishers were in this case more a reflection of frustration of resource limitation. There was a limited purse seine fishery on Sprat in 2009–2010, and no fish-ing in 2011–2012 in the Trondheimsfjord (Bakketeig et al. 2013). In 2009, the local newspaper did report fantastic Sprat catches by purse seiners from these areas, as well as bycatches of cod (Roel 2009a) though, which could be another reason that fuelled the frustration of the fishers. At the same time, however, the Ytterøya cod production facility experienced two big escape events in December 2009 and September 2010, with 25,000 and 43,000 cod specimens, respectively (Roel 2009b). Those escapes took place in the actual purse seining area those years. Thus, the purse seine bycatch of cod those years might not have affected native Trondheimsfjord cod stock as severely as the artisanal fishers perceived it to have. Thus, even though the input of local ecological knowledge from stakeholders is critical for legitimacy of policy and compliance of stakeholders to management decisions, their perceptions need to be considered within the trinity of bio-logical and economic data as well. In involving not only highly trained experts, including biologists, modellers, economists, and political scientists in determining the adaptive capacity of a given community, the stakeholders can highlight the areas of most critical emphasis to them and their perceptions of threat to their focus area. Ulti-mately, sustainable development is critical to achieve, and the participation of stakeholders is necessary to uncover possible avenues of management that are dormant in expert opinion. This would bring legitimacy to the process, thereby ensuring a higher likelihood of compliance to regu-lations by the government and, in turn, effective resource management. In the case of the cod fishers in the inner Trondheimsfjord, the serendipitous discovery was that the threat they perceived to be more imminent than jellyfish infestation in their fishing grounds was a commercially driven threat of larger ocean-going fleets with higher capacity fishing in the same area. The importance of this finding to management is that strategies should include acknowledging this de facto fear through consultations and inclusion, thereby creating more legitimacy and less frus-tration with the process.

ACKNOWLEDGMENTS

The author would like to acknowledge the Norwegian Research Council for their financial support of the JANUS project, which enabled the effectuation of the research in this article. The Gemini Center for Sustainable Fisheries is also acknowledged for laying the groundwork and encour-aging the interdisciplinary and international collaboration of this article.

REFERENCES

Arai, M. N. 2001. Pelagic coelenterates and eutrophication: a review. Hydobiologia 541:69–87.

Association of Norwegian Coastal Fishers. 2014. Fakta om Norges Kystfiskarlag. [Facts about the Association of Norwegian Coastal Fishers.] Available: http://www.norgeskystfiskarlag.no/index.php/om-nkf. (January 2015).

Bakketeig, I. E., H. Gjøsæter, M. Hauge, H. Loeng, B. H. Sunnset, and K. Ø. Toft. 2013. Havforskningsrapporten 2013. Fisken og havet, særnummer 1–2013. [The marine science report 2013. The fish and the sea, special issue 1-2013.] Institute of Marine Research, Bergen, Norway.

Båmstedt, U., J. H. Fosså, M. B. Martinussen, and A. Fosshagen. 1998. Mass occurrence of the physonect siphonophore Apolemia uvaria (Lesueur) in Norwegian waters. Sarsia 83:79–85.

Barkin, J. S., and E. R. DeSombre. 2013. Saving global fisheries: reducing fishing capacity to promote sustainability. Massachusetts Institute of Technology Press, Cambridge.

Barnard, A. 1983. Contemporary hunter-gatherers: current theoretical issues in ecology and social organization. Annual Review of Anthropology 12: 193–214.

Broch, H. 2013. Social resilience - local responses to changes in social and natural environments. Maritime Studies 12:1–17.

Brotz, L., W. L. Cheung, K. Kleisner, E. Pakhomov, and D. Pauly. 2012. Increasing jellyfish populations: trends in large marine ecosystems. Hydrobiologia 690:3–20.

Bunnefeld, N., E. Hoshino, and E. J. Milner-Gulland. 2011. Management strategy evaluation: a powerful tool for conservation? Trends in Ecology and Evolution 26:441–447.

Dahl, K. 1899. Beretning om fiskeriundersøgelser i og om Trondhjemsfjorden 1898. [Report on fisheries research in and around the Trondheimfjord 1898.] Kongelige Norske Videnskabers Selskab, Volume 10, Trondhjem, Norway.

Dong, Z., D. Iiu, Y. Wang, B. Di, S. X. and Y. Shi. 2012. A report on a moon jellyfish Aurelia aurita bloom in Shishili Bay, northern Yellow Sea of China in 2009. Aquatic Ecosystem Health and Management 15:161–167.

Dryzek, J., D. Downs, H.-K. Hernes, and D. Schlosberg. 2003. Green states and social movements: environmentalism in the United States, United Kingdom, Germany, and Norway. Oxford University Press, Oxford, UK.

FAO (Food and Agriculture Organization of the United Nations) Fisheries and Aquaculture Department. 2010. The state of world fisheries and aquaculture 2010. FAO, Rome.

Fernandes, P. 2012. Invasion of the jellyfish. The Times of India (October 22). Available: http://timesofindia.indiatimes.com. (May 2015).

Freeman, R. E. 2010. Strategic management: a stakeholder approach. Cambridge University Press, New York.

Garcia, S. M., and A. A. Rosenberg. 2010. Food security and marine capture fisheries: characteristics, trends, drivers and future perspectives. Philosophical Transactions of the Royal Society B Biological Sciences 365: 2869–2880.

Gershwin, L.-A. 2013. Stung! On jellyfish blooms and the future of the ocean. University of Chicago Press, Chicago.

Gezelius, S. 2004. Food, money, and morals: compliance among natural resource harvesters. Human Ecology 32:615–634.

Graham, W., D. Martin, D. Felder, V. Asper, and H. Perry. 2003. Ecological and economic implications of a tropical jellyfish invader in the Gulf of Mexico. Biological Invasions 5:53–69.

Halpern, B. S., S. Walbridge, K. A. Selkoe, C. V. Kappel, F. Micheli, C. D'Agrosa, J. F. Bruno, K. S. Casey, C. Ebert, H. E. Fox, R. Fujita, D. Heinemann, H. S. Lenihan, E. M. P. Madin, M. T. Perry, E. R. Selig, M. Spalding, R. Steneck, and R. Watson. 2008. A global map of human impact on marine ecosystems. Science 319:948–952.

Hay, S. 2006. Marine ecology: gelatinous bells may ring change in marine ecosystems. Current Biology 16:R679–R682.

Hay, S. J., J. R. G. Hislop, and A. M. Shanks. 1990. North Sea Scyphomedusae; summer distribution, estimated biomass and significance particularly for 0-group Gadoid fish. Netherlands Journal of Sea Research 25:113–130.

Johnson, D. S. 2006. Category, narrative, and value in the governance of small-scale fisheries. Marine Policy 30:747–756.

Kawahara, M., K. Ohtsu, and S.-I. Uye. 2013. Bloom or non-bloom in the giant jellyfish Nemopilema nomurai (Scyphozoa: Rhizostomeae): roles of dormant podocysts. Journal of Plankton Research 35: 213–217.

Kim, D.-H., J.-N. Seo, W.-D. Yoon, and Y.-S. Suh. 2012. Estimating the economic damage caused by jellyfish to fisheries in Korea. Fisheries Science 78:1147–1152.

Kjærulff, U. B., and A. L. Madsen. 2008. Bayesian networks and influence diagrams: a guide to construction and analysis. Springer, New York.

Kütting, G. 2000. Environment, society, and international relations: towards more effective international environmental agreements. Psychology Press, Routledge, London.

Le Manach, F., M. Andriamahefazafy, S. Harper, A. Harris, G. Hosch, G.-M. Lange, D. Zeller, and U. R. Sumaila. 2013. Who gets what? Developing a more equitable framework for EU fishing agreements. Marine Policy 38:257–266.

Lovdata.no. 2005. Forskrift om strukturkvoteordning mv. for havfiskeflåten: §15 Slumpfiskeordning. [Regulation about the structural quota scheme for the ocean going fleets transfer of quotas between vessels.] Ministry of Trade Industry and Fisheries, FOR-2005-03-04-193. Available: http://lovdata.no/forskrift/2005-03-04-193/§15. (May 2015).

Lynam, C. P., S. J. Hay, and A. S. Brierley. 2005. Jellyfish abundance and climatic variation: contrasting responses in oceanographically distinct regions of the North Sea, and possible implications for fisheries. Journal of the Marine Biological Association of the United Kingdom 85:435–450.

Mills, C. 2001. Jellyfish blooms: are populations increasing globally in response to changing ocean conditions? Pages 55–68 in J. E. Purcell, W. M. Graham, and H. J. Dumont, editors. Jellyfish blooms: ecological and societal importance. Kluwer Academic Publishers, Dordrecht, The Netherlands.

Ministry of Trade Industry and Fisheries. 2007. Høringer. [Consultations.] Available: https://www.regjeringen.no/nb/dokument/id2000006/?documentType=dokumenter/h%C3%B8ringer&ownerid=709. (June 2015).

Ministry of Trade Industry and Fisheries. 2013. Meld. Stg. 22: verdens fremste matnasjon. [White paper 22: the world's leading seafood nation.] Available: https://www.regjeringen.no/nb/dokumenter/meld-st-22-20122013/id718631/. (June 2015).

Mork, J., C. Reuterwall, N. Ryman, and G. Ståhl. 1982. Genetic variation in Atlantic Cod (Gadus morhua L.): a quantitative estimate from a Norwegian coastal population. Hereditas 96:55–61.

Mork, J., N. Ryman, G. Ståhl, F. Utter, and G. Sundnes. 1985. Genetic variation in Atlantic Cod (Gadus morhua) throughout its range. Canadian Journal of Fisheries and Aquatic Sciences 42:1580–1587.

Österblom, H., M. Sissenwine, D. Symes, M. Kadin, T. Daw, and C. Folke. 2011. Incentives, social–ecological feedbacks and European fisheries. Marine Policy 35:568–574.

Nagvenkar, M. 2012. Goa's fishy problem: falling haul blamed on preying jellyfish. TwoCircles.net (November 12). Available: http://twocircles.net. (May 2015).

Nastav, B., M. Malej, A. Malej Jr., and A. Malej. 2013. Is it possible to determine the economic impact of jellyfish outbreaks on fisheries? A case study–Slovenia. Mediterranean Marine Science 14:214–223.

Norwegian Fishermans Association. 2014. Welcome to Norges Fiskarlag. Available: http://www.fiskarlaget.no/index.php/fiskarlaget-engelsk. (January 2015).

Palmieri, M. G., A. Barausse, T. Luisetti, and K. Turner. 2014. Jellyfish blooms in the northern Adriatic Sea: fishermen's perceptions and economic impacts on fisheries. Fisheries Research 155:51–58.

Pauly, D. 2007. The sea around us project: documenting and communicating global fisheries impacts on marine ecosystems. AMBIO: a Journal of the Human Environment 36:290–295.

Pauly, D. 2008. Global fisheries: a brief review. Journal of Biological Research–Thessaloniki 9:3–9.

Pauly, D., V. Christensen, S. Guenette, T. J. Pitcher, U. R. Sumaila, C. J. Walters, R. Watson, and D. Zeller. 2002. Towards sustainability in world fisheries. Nature 418:689–695.

Purcell, J., and M. Arai. 2001. Interactions of pelagic cnidarians and ctenophores with fish: a review. Hydrobiologia 451:27–44.

Purcell, J. E. 2005. Climate effects on formation of jellyfish and ctenophore blooms: a review. Journal of the Marine Biological Association of the United Kingdom 85:461–476.

Purcell, J. E. 2007. Anthropogenic causes of jellyfish blooms and their direct consequences for humans : a review. Marine Ecology Progress Series 350:153–174.

Quiñones, J., A. Monroy, E. M. Acha, and H. Mianzan. 2013. Jellyfish bycatch diminishes profit in an anchovy fishery off Peru. Fisheries Research 139:47–50.

Richardson, A. J., A. Bakun, G. C. Hays, and M. J. Gibbons. 2009. The jellyfish joyride: causes, consequences and management responses to a more gelatinous future. Trends in Ecology and Evolution 24:312–322.

Richards, R., M. Sanó, A. Roiko, R. W. Carter, M. Bussey, J. Matthews, and T. F. Smith. 2013. Bayesian belief modeling of climate change impacts for informing regional adaptation options. Environmental Modelling and Software 44:113–121.

Roel, J. E. 2009a. Eventyrlig brislingfisk. [Fairytale sprat fishing.] Adresseavisen (November 18).

Roel, J. E. 2009b. Halvt tonn oppdrettstorsk i brislingnota. [Half a ton of farmed cod in the sprat net.] Adresseavisen (November 19).

Solheim, H. 2012. Population trend of *Periphylla periphylla* in inner Trondheimsfjord. Master's thesis. Norwegian University of Science and Technology, Trondheim.

Sterman, J. D. 2000. Business dynamics: systems thinking and modeling for a complex world, volume 19. Irwin/McGraw-Hill, Boston.

Tiller, R., R. Gentry, and R. Richards. 2013. Stakeholder driven future scenarios as an element of interdisciplinary management tools; the case of future offshore aquaculture development and the potential effects on fishermen in Santa Barbara, California. Ocean and Coastal Management 73:127–135.

Tiller, R. G. 2008. The Norwegian system and the distribution of claims to redfeed. Marine Policy 32:928–940.

Tiller, R. G. 2010. Regime management at the bottom of the food web. Journal of Environment and Development 19:191–214.

Tiller, R. G., J. Mork, R. Richards, L. Eisenhauer, Y. Liu, J.-F. Nakken, and Å. L. Borgersen. 2014. Something fishy: assessing stakeholder resilience to increasing jellyfish (*Periphylla periphylla*) in Trondheimsfjord, Norway. Marine Policy 46:72–83.

Uye, S.-I. 2011. Human forcing of the copepod–fish–jellyfish triangular trophic relationship. Hydrobiologia 666:71–83.

Uye, S.-I., and U. Ueta. 2004. Recent increase of jellyfish populations and their nuisance to fisheries in the Inland Sea of Japan. Bulletin of the Japanese Society of Fisheries Oceanography 68:9–19.

Worm, B., E. B. Barbier, N. Beaumont, J. E. Duffy, C. Folke, B. S. Halpern, J. B. C. Jackson, H. K. Lotze, F. Micheli, S. R. Palumbi, E. Sala, K. A. Selkoe, J. J. Stachowicz, and R. Watson. 2006. Impacts of biodiversity loss on ocean ecosystem services. Science 314:787–790.

Zürn, M. 1998. The rise of international environmental politics: a review of current research. World Politics 50:617–649.

15

Using Salmon Survey and Commercial Fishery Data to Index Nearshore Rearing Conditions and Recruitment of Alaskan Sablefish

Ellen M. Yasumiishi,* S. Kalei Shotwell, Dana H. Hanselman, Joseph A. Orsi, and Emily A. Fergusson

National Marine Fisheries Service, Alaska Fisheries Science Center, 17109 Point Lena Loop Road, Juneau, Alaska 99801, USA

Abstract

We examined physical and biological indices from Pacific salmon *Oncorhynchus* spp. surveys and commercial fisheries to index nearshore rearing habitats used by age-0 and age-1 Sablefish *Anoplopoma fimbria* in the eastern Gulf of Alaska and as indicators for their recruitment to age2 during the period 2001–2013. The best-fitting general linear model used to estimate age-2 Sablefish recruitment included chlorophyll-*a* concentration during late August and an index of juvenile Pink Salmon *O. gorbuscha* abundance during the age-0 stage of Sablefish. The model and biophysical indices from 2012 and 2013 produced a forecast of 23 million age-2 Sablefish for 2014 and a forecast of 8 million Sablefish for 2015. This study highlights the opportunity to use proxies for direct ambient physical and biological observations of rearing habitats in estimating groundfish recruitment to older ages.

Environmental forcing during early life stages often plays a critical role in determining the survival of marine fish (Cushing 1969; Houde 1987). Sablefish *Anoplopoma fimbria* in Alaska have had several high recruitment events during periods of low observed spawning biomass; therefore, recruitment does not appear to be closely related to the level of spawning biomass and is more likely related to environmental effects on survival during early life stages (Hanselman et al. 2013). Sablefish recruitment was positively correlated with the intensity of the winter Aleutian low-pressure system (hereafter, Aleutian Low; McFarlane and Beamish 1992) and associated conditions, such as cooler-than-average winter sea surface temperatures (SSTs) in the central North Pacific and warmer-than-average summer sea temperatures in the Gulf of

Alaska (GOA; Sigler et al. 2001; Shotwell et al. 2014). One hypothesized mechanism is an increase in primary productivity (PP) from phytoplankton blooms and the subsequent effects on survival of and prey availability for Sablefish in nearshore rearing habitats (Shotwell et al. 2014). In the present study, we explore the value of using salmon survey and commercial fishery data from Southeast Alaska (SEAK) to index nearshore rearing conditions and recruitment of Alaskan Sablefish.

The Sablefish is a long-lived, deep-dwelling, oily fleshed species of commercial importance. Sablefish are distributed across the North Pacific Ocean from northern Mexico to the Bering Sea, but the bulk of the Alaskan population is primarily located in the GOA, where they are subject to a major commercial fishery (Hanselman et al. 2013). Of the six major

Subject editor: Donald Noakes, Vancouver Island University, Nanaimo, British Columbia

*Corresponding author: ellen.yasumiishi@noaa.gov

groundfish groups in Alaska, Sablefish ranked fourth in ex-vessel value at US\$120.3 million and sixth in catch at 13,900 metric tons during 2012, but the price per pound (1 lb = 0.4536 kg) of Sablefish was highest at \$4.18 (\$9.22 per kg), in comparison with less than \$0.30 (\$0.66 per kg) for the other groundfish groups (Fissel et al. 2013).

Alaskan Sablefish are typically encountered at depths of 200–1,000 m along the continental slope, in shelf gullies, and in deep-sea canyons (Wolotira et al. 1993). They are thought to spawn at depths of 300–500 m along the upper continental slope during winter (Mason et al. 1983). Eggs and yolk sac larvae are initially found at depths greater than 200 m; the larvae then migrate to the surface and are neustonic for the remainder of the age-0 stage, being present in the central and eastern GOA during spring and summer (Mason et al. 1983; Moser et al. 1994; Sigler et al. 2001). During mid- to late summer, age-0 Sablefish are advected onshore by currents and settle out in bays and inlets along the coast, where they spend the next 1–2 years (Rutecki and Varosi 1997; Maloney and Sigler 2008). Consequently, the coastal waters of Alaska are important rearing habitats for age-0 and age-1 Sablefish.

Quantifying the conditions associated with nearshore rearing habitat used by age-0 and age-1 Sablefish may help in predicting the success of incoming year-classes to the fishable population at age 2 and older ages. Assessments of the Alaskan Sablefish population are based on catches of Sablefish during annual fishery surveys using longline gear along the upper continental slope from the southern tip of SEAK to 60°N in the Bering Sea (Lunsford and Rodgveller 2013). Information on catch, effort, age, length, weight, and maturity is used to estimate the abundance of age-2 and older Sablefish (Hanselman et al. 2013). Indexing of the nearshore rearing habitat used by age-0 and age-1 Sablefish may provide an early indication of the strength of the incoming year-class.

Observations of juvenile Sablefish in pelagic surveys of nearshore habitats are typically rare (Rutecki and Varosi 1997; Orsi et al. 2013b). This may be due to the settlement of age-0 Sablefish into bays, their diel vertical migrations (Wing and Kamikawa 1995), and their ability to evade capture via the standard surface-trawl survey techniques used to sample juvenile fishes in nearshore habitats—specifically, by vertically migrating to deeper depths below the trawl or by inhabiting non-strait habitats and shallower waters (Orsi et al. 2013b). Thus, a time series of juvenile Sablefish abundance is not available from surveys; however, information on co-occurring species and nearby strait habitats might provide a usable substitute for Sablefish rearing conditions.

Pink Salmon *Oncorhynchus gorbuscha* also inhabit the coastal waters of Alaska during their seaward migration and on their return to spawn in freshwater (Heard 1991). Therefore, rearing conditions for and the abundance of juvenile Pink Salmon may serve as suitable indices for age-0 Sablefish survival. Pink Salmon are considered juveniles from the time

of seawater entry in the spring and summer to the end of December; they are considered adults from January 1 to the time of freshwater entry for the spawning migration. Pink Salmon inhabit nearshore waters during the spring and summer of their juvenile stage; they move offshore during the fall, remaining there during winter and spring, and then return to nearshore waters during summer. Age-0 Sablefish and juvenile Pink Salmon reach a similar size and feed on similar prey (Sigler et al. 2001). Age-0 Sablefish (<200 mm SL) initially feed primarily on euphausiids and pelagic tunicates but later become more piscivorous as they increase in size (Laidig et al. 1997; Sigler et al. 2001). Juvenile Pink Salmon (<300 mm FL) primarily feed on amphipods, euphausiids, fishes, and copepods in northern waters of SEAK. Adult Pink Salmon feed upon and compete with Sablefish; they are known to feed on age-0 Sablefish and their prey (Kaeriyama et al. 2000). Age-0 and age-1 Sablefish also feed on salmon eggs and tissue that are washed out from rivers into saltwater (Coutré 2014). The catch of juvenile Pink Salmon in nearshore surveys acts as a valuable pre-season forecasting tool for the commercial harvest of adult Pink Salmon in SEAK during the following year (Orsi et al. 2013a; Wertheimer et al. 2013). Indices of juvenile Pink Salmon abundance may be useful as proxies for growth and survival of age-0 Sablefish and as indicators of Sablefish recruitment to age 2.

We hypothesized that physical and biological indicators from fisheries and oceanography surveys that target juvenile Pink Salmon would be useful for understanding relationships between the two co-occurring species and for predicting Sablefish recruitment. We employed a generalized linear model to evaluate the importance of summer conditions during the Sablefish age-0 stage for the subsequent recruitment of Sablefish to age 2 by using Pink Salmon returns and biophysical indices from a survey conducted in nearshore strait habitats during late July and late August. In addition, our model and biophysical data were used to forecast age-2 recruitment 2 years in advance. These conditions included (1) summer SST; (2) summer PP, as measured by chlorophyll-*a* concentration; (3) abundance of juvenile Pink Salmon; and (4) abundance of adult Pink Salmon. We expected that Sablefish recruitment to age 2 would be (1) positively related to the nearshore summer SST and PP measured during the Sablefish age-0 stage, (2) positively related to juvenile Pink Salmon (indicator species) abundance during the Sablefish age-0 stage, and (3) negatively related to adult Pink Salmon (predator and competitor) abundance during the Sablefish age-0 stage. Possible mechanisms for increased Sablefish survival included favorable ocean conditions for growth and feeding (warm nearshore summer SST), increased ocean productivity (higher chlorophyll-*a* concentration), and greater survival of co-occurring species (juvenile Pink Salmon) during the Sablefish age-0 stage; and reduced predatory and competitive interactions between adult Pink Salmon and age-0 Sablefish.

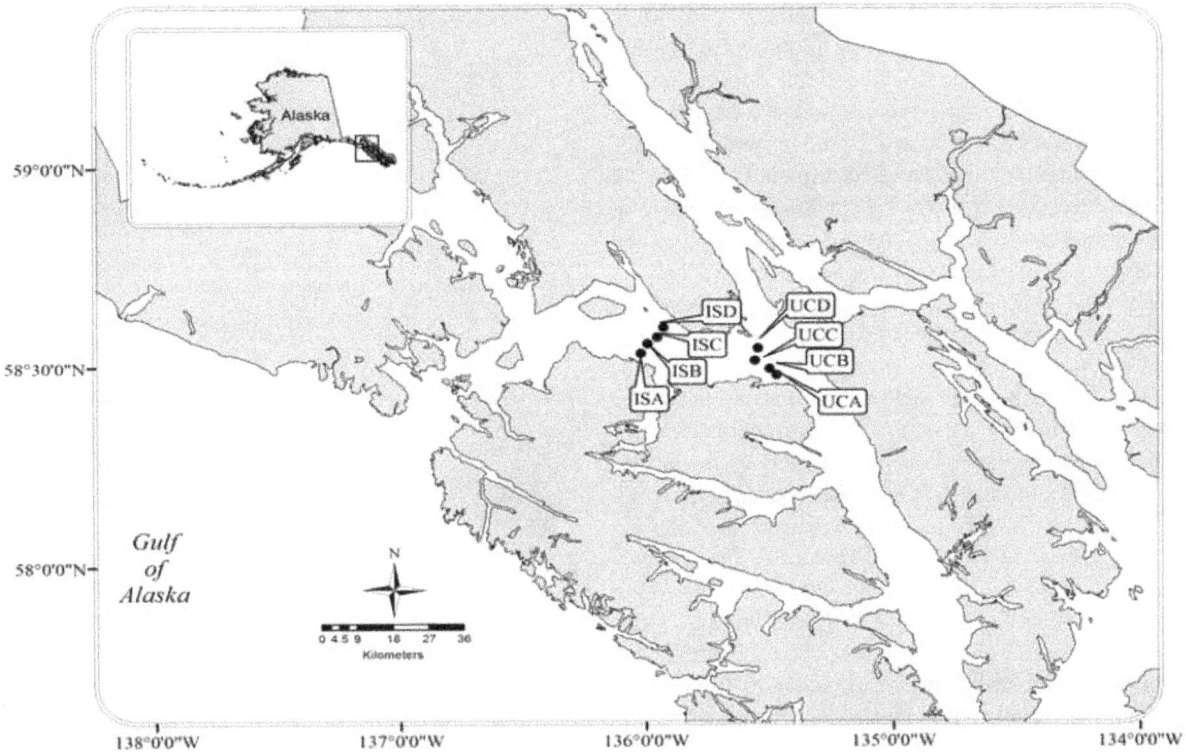

FIGURE 1.   Northern waters of Southeast Alaska, where sea surface temperature, primary productivity, and juvenile Pink Salmon abundance were sampled along two transects. Sampling occurred at four stations per transect: Icy Strait stations A–D (ISA, ISB, ISC, and ISD) and Upper Chatham Strait stations A–D (UCA, UCB, UCC, and UCD).

METHODS

We examined data from the Southeast Alaska Coastal Monitoring (SECM) project (National Marine Fisheries Service [NMFS]) in conjunction with commercial fisheries data to determine factors affecting Sablefish recruitment to age 2 in nearshore habitats. Surveys were conducted in Icy Strait and Upper Chatham Strait in the northern region of SEAK during late July and late August from 1999 to 2013 (Figure 1; Orsi et al. 2013b). Biophysical variables that were recorded during the SECM surveys and used in this study included SST, PP, and the CPUE of juvenile Pink Salmon. Information on commercial harvest and escapement was used to index juvenile and adult Pink Salmon abundances. Some age-0 Sablefish were captured, indicating that they can co-occur with juvenile and adult Pink Salmon in the survey area, but the Sablefish catches were not consistent over time and were too sporadic to be used in the time series analysis. A multiple regression model was used to examine the relationship between Sablefish recruitment and biophysical indices.

Response variable.—Abundance estimates for age-2 Sablefish in Alaskan waters were taken from the GOA Sablefish assessment (Figure 2; Hanselman et al. 2014). Fishery-independent information (annual longline and trawl surveys) included age and length compositions and survey abundance indices. Fishery-dependent information included foreign and domestic fishery catch, fishery age and length compositions, and CPUE.

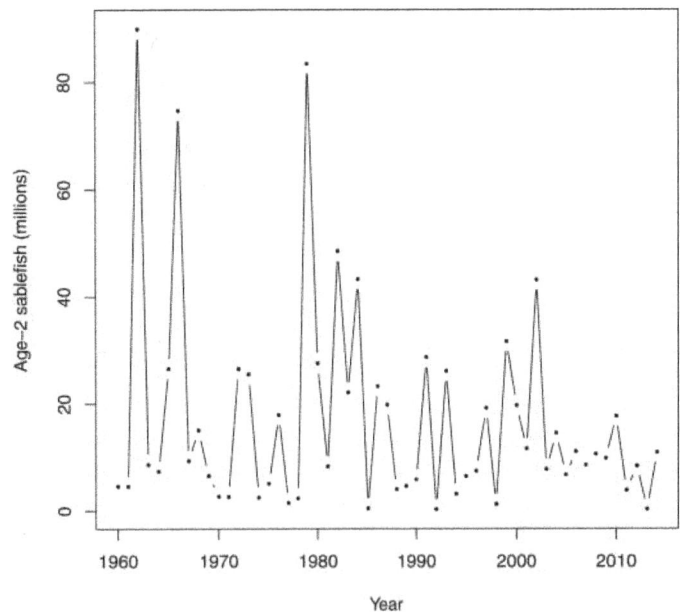

FIGURE 2.   Age-2 Sablefish recruitment estimates from the stock assessment report by Hanselman et al. (2014). Years represent age-2 recruitment years, so age-0 Sablefish occurred 2 years earlier (i.e., age-2 Sablefish that recruited in 2001 were exposed to conditions as age-0 fish in 1999).

The Sablefish stock assessment model (Hanselman et al. 2013) estimates parameters simultaneously while allowing for missing data within a penalized maximum likelihood function; the model produces spawning stock biomass, recruitment, and fishing mortality estimates as well as projections of future harvest scenarios. Recruitment is computed as mean recruitment with annual recruitment deviations rather than as a stock–recruitment relationship (Hanselman et al. 2013). Sablefish recruits are defined as 2 year olds because this is the first age at which Sablefish are caught in traditional adult surveys. As age and length compositions for each year are added to the stock assessment model, the amount of information on year-class strength increases (i.e., estimates of age-3 fish in the next year adds information on the strength of age-2 fish in the prior year). Estimates of age-2 Sablefish recruitment were available from 1960 to 2014—the period over which (1) abundance data were available for use in the stock assessment model and (2) recruitment was reasonably well estimated (Hanselman et al. 2014). Initial estimates of age-2 Sablefish are uncertain, so as more demographic data are collected the estimates of year-class strength become more precise. To formulate our model, we used stock assessment model estimates of age-2 Sablefish recruitment for the years 2001–2013, as those estimates were more recent and more certain. The model coefficients and biophysical indices from 2012 and 2013 were then used to make predictions of age-2 Sablefish abundance for 2014 and 2015.

Ecological predictors.—We used oceanographic indices (SST and PP) from Icy Strait due to the potential mixing of the water column at the Chatham Strait transect. Temperature was measured at 3-m depth by using an SBE 19plus SeaCAT CTD Profiler (Sea-Bird Electronics). Water samples taken from the surface were analyzed for chlorophyll-*a* concentration (μg/L) as an indicator of PP. Annual average and maximum values of SST and PP for July and August were calculated from the four stations at the Icy Strait transect.

A juvenile Pink Salmon abundance index, calculated as juvenile CPUE ($Jpink_{CPUE}$), was used as a proxy for ocean productivity and fish survival. Juvenile Pink Salmon were sampled with a Nordic 264 rope trawl that was modified to fish from the surface to 20 m directly astern of the trawl vessel (Orsi et al. 2013b). Station coordinates were targeted as the midpoint of the trawl haul; current, swell, and wind conditions usually dictated the setting direction. Several different vessels were used over time, and calibrations were made to standardize catch whenever possible (Wertheimer et al. 2010). The $Jpink_{CPUE}$ values for July and August were calculated as the average of juvenile Pink Salmon CPUEs at the eight stations.

An alternative index of juvenile Pink Salmon abundance ($Jpink_{adults}$) was constructed from returns (catch plus escapement) of adult Pink Salmon to SEAK (Figure 3B). The $Jpink_{adults}$ index was useful because the time series encompassed the outcome of conditions that influenced overwinter survival for juvenile Pink Salmon (i.e., from fall to the following summer)

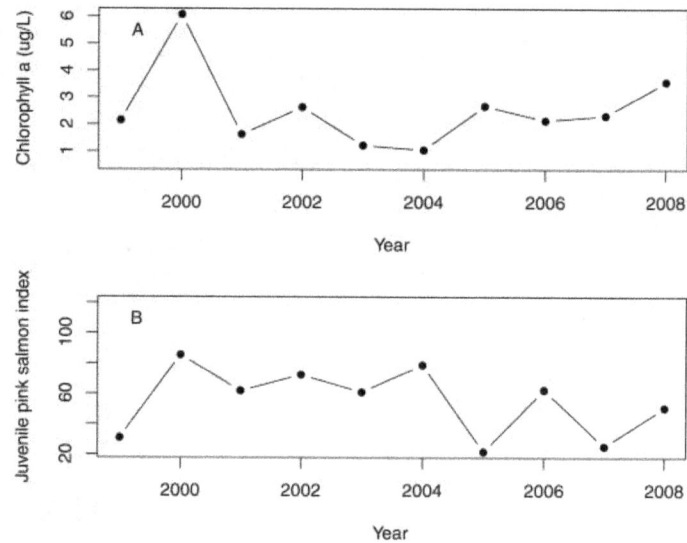

FIGURE 3. Statistically significant predictor variables and data used in estimation of model coefficients (2001–2013) and predictions (2014–2015) of Sablefish recruitment at age 2. Predictor variables are **(A)** maximum chlorophyll-*a* concentration at the Icy Strait stations during late August; and **(B)** juvenile Pink Salmon production based on Pink Salmon adult returns to Southeast Alaska. Year corresponds to the age-0 life stage of Sablefish ($t - 2$) that was related to age-2 Sablefish recruitment in year t.

during the first year at sea, a proxy for overwinter survival of Sablefish from age 0 to age 1. Adult Pink Salmon abundance (Apink) in year $t - 1$ was used to represent juvenile Pink Salmon abundance in year $t - 2$.

Abundances of adult Pink Salmon during the age-0 stage of Sablefish were also indexed as adult Pink Salmon returns (catch plus escapement) to SEAK. Data were from the Alaska Department of Fish and Game (Piston and Heinl 2013).

Correlation analysis.—A correlation analysis using Pearson's product-moment correlation was used to examine the relationships between predictor variables. A Bonferroni correction factor was applied to the α-value ($\alpha = 0.05$). Correlated predictor variables were included separately in the Sablefish regression model. Statistical analyses were conducted in R version 3.0.3 (R Development Core Team 2014).

Recruitment model.—A generalized linear regression model was used to relate age-2 Sablefish recruitment to biophysical parameters,

$$\text{Sablefish}_t = (\beta_1 \cdot \text{SST}_{t-2}) + (\beta_2 \cdot \text{PP}_{t-2}) + (\beta_3 \cdot \text{Jpink}_{t-2}) + (\beta_4 \cdot \text{Apink}_{t-2}) + e_t, \quad (1)$$

where SST is sea surface temperature, PP is chlorophyll-*a* concentration, Jpink is juvenile Pink Salmon abundance, Apink is adult Pink Salmon abundance, β is the coefficient estimate, e is error, and t is year. Error was assumed to be normally

TABLE 1.   Candidate biophysical indices (predictor variables) used in the model of age-2 Sablefish recruitment in northern waters of Southeast Alaska (SST = sea surface temperature; PP = primary productivity).

Predictor variable	Symbol
Oceanographic indices	
Mean SST at the four Icy Strait stations during late July (year $t - 2$)	SST1
Maximum SST at the four Icy Strait stations during late July (year $t - 2$)	SST2
Mean SST at the four Icy Strait stations during late August (year $t - 2$)	SST3
Maximum SST at the four Icy Strait stations during late August (year $t - 2$)	SST4
Mean chlorophyll-a concentration at the four Icy Strait stations during late July (year $t - 2$)	PP1
Maximum chlorophyll-a concentration at the four Icy Strait stations during late July (year $t - 2$)	PP2
Mean chlorophyll-a concentration at the four Icy Strait stations during late August (year $t - 2$)	PP3
Maximum chlorophyll-a concentration at the four Icy Strait stations during late August (year $t - 2$)	PP4
Indices of juvenile Pink Salmon abundance during Sablefish Age 0	
Mean juvenile Pink Salmon CPUE at Icy Strait and Chatham Strait stations during late July (year $t - 2$)	Jpink1
Mean juvenile Pink Salmon CPUE at Icy Strait and Chatham Strait stations during late August (year $t - 2$)	Jpink2
Southeast Alaska Pink Salmon production (harvest plus escapement of returning adults in year $t - 1$)	Jpink3
Predator index	
Southeast Alaska Pink Salmon production during Sablefish age 0 (harvest plus escapement of returning adults in year $t - 2$)	Apink

distributed, with a mean of zero and a constant variance. The SDs of the stock assessment estimates were added to the SEs of the fitted values from equation (1). Candidate predictor variables are listed in Table 1.

Selection of the best-fitting generalized linear regression model was based on cross-validation criteria and the Bonferroni correction factor. Cross validation was conducted to select the model with the best performance by using the "bestglm" package (version 0.34) in R. An additional penalty was imposed on the t-values of the significant predictor variables to account for the 12 possible predictors (at an α-value of 0.05) by using the Bonferroni correction factor ($P < \alpha$/[number of predictor variables]). Coefficients with P-values less than 0.004 (i.e., 0.05/12) were considered statistically significant at the 95% level. Summary statistics included the R^2, the F-statistic for R^2, and the P-value for the F-statistic. For each model, we tested the following assumptions: (1) homoscedasticity of the residuals (plot of absolute residuals versus fitted values; and a Goldfeld–Quandt test), (2) normality of the residuals (Q–Q plots; and a Shapiro–Wilk test), (3) absence of multicollinearity (Pearson's product-moment correlation), and (4) absence of autocorrelation in the residuals (correlograms of the autocorrelation function estimates).

Predictions of Sablefish recruitment to age 2 were made by using the coefficient from the best-fitting model and biophysical indices. Biophysical indices from 2012 and 2013 were entered into the model to produce estimates of age-2 Sablefish recruitment for 2014 and 2015. Model predictions for 2014 were compared with stock assessment estimates of age-2 Sablefish recruitment in 2014.

RESULTS

Correlation Analysis

Pearson's product-moment correlation coefficients relating predictor variables used in the Sablefish model are given in Table 2. Significant correlations were found between mean and maximum values of SST and PP within each month, except between the mean and maximum PP index for August. Therefore, the mean and maximum values of SST and PP were added separately to the Sablefish model. Mean August PP was correlated with mean July PP, maximum July PP, and maximum August PP, so these variables were included in separate models. The $Jpink_{CPUE}$ index for July and the $Jpink_{adults}$ index were introduced separately in the Sablefish model.

Recruitment Model

The best-fitting generalized linear model identified by cross validation included 3 of the 12 possible variables: maximum SST during late August, maximum chlorophyll-a concentration (i.e., PP) during late August, and $Jpink_{adults}$. However, the SST coefficient was not significant after Bonferroni correction (Bonferroni-corrected $\alpha = 0.004$, but $P = 0.02$), so SST was removed from the model. The final model therefore included maximum chlorophyll-a concentration during late August and $Jpink_{adults}$ (Table 3). The model residuals were normally distributed (Shapiro–Wilk test: $W = 0.9625$, $P = 0.7924$), were homoscedastic (Goldfeld–Quandt test: $GQ = 2.1155$, $P = 0.2439$), and had no serial correlation. The biophysical indices captured 79% of the variability in stock assessment estimates of age-2 Sablefish recruitment.

TABLE 2. Pearson's product-moment correlation coefficients for the predictor variables used in the Sablefish recruitment model (codes for predictor variables are defined in Table 1). Values in bold italics represent significant correlations.

Predictor variable	SST2	SST3	SST4	PP1	PP2	PP3	PP4	Jpink1	Jpink2	Jpink3	Apink
SST1	*0.91*	0.49	0.50	−0.26	−0.13	−0.53	−0.22	−0.23	−0.39	0.09	0.28
SST2		0.43	0.46	−0.30	−0.16	−0.22	−0.18	−0.41	−0.23	0.06	0.37
SST3			*0.94*	−0.39	−0.34	−0.48	−0.32	−0.08	−0.15	0.05	0.33
SST4				−0.45	−0.37	−0.51	−0.41	−0.17	−0.14	−0.12	0.40
PP1					*0.90*	*0.76*	0.55	−0.29	0.47	−0.42	0.06
PP2						*0.76*	*0.66*	−0.25	0.56	−0.31	−0.02
PP3							*0.84*	−0.46	0.34	−0.35	−0.08
PP4								−0.09	0.34	0.04	−0.30
Jpink1									0.03	*0.65*	−0.45
Jpink2										−0.29	−0.12
Jpink3											−0.47

The final model and biophysical indices from 2013 and 2014 yielded a prediction of 22.8 million (SD = 4.54 million) age-2 Sablefish for 2014 and a prediction of 8.46 million (SD = 1.43 million) age-2 Sablefish for 2015 (Figure 4). Our 2014 estimate was high due to the large return of adult Pink Salmon in 2013, indicating favorable survival conditions for Sablefish from age 0 to age 1 in 2012–2013. The 2014 estimate from our model was two times higher than the stock assessment estimate of age-2 Sablefish recruitment for 2014. Future estimates of age-2 Sablefish abundance from the stock assessment model and based on survey data will help to determine the performance of these models for predicting recruitment.

DISCUSSION

Our study focused on the use of survey and fishery data to index environmental conditions during the age-0 stage of Alaskan Sablefish and predict their recruitment to age 2. Previous studies have primarily focused on environmental conditions during the winter prior to and during the age-0 stage (McFarlane and Beamish 1992; Sigler et al. 2001; Shotwell et al. 2014). We found that significant indicators for the estimated recruitment of age-2 Sablefish included late-summer SST, late-summer PP, and a juvenile Pink Salmon abundance index during age 0.

All of the relationships were positive. Higher recruitment of Sablefish to age 2 was associated with higher chlorophyll-a concentrations in nearshore areas during late summer of the Sablefish age-0 stage and with higher Pink Salmon productivity, possibly the result of increased storms during an intensified Aleutian Low. Previous studies support this relationship. For example, McFarlane and Beamish (1992) showed that Sablefish year-class strength in British Columbia during 1965–1980 was positively correlated with the Aleutian Low index ($R^2 = 0.53$, $P < 0.001$) and copepod abundance in the central Northeast Pacific at Ocean Station Papa ($R^2 = 0.45$, $P < 0.001$), indicating that stronger year-classes were associated with climate-related changes in prey abundances. At about 40 d posthatch (usually in March), larval Sablefish generally absorb their yolk sacs and begin feeding on copepods, the most abundant zooplankton in the water column (McFarlane and Beamish 1992). McFarlane and Beamish (1992) further hypothesized that the more intense Aleutian Low resulted in (1) greater upwelling of nutrients to the surface, which improved PP; and (2) greater wind transport of nutrients and plankton from offshore to nearshore waters, which in turn increased the survival of Sablefish by increasing the

TABLE 3. Best-fitting model describing age-2 Sablefish recruitment (year t) estimates (2001–2013) as a function of biophysical indices during the age-0 life stage (1999–2011; PP_{t-2} = primary productivity [chlorophyll-a concentration] during the Sablefish age-0 life stage [year $t - 2$]; $Jpink_{adults,t-2}$ = juvenile Pink Salmon abundance index during year $t - 2$ based on Pink Salmon adult returns to Southeast Alaska in year $t - 1$). Coefficient statistics include the estimate, t-value, and associated P-value for the significant predictor variables in the generalized least-squares regression model of recruitment. Model statistics include the adjusted R^2, F-statistic, and associated P-value.

Predictor variable	Coefficient statistics			Model statistics		
	Estimate	t	P	Adjusted R^2	F	P
PP_{t-2}	0.739	5.796	0.00012	0.79	25.3	0.000077
$Jpink_{adults,t-2}$	0.497	3.895	0.00249			

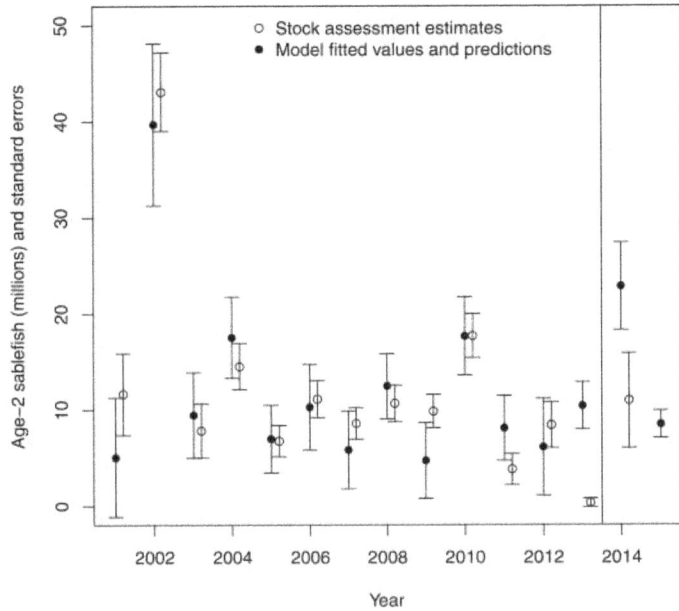

FIGURE 4. Estimates (±SE) of age-2 Sablefish recruitment from the stock assessment by Hanselman et al. (2014; open circles, 2001–2014); and estimates (solid circles, 2001–2013) and forecasts (solid circles, 2014–2015) from the generalized linear regression model describing age-2 recruitment as a function of the chlorophyll-*a* concentration and juvenile Pink Salmon abundance index during the Sablefish age-0 life stage (adjusted $R^2 = 0.79$, $P = 0.000077$).

production of copepods, their primary prey. Shotwell et al. (2014) developed a conceptual model of Sablefish recruitment (termed the ocean domain dynamic synergy [ODDS] model) that consisted of three linked mechanisms relating to intensification of the Aleutian Low. First, an increase in atmospheric and oceanic cyclonic circulation results in a match between the arrival of Sablefish larvae and the arrival of cool, productive water from the central North Pacific Ocean to the outer-shelf domain. Second, increased anti-cyclonic eddy activity in the mid-shelf domain entrains nutrients and prey that are used by Sablefish. Third, increased stratification along the coast due to warmer SSTs may result in an earlier spring phytoplankton bloom and enhanced zooplankton biomass, which could increase the growth of juvenile Sablefish in the nearshore zone and subsequently enhance survival. We explored conditions more associated with this third mechanism for their effects on the nearshore timing of Sablefish early life history and found that the late-summer SST and PP indicative of a late-summer phytoplankton bloom were particularly related to Sablefish recruitment.

Summer SST in nearshore waters during the Sablefish age-0 stage was not a significant predictor of age-2 Sablefish recruitment after we accounted for the number of predictors in our model. However, previous studies have shown that temperature can affect Sablefish recruitment via direct and indirect influences on motor activity, feeding, growth, and size (Sigler and Zenger 1989; Sigler et al. 2001). In adult Sablefish, feeding activity (number of squid consumed) was

33–71% greater at higher water temperatures (8°C) in comparison with colder water temperatures (2°C; Stoner and Sturm 2004). Higher temperatures also reduced the time required by Sablefish to attack, consume, and handle squid bait (Stoner and Sturm 2004). For Sablefish that were collected off the Oregon coast, the otolith-derived growth rate of late larvae and early juveniles was positively correlated with body size and temperature; furthermore, growth rate was positively correlated with recruitment except during the 1997 El Niño, when growth was high and survival was low—possibly due to poor feeding conditions in late summer (Sogard 2011). The late-summer maximum SSTs (10.7–14.5°C) observed during the years included in our study may have been within the optimum range favoring Alaskan Sablefish recruitment.

Late-summer PP in Icy Strait explained most of the variability in age-2 Sablefish recruitment over the 13-year time series. The two sporadic high-recruitment events for age-2 Sablefish in 2002 and 2010 corresponded with high PP levels for age-0 Sablefish in Icy Strait during late August of 2000 and 2008. A threefold increase in PP during late August suggests the occurrence of a late-summer phytoplankton bloom. We hypothesize that the high PP in late summer may increase the supply of nutrients and prey for age-0 Sablefish prior to winter, thereby increasing fish energy stores and, in turn, over-winter survival. The "critical size and critical period" hypothesis (Beamish and Mahnken 2001) states that if fish are larger and have higher energy density by fall, then they have an increased likelihood of surviving the winter. Long-term monitoring of late-summer PP and the feeding habits and whole-body energy of age-0 Sablefish could contribute to a better understanding of the influence of late-summer phytoplankton blooms on Sablefish recruitment.

Juvenile Pink Salmon abundance during the Sablefish age-0 stage, as indexed by returns of adult Pink Salmon ($Jpink_{adults}$), was a positive predictor of Sablefish recruitment to age 2. We found that $Jpink_{adults}$ was a better survival index than $Jpink_{CPUE}$. Returning adult Pink Salmon represent the portion of the population that survived the first winter at sea, whereas the CPUE of juvenile Pink Salmon represents the abundance of the population prior to overwintering mortality. The $Jpink_{CPUE}$ index is also just a snapshot of the juvenile population at one point in time and therefore may not be as accurate as juvenile abundance indices based on adult returns. Estimates of adult returns based on escapement and catch are also prone to error. As a postmortality index, the abundance of returning adult Pink Salmon provides an index of the Pink Salmon population closer to the time of Sablefish recruitment to age 2 and indicates additional survival conditions through the first winter of life. Furthermore, Pink Salmon juveniles may spend time on the continental shelf during late fall, near the area where age-0 Sablefish enter in late summer to rear and overwinter for up to 2 years. Our results indicate that the abundance of juvenile Pink Salmon is a reasonable proxy for growth and survival conditions experienced by age-0 Sablefish due to the similarity in size, growth rates, and diets between these two groups of fish.

Results of the present study demonstrate that indices of nearshore rearing habitats have potential as a means of forecasting Sablefish recruitment. The magnitude of late-summer PP and the survival of a co-occurring species were shown to correlate with Sablefish recruitment. Our present results contributed to ecosystem considerations within the Alaskan groundfish Stock Assessment and Fishery Evaluation Report (Zador and Aydin 2013). The report provides current ecosystem conditions and their impacts on fish populations. In the management of marine ecosystems, an ecosystem approach to fisheries involves incorporating ecosystem factors into fisheries stock assessments so as to support tactical management decisions (Link 2010; Idhe and Townsend 2013). Currently, our model only provides a qualitative measure of whether to expect above-average or below-average recruitment of Sablefish. A longer time series of ocean productivity will be necessary to match the time series of the stock assessment estimates; an evaluation of correlations between ocean productivity indices and uncertainty in stock assessment estimates of abundance is also needed.

There are several caveats to consider in the context of our study. First, the examined time series of biophysical indices was short (13 years) relative to the available time series of the age-2 Sablefish recruitment index (55 years, beginning in 1960). Second, we did not examine the relationship between biophysical indices and the uncertainty in recruitment estimates from the stock assessment model. Third, the present study was based on correlation analyses and was not supported by a process study directed at addressing Sablefish early life history. Although the significant correlations we identified were corroborated by previous findings in field research on Pink Salmon, the correlations do not necessarily equal causation. For example, Cooney et al. (2001) found that two co-occurring species—juvenile Pink Salmon and juvenile Pacific Herring *Clupea pallasii*—used different portions of the annual production cycle in the northern GOA. Cooney et al. (2001) also highlighted the importance of seeking mechanisms rather than a correlative understanding of complex marine ecosystems. The next logical step will be to develop a long-term monitoring project for sampling nearshore areas where age-0 Sablefish have historically occurred.

In summary, our study highlights the potential value of using fisheries and oceanography survey data and information from a commonly sampled species to index conditions experienced by a more elusive species when those species have shared rearing areas. Such data indicate status and trends in ecosystem pressures and can serve as early indicators of year-class strength for Alaskan Sablefish. Predictor variables can be useful in forecasting recruitment 2 years in advance to assist the stock assessment process. This study emphasizes the opportunity to use proxies for direct ambient physical and biological observations of rearing habitats and time series parameters in estimating groundfish recruitment to older ages.

ACKNOWLEDGMENTS

All co-authors contributed to the ideas presented in this article. We thank the anonymous reviewers for their valuable feedback; Jordan Watson (NMFS Alaska Fisheries Science Center) for assistance with R software; and Andy Piston (Alaska Department of Fish and Game, Ketchikan) for providing Pink Salmon harvest and escapement data. The findings and conclusions in this paper are those of the authors and do not necessarily represent the views of NMFS. Reference to trade names does not imply endorsement by NMFS.

REFERENCES

Beamish, R. J., and C. Mahnken. 2001. A critical size and period hypothesis to explain natural regulation of salmon abundance and the linkage to climate and climate change. Progress in Oceanography 49:423–437.

Cooney, R. T., J. R. Allen, M. A. Bishop, D. L. Eslinger, T. Kline, B. L. Norcross, C. P. McRoy, J. Milton, J. Olsen, V. Patrick, A. J. Paul, D. Salmon, D. Scheel, G. L. Thomas, S. L. Vaughan, and T. M. Willette. 2001. Ecosystem controls of juvenile Pink Salmon (*Oncorhynchus gorbuscha*) and Pacific Herring (*Clupea pallasi*) populations in Prince William Sound, Alaska. Fisheries Oceanography 10(Supplement 1):1–13.

Coutré, K. M. 2014. Feeding ecology and movement patterns of juvenile Sablefish in coastal Southeast Alaska. Master's thesis. University of Alaska, Fairbanks.

Cushing, D. H. 1969. The regularity of the spawning season of some fishes. ICES Journal of Marine Science 33:81–92.

Fissel, B., M. Dalton, R. Felthoven, B. Garber-Yonts, A. Haynie, A. Himes-Cornell, S. Kasperski, J. Lee, D. Lew, L. Pfeiffer, and C. Seung. 2013. Stock assessment and fishery evaluation report for the groundfish fisheries of the Gulf of Alaska and Bering Sea/Aleutian Islands areas: economic status of the groundfish fisheries off Alaska, 2012. North Pacific Fishery Management Council, Anchorage, Alaska.

Hanselman, D. H., C. R. Lunsford, and C. J. Rodgveller. 2013. Assessment of the Sablefish stock in Alaska. Pages 267–376 in M. Sigler and G. Thompson, editors. Stock assessment and fishery evaluation report for the groundfish resources of the Bering Sea/Aleutian Islands regions. North Pacific Fishery Management Council, Anchorage, Alaska.

Hanselman, D. H., C. R. Lunsford, and C. J. Rodgveller. 2014. Assessment of the Sablefish stock in Alaska. Pages 283–424 in Appendix B: stock assessment and fishery evaluation report for groundfish resources of the Gulf of Alaska. North Pacific Fishery Management Council, Anchorage, Alaska.

Heard, B. 1991. Life history of Pink Salmon (*Oncorhynchus gorbuscha*). Pages 119–230 in C. Groot and L. Margolis, editors. Pacific salmon life histories. University of British Columbia Press, Vancouver.

Houde, E. D. 1987. Fish early life dynamics and recruitment variability. Pages 17–29 in R. D. Hoyt, editor. 10th Annual larval fish conference. American Fisheries Society, Symposium 2, Bethesda, Maryland.

Idhe, T., and H. Townsend. 2013. Interview with Jason Link. Fisheries 38:363–369.

Kaeriyama, M., M. Nakamura, M. Yamaguchi, H. Ueda, G. Anma, S. Takagi, K. Y. Aydin, R. V. Walker, and K. W. Myers. 2000. Feeding ecology of Sockeye and Pink salmon in the Gulf of Alaska. North Pacific Anadromous Fish Commission Bulletin 2:55–63.

Laidig, T. E., P. B. Adams, and W. M. Samiere. 1997. Feeding habits of Sablefish off the coast of Oregon and California. NOAA Technical Report NMFS 130:65–79.

Link, J. 2010. Ecosystem-based fisheries management: confronting trade-offs. Cambridge University Press, New York.

Lunsford, C., and C. Rodgveller. 2013. F/V *Ocean Prowler* cruise report OP-13-01: longline survey of the Gulf of Alaska and the eastern Bering Sea, May 26–August 28, 2013. Alaska Fisheries Science Center, Juneau.

Maloney, N., and M. Sigler. 2008. Age-specific movement patterns of Sablefish *Anoplopoma fimbria* in Alaska. U.S. National Marine Fisheries Service Fishery Bulletin 106:305–316.

Mason, J. C., R. J. Beamish, and G. A. McFarlane. 1983. Sexual maturity, fecundity, spawning, and early life history of Sablefish *Anoplopoma fimbria* off the Pacific coast of Canada. Canadian Journal of Fisheries and Aquatic Sciences 40:2126–2134.

McFarlane, G. A., and R. J. Beamish. 1992. Climatic influence linking copepod production with strong year-class in Sablefish, *Anoplopoma fimbria*. Canadian Journal of Fisheries and Aquatic Sciences 49:743–753.

Moser, H. G., R. L. Charter, P. E. Smith, N. C. H. Lo, D. A. Ambrose, C. A. Meyer, E. M. Sandknop, and W. Watson. 1994. Early life history of Sablefish, *Anoplopoma fimbria*, off Washington, Oregon, and California, with application to biomass estimation. California Cooperative Oceanic Fishery Investigations Reports 35:144–159.

Orsi, J., E. Fergusson, M. Sturdevant, and A. Wertheimer. 2013a. Forecasting Pink Salmon harvest in Southeast Alaska. Pages 131–134 *in* S. Zador, editor. Ecosystem considerations for 2013, appendix C of the Bering Sea/Aleutian Islands/Gulf of Alaska stock assessment and fishery evaluation report. North Pacific Fishery Management Council, Anchorage, Alaska.

Orsi, J. A., E. A. Fergusson, M. V. Sturdevant, E. V. Farley Jr., and R. A. Heintz. 2013b. Annual survey of juvenile salmon, ecologically related species, and biophysical factors in the marine waters of southeastern Alaska, May–August 2012. North Pacific Anadromous Fish Commission, Document 1485, Vancouver.

Piston, A. W., and S. C. Heinl. 2013. Pink Salmon stock status and escapement goals in Southeast Alaska. Alaska Department of Fish and Game, Special Publication 11-18, Anchorage.

R Development Core Team. 2014. R: a language and environment for statistical computing. R Foundation for Statistical Computing, Vienna.

Rutecki, T. L., and E. R. Varosi. 1997. Distribution, age, and growth of juvenile Sablefish, *Anoplopoma fimbria*, in Southeast Alaska. NOAA Technical Report NMFS 130:45–54.

Shotwell, S. K., D. H. Hanselman, and I. M. Belkin. 2014. Toward biophysical synergy: investigating advection along the Polar Front to identify factors influencing Alaska Sablefish recruitment. Deep-Sea Research Part II Topical Studies in Oceanography 107:40–53.

Sigler, M. F., T. L. Rutecki, D. L. Courtney, J. F. Karinen, and M.-S. Yang. 2001. Young-of-the-year Sablefish abundance, growth, and diet. Alaska Fishery Research Bulletin 8:57–70.

Sigler, M. F., and H. H. Zenger Jr. 1989. Assessment of Gulf of Alaska Sablefish and other groundfish based on the domestic longline survey, 1987. NOAA Technical Memorandum NMFS-F/NWC-169.

Sogard, S. M. 2011. Interannual variability in growth rates of early juvenile Sablefish and the role of environmental factors. Bulletin of Marine Science 87:857–872.

Stoner, A. W., and E. A. Sturm. 2004. Temperature and hunger mediate Sablefish *Anoplopoma fimbria* feeding motivation: implication for stock assessment. Environmental Biology of Fishes 61:238–246.

Wertheimer, A. C., J. A. Orsi, E. A. Fergusson, and M. V. Sturdevant. 2010. Calibration of juvenile salmon catches using paired comparisons between two research vessels fishing Nordic 264 surface trawls in Southeast Alaska, July 2009. North Pacific Anadromous Fish Commission, Document 1277, Vancouver.

Wertheimer, A. C., J. A. Orsi, E. A. Fergusson, and M. V. Sturdevant. 2013. Forecasting Pink Salmon harvest in Southeast Alaska from juvenile salmon abundance and associated biophysical parameters: 2012 returns and 2013 forecast. North Pacific Anadromous Fish Commission, Document 1486, Vancouver.

Wing, B. L., and D. J. Kamikawa. 1995. Distribution of neustonic Sablefish larvae and associated ichthyoplankton in the eastern Gulf of Alaska, May 1990. NOAA Technical Memorandum NMFS-AFSC-53.

Wolotira, R. J. J., T. M. Sample, S. F. Noel, and C. R. Iten. 1993. Geographic and bathymetric distributions for many commercially important fishes and shellfishes off the West Coast of North America, based on research survey and commercial catch data, 1912–1984. NOAA Technical Memorandum NMFS-AFSC-6.

Zador, S., and K. Aydin. 2013. Ecosystem assessment. Pages 1–235 *in* S. Zador, editor. Ecosystem considerations for 2013; appendix C of the Bering Sea/Aleutian Islands/Gulf of Alaska stock assessment and fishery evaluation report. North Pacific Fishery Management Council, Anchorage, Alaska.

16

Modeling Larval Transport and Settlement of Pink Shrimp in South Florida: Dynamics of Behavior and Tides

Maria M. Criales*
Rosenstiel School of Marine and Atmospheric Science, University of Miami,
4600 Rickenbacker Causeway, Miami, Florida 33149, USA

Laurent M. Cherubin
Harbor Branch Oceanographic Institute, Florida Atlantic University, 5600 U.S. 1 North, Fort Pierce,
Florida 34946, USA

Joan A. Browder
National Oceanic and Atmospheric Administration, National Marine Fisheries Service,
Southeast Fisheries Science Center, 75 Virginia Beach Drive, Miami, Florida 33149, USA

Abstract

The pink shrimp *Farfantepenaeus duorarum*, one of the commercially important Penaeidae, reproduces offshore of the southwest Florida (SWF) shelf. Larvae migrate to nursery grounds in estuarine Florida Bay. Using a numerical approach, we investigated the role of spawning location, larval traits, and physical forces on the transport of pink shrimp larvae. First, the Regional Oceanic Modeling System that is based on tides, air–ocean fluxes, and freshwater flows was used to simulate the SWF shelf oceanography. The model replicates the tides, winds, salinity, currents, and seasonality of the shelf. Secondly, the Regional Oceanic Modeling System was coupled offline with the Connectivity Modeling System, in which virtual larvae were released near the surface from two spawning sites, Dry Tortugas and Marquesas, and tracked until the time for settlement (about 28–30 d). Virtual larvae moved vertically in the water column following ontogenetic behaviors previously observed in the field: diel vertical migration (DVM) and selective tidal stream transport (STST). Lagrangian trajectories indicated that migration paths changed radically between summer and winter during model years (1995–1997). Maximum settlements occurred in summer by larvae crossing the SWF shelf, while the lowest settlement occurred in winter by larvae moving through passes in the Florida Keys. Modeling results demonstrated an effective east-northeast transport across the SWF shelf during summer as a result of the tidal currents, the subtidal currents, and the combined DVM and STST behaviors. The current phase captured during the initial DVM period was critical to determine the direction in which larvae move, favorable (east and northward) or unfavorable (south and westward), before the STST behavior captures the eastward tidal current that brings larvae to the nursery grounds. Unfavorable currents were driven by the summer easterlies and low salinities at the coast. Results indicated that Marquesas is the more effective spawning ground, with 4.5 times more likely settlement of originating larvae compared with Dry Tortugas. Model-estimated seasonal settlement patterns concurred with postlarval influxes previously observed at Florida Bay boundaries.

Subject editor: Suam Kim, Pukyong National University, Busan, South Korea

*Corresponding author: mcriales@rsmas.miami.edu

Most coastal species of fish and marine invertebrates spawn offshore in shelf waters, and early life stages migrate to coastal estuarine nursery grounds. Coastal oceanographic processes greatly influence transport and settlement of these estuarine-dependent species. The dominant physical mechanisms that may yield successful transport include wind-driven Ekman transport, upwelling fronts, moving frontal systems, counter-currents generated by coastal eddies, coastal boundary layers, and nonlinear internal tides and bores (e.g., Shanks 1995, 2006; Queiroga and Blanton 2004). Larval behavior in the form of diel or tidal ontogenetic vertical migrations is an important mechanism that influences horizontal dispersal (Cowen and Castro 1994; Paris et al. 2002) and enhances local retention in nearshore zones (Paris and Cowen 2004; Paris et al. 2007; Butler et al. 2011) and in upwelling systems (Peterson 1998; Almeida et al. 2006). In this study, we used a modeling approach to determine the effect of physical factors and larval behaviors on the transport and settlement of the important penaeid pink shrimp *Farfantepenaeus duorarum* in southern Florida.

Tropical penaeid shrimps are highly fecund species with short-lived, complex life cycles and highly variable patterns of recruitment. Females usually spawn offshore, and larval stages migrate to estuaries, where juveniles develop and grow before returning to the fishery grounds (e.g., Garcia and Le Reste 1981; Dall et al. 1990). The larvae employ diurnally and tidally modulated behaviors to successfully migrate to the nursery ground. Diel vertical migration (DVM), a widespread behavior in crustacean decapod larvae, has been observed in the early protozoea and mysis stages of several penaeid species (Jones et al. 1970; Rothlisberg 1982; Rothlisberg et al. 1983; Criales et al. 2007). These early larvae, with very limited ability to swim, coordinate their migrations with the day–night cycle, rising in the water column exclusively at night. Later in their development, as they approach the nursery grounds, postlarvae change activity from a diurnal to a tidal rhythm (Rothlisberg et al. 1995, 1996; Criales et al. 2010, 2011). Prompted by environmental cues associated with the tides, postlarvae swim in the water column during the dark flood tide and rest near the bottom during the ebb tide. This general mechanism is known as selective tidal stream transport (STST), and more than one behavior has been associated with it (for review see Shanks 1995; Forward and Tankersley 2001; Queiroga and Blanton 2004).

Pink Shrimp Larval Ecology

The pink shrimp is one of the most economically and ecologically important species in southern Florida and in Campeche Sound, Gulf of Mexico (Sheridan 1996; Ramírez-Rodríguez et al. 2003). This species supports an important year-round fishery on the lower southwest Florida (SWF) shelf between Dry Tortugas and Key West (Klima et al. 1986; Hart 2012). The spawning center and major fishery are located near the Dry Tortugas and Marquesas islands (Cummings 1961; Roberts 1986), and the main nursery grounds are in Florida Bay and other coastal estuaries of the lower southwestern Florida coast (Tabb et al. 1962; Costello and Allen 1966; Browder and Robblee 2009). The Dry Tortugas fishery is directly dependent on young shrimp that migrate from inshore nursery areas onto the offshore fishing grounds (Ehrhardt and Legault 1999; Browder et al. 2002). Larvae develop quickly, passing through several changes in feeding habitats, behavior, and morphologic stages (nauplii, zoeae, myses, postlarvae) and needing only about 30 d to become postlarvae ready to settle (Dobkin 1961; Ewald 1965). Larval development and ocean hydrodynamics must be well synchronized to successfully bring planktonic stages to their coastal nursery grounds. Because Florida Bay interacts with two different continental shelves—i.e., through its western border with the SWF shelf of the Gulf of Mexico and through the eastern and southeastern borders with the Florida Keys coastal zone in the Atlantic Ocean—pink shrimp larvae arriving from Dry Tortugas have alternative routes to recruit to the nursery grounds. The two main hypothesized transport pathways are as follows: (1) larvae may drift south–southeast downstream with the Florida Current front and enter Florida Bay through the tidal inlets of the Lower and Middle Florida Keys (Munro et al. 1968; Criales et al. 2003) and (2) larvae may move east–northeast across the SWF shelf and enter the bay at its northwestern boundary (Jones et al. 1970; Criales et al. 2006).

Simulations of larval transport that include larval behavior have been conducted for penaeid species in the Gulf of Carpentaria, in the Indo-Pacific (Rothlisberg et al. 1995, 1996; Condie et al. 1999), and in the Gulf of California in the eastern Pacific Ocean (Calderon-Aquilera et al. 2003; Marinone et al. 2004). Rothlisberg et al. (1996) and Condie et al. (1999) developed larval behavior models for penaeid species (*Penaeus semisulcatus, P. merguiensis,* and *P. esculentus*) and coupled them to hydrodynamic models from the Gulf of Carpentaria to determine the distribution of effective spawning grounds and the advection of larvae to the nursery grounds. The modeling approach of these two studies was similar in using a transition depth as the point where larvae switch from vertical migration cued by the diel cycle to that cued by the tides. Marinone et al. (2004) studied larval transport from a Lagrangian point of view by including larval behavior as part of the biological component of the model to examine the potential location of spawning grounds for the blue shrimp *Litopenaeus stylirostris* and yellowleg shrimp *Farfantepenaeus californiensis*. In the present study, a biophysical transport model that resolves the ontogenetic behaviors of the pink shrimp larvae was used to determine larval migration of pink shrimp larvae from their spawning grounds to their nursery grounds, the most frequent recruitment pathways, and the larval behaviors and physical forces necessary to successful transport and settlement.

Southwest Florida Shelf Dynamics

The SWF shelf is a wide and shallow shelf with a smoothly varying topography aligned northwest–southeast bounded by Gulf of Mexico waters to the west and Atlantic Ocean waters near the Florida Keys and the Dry Tortugas to the south (Figure 1). Previous studies by Li and Weisberg (1999a, 1999b), Weisberg et al. (2001), Weisberg et al. (2009), and Liu and Weisberg (2012) suggested a dynamically based partition of the shelf that leads to the distinction of outer, middle, and inner shelf regions, plus a nearshore region embedded within the inner shelf. The outer shelf extends an internal Rossby radius of deformation inward from the shelf break, providing a buffer zone between the deep ocean and the shelf and reducing the seasonality of the circulation. The outer shelf circulation is mainly under the influence of the Loop Current and Florida Current (LFC) system and warm core (anticyclonic) and cold core (cyclonic) eddies that propagate southward along the Loop Current front and continue eastward between the shelf and the Florida Straits (Lee et al. 1992; Fratantoni et al. 1998; Le Hénaff et al. 2012).

The inner shelf is the shallow region controlled by friction, where surface and bottom Ekman layers are likely to either interact or be separated from one another by stratification. This region is mainly under the influence of baroclinic, heat-flux-driven and barotropic, wind-and-tide-driven circulations. Using long-term measurements, Liu and Weisberg (2012) found a robust seasonal cycle in velocity that varies across the shelf. The inner shelf circulation is predominantly upwelling favorable from October to April, and the circulation consists of a southward alongshore current on the central and southwestern shelf. From June to September, the circulation is predominantly downwelling favorable and consists of a strong coastal northward alongshore flow on the central shelf. On the southern and southwestern sides of the SWF shelf in summer, the circulation is mostly southward, although the standard deviation of the flow speed and direction is much larger than during other times.

Estuarine-imposed salinity gradients may add an additional baroclinic effect to the nearshore region of the inner shelf that sometimes causes a southward flow in summer. The western Florida coast is under the influence of a significant number of freshwater inputs (Figure 1). They drain freshwater from precipitation (direct or delayed, local or regional) from rivers, streams, lakes, and canals into the near shore, majorly contributing to the estuarine properties of SWF shelf coastal waters. Lee et al. (2002) observed salinity patterns associated with the Shark River Slough low-salinity plume in June 1998. The plume was elongated toward the southeast with little spreading offshore, suggesting advection by a background current. The general movement of the plume was to the southeast through western Florida Bay and middle Florida Keys passages. Extreme discharges or large nutrient inputs have been associated with harmful algal blooms (Zhao et al. 2013) and "black water" events (Hu et al. 2004; Zhao et al. 2013). Zhao et al.

(2013) showed that freshwater coastal plumes tend to move south along the southwestern Florida coast and slowly disperse to the southwest, as well as approaching and moving through the Florida Keys. The complete clearing of low-salinity water from the inner shelf following heavy rainfall events can take several months (Hu et al. 2004; Zhao et al. 2013).

The west Florida shelf (WFS) displays a strong coherent response to synoptic-scale alongshore wind forcing (Li and Weisberg 1999a, 1999b; Weisberg et al. 2000, 2005) that is also observed on the inner SWF shelf (Lee et al. 2002). According to Lee and Smith (2002), synoptic-scale upwelling (downwelling) alongshore winds, together with the coastal constraint, results in barotropic alongshore currents over the shelf aligned with the wind direction at the surface and balanced by set-up (set-down) of the coastal sea level. Using multiyear remotely sensed and sea-level wind measurements, Liu and Weisberg (2012) showed that during the boreal fall, winter, and spring seasons (from October through April), the wind tends to be northeasterly and upwelling favorable. During boreal summer (June through August), the winds tend to be southerly and southeasterly. Monthly mean winds are stronger in winter than in summer on the WFS, including the SWF shelf.

The circulation of the southern portion of the SWF shelf is complex. This area experiences mixed semidiurnal–diurnal tides, whereas tides in the Florida Keys zone are primarily semidiurnal (Wang et al. 1994; Smith 1997, 2000; Wang 1998). Smith (2000) measured the flow exchange between Florida Bay and the SWF shelf and found that the northern two-thirds of the western boundary of the bay is likely a region of inflow, whereas the southern third emerged as a region of weak outflow. According to Smith (2000), westward flow from the southern bay is a logical consequence of the north–south decrease in tidal amplitudes (Smith 1997) and the clockwise rotation of tidal ellipses. The magnitude of flow across the open western boundary related to the oscillatory ebb and flood of the tide can be substantial, even if there is no net volume transport. Moreover, in addition to the tide, larger-scale wind set-up on the inner shelf is likely to drive the net coastal circulation (Wang et al. 1994; Liu and Weisberg 2007).

The southernmost part of the Florida shelf has unique geographic and oceanographic characteristics that present special challenges and opportunities to organisms seeking transport using currents. The presence of the pink shrimp stock and fishery on the SWF shelf provides an especially interesting opportunity to explore the larval behavior and oceanographic and meteorological processes enabling larval migration from spawning to nursery grounds in this complex setting. In the present study, a three dimensional high-resolution, region-wide hydrodynamic model forced by tides and air–ocean fluxes is coupled to a biophysical model that simulates the ontogenetic larval behaviors observed for this species to determine the pink shrimp larval migration pathways from their spawning grounds to their nursery grounds and to explore spatial, seasonal, and interannual contrasts.

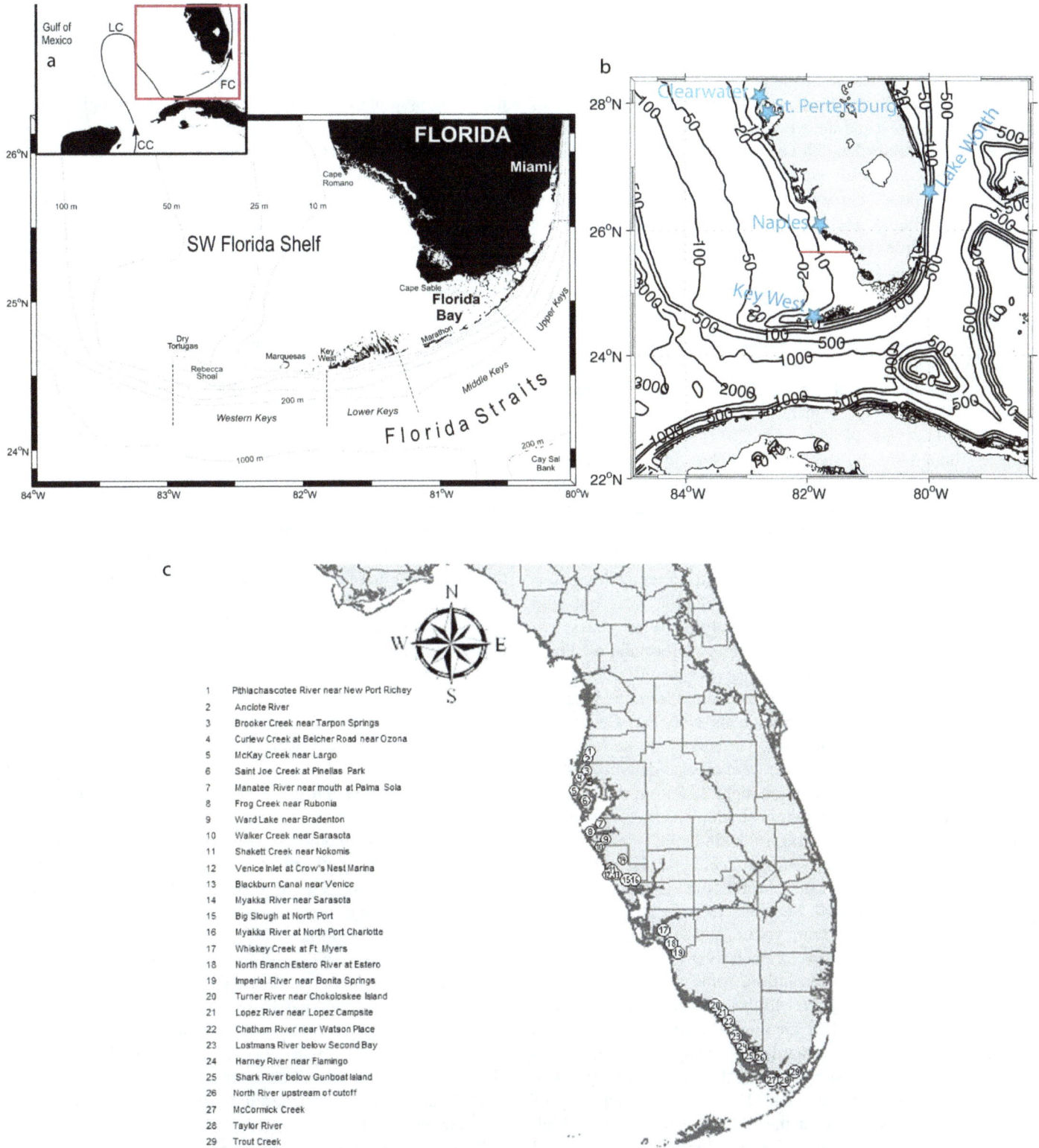

FIGURE 1. Map of (a) the southwest Florida shelf study area. The red box on the small map indicates the domain of the Regional Oceanic Modeling System (ROMS) model. Abbreviations are as follows: CC = Caribbean Current, LC = Loop Current, FC = Florida Current, and GS = Gulf Stream. Map showing (b) model domain and bathymetry on which blue stars show the location of tidal gauges used for model verification. The horizontal red line at 25.7°N indicates the section used to calculate cross-shore salinity gradient. Map of (c) the watersheds and rivers on Florida's west coast that were included in the model. Open circles with numbers denote U.S Geological Survey hydrologic stations, providing flow, temperature, and salinity data.

METHODS

Hydrodynamic model description.—The three-dimensional Regional Oceanic Modeling System (ROMS; Shchepetkin and McWilliams 2005) was implemented to simulate the circulation that connects the lower Florida shelf with larger-scale flows of the Gulf of Mexico and the western subtropical North Atlantic, while incorporating detailed coastal dynamics around Florida Bay and the SWF shelf and freshwater discharge from west coast Florida point sources. The model domain is centered on Florida Bay (Figure 1a), encompasses the eastern side of the Loop Current area and the Florida Straits all the way to Cuba to the South and the Great Bahamas Banks to the East (22.000–28.405°N; 84.900–78.317°W; Figure 1b). The horizontal resolution is 2.862 km, the minimum depth is 3 m, and the model depth is discretized over 25 terrain-following *s*-layers. The model was started on January 3, 1994, and run until 2001. The model ocean state variables were relaxed weekly to the high-resolution (1/12°) TOPAZ reanalysis of the North Atlantic for the period 1979 to present (Sakov et al. 2012). Tides were set at the model's boundaries by the TPXO6 global tide model (Egbert and Erofeeva, 2002). Six tidal components, M2 (principal lunar, semi diurnal), S2 (principal solar, semidiurnal), N2 (larger lunar elliptic, semidiurnal), K2 (lunisolar, semidiurnal), K1 (lunisolar, diurnal), and O1 (principal lunar diurnal), were included. The model surface was forced by 3-hourly wind, air temperature, relative humidity, evaporation, and precipitation rates obtained from North American Regional Reanalysis. Wind stress was calculated by the model's air–sea fluxes bulk formulation. Four-hourly net surface shortwave and longwave heat fluxes, as well as the net shortwave radiation, were obtained from the National Center for Environmental Predictions–National Center for Atmospheric Research reanalysis. The model sea surface temperature (SST) and salinity (SSS) were relaxed daily to, respectively, the coarser 4-km-resolution night SST from the National Oceanic and Atmospheric Administration (NOAA) Advanced Very High Resolution Radiometer Pathfinder SST version 5 and to monthly SSS from the National Center for Environmental Predictions Global Ocean Data Assimilation System.

Rivers were simulated as point sources in the model. Monthly river discharge, temperature, and salinity were calculated from in situ time series for each river listed in Figure 1c. Because of the model's resolution, some rivers were joined together and their respective discharge summed. River data were obtained from the U.S. Geological Survey. Not all rivers had a complete record of input data. The gap was filled by calculating a monthly climatology from the existing records. The model was spun-up for one year (1994) and the frequency of model output was based on the release experiments, which took place in boreal winter (January–March) and in boreal summer (June–September). During the release periods, the output frequency was every 3 h in order to resolve semidiurnal tides, whereas every 8 h the rest of the year. The winter and summer 3-hourly inputs necessary to the larval transport study were produced for only the years 1995–1997.

The connectivity modeling system.—We used the connectivity modeling system (CMS; version 1.0; Paris et al. 2013) to investigate larval migration and the behavior of pink shrimp in south Florida. The CMS is an open access biophysical modeling system developed in a stochastic Lagrangian framework. The CMS consists of a spatially explicit individual-based model composed of different modules that are coupled. Each of these modules can be turned on or off within the configuration files. A Lagrangian stochastic model integrates, along each individual particle path, information derived from the other modules and tracks the trajectories of individual larval attributes (Grimm et al. 2006). In the Lagrangian stochastic model, larvae are moved at each time step ($dt = 7,200$ s in our application) by a fourth-order Runge-Kutta integration of the ordinary differential equation $d\mathbf{X} = \mathbf{u} \cdot dt$, where $d\mathbf{X}$ is the displacement vector and $\mathbf{u}$ the velocity vector (Griffa 1996), that is applied in both time and space (Paris et al. 2013). The latter is based on a time series of three-dimensional velocity fields from the oceanographic module. In addition to advection by ROMS velocities and stochastic diffusion ($K_h = 20$ m^2/s^2), larvae are moved vertically following the ontogenetic vertical migrations prescribed in the biological module and are settled in suitable settlement habitat based on information derived from the seascape module, provided that they arrive when it is their ontogenetically ordained time to settle (also prescribed by the biological module). The CMS allows us to track the source and destination of each larva. The CMS previously was used to investigate the interaction between the life history characteristics and oceanography of coral reef fishes and the spiny lobster in the Gulf of Mexico and the Caribbean Sea (Paris et al. 2005, 2007; Cowen et al. 2006; Butler et al. 2011; Sponaugle et al. 2012; Kough et al. 2013; Holstein et al. 2014). More details on the coupled biophysical algorithms and modeling approach can be found in Paris et al. (2013) (http://code.google.com/p/connectivity-modeling-system/).

The biological module.—The biological module of the CMS accounted for pink shrimp early life history traits, specifically, pelagic larval duration and larval behaviors (ontogenetic vertical migrations consisting of DVM and STST behavior). Other important biological variables such as egg production, larval growth, and larval mortality were excluded from the model in order to isolate the effect of circulation and larval behavior on the spatial trajectories. The pelagic larval duration in the model was set at 30 d based on laboratory studies of the species (Dobkin 1961; Ewald 1965). The five nauplius, three protozoea, and three mysis stages were completed in 15 d, and additional planktonic postlarval stages were completed in another 15 d.

The larval behavior of pink shrimp was parameterized in the model using results derived from repetitive vertically stratified plankton surveys conducted by Jones et al. (1970) and Criales et al. (2007). Larval vertical distribution patterns observed from plankton surveys were used to create the vertical matrix (z, Δt), in which the probability density distributions in the water column (z) were calculated through time intervals (Δt) for each larval developmental stage (Table 1). Larval stage durations were assigned according to the studies of Dobkin (1961) and Ewald (1965). As Table 1 indicates, pink shrimp larvae perform vertical migrations and the specific behavior changes ontogenetically; protozoeae were found deeper than myses and myses deeper than postlarvae. The relative concentrations of protozoeae in the upper, middle, and bottom layers were consistent with a DVM behavior, whereas that of late myses and postlarvae were consistent with a STST behavior in phase with the flood tide (Criales et al. 2007). The probability vertical matrix was set up for 12-h periods to simulate the DVM behavior. The STST behavior was reproduced in the model by creating a module in which particles move with the current when the sea surface height rises (during flood tide) and stop moving when the sea surface height falls (during ebb tide). The STST behavior was added into the model at day 15, which corresponds to the time reported in the literature (Rothlisberg et al. 1995; Criales et al. 2006).

The benthic habitat module.—The benthic habitat module accounted for pink shrimp spawning ground aggregations and habitats suitable for settlement. The two selected spawning locations were Dry Tortugas and Marquesas. The Dry Tortugas site has been considered the main spawning ground of this species (Roberts 1986) and is located on the outer SWF shelf about 40 km northeast of the Dry Tortugas Islands at a depth of 35 m. The Marquesas site was selected based on the high percentages of early protozoeae collected during the summer months by Jones et al. (1970) and by Criales et al. (2007). The Marquesas site is located about 30 km north of the Marquesas Islands at a depth of 25 m.

The suitable habitats for settlement were represented as polygons of settlement created in ArcMap and coupled to the CMS benthic habitat module as (X, Y, Z), where X and Y were the coordinates of the vertices and Z the polygon number. Thirty-one 10-km × 10-km square polygons were created covering the Florida Bay area and adjacent coastal shelf. A variety of distinct bottom habitats are contained within the area

TABLE 1. Matrix table of the distribution of pink shrimp larvae in the water column used in the biological module of the CMS model, showing the density probabilities (%) of occurrence of the larval stages at each depth and the duration of larval stages (time). The data is from Criales et al. (2007) and is based on vertically stratified plankton sampling.

Time and depth	Planktonic stages		
	Nauplii–protozoea	Mysis	Postlarvae
Time (d)	7	7	15
Depth (m)			
5	0	0	50
15	25	54	25
30	35	26	20
45	40	20	5

covered by the polygons, including submerged aquatic vegetation (mainly seagrass beds of turtle grass *Thalassia testudinium*, manatee grass *Syringodiun filiforme*, and shoal grass *Halodule wrightii*) and different types of unvegetated bottom (mainly soft and muddy bottom) (Zieman et al. 1989). Pink shrimp postlarvae usually settle in areas covered with dense mixed species of seagrasses (Costello and Allen 1966), therefore the presence of seagrasses may be another factor determining settlement. One major simplification in the settlement module was that particles within the polygon boundaries were considered successful regardless of the type of bottom habitat.

Simulations and analysis of model output.—Ten virtual larvae were released in the surface layer of ROMS daily for 28–31 d, starting at day 1 of each month at midnight, from the two selected spawning sites (Dry Tortugas and Marquesas). Moon phase and stage of the tide were not taken into account at the moment of releasing larvae. Between 280 and 310 virtual larvae were released from each spawning site every month. Virtual larvae were tracked until they settled in the coastal nursery habitats represented by the polygons. Larvae were recorded as "settled" if they arrived within any of the habitat polygons. For each simulation, the CMS model generates two types of files. One output file provides information on the distance traveled, geographic position, and status of each particle. The other file is the connectivity matrix, in which each cell contains the number of larvae from each spawning site settling in each polygon. The monthly percentage of settlement was calculated by dividing the output number of larvae settled at the polygons by the total number of larvae released from each spawning site (i.e., Dry Tortugas and Marquesas).

Simulations conducted during the month of July 1996 were used to explore the contribution of a passive versus active (DVM and the STST behavior) drift on larval dispersal between offshore spawning and onshore nursery grounds. We analyzed three behavior scenarios: (1) passive dispersion or particles purely advected by the current for 30 d, (2) particles drifting with a DVM behavior for 30 d, and (3) particles drifting with a DVM behavior for 30 d plus an additional STST behavior starting at day 15. Another set of simulations was conducted during the summer months of June, July, August, and September 1995–1997 and the winter months of January, February, and March 1995–1997 to determine the effect of seasonality on dispersal and settlement. Simulations were conducted with behavior scenario 3. A two-way ANOVA was carried out on the percentages of settlement to determine the effect of the release location (Dry Tortugas and Marquesas) and the season (summer and winter) on transport and settlement. The percentages were arcsine transformed to resolve the nonnormal distribution of percentages (Zar 2010).

The probabilities of settlement from each spawning location (Dry Tortugas and Marquesas) were examined together with the Lagrangian trajectories to determine the main migration routes that larvae utilize to reach the settlement polygons. Data were separated by season (summer versus winter) and by two main migration routes: (1) larvae that arrive at the settlement polygons moving east–northeast across the SWF shelf and (2) larvae that arrive at the polygons moving south–southeast via the Florida Current and Florida Keys inlets and entering through Rebecca Shoal, Marquesas, Key West, and tidal channels of the Upper, Middle, or Lower Florida Keys.

Evaluation of the hydrodynamic model—The hydrodynamic model was evaluated primarily by comparing its representation of the tide to actual observations. The first approach was a statistical comparison of amplitude and periodicity in modeled time series to actual time series of tidal stages at several locations on the Florida coast and in the Florida Keys (Figure 1b). Statistical parameters were the linear correlation coefficient (R), its normalized standard deviation (σ; which indicates agreement in the amplitude of the signal), and the root-mean-square difference (RMSD) between the modeled and corresponding in situ values. The second approach was a graphical comparison of monthly mean modeled flow at seven observation sites on the SWF shelf to observed flow from acoustic Doppler current profiler (ADCP) data at those sites. The third approach was to compare model output fields to the general features of the seasonal circulation pattern and associated patterns of SST observed by Liu and Weisberg (2012). Equations for the statistical parameters are as follows:

For the correlation coefficient, R:

$$R = \frac{\frac{1}{N} \sum_{n=1}^{N} (m_n - \overline{m})(r_n - \overline{r})}{\sigma_m \sigma_r},$$

where N is the number of elements, n is the index, m indicates the model field, r is the reference field, the overbar is the average, and σ is the standard deviation. The correlation coefficient is in the range (-1, $+1$) and, if the two signals are perfectly in phase, $R = 1$.

For the standard deviation, σ:

$$\sigma^* = \frac{\sigma_m}{\sigma_r},$$

where $\sigma = 1$ indicates complete agreement in amplitude of the signal.

For RMSD:

$$RMSD = \left[\frac{1}{N} \sum_{n=1}^{N} (m_n - r_n)^2 \right]^{0.5}.$$

Partitioning components of the current.—Modeled current fields were examined for the effect on larval transport of

separated driving currents in order to better understand the main forces affecting variation in larval transport pathways and settlement success between seasons and spawning ground. The main forces driving current patterns in the model were wind, tide, and geostrophic, the latter of which arose mainly from density gradients formed by the combination of temperature and salinity. The wind- and tidal-driven current velocity and direction were obtained by first conducting the simulations with and without each of the two and then subtracting both forced and unforced fields from each other. To understand the effect of the background flow, which only includes the geostrophic and wind-driven current, we calculated what we called the subtidal flow. Because of the critical role of the 12-h DVM period in the shrimp larvae transport, the mean 12-h DVM flow was used to understand its role in the fate of larval transport. Then because of the 15-d time period that characterizes the ontogenetic behavior, the mean, subtidal, wind-driven, tidal, and 12-h DVM flow were averaged biweekly to understand the balance of forces on the fate of larval transport over a 15-d time window. Finally, in order to understand the effect of the wind on the SWF shelf circulation, we projected the wind vectors on the axis of greatest variance as defined by van Aken et al. (2007) in order to capture the direction of most significant influence on the circulation.

RESULTS

Hydrodynamic Model Evaluation

Tides and tidal flow.—The model tides during periods of simulated larval releases were compared with data of the same period from in situ tidal gauges along the southern Florida coast and in the Florida Keys. In Figure 2, the model mean sea levels are overlaid on the ones observed in Lake Worth, Naples, Clearwater, St. Petersburg, and Key West. Beside the relatively coarse resolution of the coastal model, the site specificities of each tidal location shown here are very well reproduced by the model.

Results in Table 2 indicate that the model tide is in phase with the observed tide at all tide gauge locations shown in Figure 1b (i.e., $R \geq 0.75$ at all times and all locations). In Table 2, σ^* is on average between 1 and 1.5, which indicates that the model tends to overestimate the tidal amplitude and particularly at locations that are in bays (Naples, St. Petersburg) and near a tidal channel (Key West). The RMSD values (also in Table 2) are higher and the R-values are lower in winter than in summer, which may be explained by the stronger winds in winter that yield significant coastal sea level set-up or set-down (Lee et al. 2002; Liu and Weisberg 2012). Smooth bathymetry and coarse resolution are likely to affect the model solution.

Further model flow evaluation on the shelf and in the vicinity of the spawning grounds is presented in Figure 3. The biweekly mean current halfway through the incoming tide, as it is in the tidal phase ridden by pink shrimp larvae during the

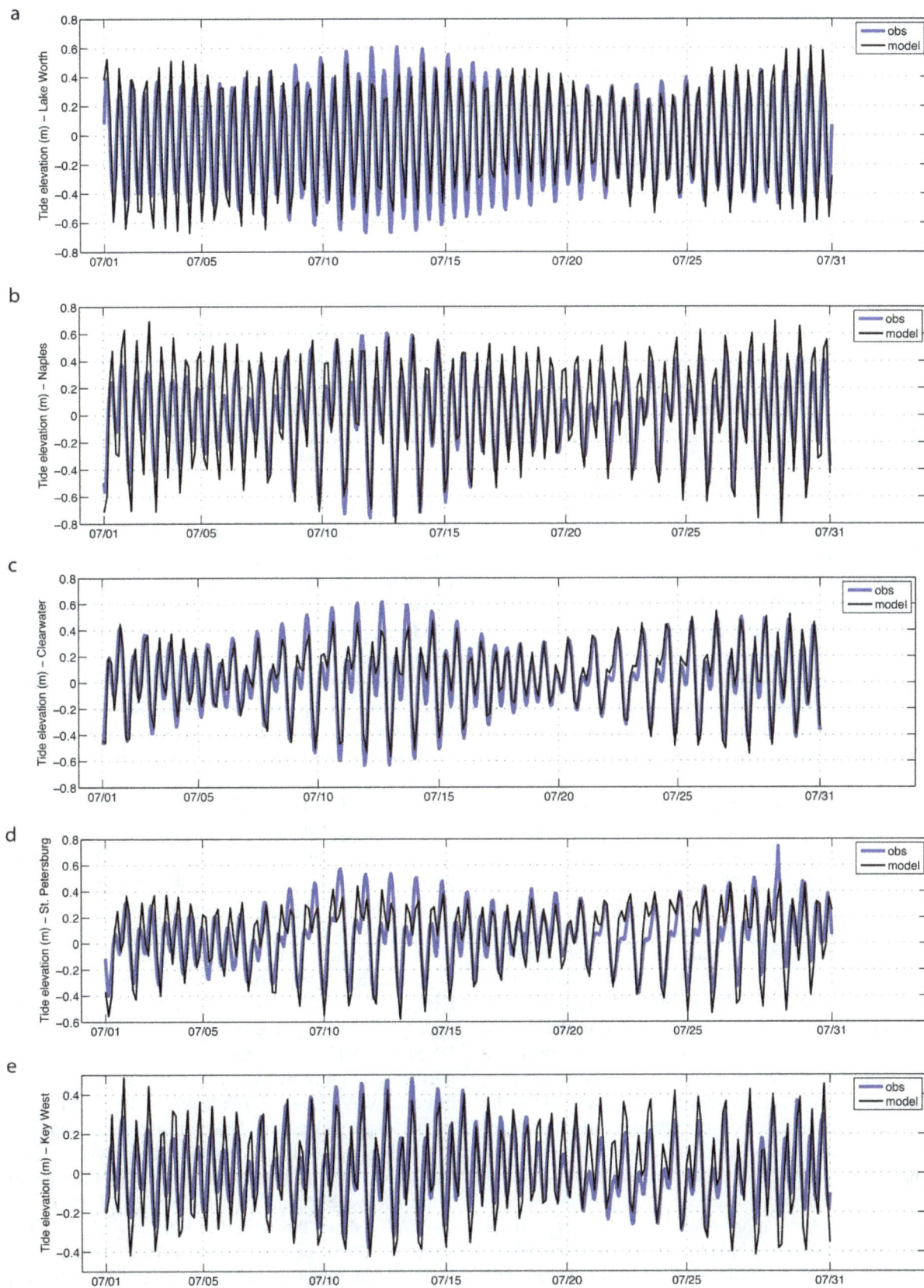

FIGURE 2. Time series of mean sea level in model month July 1995 at the following sites in Florida: (a) Lake Worth, (b) Naples, (c) Clearwater, (d) St. Petersburg, and (e) Key West. The thick blue line shows the tidal gauge measurements and the thin black line shows the model's water level; obs = observed.

TABLE 2.　Pattern statistics of model sea-surface height during months of Lagrangian experiments at tidal gauge locations in Florida. Gauge data were obtained from http://tidesandcurrents.noaa.gov. Abbreviations are as follows: R is the linear correlation coefficient, σ^* the standard deviation ratio, RMSD the root-mean-square difference, S the summer months, and W the winter months.

Statistics	Lake Worth		Naples		Clearwater		St. Petersburg		Key West	
	S	W	S	W	S	W	S	W	S	W
R	0.83	0.77	0.90	0.83	0.93	0.81	0.81	0.77	0.83	0.74
σ^*	1.12	1.05	1.41	1.51	1.04	1.15	1.24	1.32	1.40	1.50
RSMD	0.19	0.22	0.18	0.23	0.09	0.16	0.15	0.18	0.13	0.16

STST transport, is shown. Both the ADCP measurements (Table 3) and the model are interannually consistent in flow amplitude and direction and have limited monthly variability (Figure 3a, b). The model is in agreement with the flow speed and direction at most locations on the shelf, although discrepancies exist at the mouth of Florida Bay.

Mean flow, SST, and SSS.—In order to verify the model mean flow, we calculated the biweekly mean flow both in winter (Figure 4) and in summer (Figure 5) of 1995 and 1997 without de-tiding the flow before averaging. We chose a biweekly average because the larval behavior is divided into two types of behavior: DVM that operates exclusively during the first 2 weeks of development and STST that starts after the first 2 weeks. Therefore, larvae may not face the same mean flow from one fortnight to another.

The modeled circulation in winter (January 1995 and 1997) is characteristic of an upwelling circulation and temperature pattern (Figure 4). Coastal waters on the inner shelf are cooler than offshore. The mean flow is parallel to the coastline, directed mainly southward and intensified at the edge of cool waters. In the Dry Tortugas and Marquesas region, the mean flow is to the southwest. This state of the model is fairly similar to the finding of Liu and Weisberg (2012).

In summer (July 1995 and 1997), the cross-shelf temperature gradient is reversed, as reflected in modeled mean flow (Figure 5). Warmer waters are found inshore and sustain a downwelling circulation pattern, which appears to be unstable. Indeed, the flow is southward on the inner shelf during the first fortnight of July 1995 and turns northward during the second fortnight. Overall, the monthly mean flow has a consistent downwelling circulation pattern (not shown). The mean flow at the spawning grounds is to the northwest. In July 1997, the downwelling circulation only occurs in the second fortnight of July and is very weak. The biweekly mean flow is very weak to the south in coastal waters and to the southwest at the spawning grounds. Overall, the model downwelling regime agrees well with the observations of Liu and Weisberg (2012). As expected, the circulation is in balance with the wind, whose axis of greatest variance is downwelling favorable in July 1997 (Figure 6).

Summer is the rainy season in southern Florida. There is significant variability between years in the input of freshwater into coastal waters as well as shelf waters, as indicated in our analysis of modeled biweekly salinity (Figure 7). The salinity in coastal waters is about 0.75‰ lower on the SWF shelf in July 1997 than in July 1995. The plume of low-salinity water extends further offshore all along the western Florida coast in 1997. Lee et al. (2002) suggested that the freshwater plume is likely to drift southeast along the coast and southwest as it nears the Florida Keys. This motion would oppose the northward coastal current and yield the weaker downwelling circulation in July 1997, even though wind magnitude and direction are similar to July 1995 (Figure 6).

Passive versus active larval transport.—Larvae deployed at the Marquesas and Dry Tortugas spawning stations in July 1996 as passive particles (behavior scenario 1) moved mainly north-northeast along the SWF shelf (Figure 8a, b). From the Dry Tortugas station, some larvae also moved west and entered the LFC system, while the LFC had no effect on larval dispersal from Marquesas. Larvae deployed at Marquesas and Dry Tortugas with a DVM behavior (scenario 2) showed a similar north-northeast drift to those of passive larvae (Figure 8c, d). However, the northern distance traveled with the DVM was greater than that of the passive larvae, suggesting that the DVM behavior may improve the northward drift. The eastward distance traveled was only a few kilometers per larvae. Larvae deployed at the Marquesas and Dry Tortugas sites with DVM and STST behavior (scenario 3) moved mainly east-northeast (Figure 8e, f). The maximum eastward distance traveled was about 200 km, and ~30% and ~63% of the larvae deployed at the Dry Tortugas and Marquesas sites, respectively, reached the settlement polygons located north of Florida Bay. A small number of larvae deployed at Dry Tortugas were caught up in the LFC system and entered the SWF shelf through the Rebecca Shoal Channel, but none of them entered Florida Bay through the Florida Keys. Trajectories of larvae were similar for both spawning grounds, suggesting that tidal currents may be strong

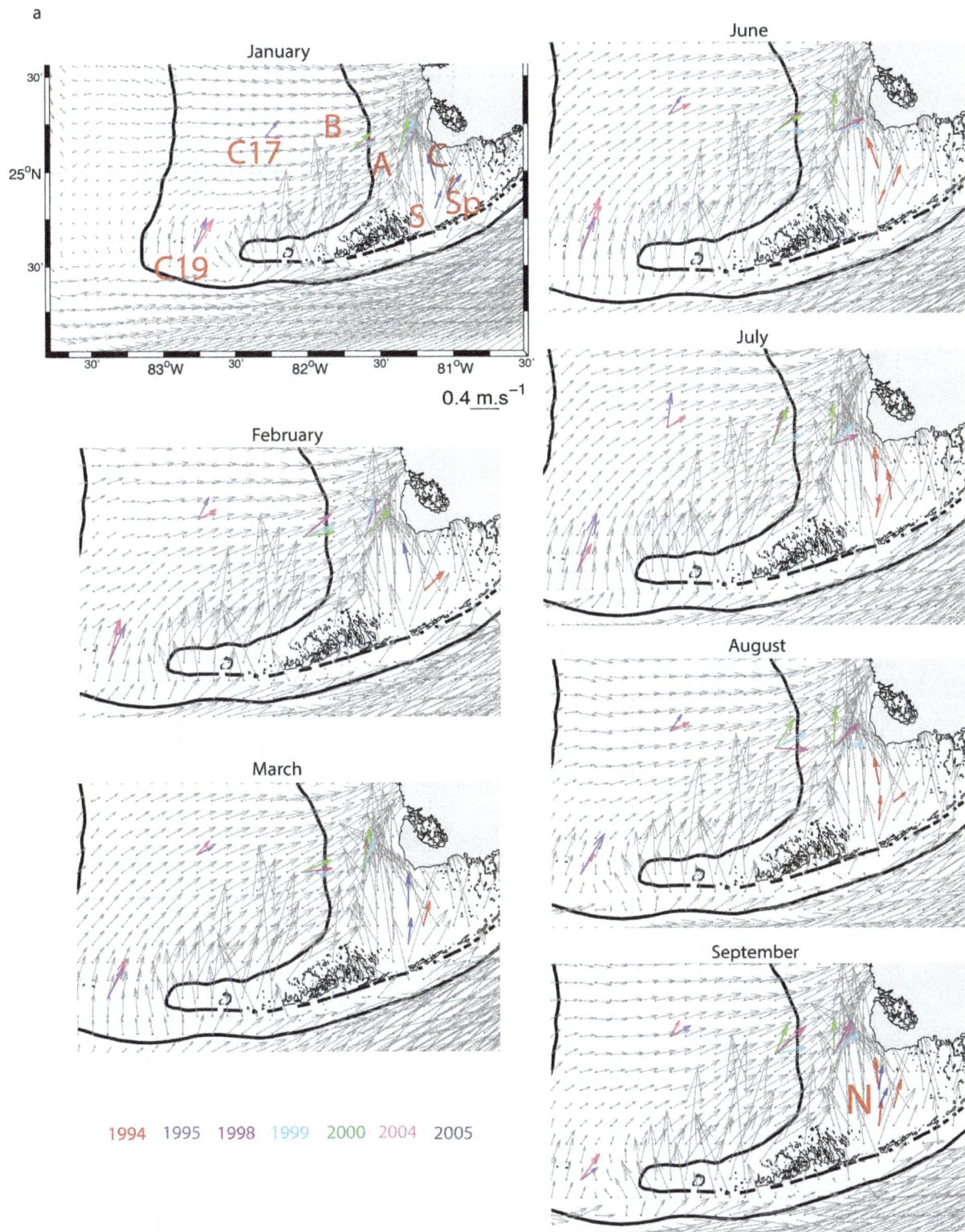

FIGURE 3. (a) In situ current vectors overlaid on model year 1995 current vector maps and (b) the model current vectors for 1996 and 1997 at the measurement locations. The vectors shown correspond to the 2-week average of the flow halfway through the incoming tide on the southwest Florida shelf. Each color corresponds to a year. Only months when virtual larvae were released are shown. The names of the moorings listed in Table 3 are shown in red letters on the January and September panels. (*Continued on next page*)

b

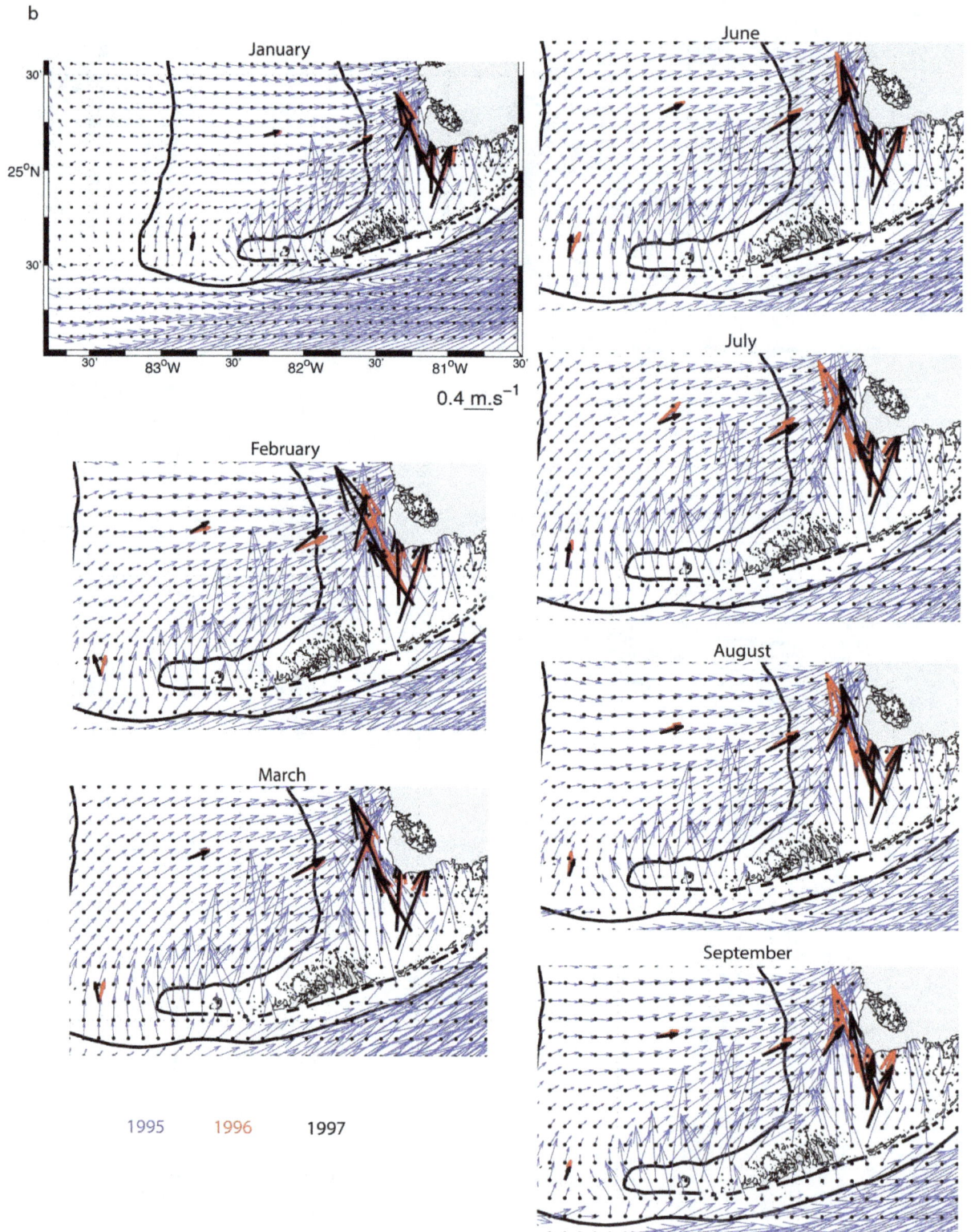

FIGURE 3.   Continued.

TABLE 3. Acoustic Doppler current profiler data used for model verification, showing the recorded time period (X indicates data for the whole year). Each location of data collection is shown on Figure 3a. The NOAA South Florida Program (SFP) data are available online at http://www.aoml.noaa.gov/phod/sfp/data/index.php. The Harbor Branch Oceanic Institute (HBOI) Florida Bay (FB) data were made available by N. Smith to the National Oceanographic Data Center (NODC Accession 582, http://www.nodc.noaa.gov/archive/arc0001/0000582/). The University of South Florida (USF) data were kindly made available to us by B. Weisberg and L. Zheng.

Data source	Abbreviation	1994	1995	1997	1998	1999	2000	2004	2005
Model			X	X	X	X	X		
NOAA SFP southeast A	A			Sep–Dec	X	X	X		
NOAA SFP southwest B	B			Sep–Dec	X	X	X		
HBOI Sprigger	Sp	X	Jan						
HBOI FB mouth south	S	X	Jan–Apr	Aug–Nov					
HBOI FB central	C	X	Jan–Apr						
HBOI FB north	N			Aug–Nov					
USF C17	C17							X	X
USF C19	C19							X	X

enough to facilitate STST transport from as far as the Dry Tortugas site.

Variability in Settlement and its Drivers

Variability in settlement.—Simulated settlement data indicated that there is a high interannual and intermonthly variability in settlement at the coastal nursery habitats (Figure 9), as well as substantial differences in the percent arriving from the two spawning grounds. The highest percentage of settlement occurred during 1996 in almost every month, independent of whether the origin of deployment was from the Dry Tortugas or Marquesas site (Figure 9). July 1996 had the highest settlement rate, and July was the consistently highest settlement month each year. An ANOVA conducted on the percentage of settlements indicated that season (summer versus winter) and location of spawning have a significant effect on the success of larvae settling in the coastal habitats; however, the interaction of these two factors did not have a significant effect (two-way ANOVA for location and season: $P < 0.001$; interaction $P = 0.07$). The larvae released at the Marquesas site during the summer months had the highest larval settlement rate, 15.3 ± 3.1 (mean $\pm$ SD), which constitutes ~81% of the total settlement. Larvae released at the Dry Tortugas site during the summer had a settlement rate of 3.4 ± 3.1. Settlement rates during the winter months were about one-fifth the summer rates. The settlement rate of larvae released at Marquesas during the winter months was 2.4 ± 3.7, while that of larvae released at Dry Tortugas during the winter was 0.9 ± 3.7.

Variability in recruitment pathways.—The probabilities of settlement indicated that larvae recruiting to the settlement habitats might arrive from different routes depending on release location, season of the year, and predominant oceanographic conditions (Figures 10, 11). The examples shown in Figure 10 indicate that larvae released at Dry Tortugas and

Marquesas in the summer months of July 1995 and August 1996 tended to remain on the SWF shelf, moving northeast, and a large percentage of larvae reached the settlement habitats. In three of the four illustrated periods, a part of larvae released at Dry Tortugas were pushed southward and later caught within the LFC system. Especially in July 1995, some of these larvae entered the shelf through the Rebecca Shoal Channel and crossed the tidal channels of the Florida Keys to reach settlement habitat. The rest of the larvae were flushed further north with the LFC. The example from August 1996 indicated different oceanographic conditions at the Dry Tortugas–Marquesas region during the first days of deployment, when larvae were migrating with only a DVM behavior. Some larvae remained in the vicinity of the spawning ground for several days, while some drifted north and others south. After that, during the STST period, some larvae moved eastward across the SWF shelf, but the ones who went south were entrained by the LFC flow and a few entered the shelf through the Rebecca Shoal Channel and the tidal channels of the Florida Keys.

The trajectories from winter showed different patterns from those of summer (Figure 10). In the January and March 1997 examples, none of the larvae released at Marquesas reached the settlement habitats by traveling across the SWF shelf. Those released at the Dry Tortugas in January 1997 drifted westward from the spawning grounds during the DVM period and then were flushed into the LFC system, and only one larva entered the nursery ground area, moving through the Middle Florida Keys. In the March 1997 example, the number of larvae moving with the LFC system was larger and the number of settled larvae was increased in comparison to January 1997.

Probability analysis of the settlement rates at each spawning site indicated that settlement rates and recruitment paths change radically between summer and winter (Figure 11). During summer the majority of larvae recruited to the coastal

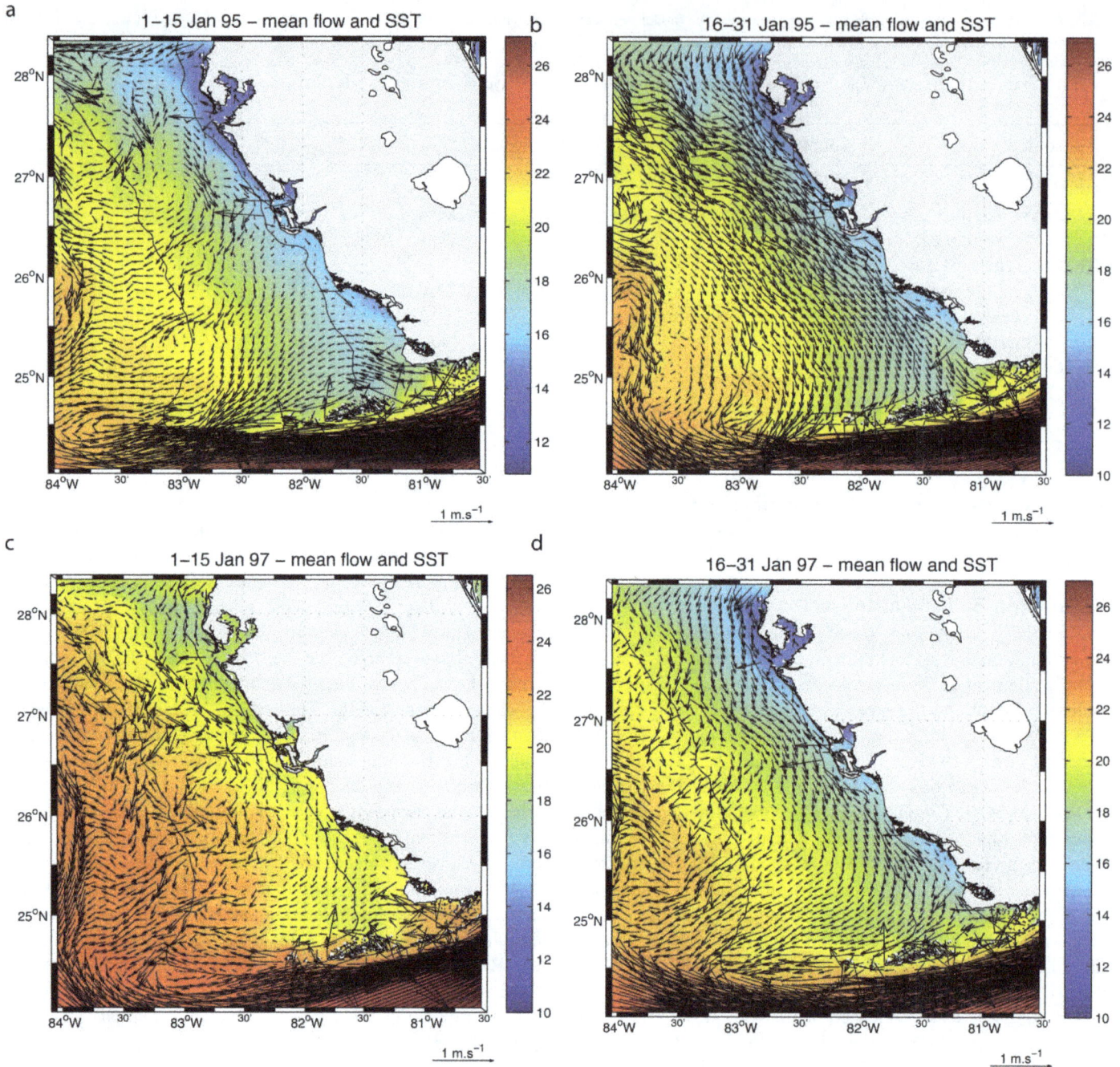

FIGURE 4. Model sea surface temperature (SST; colored bars in °C) and current vectors during winter 1995 and 1997, showing the (a) first and (b) second fortnight in January 1995 and the (c) first and (d) second fortnight in January 1997.

areas moving across the SWF shelf, while recruitment through the Florida Keys was more often in winter months. In summer, about 15% of the larvae originating from Marquesas and 2% of the larvae originating from Dry Tortugas arrived at settlement habitats moving across the shelf, while recruitment through the Florida Keys was 0.1% of larvae originating from Marquesas and 1% of larvae originating from Dry Tortugas. The pattern changed substantially during the winter months. About 1.5% of recruited larvae originating from Dry Tortugas

and 0.9% of larvae originating from Marquesas arrived through the Florida Keys channels during winter months. Recruiting larvae moving across the shelf during winter months made up only 1% of those originating from Marquesas, and there was no recruitment across the shelf from larvae originating from Dry Tortugas during winter months.

Seasonal and interannual variability affecting recruitment.— We previously showed that the seasonal variability on the WFS consists mostly of a current reversal associated with an

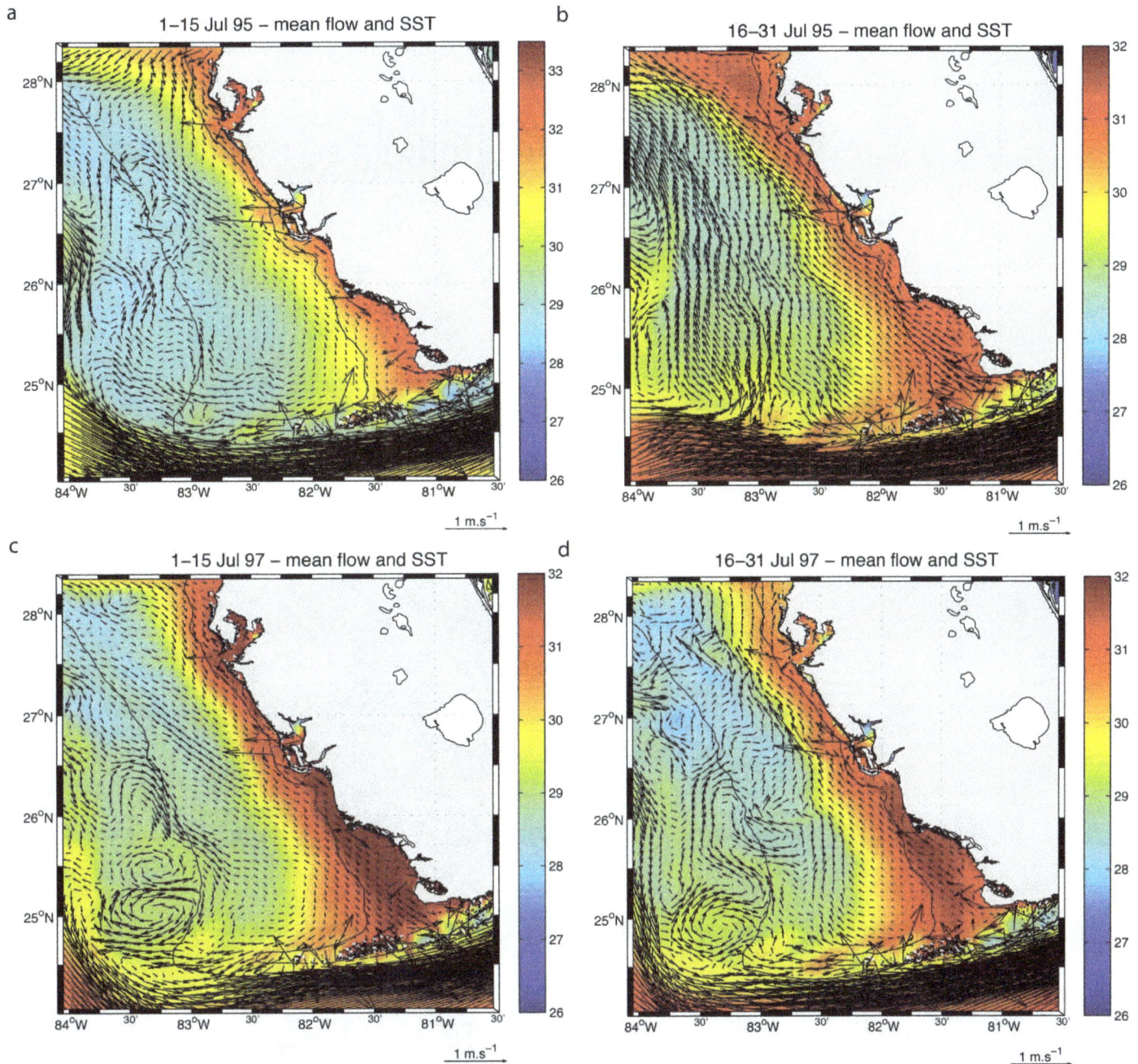

FIGURE 5.   Model sea surface temperature (SST; colored bars in °C) and current vectors during summer 1995 and 1997, showing the (a) first and (b) second fortnight in July 1995 and the (c) first and (d) second fortnight in July 1997.

upwelling (winter) circulation and a downwelling (summer) circulation. To reach the nursery grounds, larval behavior changes from DVM to STST, and the forces affecting larval transport pathways and recruitment may be different during each of these two behavioral phases.

Diel vertical migration consists of the vertical migration of marine planktonic organisms from deep waters to near surface approximately every 12 h, usually synchronized with the local diurnal cycle occurring at night (e.g., Dall et al. 1990). We,

therefore, examined the potential influence of the direction and magnitude of modeled background flow on the direction and magnitude of the modeled 12-h DVM mean flow and percent of larvae reaching the nursery grounds in the winter and summer months of 1995, 1996, and 1997 (Figure 12). During the winter months of all 3 years, the three flows are toward the same direction (southeast to southwest, i.e., $> 90°$ to $< 270°$), which is away from the coast, resulting in the lowest settlement levels. Transport through Florida Keys channels is the

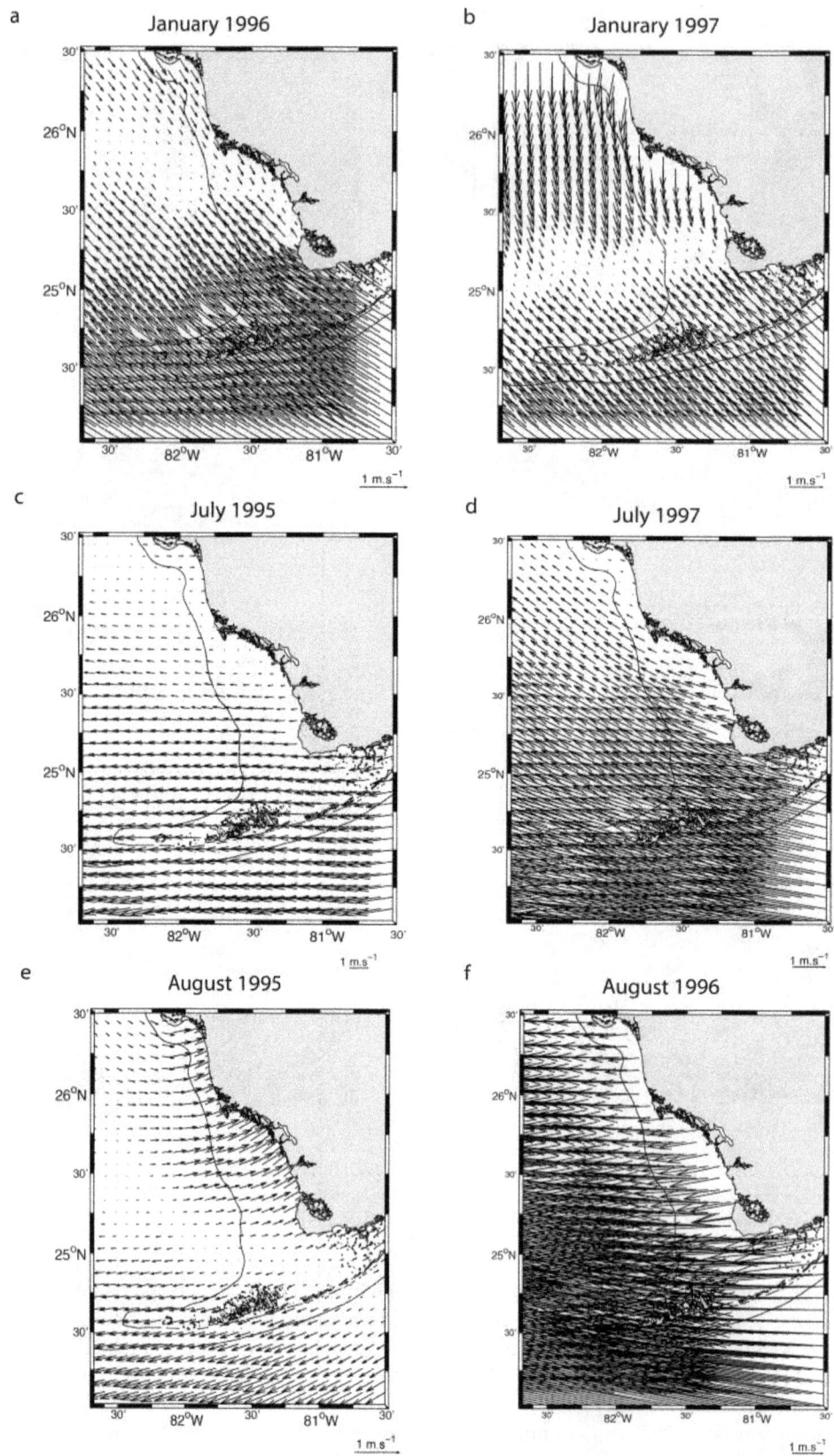

FIGURE 6.   Monthly wind model vectors map along the axis of greater variance for **(a)** January 1996, **(b)** January 1997, **(c)** July 1995, **(d)** July 1997, **(e)** August 1995, and **(f)** August 1996.

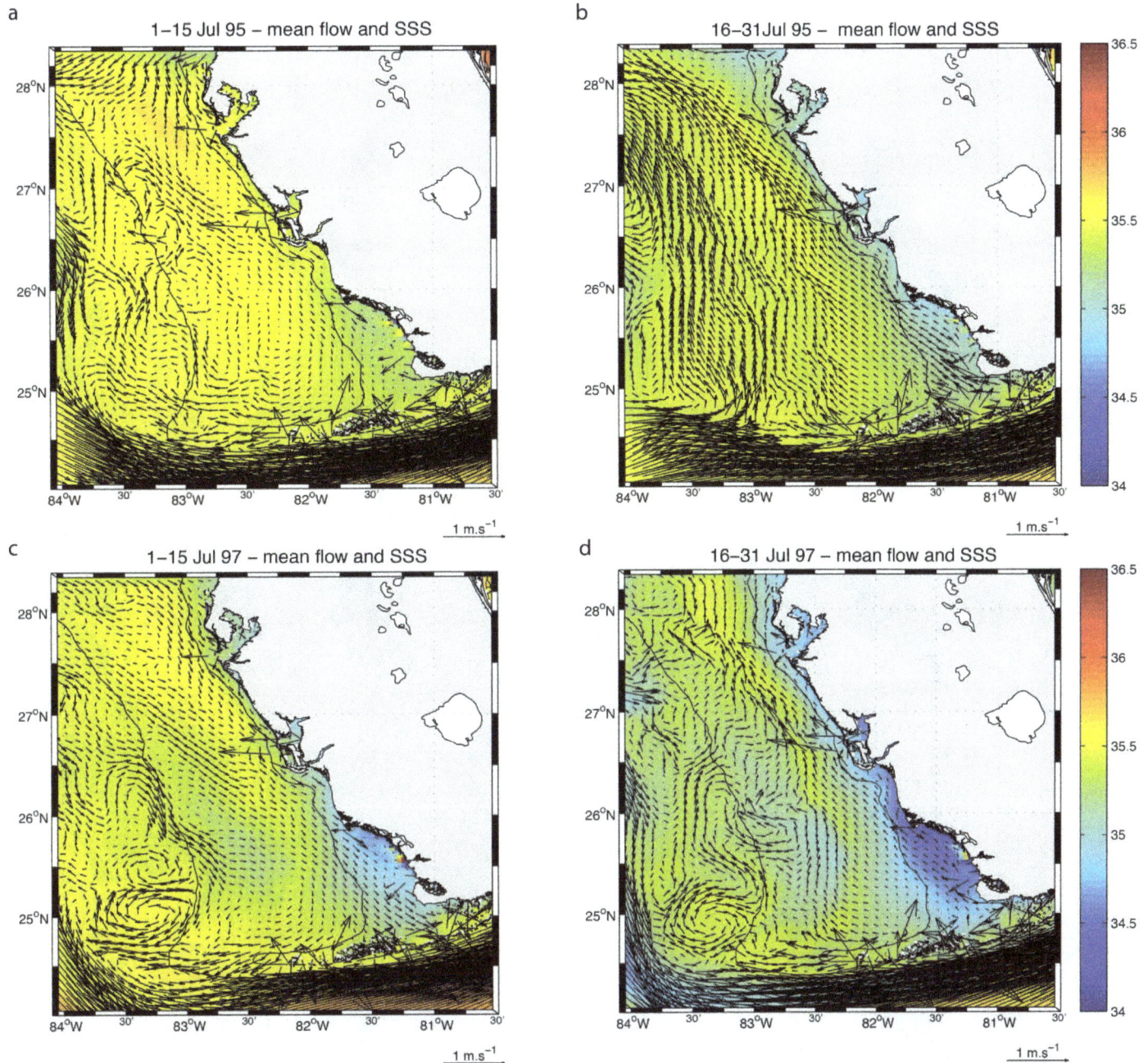

FIGURE 7. Fortnightly model of sea surface salinity (SSS; colored bars in ‰) and current vectors during summer 1995 and 1997, showing the (a) first and (b) second fortnight in July 1995 and the (c) first and (d) second fortnight in July 1997.

main pathway for recruiting larvae in the winter months (Figure 11). During the summer months, there is more deviation between the three flows. The highest levels of settlement were obtained in summer months when the 12-h mean DVM flow has an eastward or northward component at both spawning grounds, Dry Tortugas and Marquesas. In June 1996, the 12-h mean DVM flow was to the southwest at Marquesas but to the southeast at Dry Tortugas. The 12-h mean in June 1996 revealed a cyclonic gyre in the vicinity of the spawning

ground, which biased the mean toward the southwest, although particles remained in the vicinity of the spawning location (Figure 13).

Transport during the DVM period will carry larvae a considerable distance from the spawning grounds to their nursery grounds, even when current patterns are not ideal, as in June 1996 (Figure 13a, b). Some of the larvae originating from Dry Tortugas advanced northeast during the DVM period despite the confusion of currents associated with the previously

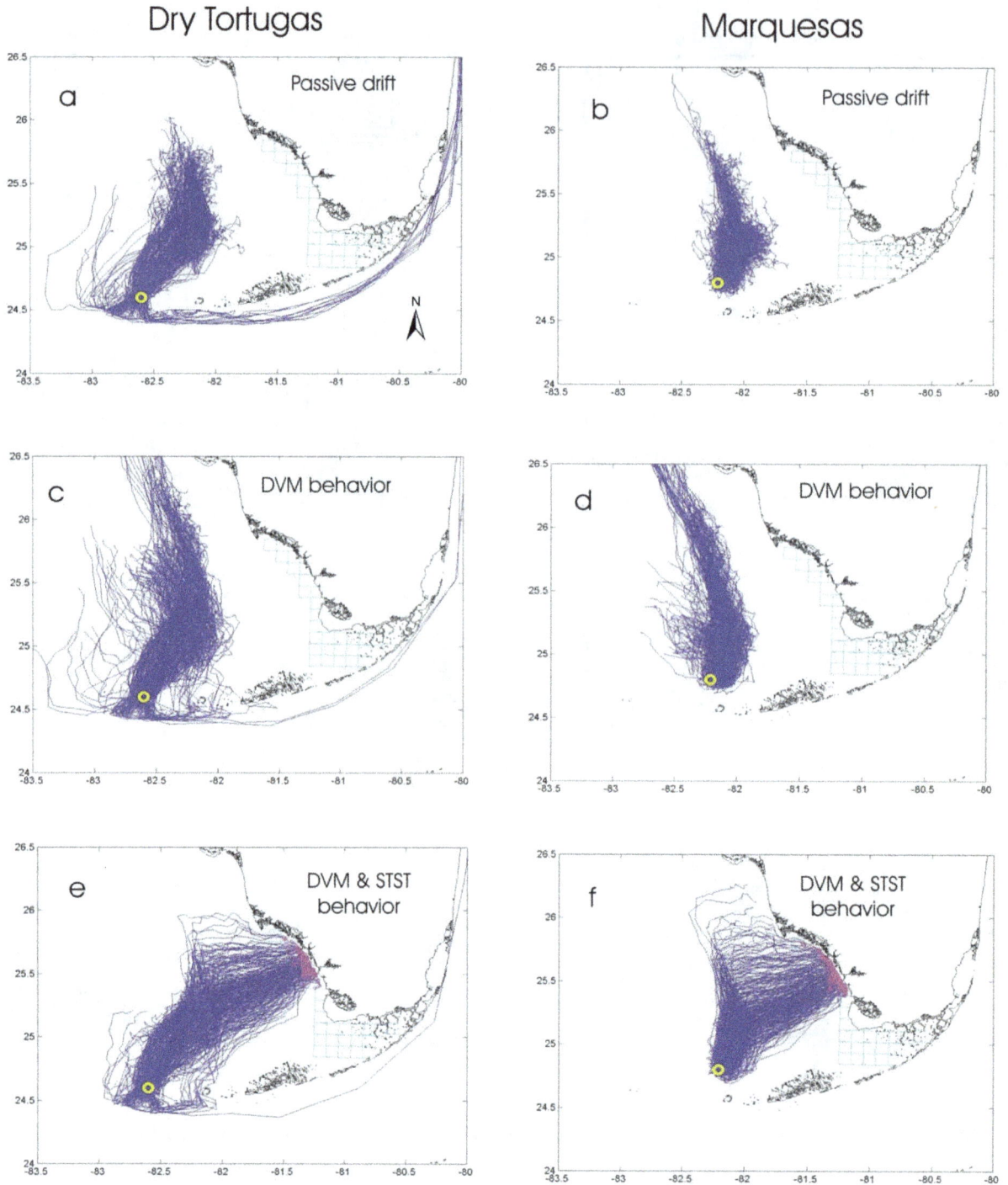

FIGURE 8.   Simulated trajectories of pink shrimp larvae under different scenarios of behavior: **(a, b)** passive transport, **(c, d)** active transport with diel vertical migratory (DVM) behavior, and **(e, f)** active transport with DVM and selective tidal stream transport (STST) behavior. Virtual larvae were released daily from July 1–31, 1996, at two spawning locations, Dry Tortugas and Marquesas (yellow circles). Purple dots indicate the larvae that reached the settlement habitats, which are represented by the cyan-colored polygons.

mentioned cyclonic gyre of that period, which is reflected in the fortnightly mean 12-h DVM flow (Figure 13c). The match between trajectories during the DVM period and the 15-day mean of the 12-h mean DVM flow may be inferred by comparing panel c with panels a and b in Figure 13. Nonetheless,

there is a strong seasonal variability in the DVM mean flow, which is mostly westward in winter and northeastward or weak in summer (Figure 12).

Interestingly, the STST movement is always eastward, independent of the time of the year (Figures 3, 13). The tidal

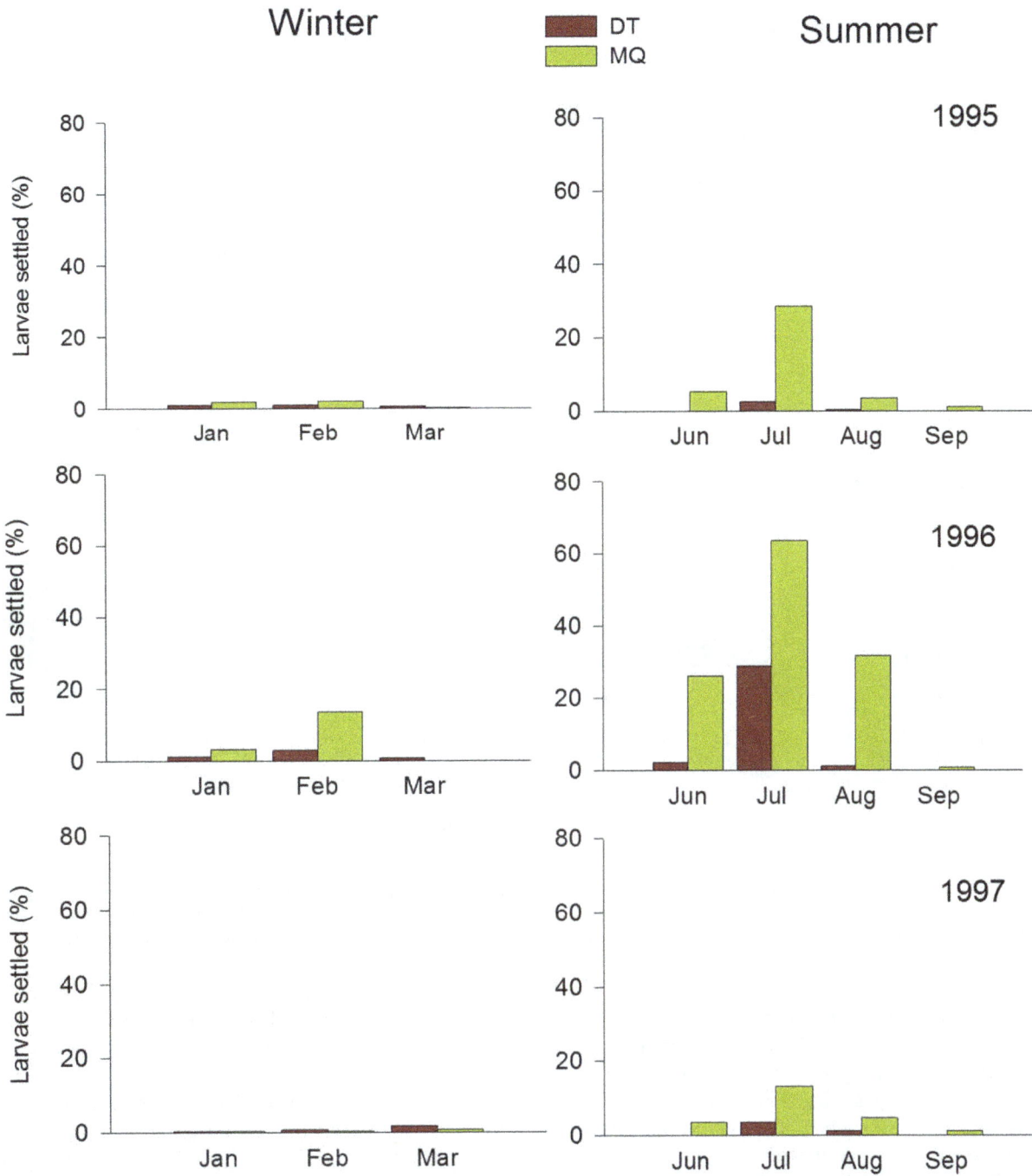

FIGURE 9.   Monthly percentages of simulated pink shrimp larvae released at the Dry Tortugas (DT) and Marquesas (MQ) spawning locations (on the first day of the month at midnight of the indicated date) that settled in the nursery habitat. Virtual larvae were moved with diel vertical migration (DVM) and selective tidal stream transport (STST) behaviors during the summer and winter months of 1995–1997.

incoming flow is always directed toward the coast on the SWF shelf and toward Florida Bay in the channels through the Florida Keys reef tract. Therefore the recruitment variability of larvae transport across the SWF shelf is mostly dependent upon the fate of larvae during the DVM period, which favors successful recruitment in summer. The lack of consistency among years in the summer DVM mean flow likely drives the interannual variability of summer recruitment levels. If the mean flow during the DVM period is critical for the fate of pink shrimp larvae, it is important to understand the drivers of the DVM flow, which are described in the following sections.

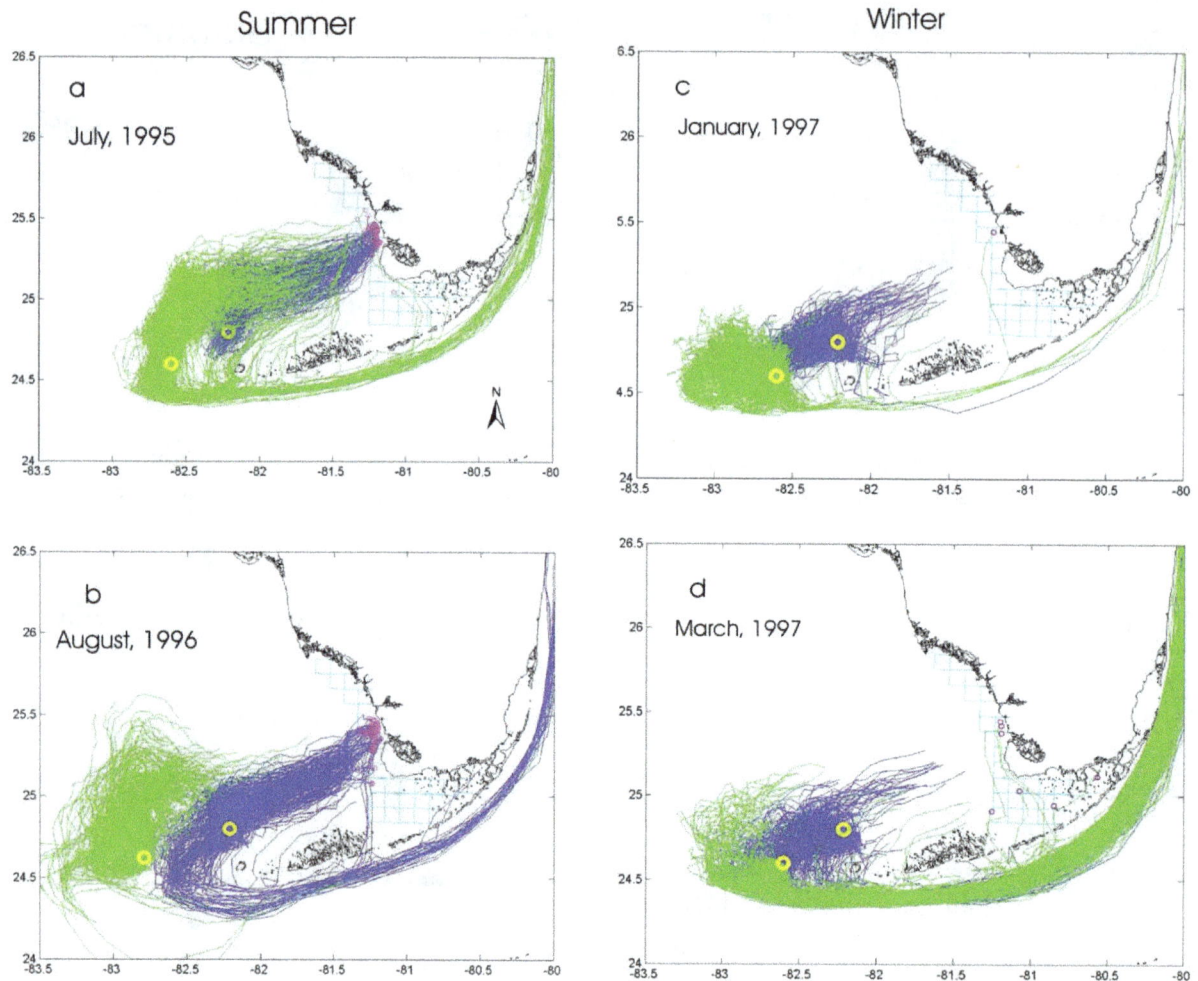

FIGURE 10.   Simulated trajectories of pink shrimp larvae released at the Dry Tortugas and Marquesas spawning sites (yellow circles) and moving with diel vertical migratory behavior and selective tidal stream transport behavior during the summer months of **(a)** July 1995 and **(b)** August 1996 and the winter months of **(c)** January and **(d)** March 1997. The purple dots represent the larvae that settled in the settlement habitats, which are represented by the cyan-colored polygons.

Ekman transport during the DVM window.—As shown by Liu and Weisberg (2012), the WFS exhibits a robust seasonal cycle in velocity, dominantly driven by the wind, but also in temperature and salinity gradients across the shelf. Rather than the arithmetic mean wind direction, we calculated the characteristic mean direction of the wind on the shelf identified by the direction in which the total variance of the velocity component is maximized (van Aken et al. 2007). This direction is the one that has the most influence on the wind-driven circulation. From October to April, the wind is predominantly blowing from the northeast, with a stronger southward component in winter than in fall in the central WFS and a stronger eastward component on the SWF shelf overall. However, year-to-year variability can be significant and wind patterns inconsistent, not only from year to year but spatially within the same timeframe. For example, in January 1996 the monthly wind direction consisted of strong winds from the south-southeast on the SWF shelf and winds from the northwest on the central WFS (Figure 6a). Similar inconsistencies were also seen in January 1997 (Figure 6b). These wind features could set up opposing circulation cells on the SWF shelf. From June to August, the wind blows from the southeast on the WFS, with a stronger northward component on the central WFS than on the SWF shelf, where it is more west-northwest (Figure 6c, d). Winds during summer are weaker than those of the rest of the year, unless a hurricane is in the vicinity. Again, wind anomalies exist from one year to another and are characterized by opposing wind directions between the central WFS and the SWF shelf, as in August 1995, or by stronger trade winds, as seen on the SWF shelf in August 1996 (Figure 6f).

In order to estimate the wind-driven current on the SWF shelf, we subtracted two simulations that were identical except that one was wind forced. This revealed, for example, that in model month July 1995 the surface flow in the vicinity of the spawning grounds during the 2-week DVM period was

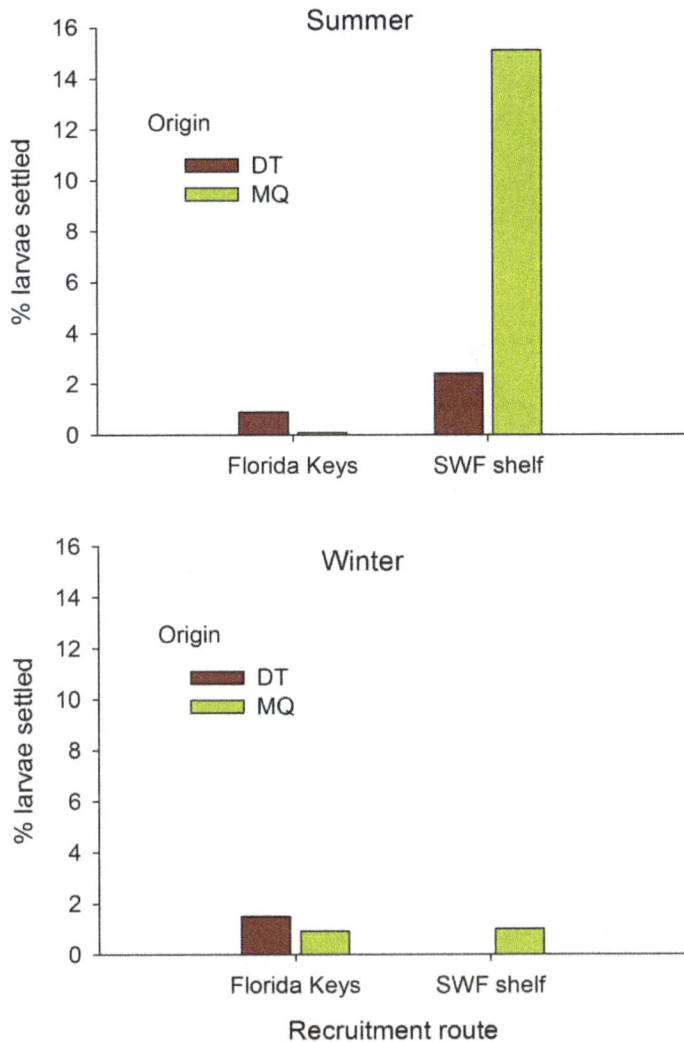

FIGURE 11. Mean percentages of pink shrimp larvae settled within the nursery habitat in summer (upper figure) and winter (lower figure) calculated from the connectivity matrix data based on 3 years of simulations (1995–1997). Larvae originated at the spawning sites of Dry Tortugas (DT) and Marquesas (MQ) and migrated for 30 d with diel vertical migration and selective tidal stream transport behaviors. Settlement percentages were separated into two recruitment routes: larvae that arrived to the settlement habitat moving across the southwest Florida (SWF) shelf and larvae that arrived via the Florida Current across the Florida Keys channels.

westward (i.e., easterlies), the same as the direction of greatest variance in the wind (Figure 14a). This suggests that sustained easterlies are unfavorable to cross-shelf transport.

Tidal-phase-induced variability and DVM mean flow.— Tidal currents are interannually consistent and are always in the same direction at a given stage. An example of tidal-driven flow during the DVM period was obtained for July 1995 by calculating the difference between two simulation fields that are identical except that one was forced by tide (Figure 14b). The fortnightly 12-h DVM tidal current was to the northeast at both spawning grounds when the DVM was initiated at midnight on the first day of release. But the DVM mean flow is

likely to respond to timing in relation to the phase of the tide as shown by the different flow fields obtained for different onsets of the DVM period (Figure 15). Each DVM mean flow shows markedly different circulation patterns over the SWF shelf and, in particular, at the spawning sites. The DVM mean flow can even be in opposite directions, depending on the phase of the tide at the time of the particle's release. This suggests a substantial contribution of the tidal flow to the DVM mean flow.

The sensitivity of the timing of the 12-h mean DVM flow direction to the phase of the tide was tested in order to estimate the optimum release time and tidal cycle that adult shrimp should choose to maximize the eastward flow of their larvae, which would potentially increase their recruitment level. For example, at the Marquesas spawning ground, if the DVM behavior would be started at midnight on July 1, 1995, the mean resultant tidal flow is to the northeast and the incoming flow is synchronized with the nighttime during the 2-week duration of the DVM period (Figure 16a). If the beginning of the 12-h DVM window is shifted forward in time by 6 h, the resultant mean flow is to the southwest (or northwest for a 12-h shift) (Figure 16b, c). Moving the DVM windowing further forward in time, we find that the tidal phase and cycle that leads to the strongest northeastward current is obtained when the 12-h DVM window is started 2 d before the full moon in July 1995. The beginning of the DVM window falls within the incoming tide, which starts at sunset (Figure 16d). The same effect is obtained at the Dry Tortugas spawning ground.

Subtidal flow and coastal salinity influence.—Baroclinicity from freshwater discharge into the nearshore region can locally modify the local advection of riverine water as shown by Lee and Smith (2002), Hu et al. (2004), and Zhao et al. (2013). Although the wind stress was stronger in July 1997 than in July 1995 (Figure 6), the northward flow was much weaker during the second fortnight of July 1997 than in 1995, while the water was fresher at the coast in 1997 than in 1995 (Figure 7b, d). The downwelling-driven northward current can be turned southward by the riverine water in the nearshore region (Lee and Smith, 2002). With increased discharge, the freshwater flow would not only be flushed southward through the Florida Keys channels (Lee and Smith 2002) but also be pushed westward toward the Marquesas and Dry Tortugas sites, as shown by Hu et al. (2004) and Zhao et al. (2013). In order to verify this effect in the model, we calculated the salinity gradient across the SWF shelf and the corresponding monthly subtidal current direction and amplitude for every month of July from 1995 to 2002 (Figure 17). The model shows that large negative salinity gradients, with the lowest salinity inshore and increasing salinity offshore (as in July 1997, 1998, 2001 and 2002), are associated with a southwestward-westward flow at Marquesas and Dry Tortugas, which confirms the westward transport of riverine water north of the Florida Keys. During dry summers, the flow is either

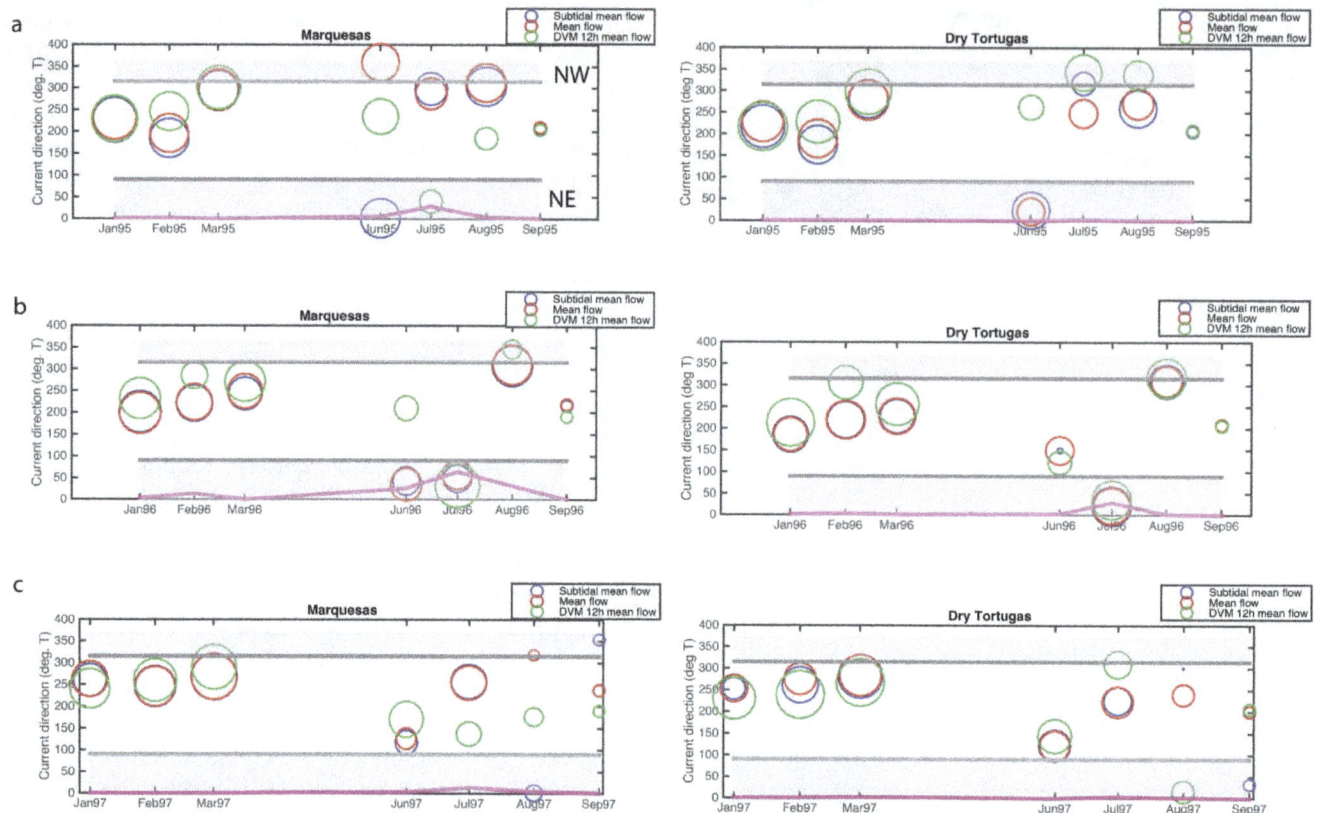

FIGURE 12.   Discrete flow magnitude (circle diameter) and direction (y-axis scale gives compass coordinate, 0° to 360°, of the center of the circle) of the first fortnight subtidal mean flow (blue circles), mean flow (red circles), and 12-h mean diel vertical migration (DVM; green circles) timed with each particle release for (a) 1995, (b) 1996, and (c) 1997 at the Marquesas and Dry Tortugas spawning grounds. The legend circle diameter is equivalent to 5 cm/s. Only the months of pink shrimp transport simulation are shown. Shaded regions indicate a northward component in the flow direction. The thin pink line shows the recruitment percentage on the y-axis, where label 100 on the y-axis corresponds to 100%. Mean flow was calculated in a box around the release location representing the spawning ground.

southward or northward, with an eastward component most of the time. Therefore, independently of the SWF shelf seasonal circulation, the flow at the spawning grounds seems to be under the influence of the freshwater discharge at the coast, which is influenced by both rainfall and water management that controls freshwater inflows to Everglades National Park. Because the influence of tide is removed, the flow direction remains the same if the DVM windowing is applied (i.e., the timing of the 12-h DVM period relative to the tidal phase would not matter), yielding that the subtidal flow during the DVM period would tend to have an eastward component during dry summers.

DISCUSSION

Overall, the ROMS simulations reproduce the dynamics of the WFS as observed by Liu and Weisberg (2012) for both seasons. However, some discrepancies exist in Florida Bay and shallow areas, where the model overestimates velocity magnitudes as shown by Figure 3. Because of resolution-dependent numerical stability constraints, a minimum depth (here 3 m) was set in shallow areas. Therefore, the shallow banks that

slow down the flow at the mouth of Florida Bay are not accounted for in the model, which explains the relatively high velocities of the model in regards to the ones observed, and the water is much deeper in the model throughout the bay. The model accurately simulates the SST and SSS seasonal cycles on the outer and inner shelf. It also roughly simulates the low-salinity patterns as observed by Lee and Smith (2002) in the nearshore region. Salinity values were realistic in a general sense in terms of seasonality, distribution, and magnitude. Values were relatively lower on the inner shelf than off-shelf in the Gulf of Mexico. Salinities are very susceptible to the freshwater input and are highly variable interannually, as shown by the synoptic shipboard survey of NOAA's Atlantic Oceanographic and Meteorological Laboratory South Florida Program (http://www.aoml.noaa.gov/phod/sfp/index.php).

Biophysical simulations indicated that a successful settlement of pink shrimp larvae would require not only the appropriate circulation conditions for transport but also the right combination of DVM and tidal rhythm activity (STST) behavior. Larvae deployed at the Dry Tortugas and Marquesas spawning grounds as passive particles or with only a DVM behavior cannot reach the coastal nursery grounds in 30 d, the

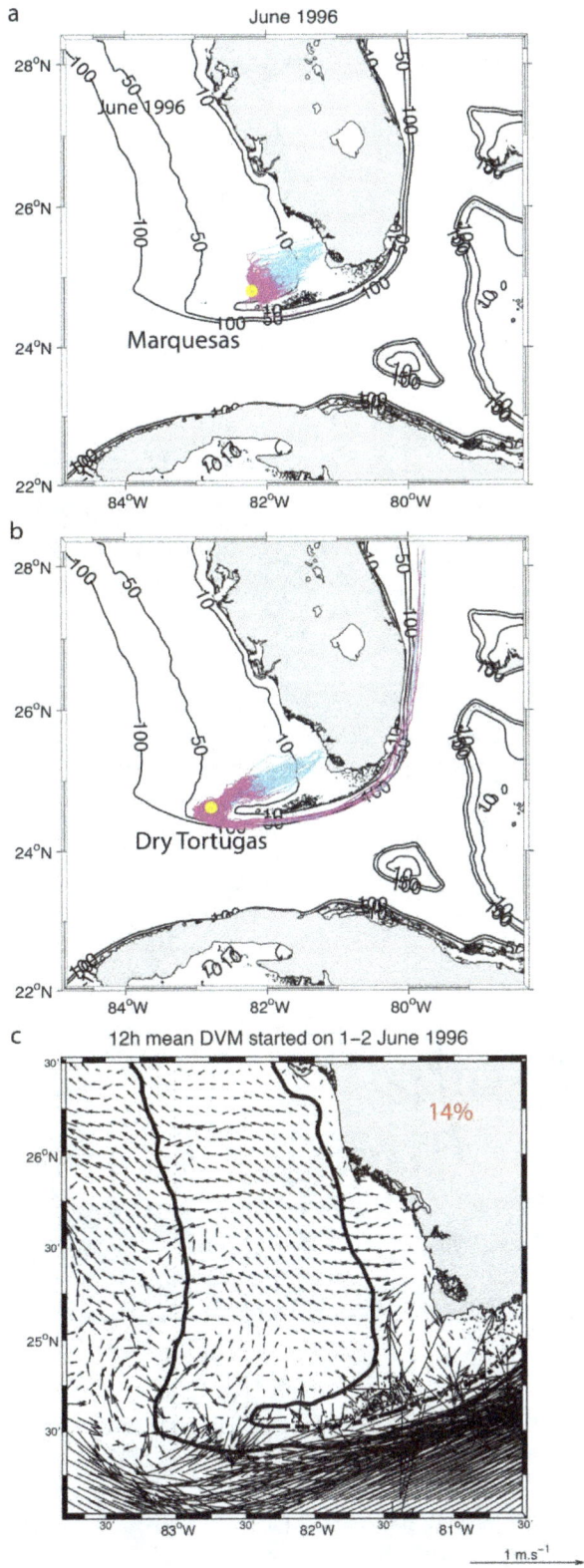

FIGURE 13. Particle trajectories, including diel vertical migration (DVM; magenta) and selective tidal stream transport (cyan), from their release locations in June 1996 at (a) Marquesas and (b) Dry Tortugas and (c) the corresponding 15-d mean of 12-h DVM mean flow started on June 1, 1996. The total recruitment percentage is indicated in red.

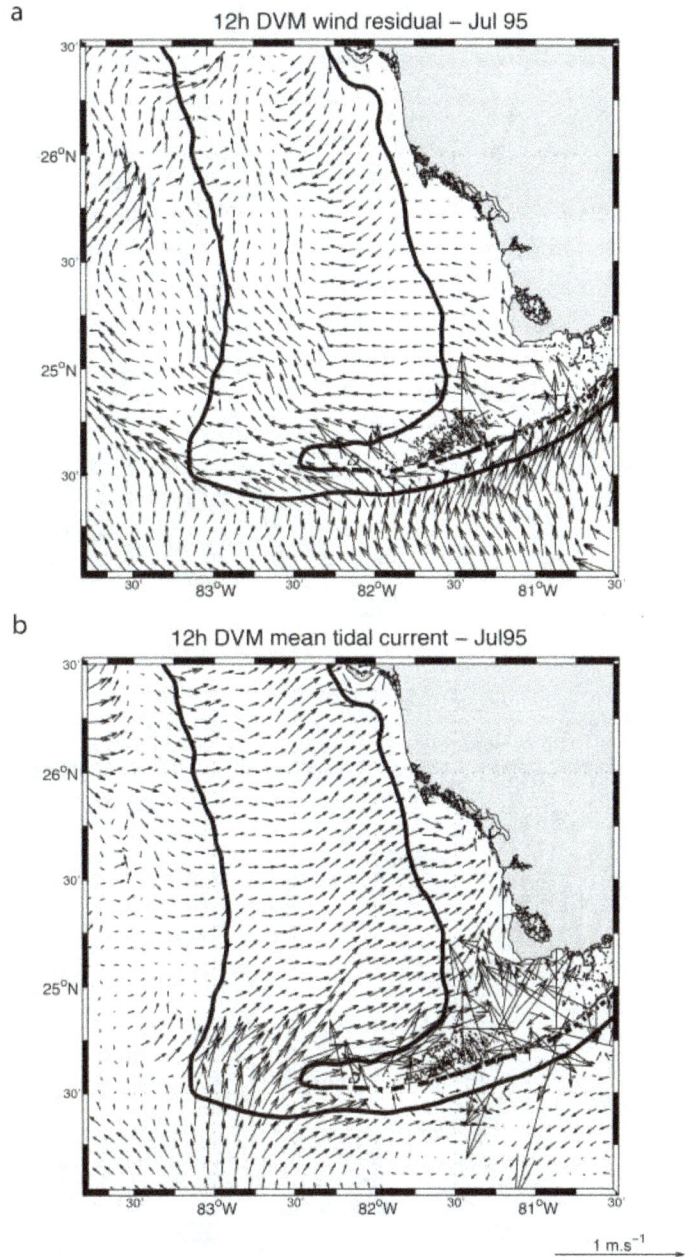

FIGURE 14. (a) First fortnightly 12-h diel vertical migration (DVM) mean wind-driven surface current in July 1995 and (b) the first fortnightly 12-h DVM mean tidal surface current for a DVM initiated at midnight of the first day of particle release in July 1995.

estimated period of larval transport (Figure 8). By adding an STST behavior at day 15 of migration, ~15% of the larvae released at Marquesas and 2% of the larvae released at Dry Tortugas reached the nursery grounds, moving north-northeast across the SWF shelf during summer. The maximum eastward distance traveled by larvae with DVM and STST behavior depends on the strength of the circulation, the season of the year, the release location, and the stage of the tidal current at the time of initiating the larval DVM period.

a

Nights only of first fortnight – Jul 95

b

12h mean started 12 am 1 July 1995

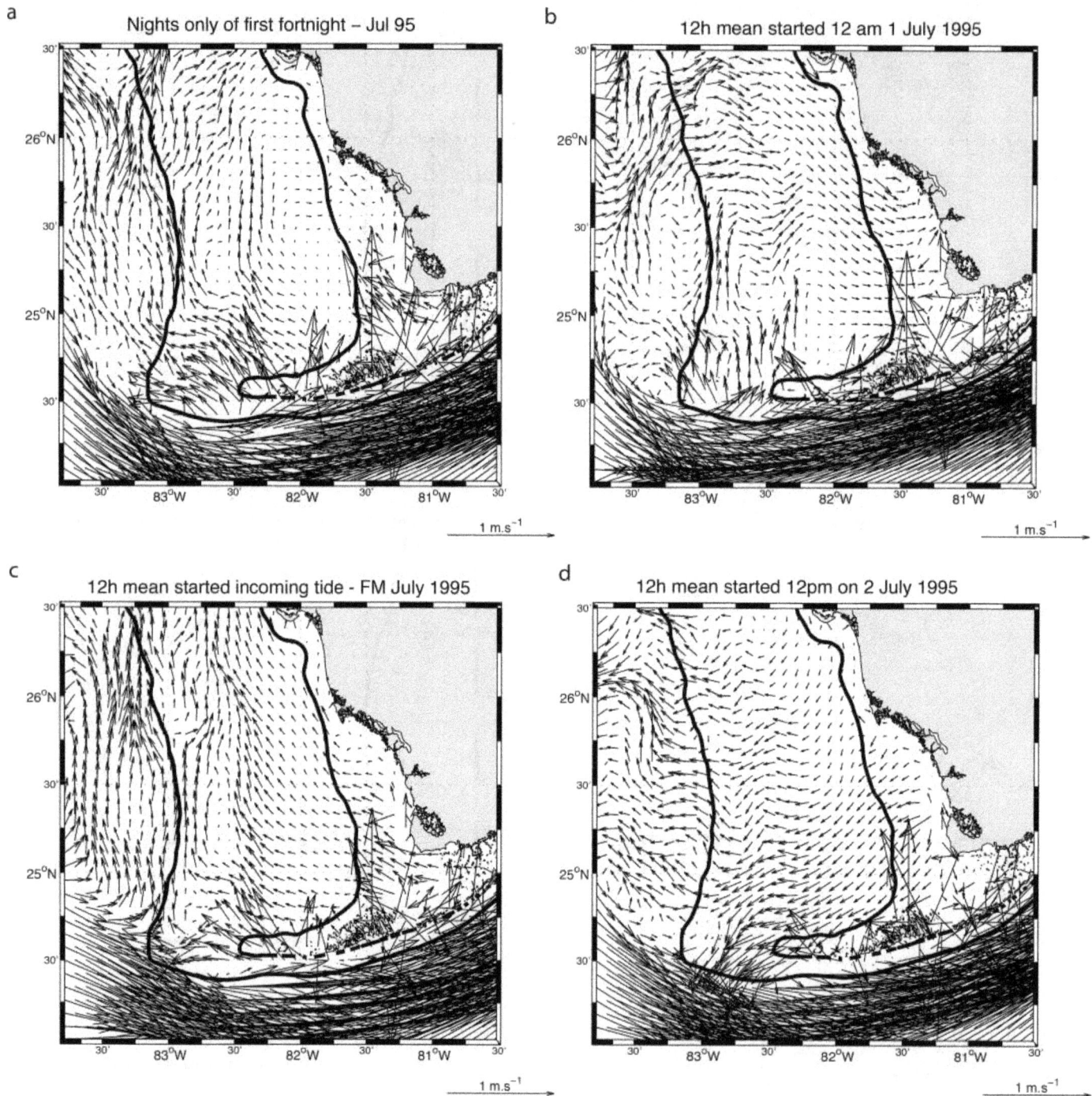

1 m.s^{-1}

c

12h mean started incoming tide - FM July 1995

d

12h mean started 12pm on 2 July 1995

1 m.s^{-1}

FIGURE 15. Fortnightly 12-h mean diel vertical migration (DVM) current vector field (a) based on model nights from only the first fortnight of July 1995, (b) started at midnight on July 1, 1995, (c) started with the incoming tide on the full moon of July 1995, and (d) started at noon on July 2, 1995.

Biophysical simulations revealed substantially different seasonal patterns in larval migration and in settlement. These different migration patterns were in accordance with the general circulation patterns described for the region, partly dependent on synoptic-scale winds, coupled with a strong seasonality of the subtidal flow and strong tidal currents (e.g., Weisberg et al. 1996; Smith 2000; Lee et al. 2002). The

transport of passive particles or particles with a DVM behavior from the spawning grounds to the coastal nursery grounds appeared to be controlled by a combination of four factors: (1) the subtidal flow, namely the geostrophic flow, which is in general very weak in summer and southward in the vicinity of the spawning grounds; (2) the coastal freshwater-driven density current, mostly controlled by watershed discharge

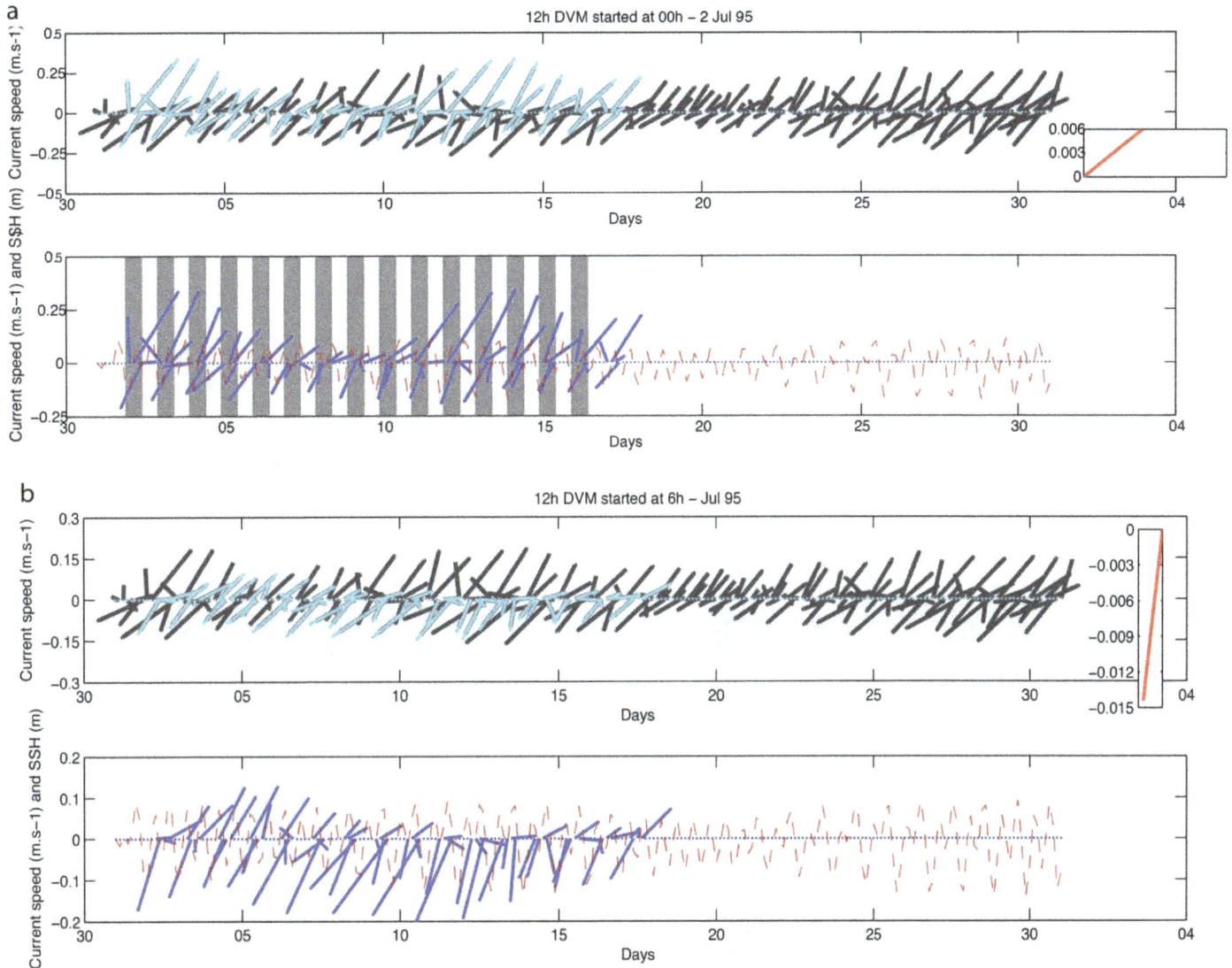

FIGURE 16. Current stick diagram at Marquesas spawning ground for the month of July 1995 (black sticks), showing (a) the 12-h DVM started at 0000 hours on July 2, (b) the DVM started at 0600 hours on July 2, (c) the DVM started at 1200 hours on July 2, and (d) the DVM started on July 10, 2 d before the full moon, at 1800 hours. The cyan-colored sticks in each of the upper diagrams show the current vectors selected by the 12-h DVM. The red stick in the small box shows the corresponding mean current direction. In each of the lower diagrams, the sticks selected by the 12-h DVM are overlaid with the mean sea level (red dashed line). The shaded gray bars show the nights. (*Extended on next page*)

(Hu et al. 2004; Zhao et al. 2013); (3) the Ekman transport, which is on average northwestward (Liu and Weisberg 2012); and (4) the tidal current, which is highly consistent. If both the wind-driven and the geostrophic current are weaker than the tidal current, the 12-h DVM mean flow can be northeastward, which increases the chances of larvae reaching the western Florida coast by crossing the SWF shelf. The strength of the geostrophic flow is partly controlled by the salinity at the coast, and it can overcome the tidal flow, as happened in July 1997. The subtidal flow points to the southwest as shown in Figure 7c, and the 12-h mean DVM flow is directed to the northwest at Dry Tortugas and to the south at Marquesas (Figure 12), which suggests that the geostrophic current was

against the northeastward tidal current at both Dry Tortugas and Marquesas. In July 1995, the relatively low freshwater discharge in inshore waters (Figure 7a) yielded a geostrophic flow directed to the northwest at Dry Tortugas and Marquesas (Figure 12). Combined with the selected tidal flow, the resulting 12-h DVM current was to the northeast at Marquesas and to the northwest at Dry Tortugas, yielding a much higher recruitment percentage in July 1995 than in July 1997 (Figure 9). In July 1996, the subtidal flow and the selected tidal flow were in the same direction and contributed to a strong northeastward 12-h mean DVM flow (Figure 12), yielding the largest recruitment percentage among the 3 years we simulated. This analysis shows that the current balance

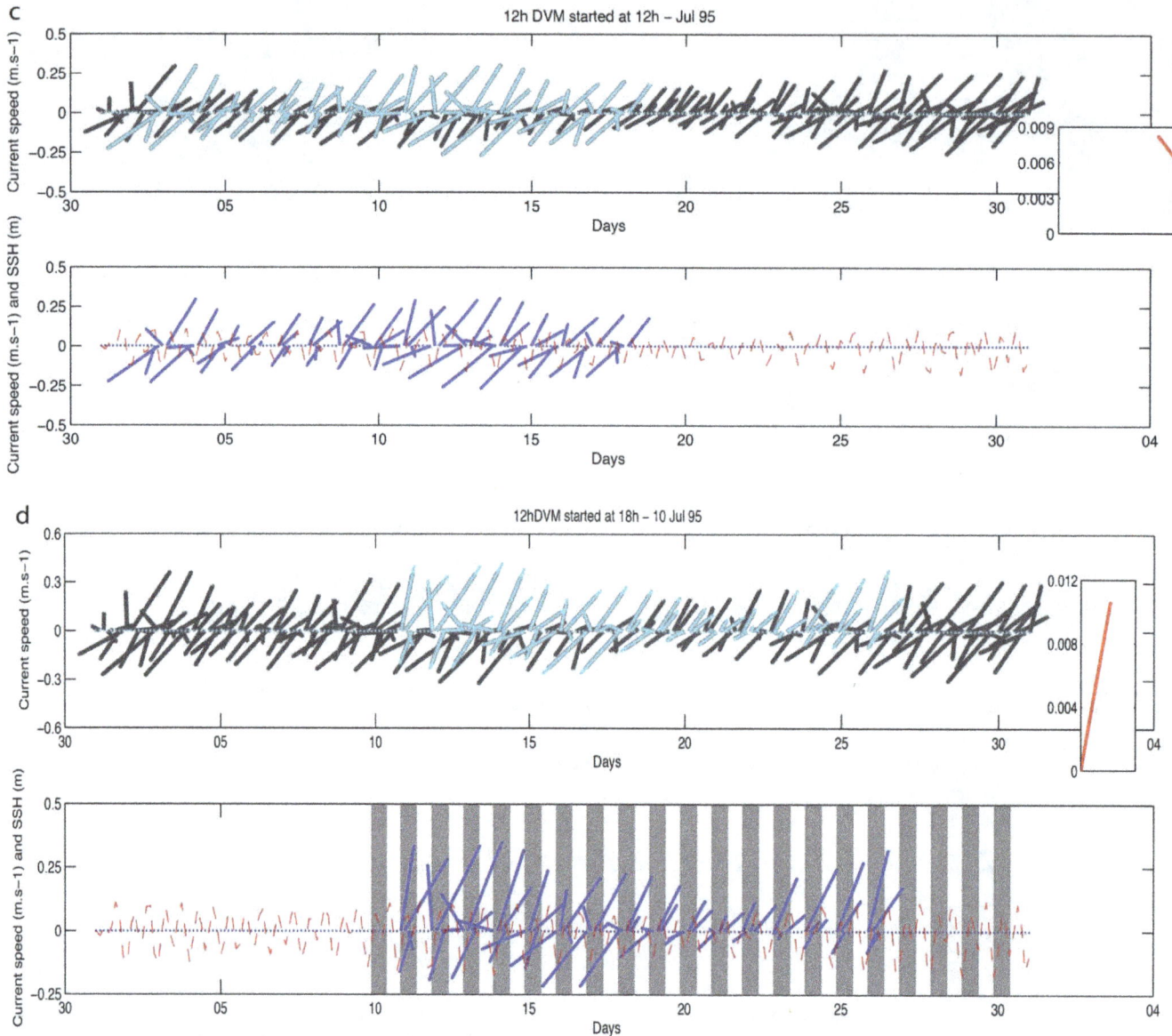

FIGURE 16.   Extended.

captured during the DVM period is critical to determine the direction in which larvae initially move, whether it is favorable (east or northward) or unfavorable (south or westward) to larval transport toward the nursery grounds.

The migration pattern changed radically during the winter months. Larvae moving with DVM and STST behavior during winter advanced only a short distance east-northeast across the SWF shelf because the DVM flow was westward (Figures 10 and 12). Some of the larvae originating at Dry Tortugas may move to the northwest, others may be caught by the LFC and associated gyre circulation, such as cyclonic eddies that originate westward of the Dry Tortugas and move downstream along the edge of the shelf (Fratantoni et al. 1998; Yeung et al. 2001; Yeung and Lee 2002). The different oceanographic conditions observed between summer and winter produce different migration routes that larvae use to reach the settlement grounds. During summer, most larvae released at Marquesas and at Dry Tortugas that recruited to settlement habitat arrived there by moving across the SWF shelf. During winter the settlement rates were about 5 times lower than during summer, and most larvae arrived at the settlement grounds via the Florida Current through the Florida Keys channels. This seasonal pattern in transport and settlement observed in our simulations agrees with previous field observations on postlarval influxes of pink shrimp into Florida Bay (Criales et al. 2006). The influx of postlarvae entering Florida Bay through its northwestern border, having crossed the SWF shelf, was about seven times greater than the influx through the channel of the

a

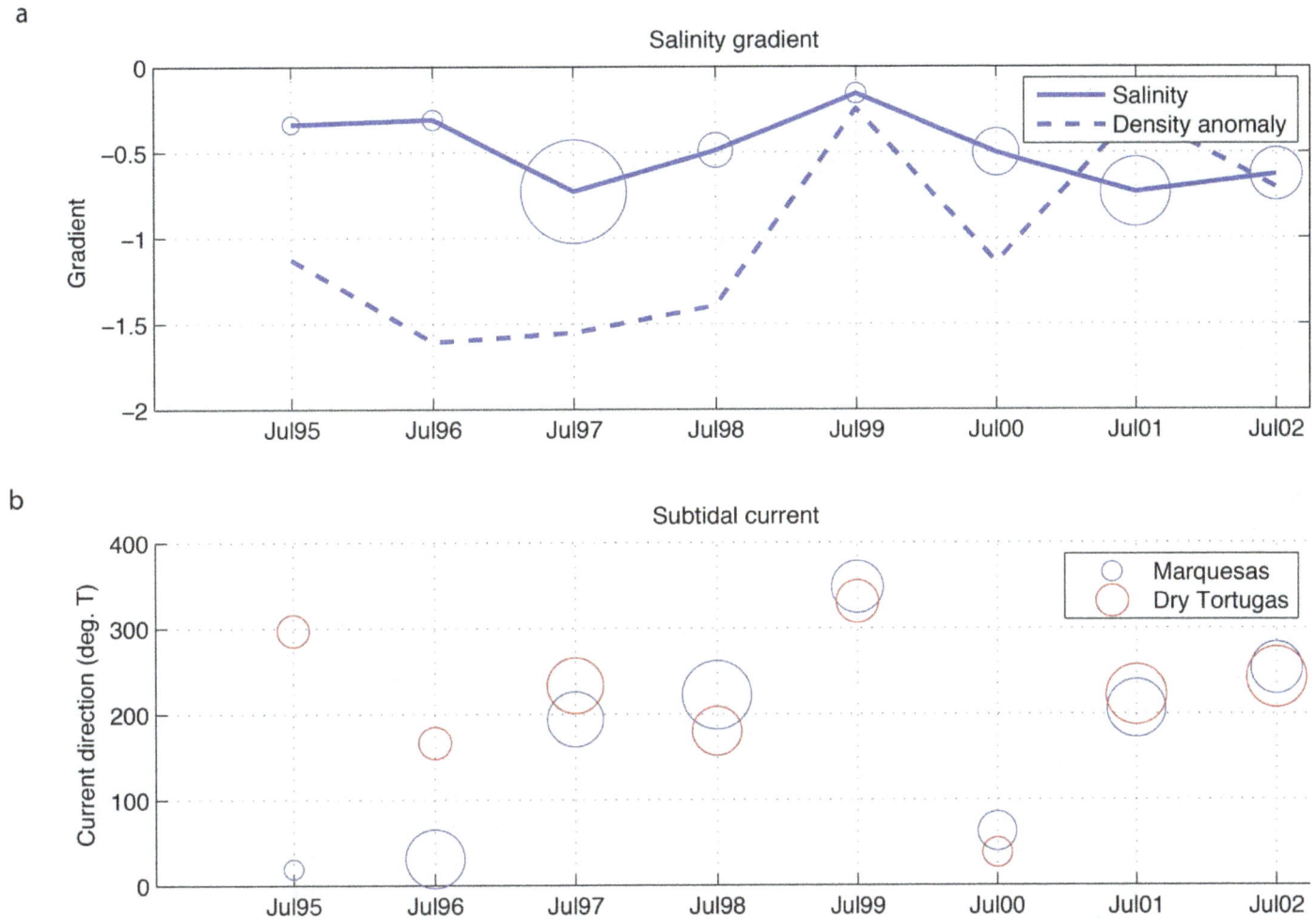

b

FIGURE 17.   Graphs of salinity and subtidal flow, showing an **(a)** 8-year time series of model monthly salinity gradient (solid line) and density anomaly (dashed line) in July along a zonal section at 25.7°N (shown in Figure 1b). Circles show the gradient concavity, which indicates the width of the gradient. Also shown is **(b)** subtidal monthly mean flow direction (*y*-axis coordinate of the center of the circle) and magnitude (circle diameter) at the Marquesas and the Dry Tortugas sites.

Middle Florida Keys. Furthermore, the peaks of abundance at the northwestern Florida Bay border were centered in summer for four consecutive years, whereas the influx of postlarvae through the Middle Florida Keys channels exhibited a highly variable seasonal pattern from year to year.

The highest settlement rates of pink shrimp larvae were observed in summer. Besides the favorable balance of the forcing on currents previously described, these settlements may have been enhanced by a horizontal displacement caused by the interaction between the periods of the DVM and the tidal constituents (e.g., Hill 1991a, 1991b; Smith and Stoner 1993; Power 1997; Criales et al. 2005). Organisms that use tidal currents with a DVM behavior for onshore transport may take advantage of an annual tidal cycle to improve their chances of reaching coastal nursery habitats. However, there are significant geographic variations in the seasonal diel transport. Using a simple Lagrangian model for the SWF shelf, Criales et al. (2005) estimated that particles moving at night with a STST behavior could enhance their eastward-traveled distance by up to 65 km due to the

superposition of the phase of the tidal constituents (solar semidiurnal S_2 and diurnal K_1) with the 12-h DVM period during long summer days. We found in the DVM onset sensitivity analysis that in July the strongest northeastward tidal current was synchronized with a 12-h DVM windowing starting with the incoming tide at night. The strongest northeastward tidal current at Dry Tortugas and Marquesas spawning grounds was obtained when the DVM started on July 10, 1995, 2 d before the full moon (Figure 16d). Pink shrimp females spawn all year round, but the highest frequency of ripe females (Cummings 1961) and the highest concentrations of larvae (Munro et al. 1968; Jones et al. 1970) occur in summer. The fact that the highest settlement rates in our simulation also occurred during summer may suggest that the reproductive cycle of this species in southern Florida is synchronized with the summer months when the oceanographic conditions (currents, tidal currents, tidal phase, and wind) are the most favorable for the DVM and the STST behavior.

The main nursery grounds of pink shrimp are located on the seagrass banks of western Florida Bay (Costello and Allen

1966), but other coastal estuaries of the lower southwestern Florida coast, such as Lostman's River, Oyster Bay, Ponce de Leon Bay, and Whitewater Bay, are also nursery grounds (Browder and Robblee 2009). Our model results indicated that, during summer, the vast majority of pink shrimp postlarvae settled at a location between the mouth of Lostman's River and Ponce de Leon Bay, north of Florida Bay (Figures 8, 10). Some of these postlarvae may actually settle in these small bays, but the great majority of them must move south along the coast from this location to reach Florida Bay. The connection between Florida Bay and interior bays was closed with the construction of the man-made Buttonwood Canal in Everglades National Park in 1982. Trajectories from surface drifters deployed at Shark River by Lee et al. (2002) during summer showed a northwestward displacement along the coast; however, Lee et al. (2002) also noted that salinity patterns associated with the Shark River low-salinity plume are generally oriented toward the southeast, with little spreading offshore of the river mouth. As previously shown, freshwater from the rivers that constitute the nursery grounds sustain a southward flow along the coast. Postlarvae arriving at the coast north of Florida Bay could use this southeastward flow to reach the seagrass banks inside of Florida Bay.

The number of ripe females reported from pink shrimp fishery data suggested that the main spawning ground of the species in southern Florida is located northeast of the Dry Tortugas (Cummings 1961; Roberts 1986). Based on the number of early planktonic stages (protozoeal stages), a location northward of Marquesas has also been suggested as spawning grounds (Jones et al. 1970; Criales et al. 2007). Our model results support Marquesas as the most effective spawning ground for pink shrimp, since larvae released at Marquesas were 4.5 times more likely to recruit to settlement habitats than larvae released at Dry Tortugas. Factors in the model that may have contributed to making Marquesas the more favorable spawning ground are as follows: (1) the distance to the nursery grounds, which is about half of that from Dry Tortugas, and (2) the circulation around Dry Tortugas, which is more variable than at Marquesas and larvae could be caught by the LFC and associated gyres, which usually hug the shelf edge southwest of the Dry Tortugas, and travel downstream with the Florida Current (Lee et al. 2002). Results of simulations conducted for other penaeid species have reached similar conclusions to ours. The estimated distance between effective spawning and nursery grounds for penaeid species in the Gulf of Carpentaria and the Gulf of California was closer to shore than the original estimates (Condie et al. 1999; Marinone et al. 2004), suggesting that the contributions of nearshore spawning grounds should not be underestimated.

In summary, results of the biophysical model clearly demonstrated the seasonality in larval settlement and in the migration routes to reach the nursery grounds. An effective transport mechanism promoting travel east-northeast across

the SWF shelf occurs during summer as the result of tidal currents, favorable wind, geostrophic flow and a combined DVM and STST behavior. The current phase captured during the DVM period appeared to be critical at selecting how far from the nursery ground larvae will be moved before the STST behavior enables the larvae to capture the eastward phase of the tide, which will move them to the nursery grounds. Interannual variability of the DVM was shown to be driven by the wind and the salinity at the nursery grounds, which is driven by rainfall and freshwater discharge. Therefore, salinity may be an important environmental factor controlling pink shrimp recruitment. Results suggest that adult pink shrimp may select spawning locations where one of the flow drivers is always favorable to their transport across the SWF shelf toward their nursery grounds, as tide is at the shrimp spawning grounds. The migration pattern changed in winter months when the alongshore winds reversed direction and the resultant flow component at the spawning grounds was westward. A few larvae that originated at Dry Tortugas and Marquesas reached the nursery grounds via the Florida Current and Florida Keys channels, but none of the larvae that originated at Dry Tortugas arrived at the nursery grounds via the SWF shelf. The Marquesas ground may be a more effective spawning ground than the Dry Tortugas ground, since larvae released at Marquesas were 4.5 times more likely to recruit to settlement habitats than larvae released at Dry Tortugas. These results will enable more effective management of this important fishery species.

An obvious next step of the study should include in the model other important nonlinear biological processes such as egg production, larval growth, and mortality to improve predictions of spawning ground locations and spatial distributions of settlement. New modeling should also determine the behavior and timing necessary to allow some of the postlarval pink shrimp that reach the southwestern Florida coast to finish the journey to Florida Bay. This model should also cover the flow of larvae among subpopulations of pink shrimp in Florida such as those in Biscayne Bay, Sanibel, and St. Augustine.

ACKNOWLEDGMENTS

This study was supported in part through NOAA grants, Fisheries and the Environment, and Coastal and Ocean Climate Applications NA12OAR4310105. Support was also provided through the NOAA South Florida Ecosystem Restoration Program and the Habitat Program. We thank Claire B. Paris for implementing the DVM and STST module in the CMS and for her valuable help and guidance with the use of the CMS model. The authors are also thankful to Andrew S. Kough and Chris Kelble for valuable help and comments. We also thank Silvia Gremes-Cordero for her assistance with the river data analysis.

REFERENCES

Almeida, M. M., J. Duberti, J., A. Peliz, and H. Queiroga. 2006. Influence of vertical migration pattern on retention of crab larvae in a seasonal upwelling system. Marine Ecology Progress Series 307:1–19.

Browder, J. A., and M. B. Robblee. 2009. Pink shrimp as an indicator for restoration of Everglades Ecosystems. Ecology Indicators 9(Supplement):S17–S28.

Browder, J. A., Z. Zein-Eldin, M. M. Criales, M. B. Robblee, S. Wong, T. L. Jackson, and D. Johnson. 2002. Dynamics of pink shrimp (*Farfantepenaeus duorarum*) recruitment potential in relation to salinity and temperature in Florida Bay. Estuaries 25:1355–71.

Butler, M. J., C. B. Paris, J. S. Goldstein, H. Matsuda, and R. K. Cowen. 2011. Behavior constrains the dispersal of long-lived spiny lobster larvae. Marine Ecology Progress Series 422:223–237.

Calderon-Aquilera, L. E., S. G. Marinone, and E. A. Aragon-Noriega. 2003. Influence of oceanographic processes on the early life stages of the blue shrimp (*Litopenaeus stylirostris*) in the upper Gulf of California. Journal of Marine Systems 39:117–128.

Condie, S. A., N. R. Loneragan, and D. J. Die. 1999. Modelling the recruitment of tiger prawns *Penaeus esculentus* and *P. semisulcatus* to nursery grounds in the Gulf of Carpentaria, northern Australia: implications for assessing stock-recruitment relationships. Marine Ecology Progress Series 178:55–68.

Costello, T. J., and D. M. Allen. 1966. Migrations and geographical distribution of pink shrimp, *Penaeus duorarum*, of the Tortugas and Sanibel grounds, Florida. U.S. Fish and Wildlife Service Fishery Bulletin 65:449–459.

Cowen, R. K., and L. R. Castro. 1994. Relation of coral reef fish larval distributions to island scale circulation around Barbados, West Indies. Bulletin Marine Science 54:228–244.

Cowen, R. K., C. B. Paris, and A. Srinivasan. 2006. Scaling of connectivity in marine populations. Science 311:522–527.

Criales, M. M., J. A. Browder, C. N. K. Mooers, M. B. Robblee, H. Cardenas, and T. L. Jackson. 2007. Cross-shelf transport of pink shrimp larvae: interactions of tidal currents, larval vertical migrations and internal tides. Marine Ecology Progress Series 345:167–184.

Criales, M. M., M. B. Robblee, J. A. Browder, H. Cardenas, and T. Jackson. 2010. Nearshore concentration of pink shrimp *Farfantepenaeus duorarum* postlarvae in northern Florida Bay in relation to the nocturnal flood tide. Bulletin of Marine Science 86:51–72.

Criales, M. M, M. B. Robblee, J. A. Browder, H. Cardenas, and T. Jackson. 2011. Field observations on selective tidal stream transport for postlarval and juvenile pink shrimp in Florida Bay. Journal of Crustacean Biology 31:26–33.

Criales, M. M., J. Wang, J. A. Browder, and M. Robblee. 2005. Tidal and seasonal effect on transport of pink shrimp postlarvae. Marine Ecology Progress Series 286:231–238.

Criales, M. M., J. Wang, J. A. Browder, M. B. Robblee, T. L. Jackson, and C. Hittle. 2006. Variability in supply and cross-shelf transport of pink shrimp (*Farfantepenaeus duorarum*) postlarvae into western Florida Bay. U.S. National Marine Fisheries Service Fishery Bulletin 104:60–74.

Criales, M. M., C. Yeung, D. Jones, T. L. Jackson, and W. J. Richards. 2003. Variation of oceanographic processes affecting the size of pink shrimp (*Farfantepenaeus duorarum*) postlarvae and their supply to Florida Bay. Estuarine Coastal and Shelf Science 57:457–468.

Cummings, D. C. 1961. Maturation and spawning of pink shrimp, *Penaeus duorarum* Burkenroad. Transactions of the American Fishery Society 90:462–468.

Dall, W., B. J. Hill, P. C. Rothlisberg, and D. J. Staples. 1990. The biology of Penaeidae. Advances in Marine Biology 27.

Dobkin, S. 1961. Early development stages of pink shrimp, *Penaeus duorarum*, from Florida waters. U.S. Fish and Wildlife Service Fishery Bulletin 61:321–349.

Egbert, G. D., and S. Y. Erofeeva. 2002: Efficient inverse modeling of barotropic ocean tides. Journal of Atmospheric Oceanic Technology 19: 183–204.

Ehrhardt, N., and C. Legault. 1999. Pink shrimp, *Farfantepenaeus duorarum*, recruitment variability as an indicator of Florida Bay dynamics. Estuaries 22:471–483.

Ewald, J. J. 1965. The laboratory rearing of pink shrimp, *Penaeus duorarum*, Burkenroad. Bulletin of Marine Science 15:436–449.

Fratantoni, P. S., T. N. Lee, G. P. Podesta, and F. Müller-Karger. 1998. The influence of Loop Current perturbations on the formation and evolution of Tortugas eddies in the southern Straits of Florida. Journal of Geophysical Research 103:24759–79.

Forward, R. B. Jr., and R. A. Tankersley. 2001. Selective tidal-stream transport of marine animals. Oceanography and Marine Biology 39:305–353.

Garcia, S., and L. Le Reste. 1981. Life cycles, dynamics, exploitation and management of coastal penaeid shrimp stocks. FAO (Food and Agriculture Organization of the United Nations) Fisheries Technical Paper 203.

Griffa, A. 1996. Applications of stochastic particle models to oceanographic problems. Page 114–140 *in* R. J. Adler, R. Muller, and B. L. Rozovskii, editors. Stochatic modelling in physical oceanography. Birkhäuser, Boston.

Grimm, V., U. Berger, F. Bastiansen, S. Eliassen, and V. Ginot. 2006. A standard protocol for describing individual-based and agent-based models. Ecological Modelling 198:115–126.

Hart, R. A. 2012. Stock assessment of pink shrimp (*Farfantepenaeus duorarum*) in the U.S. Gulf of Mexico for 2011. NOAA Technical Memorandum NMFS-SEFSC 639.

Hill, A. E. 1991a. A mechanism for horizontal zooplankton transport by vertical migration in tidal currents. Marine Biology 111:485–492.

Hill, A. E. 1991b. Vertical migration in tidal currents. Marine Ecology Progress Series 75:39–54.

Holstein, D., C. B. Paris, and P. J. Mumby. 2014. Consistency and inconsistency in multispecies population network dynamics of coral reef ecosystems. Marine Ecology Progress Series 499:1–18.

Hu, C. M., F. E. Muller-Karger, G. A. Vargo, M. B. Neely, and E. Johns. 2004. Linkages between coastal runoff and the Florida Keys ecosystem: a study of a dark plume event. Geophysical Research Letters [online serial] 31: L15307.

Jones, A. C., D. E. Dimitriou, J. J. Ewald, and J. H. Tweedy. 1970. Distribution of early developmental stages of pink shrimp, *Penaeus duorarum*, in Florida water. Bulletin of Marine Science 20:634–661.

Klima, E. F., G. A. Matthews, and F. J. Patella. 1986. Synopsis of the Tortugas pink shrimp fishery, 1960–1983, and the impact of the Tortugas Sanctuary. North American Journal Fishery and Management 6:301–310.

Kough, A. S., C. B. Paris, and M. J. IV Butler. 2013. Larval connectivity and the international management of fisheries. PLoS (Public Library of Science) ONE [online serial] 8(6):e64970.

Le Hénaff, M., V. H. Kourafalou, Y. Morel, and A. Srinivasan. 2012. Simulating the dynamics and intensification of cyclonic Loop Current frontal eddies in the Gulf of Mexico. Journal of Geophysical Research [online serial] 117: C02034.

Lee, T. N., E. Johns, D. Wilson, E. Williams, and N. P. Smith. 2002. Transport processes linking south Florida coastal ecosystems. Pages 309–342 *in* J. Porter and K. Porter, editors. Linkages between ecosystems in the South Florida hydroscape. CRC Press, Boca Raton, Florida.

Lee, T. N., C. Rooth, E. Weilaims, M. Mc Gowan, A. F. Szmant, and M. E. Clarke, 1992. Influence of Florida Current, gyres and wind-driven circulation on transport of larvae and recruitment in the Florida Keys coral reefs. Continental Shelf Research 12:971–1002.

Lee, T. N., and N. Smith. 2002. Volume transport variability through the Florida Keys tidal channels. Continental Shelf Research 22:1361–1377.

Li, Z., and R. H. Weisberg. 1999a. West Florida shelf response to upwelling favorable wind forcing: kinematics. Journal of Geophysical Research 104:13507–13527.

Li, Z., and R. H. Weisberg. 1999b. West Florida continental shelf response to upwelling favorable wind forcing, part II: dynamical analyses. Journal of Geophysical Research 104:23427–23442.

Liu, Y., and R. H. Weisberg. 2007. Ocean currents and sea surface heights estimated across the west Florida shelf. Journal of Physical Oceanography 37:1697–1713.

Liu, Y., and R. H. Weisberg. 2012. Seasonal variability on the west Florida shelf. Progress in Oceanography 104:80–98.

Marinone, S. G., O. Q. Gutierrez, and A. Pares-Sierra. 2004. Numerical simulation of larval shrimp dispersion in the northern region of the Gulf of California. Estuarine Coastal and Shelf Science 60:611–617.

Munro, J. L., A. C. Jones, and D. Dimitriou. 1968. Abundance and distribution of the larvae of the pink shrimp (*Penaeus duorarum*) on the Tortugas shelf of Florida, August 1962–October 1964. U.S. Fish and Wildlife Service Fishery Bulletin 67:165–181.

Paris, C. B., and R. K. Cowen. 2004. Direct evidence of a biophysical retention mechanism for coral reef fish larvae. Limnology and Oceanography 49:1964–1979.

Paris, C. B., R. K. Cowen, R. Claro, and K. C. Lindeman. 2005. Larval transport pathways from Cuban spawning aggregations (Snappers; Lutjanidae) based on biophysical modeling. Marine Ecology Progress Series 296:93–106.

Paris, C. B., R. K. Cowen, K. M. M. Lwiza, D. P. Wang, and D. B. Olson. 2002. Objective analysis of three-dimensional circulation in the vicinity of Barbados, West Indies: implication for larval transport. Deep-Sea Research 49:1363–1386.

Paris, C. B., L. M. Cherubin, and R. K. Cowen. 2007. Surfing, spinning, or diving from reef to reef: effects on population connectivity. Marine Ecology Progress Series 347:285–300.

Paris, C. B., J. Helgers, E. van Sebille, and A. Srinivasan. 2013. Connectivity modeling system: a probalistic modeling tool for the multi-scale tracking of biotic and abiotic variability in the ocean. Environmental Modelling and Software 42:47–54.

Peterson, W. 1998. Life cycle strategies of copepods in coastal upwelling zones. Journal of Marine Systems 15:313–326.

Power, J. H. 1997. Time and tide wait for no animals: seasonal and regional opportunities for tidal stream transport and retention. Estuaries 20:312–318.

Queiroga, H., and J. Blanton. 2004. Interactions between behavior and physical forcing in the control of horizontal transport of decapod crustacean larvae. Advances in Marine Biology 47:107–214.

Ramírez-Rodríguez, M., F. Arreguín-Sánchez, and D. Lluch-Belda. 2003. Recruitment patterns of the pink shrimp *Farfantepenaeus duorarum* in the southern Gulf of Mexico. Fisheries Research 65:81–82.

Roberts, T. W. 1986. Abundance and distribution of pink shrimp in and around the Tortugas Sanctuary, 1981–1983. North American Journal of Fisheries and Management 6:311–327.

Rothlisberg, P. C. 1982. Vertical migration and its effect on dispersal of penaeid shrimp larvae in the Gulf of Carpentaria, Australia. U.S. National Marine Fisheries Service Fishery Bulletin 80:541–554.

Rothlisberg, P. C., J. A. Church, and C. Fandry. 1995. A mechanism for nearshore density and estuarine recruitment of post-larval *Penaeus plebejus* Hess (Decapoda, Penaeidae). Estuarine Coastal and Shelf Science 40:115–138.

Rothlisberg, P. C., J. A. Church, and A. M. G. Forbes. 1983. Modelling advection of vertically migrating shrimp larvae. Journal Marine Research 41:511–538.

Rothlisberg, P. C., P. D. Craig, and J. R. Andrewartha. 1996. Modelling penaeid prawn larval advection in Albatross Bay, Australia: defining the effective spawning population. Marine and Freshwater Research 47:157–168.

Sakov, P., F. Counillon, L. Bertino, K. A. Lisaeter, P. R. Oke, and A. Korablev. 2012. TOPAZ4: an ocean-sea ice data assimilation system for the North Atlantic and Arctic. Ocean Science 8:633–656.

Shanks, A. L. 1995. Mechanisms of cross-shelf dispersal of larval invertebrates and fish. Pages 323–367 *in* L. R. McEdward, editor. Ecology of marine invertebrate larvae. CRC Press, Boca Raton, Florida.

Shanks, A. L. 2006. Mechanisms of cross-shelf transport of crab megalopae inferred from a time series of daily abundance. Marine Biology 148:1383–1398.

Shchepetkin, A. F., and J. C. McWilliams. 2005. Regional ocean model system: a split-explicit ocean model with a free-surface and topography-following vertical coordinate. Ocean Modelling 9:347–404.

Sheridan, P. 1996. Forecasting the fishery for pink shrimp *Penaeus duorarum*, on the Tortugas Grounds, Florida. U.S. National Marine Fisheries Service Fishery Bulletin 94:743–755.

Smith, N. P. 1997. An introduction to the tides of Florida Bay. Florida Scientist 60:53–67.

Smith, N. P. 2000. Transport across the western boundary of Florida Bay. Bulletin of Marine Science 66:291–304.

Smith, N. P., and A. W. Stoner. 1993. Computer simulation of larval transport through tidal channels: role of vertical migration. Estuarine Coastal and Shelf Science 37:43–58.

Sponaugle, S., C. Paris, K. D. Walter, V. Koourafalou, and E. D'Alessandro. 2012. Observed and modeled larval settlement of a reef fish to the Florida Keys. Marine Ecology Progress Series 453:201–212.

Tabb, D. C., D. L Dubrow, and A. E. Jones. 1962. Studies on the biology of pink shrimp *Penaeus duorarum* Burkenroad, in Everglades National Park, Florida. Florida Board of Conservation Marine Research Laboratory Technical Series 37.

van Aken, H. M., H. van Haren, and L. R. M. Maas. 2007. The high-resolution vertical structure of internal tides and near inertial waves measured with ADCP over the continental slope in the Bay of Biscay. Deep-Sea Research Part I Oceanographic Research Papers 54:533–556.

Wang, J. 1998. Subtidal flow patterns in western Florida Bay. Estuarine Coastal Shelf Science 46:901–15.

Wang, J. D., J. van de Kreeke, N. Krishnan, and D. Smith. 1994. Wind and tide response in Florida Bay. Bulletin of Marine Science 54:579–601.

Weisberg, R. H., A. Barth, A. Alvera-Azcárate, and L. Zheng. 2009. A coordinated coastal ocean observing and modeling system for the west Florida shelf. Harmful Algae 8:585–598.

Weisberg, R. H., B. D. Black, and Z. Li. 2000. An upwelling case study on Florida's west coast. Journal of Geophysical Research 105: 11459–11469.

Weisberg, R. H., B. D. Black, and J. Yang. 1996. Seasonal modulation of the west Florida continental shelf circulation. Journal of Geophysical Research 23:2247–2250.

Weisberg, R. H., R. He, Y. Liu, and J. Virmani. 2005. West Florida shelf circulation on synoptic, seasonal, and inter-annual time scales. Geophysical Monograph 161:325–347.

Weisberg, R. H., Z. Li, and F. E. Muller-Karger. 2001. West Florida shelf response to local wind forcing: April 1998. Journal of Geophysical Research 106:31239–31262.

Yeung, C., D. L. Jones, M. M. Criales, T. L. Jackson, and W. J. Richards. 2001. Influence of coastal eddies and counter-currents on the influx of spiny lobster, *Panulirus argus*, postlarvae into Florida Bay: influence of eddy transport. Marine and Freshwater Research 52:1217–1232.

Yeung, C., and T. L. Lee. 2002. Larval transport and retention of the spiny lobster, *Panulirus argus*, in the coastal zone of the Florida Keys, USA. Fishery Oceanography 11:86–309.

Zar, J. H. 2010. Biostatistical analysis, 5th edition. Prentice Hall, Upper Saddle River, New Jersey.

Zhao, J., H. Chuanmin, B. Lapointe, N. Melo, E. M. Johns, and R. H. Smith. 2013. Satellite-observed black water events off southwest Florida: implications for coral reef health in the Florida Keys National Marine Sanctuary. Remote Sensing 5:415–431.

Zieman, J. C., J. H. Fourqurean, and R. L. Iverson. 1989. Distribution, abundance, and productivity of seagrasses and macroalgae in Florida Bay. Bulletin of Marine Science 44:292–311.

17

Population Genetic Structure of Southern Flounder Inferred from Multilocus DNA Profiles

Verena H. Wang,* Michael A. McCartney,[1] and Frederick S. Scharf
Department of Biology and Marine Biology, Center for Marine Science,
University of North Carolina Wilmington, 601 South College Road,
Wilmington, North Carolina 28403, USA

Abstract

   Determination of stock structure is an important component of fisheries management; incorporation of molecular genetic data is an effective method for assessing differentiation among putative populations. We examined genetic variation in Southern Flounder *Paralichthys lethostigma* within and between the U.S. South Atlantic and Gulf of Mexico basins to improve our understanding of the scale of population structure in this wide-ranging species. Analysis of amplified fragment length polymorphism (AFLP) fingerprints and analysis of mitochondrial DNA (mtDNA) control region sequences found clear divergence between ocean basins. Based on mtDNA sequences, no genetic differentiation was detected within the U.S. South Atlantic at spatial scales that were broad (among states: North Carolina, South Carolina, Georgia, and Florida) or fine (among estuarine regions within North Carolina). Increased genetic resolution was observed with AFLP fingerprint data, and we found significant subdivision between nearly all Southern Flounder geographic populations, suggesting the presence of finer-scale genetic population structure within the U.S. South Atlantic. However, AFLP genetic cluster analysis also revealed evidence for a high degree of mixing within the Atlantic basin; patterns of variation, which included genetic similarity between South Carolina and Gulf of Mexico samples, were not aligned closely with geography. We examined the partitioning of genetic variation among groups by using analyses of molecular variance and found no evidence that North Carolina Southern Flounder, which are managed on the state level as a unit stock, are differentiated from the remainder of U.S. South Atlantic Southern Flounder. Our findings indicate only weak structure and the potential for basinwide mixing among Atlantic Southern Flounder, suggesting that cooperation among U.S. South Atlantic states will be essential for the effective assessment of stock dynamics and future management plans.

Accurate identification of stock structure is essential for effective fisheries management but remains a challenge for both managers and scientists (Begg et al. 1999; NCDMF 2005, 2013). Examination of genetic variation across a species' range is a commonly used approach for defining interbreeding populations, and recent advances in molecular techniques have contributed to an expanded interest in the field of fisheries genetics (Ward 2000; Hauser and Carvalho 2008; Waples et al. 2008). Marine species were once assumed to be panmictic, with long-distance dispersal capability and a lack

Subject editor: Kristina Miller, Fisheries and Oceans Canada, Pacific Biological Station, Nanaimo, British Columbia, Canada

*Corresponding author: vhw6375@uncw.edu
[1]Present address: Minnesota Aquatic Invasive Species Research Center, Department of Fisheries, Wildlife, and Conservation Biology, University of Minnesota, 2003 Upper Buford Circle, St. Paul, Minnesota 55108, USA.

of obvious barriers to gene flow in the ocean resulting in low genetic differentiation among putative populations (Ward et al. 1994; Waples 1998). However, highly polymorphic molecular markers have revealed significant genetic structure in marine fish at multiple geographic scales (Ruzzante et al. 1999; Knutsen et al. 2003; O'Reilly et al. 2004; McCairns and Bernatchez 2008), and Knutsen et al. (2011) suggested that even low levels of population subdivision in marine organisms can be biologically meaningful. Maintenance of subtle genetic stock differentiation may be important in conserving local diversity and adaptive variation. Consequently, exploitation of fish populations can lead to reductions in genetic diversity, which may impact the ability of depleted stocks to recover (Hutchinson et al. 2003). Incorporating population genetic analyses into fisheries management and considering both historical and contemporary patterns of molecular diversity can help to promote sustainability and preserve evolutionary processes in the face of increasing anthropogenic impacts (Avise 1992; Crandall et al. 2000; Conover et al. 2006; Hauser and Carvalho 2008).

The Southern Flounder *Paralichthys lethostigma* is an economically important flatfish that occurs throughout estuarine and coastal ocean waters in the northern Gulf of Mexico (hereafter, Gulf) from northern Mexico to Florida (FL) and in the U.S. South Atlantic (hereafter, Atlantic) from FL to North Carolina (NC). The species supports a thriving recreational fishery across its range and is also a valuable commercial finfish resource in some areas, particularly in NC, where commercial landings averaged 1.5 million kg (3.3 million lb) annually between 1991 and 2007 (NCDMF 2013). In fact, the NC stock has experienced elevated commercial harvest rates since at least 1991; although a fishery management plan (NCDMF 2005, 2013) was first established in 2005, the most recent stock assessment concluded that NC Southern Flounder remain overfished and are still undergoing overfishing (Takade-Heumacher and Batsavage 2009).

Mature Southern Flounder inhabiting estuarine waters migrate offshore annually to spawn during the late-fall and winter months, with recruitment of larvae into estuarine nursery grounds occurring in late winter and early spring (Wenner et al. 1990). Adults typically return to estuarine and nearshore coastal waters after spawning (Monaghan 1996), but recent studies have found that Southern Flounder migration patterns are more variable, as some adults remain offshore through the late-spring and summer months (Watterson and Alexander 2004; Taylor et al. 2008). In addition, previous tagging studies indicate that most individuals display limited movement when occupying estuarine habitats and remain in close proximity to their tagging sites (Wenner et al. 1990; Monaghan 1996; Craig and Rice 2008). Based mainly on the estuarine tagging results, the current fisheries management plan considers Southern Flounder in NC to be a unit stock (NCDMF 2013). However, there is also evidence of extensive southward movement, as some individuals tagged in NC estuaries were found in other

Atlantic states (Monaghan 1996; Craig and Rice 2008). Prior to estuarine recruitment, Southern Flounder remain in a pelagic larval stage for a moderately long duration (~45 d), which increases the potential for long-range dispersal. The unknown level of adult mixing in the offshore environment combined with the long duration of the larval stage implies that the genetic population structure of Southern Flounder could be more homogeneous than currently assumed. The use of molecular methods should provide increased knowledge of Southern Flounder population structure and connectivity, thereby contributing to effective management decisions for the long-term sustainability of the species.

Previous population genetics studies of Southern Flounder have been focused on the Gulf, but there has been no comprehensive study spanning the species' range within the Atlantic. When examining allozyme variation in Southern Flounder, Blandon et al. (2001) found that Gulf populations were significantly differentiated from Atlantic populations, and a break in allele frequencies on the coast of Texas between Galveston Bay and Matagorda Bay indicated the presence of population structure within the Gulf. Updated studies using mitochondrial DNA (mtDNA) and microsatellites again revealed high divergence between Gulf and Atlantic populations of Southern Flounder, but in contrast to the allozyme survey, little differentiation was detected within the Gulf (Anderson and Karel 2012; Anderson et al. 2012). Given the low levels of within-basin variability detected in previous studies, we applied a powerful genetic tool in an attempt to better characterize diversity and stock structure of Southern Flounder in the Atlantic basin.

We examined Southern Flounder populations by using the amplified fragment length polymorphism (AFLP) technique (Vos et al. 1995). This method rapidly generates a genetic fingerprint consisting of hundreds of highly reproducible DNA fragments that are scattered across the genome and that can be compared among individuals and populations without the need for prior characterization of an organism's genome (Bensch and Åkesson 2005). The large number of AFLPs produced allows for assessment of genome-wide variation when analyzing population structure, which can be more informative than targeting mtDNA markers (which are inherited as a single haploid locus). It has been claimed that the information content of an AFLP study is comparable to that obtained from a small number of highly informative microsatellite loci, which are commonly used but require substantial development time and cost (Campbell et al. 2003; Meudt and Clarke 2007; Sønstebø et al. 2007). Although previously underutilized in studies of animals (reviewed by Bensch and Åkesson 2005), AFLPs have become increasingly common in studies of aquatic populations, including commercially important fish and invertebrate species both in aquaculture (e.g., Channel Catfish *Ictalurus punctatus*: Mickett et al. 2003; Olive Flounder *Paralichthys olivaceus*: Liu et al. 2005) and in natural populations (e.g., common shrimp *Crangon crangon*: Weetman et al. 2007; Striped Mullet *Mugil cephalus*: Liu et al. 2009). For comparison with a more established method, we also sequenced a fragment of the mtDNA control region, which is a commonly used molecular marker. The control region has been identified as highly variable in teleosts (Lee et al. 1995) and was previously used by Anderson et al. (2012) to examine Southern Flounder population structure.

In this study, our main objective was to genotype Southern Flounder individuals from across the species' range by using both AFLPs and mtDNA to address whether distinct genetic populations are present within the Atlantic, particularly whether Southern Flounder in NC represent a unique stock. As evidenced by previous studies, we expected divergence between Gulf and Atlantic populations, but we attempted to determine whether finer structure exists within ocean basins and whether this structure alters our assumptions regarding the level of mixing and warrants a change in the spatial scale of management. In addition, we assessed the viability of the AFLP method in comparison with more traditional mtDNA sequencing for determining stock structure in a widely distributed marine species.

METHODS

Southern Flounder tissue samples were collected from multiple estuarine sites across four states (NC, South Carolina [SC], Georgia [GA], and FL) spanning the current range of the species in the Atlantic (Figure 1). North Carolina was subdivided into three regions (north [NCN], central [NCC], and south [NCS]) to examine genetic population structure within state waters. Samples were obtained from both fishery-independent and fishery-dependent sources in partnership with state fishery management agencies and commercial fishers within each state and were also obtained from licensed seafood dealers. Sampling was conducted from 2010 to 2013, and sampled fish ranged from age 0 to age 3 (96% were age 0–1), with an average TL of 360 mm (range = 166–622 mm). Additional tissue samples, provided by J. D. Anderson (Texas Parks and Wildlife Department), were collected from two sites (Figure 1) in the Gulf (Apalachicola, FL [APFL]; and San Antonio Bay, Texas [SBTX]). Sampling methods for Gulf collections are described by Anderson et al. (2012). Fin clips and gill tissue were excised and preserved in either a salt-saturated dimethyl sulfoxide solution (Atlantic samples) or a 70%

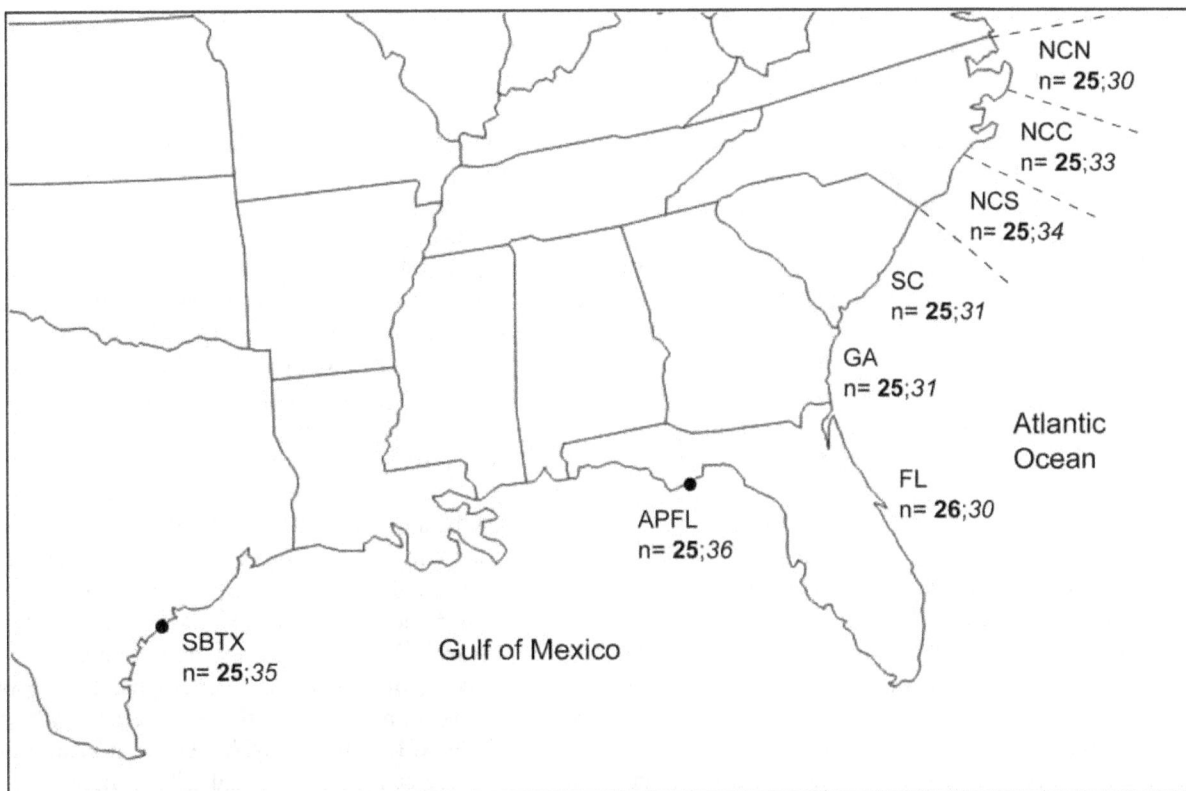

FIGURE 1. Map of Southern Flounder sampling locations in the U.S. South Atlantic and the Gulf of Mexico. Atlantic locations are represented here on a state scale, although multiple sites were sampled within each state (NC = North Carolina; SC = South Carolina; GA = Georgia; FL = Florida). North Carolina was subdivided into three regions (separated by dashed lines; NCN = NC north [Albemarle Sound]; NCC = NC central [Neuse River–Pamlico River estuary]; NCS = NC south [New River and Cape Fear River estuaries]). Gulf of Mexico samples originated from two specific locations (black circles; APFL = Apalachicola, Florida; SBTX = San Antonio Bay, Texas). Sample size (*n*) for amplified fragment length polymorphisms is shown in bold; *n* for mitochondrial DNA control region sequences is shown in italics.

solution of ethanol (Gulf samples) and were stored at 4°C. Genomic DNA was extracted from fin clips by using the DNeasy Blood and Tissue Kit (Qiagen, Valencia, California) in accordance with the manufacturer's protocols. Gill tissue was extracted only occasionally, when fin clips were damaged or not readily available.

The AFLPs were generated by using the AFLP Plant Mapping Protocol (Applied Biosystems, Inc. [ABI], Foster City, California) for regular plant genomes. We produced DNA fingerprints for 12 selective primer pairs (chosen from an earlier screen of 24 primer pairs): EcoRI-ACA and MseI-CAA; EcoRI-ACC and MseI-CAC; EcoRI-ACC and MseI-CAG; EcoRI-ACC and MseI-CAT; EcoRI-ACT and MseI-CAA; EcoRI-ACT and MseI-CAC; EcoRI-ACT and MseI-CAT; EcoRI-ACT and MseI-CTC; EcoRI-AGC and MseI-CAA; EcoRI-AGC and MseI-CTT; EcoRI-AGG and MseI-CTA; and EcoRI-AGG and MseI-CTG. Fragments were sized along with GeneScan 500 ROX size standard (Life Technologies, Carlsbad, California) on an ABI 3130XL Genetic Analyzer at the DNA Analysis Core Facility (Center for Marine Science, University of North Carolina, Wilmington). To test for reproducibility of AFLP fingerprints, one individual from each sampling location was subjected to repeat analysis across all 12 selective primer pairs beginning from genomic DNA. Fragments between 75–500 bp were scored automatically for presence/absence with a default peak calling threshold of 50 relative fluorescence units (rfu) by using ABI GeneMapper version 4.0. Peak heights were normalized across all samples for each primer pair, with only polymorphic loci considered. To further reduce noise and genotyping error, the individual loci in the AFLP profiles were filtered to a locus selection threshold of 200 rfu and a phenotype calling threshold of 20% following Whitlock et al. (2008). In this AFLP profile optimization procedure, loci with an average peak height below 200 rfu are eliminated from analysis, and presence/absence scoring is based on a threshold of 20% of the average peak height for each locus. The complete AFLP profiles (i.e., prior to optimization) were also considered in this analysis but provided similar results and are not reported here.

A fragment of the mtDNA control region was amplified by using the primers L-PROF (5′-AACTCTCACCCC-TAGCTCCCAAAG-3′; Meyer et al. 1994) and PL526H (5′-AAAAGAGAACCCCTTACCCG-3′), the latter of which is a Southern Flounder-specific primer designed by one of us (M. A. McCartney, unpublished). The PCR was conducted as follows: an initial denaturation step of 94°C for 1 min, followed by 35 amplification cycles (94°C for 1 min; 58°C for 30 s; and 72°C for 2 min) and a final extension of 72°C for 5 min. Each 25-μL reaction volume contained 1.25 units of MyTaq DNA Polymerase (Bioline, London, UK), 1 × MyTaq Red Reaction Buffer (Bioline), 0.5 μM of each primer, and 1 μL of genomic DNA (5–900 ng). The PCR products were purified by using a StratPrep PCR Purification Kit (Agilent Technologies, Santa Clara, California) and were cycle sequenced in the

forward direction (L-PROF) with ABI BigDye Terminator version 3.1. Sequencing was conducted on the same equipment as described for AFLPs. Sequences were edited, trimmed, and aligned manually in Sequencher version 4.10.1 (Gene Codes Corporation, Ann Arbor, Michigan). Control region sequences were deposited in GenBank under accession numbers KP331548–KP331677.

STRUCTURE version 2.3.4 (Pritchard et al. 2000) was used to assign individual Southern Flounder to an inferred number of genetically distinct clusters (K) present in the AFLP data set. Because AFLPs are dominant markers that are scored as the presence or absence of peaks, there is no distinction between homozygous and heterozygous states. To account for this ambiguity, we implemented the model by Falush et al. (2007) for analyzing dominant markers in STRUCTURE. A recessive AFLP profile (consisting of all peak absences) was input into the data set, and the model evaluated all possible genotypes (i.e., for a peak presence phenotype, the genotypes presence/presence and presence/absence are both considered). Based on previous studies that demonstrated a lack of structure among Southern Flounder populations within ocean basins (Anderson and Karel 2012; Anderson et al. 2012), we assumed that populations were admixed and that allele frequencies across populations were correlated. We did not provide the geographic origin of samples a priori. Five replicates for each K-value from 1 to 8 were tested, with 10,000 burn-in iterations followed by 50,000 Markov-chain Monte Carlo iterations. STRUCTURE HARVESTER (Earl and vonHoldt 2012) was used to implement the Evanno et al. (2005) method for identifying the most likely number of clusters (i.e., K) in a data set. All replicates from each K-value identified by this method were then combined using the "FullSearch" algorithm in CLUMPP version 1.1.2 (Jakobsson and Rosenberg 2007). DISTRUCT version 1.1 (Rosenberg 2004) was used to convert CLUMPP outputs into bar plots that showed each individual's probability of assignment to the genetic clusters.

We used Arlequin version 3.5.1.2 (Excoffier and Lischer 2010) to compute the genetic differentiation index F_{ST} between all pairs of geographic populations for both AFLP and mtDNA data sets. We applied a Bonferroni adjustment to correct for multiple F_{ST} comparisons by using software from Lesack and Naugler (2011). Arlequin was also used to conduct analyses of molecular variance (AMOVAs; Excoffier et al. 1992). An AMOVA examines the partitioning of genetic differentiation within and among geographical subpopulations. We used this framework to test two different regional groupings for both AFLPs and mtDNA: (1) ocean-basin-scale structure (Gulf versus Atlantic) and (2) structure within the Atlantic (NCN, NCC, and NCS versus SC, GA, and FL). To further investigate the patterns observed from our STRUCTURE results, we also considered a third grouping for AMOVA: subdivision within NC (NCN and NCC versus NCS, SC, GA, and FL). For all analyses conducted in Arlequin, AFLPs were coded as haplotypic data. Arlequin does not account for dominant marker systems, so the calculated fixation indices for AFLP data are based on peak phenotype (presence/absence) frequencies rather than allele frequencies. Despite this, Arlequin is often used for analysis of AFLPs (Svensson et al. 2004; Bensch and Åkesson 2005; Timmermans et al. 2005; Parchman et al. 2006; Toews and Irwin 2008); the degree of differentiation and their significance values are meaningful, but direct comparisons with F_{ST} values from co-dominant data sets should be avoided.

Mitochondrial DNA control region phylogeny in Southern Flounder was inferred with NETWORK version 4.6.1.2 (Fluxus Engineering; www.fluxus-engineering.com) by using the median joining method (Bandelt et al. 1999) to generate a haplotype network. Network branching patterns were postprocessed with maximum parsimony (Polzin and Daneshmand 2003). The mtDNA control region nucleotide composition and pairwise differences were calculated using MEGA version 6 (Tamura et al. 2013). Pairwise differences for the AFLP data set were also determined with MEGA. DnaSP version 5.10 (Librado and Rozas 2009) was used to assign mtDNA control region haplotypes and to calculate molecular diversity indices and Tajima's D-statistic (Tajima 1989). Sites with gaps were not considered for diversity calculations but were included in all other analyses.

RESULTS

When scored for all 201 samples across the 12 primer pair combinations, 671 polymorphic AFLP loci were identified. In the eight individuals for which the AFLP procedure was repeated, the AFLP profiles were 98.8% reproducible across all loci. Using the method of Evanno et al. (2005) to interpret the AFLP results from STRUCTURE, we identified a strong peak in ΔK at a K-value of 2 and a smaller but distinct secondary peak in ΔK at a K-value of 5. At a K of 2, we observed that each geographic population showed strong membership to one of two distinct genetic clusters representing either the Gulf or the Atlantic (Figure 2). In addition, the Atlantic samples from NCS and southward, particularly SC, displayed partial assignment to the genetic cluster dominating the Gulf locales. At a K of 5, the assignments remained similar to those produced at a K-value of 2, but SC ancestry was instead shared specifically with the western Gulf, and NCS samples displayed some shared assignment with the major genetic cluster found in the Gulf.

Because Southern Flounder from SC unexpectedly showed common ancestry with those sampled in the western Gulf (SBTX), we examined population pairwise distances for individual AFLP primer pairs in an attempt to identify the source of these similarities. For each primer pair, we compared

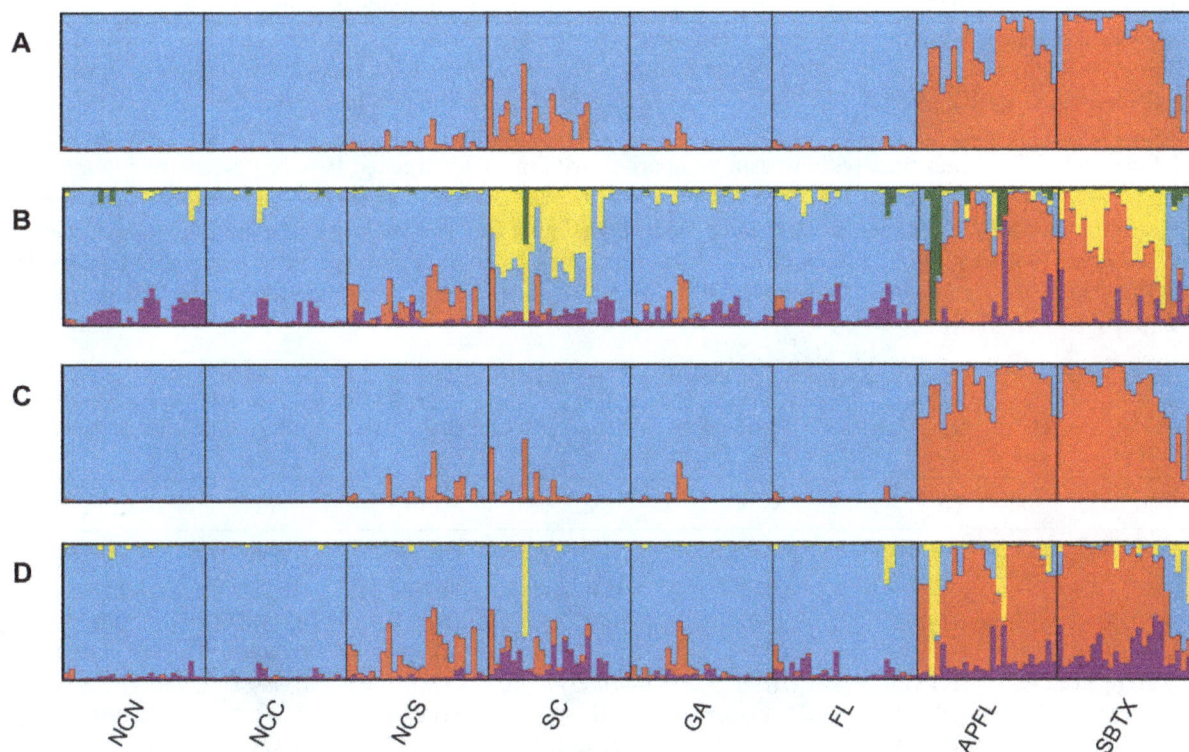

FIGURE 2.   Population structure for Southern Flounder as estimated with amplified fragment length polymorphisms (AFLPs). Each individual is represented as a vertical line along the x-axis, and individuals are grouped into geographic populations. Each color represents a different inferred genetic cluster, and the probability of assignment to each of the K genetic clusters is indicated by the proportion of color along the y-axis: (**A**) K-value of 2 and (**B**) K-value of 5. Results for the AFLP data set after removing the primer pair *EcoRI*-AGC–*MseI*-CTT are also presented: (**C**) K-value of 2 and (**D**) K-value of 4. Location codes are defined in Figure 1.

(1) the average pairwise distance among geographic population pairs that included either SC or SBTX to (2) the average pairwise distance among all other geographic population pairs (i.e., those without SC or SBTX). The percent difference between these two metrics was used to assess the unusual levels of genetic differentiation in SC and the western Gulf. The average percent difference was 6.09% (SD = 5.93) across 11 of the 12 primer pairs. The remaining primer pair (*Eco*RI-AGC–*Mse*I-CTT) displayed higher levels of differentiation, with a percent difference of 56.77%, and was likely responsible for the similarities observed between SC and SBTX samples. Subsequently, we conducted a second STRUCTURE analysis of the AFLP data set by using the same methods as previously described, except that we excluded the 38 loci from the primer pair *Eco*RI-AGC–*Mse*I-CTT. A strong peak in ΔK was identified at a K of 2, and a smaller but distinct peak was observed at a K of 4. At the K-value of 2, the pronounced similarity between SC and SBTX was no longer present, but at the K of 4, a shared partial assignment among SC samples and Gulf samples was again detected (Figure 2). Overall patterns of genetic structure were similar to those observed prior to removal of the *Eco*RI-AGC–*Mse*I-CTT primer pair from the AFLP data set.

Comparisons of F_{ST} for Southern Flounder AFLPs showed significant differentiation between all but four pairs of geographic populations ($F_{ST} = 0.020$–0.133, $P < 0.01$; Table 1). The four population pairs that were not significantly differentiated after Bonferroni adjustment were all within-Atlantic comparisons (NCN versus GA; NCN versus FL; NCC versus FL; and GA versus FL). In general, F_{ST} values for Atlantic versus Gulf population pairs were larger than within-basin comparisons, although F_{ST} values for SC versus NCC and for SC versus NCS were comparable to between-basin values. When comparing Southern Flounder from the Gulf and Atlantic basins, AMOVA identified significant differences at all levels

of hierarchical population structure (Table 2). However, within-population variation accounted for 89.9% of the differences ($F_{ST} = 0.101$, $P < 0.0001$). When partitioning differentiation in the Atlantic alone (NCN, NCC, and NCS versus SC, GA, and FL), 96.6% of the variation was due to within-population differences ($F_{ST} = 0.034$, $P < 0.0001$; Table 2). There was no differentiation between these two Atlantic regions. There was also no differentiation between regions when considering NCN and NCC as a separate group relative to NCS, SC, GA, and FL.

Across 260 samples, the final mtDNA control region alignment for Southern Flounder was 480 bp, with 88 segregating sites and a total of 130 unique haplotypes. This included a single-base-pair insertion/deletion that was present in two haplotypes, each composed of one individual from different sites. Excluding sites with gaps, overall haplotype diversity was 0.977 (SD = 0.0036) and nucleotide diversity was 0.011 (SD = 0.0003), with an average of 5.06 nucleotide changes between individuals (Table 3). Tajima's D-statistic was estimated to test for neutrality of nucleotide substitutions; the values were significantly negative for both ocean basins (Atlantic: $D = -2.08$, $P < 0.05$; Gulf: $D = -1.97$, $P < 0.05$), which may be indicative of Southern Flounder population expansion (Fu 1997).

Twenty-nine mtDNA control region haplotypes were common among individual Southern Flounder, and three of the haplotypes were shared between the Gulf and Atlantic. The majority of the haplotypes were singletons, most of which differed from common haplotypes by only one or two nucleotide substitutions. The control region haplotype network showed a pattern of divergence between Southern Flounder in the two ocean basins, although separation was not complete (Figure 3). The majority of Gulf haplotypes were separated from Atlantic haplotypes by at least three nucleotide changes, but there were Atlantic haplotypes present in the Gulf cluster

TABLE 1. Population pairwise F_{ST} comparisons based on amplified fragment length polymorphism profiles (above diagonal) and mitochondrial DNA control region haplotypes (below diagonal) from Southern Flounder sampled in the U.S. South Atlantic (NCN = North Carolina north; NCC = North Carolina central; NCS = North Carolina south; SC = South Carolina; GA = Georgia; FL = Florida) and in the Gulf of Mexico (APFL = Apalachicola, Florida; SBTX = San Antonio Bay, Texas). Values without superscript letters were not significant.

Sample	NCN	NCC	NCS	SC	GA	FL	APFL	SBTX
NCN		0.0218^c	0.0436^c	0.0506^c	0.0122^{bd}	0.0062	0.0969^c	0.1133^c
NCC	0.0049		0.0519^c	0.0724^c	0.0199^b	0.0133^{bd}	0.1119^c	0.1330^c
NCS	0.0052	−0.0002		0.0722^c	0.0215^c	0.0314^c	0.0652^c	0.1054^c
SC	0.0369^{ad}	−0.0079	0.0231		0.0526^c	0.0420^c	0.1037^c	0.0694^c
GA	0.0521^{ad}	0.0102	0.0367^{ad}	−0.0074		0.0099^{bd}	0.0771^c	0.1105^c
FL	0.0215	−0.0095	−0.0029	−0.0036	0.0003		0.0869^c	0.1100^c
APFL	0.4234^c	0.3993^c	0.4558^c	0.3841^c	0.3962^c	0.4165^c		0.0304^c
SBTX	0.3993^c	0.3736^c	0.4368^c	0.3582^c	0.3715^c	0.3949^c	0.0058	

$^aP < 0.05$.
$^bP < 0.01$.
$^cP < 0.0001$.
^{d}Not significant after Bonferroni correction.

TABLE 2. Results from analyses of molecular variance based on amplified fragment length polymorphisms in Southern Flounder. Populations are grouped at two different geographic scales: (1) ocean basin (Gulf of Mexico versus U.S. South Atlantic) and (2) U.S. South Atlantic (North Carolina versus South Carolina, Georgia, and Florida; *$P < 0.05$; **$P < 0.0001$).

Source of variation	df	Sum of squares	Variance component	Percentage of variation	Fixation index
		Ocean basin scale			
Among groups	1	401.326	4.042	6.98	$F_{CT} = 0.0698$*
Among populations within groups	6	587.121	1.822	3.15	$F_{SC} = 0.0338$**
Within populations	193	10,045.97	52.052	89.88	$F_{ST} = 0.1013$**
Total	200	11,034.42	57.916		
		U.S. South Atlantic			
Among groups	1	87.036	−0.179	−0.34	$F_{CT} = -0.0034$
Among populations within groups	4	402.025	1.964	3.71	$F_{SC} = 0.0370$**
Within populations	145	7,407.171	51.084	96.62	$F_{ST} = 0.0338$**
Total	150	7,896.232	52.869		

and vice versa. There was no clear pattern of finer-scale structure within the Atlantic cluster based on the haplotype network.

Pairwise population comparisons using the mtDNA control region showed strong differentiation ($F_{ST} = 0.37$–0.45, $P < 0.0001$) between all pairs of populations compared across basins (Table 1). Three pairs of populations within the Atlantic (GA and NCS; GA and NCN; and SC and NCN) were also significantly differentiated, although F_{ST} values were much smaller ($F_{ST} = 0.037$–0.052, $P < 0.05$). None of these within-Atlantic comparisons remained significant after a Bonferroni correction was applied, and no other pairs of populations within

TABLE 3. Molecular diversity measures for populations of Southern Flounder sequenced at the mitochondrial DNA control region (excluding sites with gaps). Number of samples (n), number of polymorphic sites excluding gaps (N_p), number of haplotypes (N_h), haplotype diversity (H_d), and nucleotide diversity (π) are given for each site. Location codes are defined in Table 1.

Location	n	N_p	N_h	H_d	π
	U.S. South Atlantic				
NCN	30	28	19	0.910	0.0082
NCC	33	30	25	0.981	0.0087
NCS	34	26	22	0.934	0.0073
SC	31	31	25	0.981	0.0085
GA	31	28	23	0.974	0.0079
FL	30	28	23	0.972	0.0080
Overall (Atlantic)	189	70	93	0.967	0.0082
	Gulf of Mexico				
APFL	36	35	25	0.946	0.0094
SBTX	35	28	19	0.909	0.0080
Overall (Gulf)	71	47	39	0.931	0.0087
Total (Atlantic plus Gulf)	260	87	129	0.977	0.0106

the Atlantic or Gulf were found to be significantly differentiated with any test. When comparing the Gulf and Atlantic, the AMOVA for the mtDNA control region showed that the majority of variation was explained by differences within populations (59.3%; $F_{ST} = 0.407$, $P < 0.0001$), although a large portion of the differentiation was due to differences between ocean basins (40.2%; F_{CT} [genetic differences among groups] $= 0.402$, $P < 0.05$; Table 4). Within the Atlantic, almost all of the variation was explained by differences within populations (98.2%; $F_{ST} = 0.018$, $P > 0.05$), and only a small portion of the variation was explained by differences between NC and the other Atlantic populations (1.8%; $F_{CT} = 0.018$, $P > 0.05$), but neither of these differences was significant (Table 4).

DISCUSSION

We observed divergence between Gulf and Atlantic Southern Flounder populations based on both mtDNA control region sequences and AFLP fingerprints. All population pairwise F_{ST} and AMOVA F_{CT} values were significant when comparing between ocean basins, regardless of the molecular marker used (Tables 1, 2, 4). The same result was observed with AFLP population assignments (Figure 2) and in the mtDNA haplotype network (Figure 3). These findings are in agreement with previous Southern Flounder studies that examined allozymes (Blandon et al. 2001), mtDNA (Anderson et al. 2012), and microsatellites (Anderson and Karel 2012), and they correspond with a well-known biogeographic break in marine fauna between the Gulf and Atlantic (Avise 1992). Southern Flounder are absent from the southern tip of Florida (Ginsburg 1952; Gilbert 1986), so a transition in genetic composition between ocean basins appears to correspond with the habitat discontinuity that is responsible for the disjunct distribution of Southern Flounder across south Florida.

We predicted that both markers would perform well for evaluating deep divergence, but we also wanted to assess

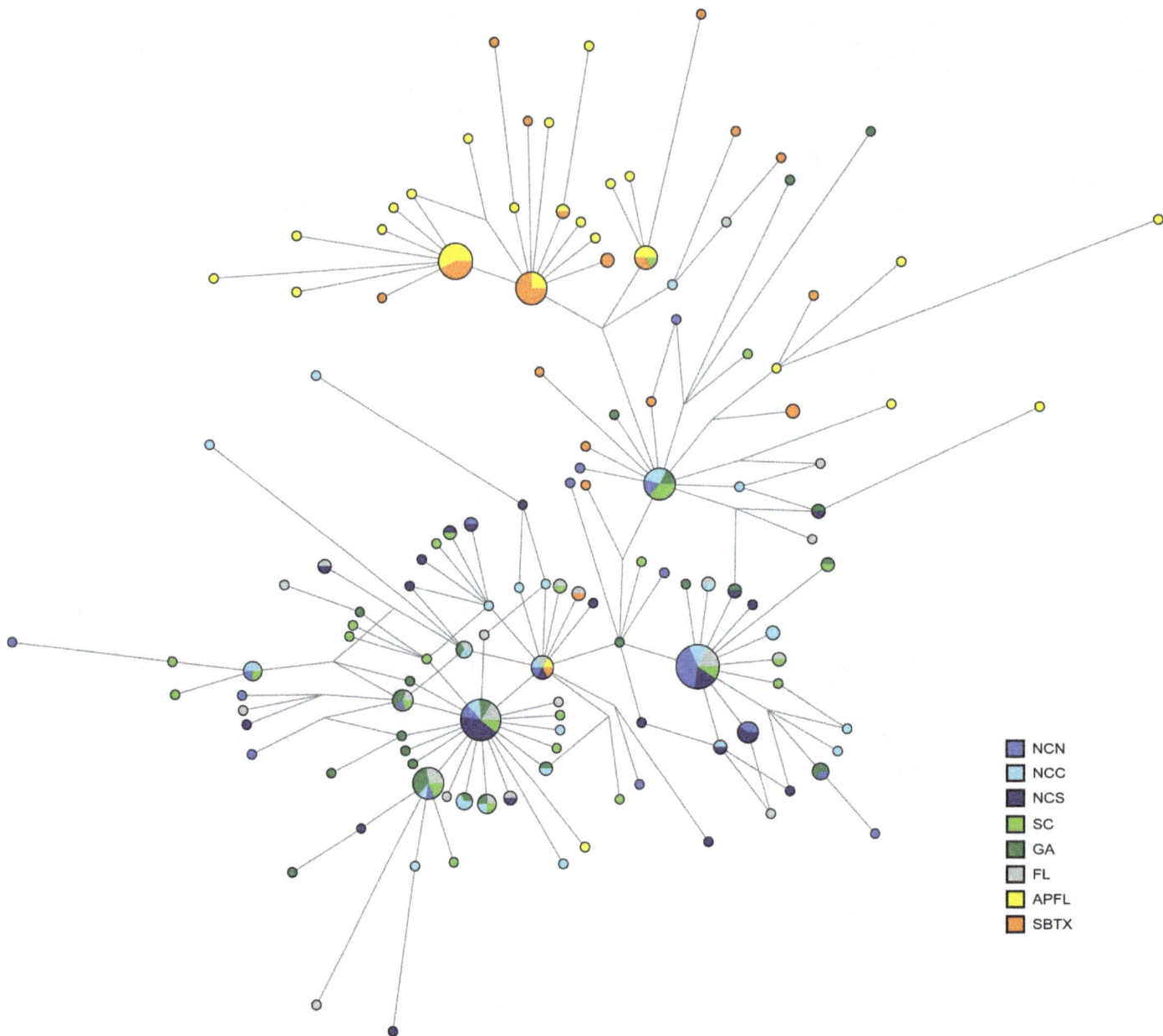

FIGURE 3.   Median joining network of mitochondrial DNA control region haplotypes identified in Southern Flounder. Circles represent haplotypes; colored wedges indicate geographic populations. Circles and wedges are proportional to the sample size, and the length of the connection is proportional to the number of nucleotide changes between haplotypes. Location codes are defined in Figure 1.

how AFLPs and mtDNA would compare when examining subtle structure, as we expected to see in the Atlantic Southern Flounder populations. Although mtDNA markers are commonly used in population genetic studies because of their high mutation rates and purportedly neutral evolution, the mitochondrial genome is also haploid (maternally inherited, effectively as a single locus) and is very small in comparison with the nuclear genome. Thus, studies that rely solely on mtDNA markers may draw incomplete conclusions when attempting to make inferences about organisms and populations as a whole (reviewed by Ballard and

Whitlock 2004). Amplified fragment length polymorphism loci are scattered across the genome and may better reflect demographic, historical, and selective events for which influences on diversity patterns can vary across loci, so we expected to obtain increased genetic resolution with AFLP analysis.

The AFLP and mtDNA results were generally in agreement when assessing Southern Flounder differentiation between the two ocean basins, but the results began to diverge when we examined structure within ocean basins. The magnitude of differentiation for pairwise comparisons between ocean basins

TABLE 4. Results from analyses of molecular variance based on mitochondrial DNA control region sequences in Southern Flounder. Populations are grouped at two different geographic scales: (1) ocean basin (Gulf of Mexico versus U.S. South Atlantic) and (2) U.S. South Atlantic (North Carolina versus South Carolina, Georgia, and Florida; *$P < 0.05$; **$P < 0.0001$).

Source of variation	df	Sum of squares	Variance component	Percentage of variation	Fixation index
Ocean basin scale					
Among groups	1	141.313	1.344	40.15	$F_{CT} = 0.4016$*
Among populations within groups	6	15.532	0.019	0.56	$F_{SC} = 0.0094$
Within populations	252	499.841	1.983	59.28	$F_{ST} = 0.4072$**
Total	259	656.685	3.346		
U.S. South Atlantic					
Among groups	1	5.263	0.035	1.78	$F_{CT} = 0.0178$
Among populations within groups	4	7.755	0.000	−0.01	$F_{SC} = -0.0001$
Within populations	183	356.284	1.947	98.24	$F_{ST} = 0.0176$
Total	188	369.302	1.982		

was much higher with mtDNA ($F_{ST} = 0.358$–0.456) than with AFLPs ($F_{ST} = 0.065$–0.133), although both were highly significant. The opposite was seen for population comparisons within the Atlantic, as mtDNA differentiation values ($F_{ST} = -0.010$ to 0.052) were nonsignificant and lower than AFLP values ($F_{ST} = 0.006$–0.072), which showed significant differentiation for nearly all within-Atlantic comparisons. If we were to interpret the degree of population structure as proposed by Wright (1978), genetic differentiation among Southern Flounder populations when comparing ocean basins would be characterized as "very great" (mtDNA) or "moderate" (AFLPs), whereas comparisons within the Atlantic basin showed "little" to "moderate" genetic differentiation based on AFLPs.

The levels of genetic differentiation detected in this study were consistent with previously reported F_{ST} values for Southern Flounder mtDNA (Anderson et al. 2012), but we observed markedly higher differentiation using AFLPs than was previously found with microsatellites (within-basin $F_{ST} = -0.002$ to 0.007; between-basin $F_{ST} = 0.025$–0.046; Anderson and Karel 2012). Similar discrepancies between markers have also been observed in recent studies of other species, for which differentiation was both higher (Mock et al. 2002; Whitehead et al. 2003; Gruenthal et al. 2007) and lower (Maguire et al. 2002) when comparing AFLPs to microsatellites. For analyses of highly variable multi-allelic loci, as in the Anderson et al. (2012) study where there were up to 53 alleles per microsatellite locus, very high levels of within-population heterozygosity can result in reduced magnitudes of differentiation between populations (Hedrick 1999). Different molecular markers are subject to different evolutionary processes, which can affect levels of diversity within and between populations (reviewed by Zink and Barrowclough 2008; Toews and Brelsford 2012). Therefore, the lack of correspondence in the magnitude of differentiation estimated with the different markers used to assess Southern Flounder genetic structure thus far is not surprising.

In this study, nearly all Southern Flounder geographic populations within the Atlantic were significantly differentiated (after Bonferroni correction for multiple comparisons) when using the AFLP data set (Table 1), thus contradicting the mtDNA results. The power to detect differentiation increases with the number of loci, and even very small values of F_{ST} can become statistically significant when enough markers are used (Ryman et al. 2006). We compared a single mtDNA marker to an AFLP profile comprising 671 loci, so we expected increased resolution. However, the levels of within-Atlantic AFLP differentiation we detected are not considered low for marine populations. Even very weak but significant levels of genetic differentiation can be biologically relevant; for example, Knutsen et al. (2011) used empirical mark–recapture data to confirm the existence of distinct, localized populations of coastal Atlantic Cod *Gadus morhua* that produced low levels of genetic differentiation. Based on AFLP differentiation, genetic structure appears to be present among geographic populations of Southern Flounder in the Atlantic, although this does not necessarily mean that each locale represents a discrete stock.

Waples et al. (2008) cautioned that statistical significance is not always correlated with biological significance and that the life history and biology of an organism must be taken into consideration when determining stocks. Thus far, we have no evidence that Southern Flounder from different locales within the Atlantic are biologically distinct. Based on otolith morphometric analysis, Midway et al. (2014) also detected only weak evidence for differentiation among Southern Flounder populations within the Atlantic, including within NC. These results are consistent with our findings. Despite the possibility that many AFLP loci should allow for greater power in the detection of population subdivision, AMOVA did not find significant genetic structure when comparing NC (or portions of NC) to the remaining Atlantic locales. The same lack of NC regional structure was found with AMOVA based on mtDNA.

These results do not support the consideration of NC Southern Flounder as a distinct stock within the Atlantic.

In addition to evaluating patterns of differentiation, we also assessed the genetic composition of each Southern Flounder population. When sufficient numbers of loci are used, AFLPs are highly accurate for population assignment tests and can be particularly informative for systems with weak genetic structure (Campbell et al. 2003). We saw similar patterns in individual assignments to inferred genetic clusters (i.e., K) across the most likely values of K for the AFLP profiles (Figure 2). Although we examined patterns of variation at K-values of 2 and 5, the major genetic patterns remained the same. The two summary plots both showed a strong genetic distinction between Gulf and Atlantic Southern Flounder. Within NC, the NCN and NCC samples appeared to be homogeneous, whereas NCS displayed a small difference in assignment probabilities, suggesting that a mild barrier to gene flow may exist between these regions. The most unexpected observation was the shared clustering of SC samples—obtained near the center of the Atlantic populations—with the Gulf samples, particularly those from SBTX.

Overall, the AFLP genetic cluster analyses suggested high gene flow within the Atlantic, with the possible exception of SC. Of the three mtDNA haplotypes shared between ocean basins, one consisted of individuals from SC and both Gulf locales, although this was a low-frequency haplotype (Figure 3). In addition, the SC samples were differentiated from one of the NC geographic populations in the mtDNA F_{ST} analysis, but this subdivision was no longer significant after Bonferroni adjustment. In contrast, SC was always highly significantly differentiated from both Gulf locales in pairwise F_{ST} comparisons, whether based on mtDNA or AFLPs. With these varying results, the potential distinction of SC genotypes from other Atlantic samples remains inconclusive.

By examining pairwise differences in AFLP profiles between geographic populations, we were able to identify a single primer pair (*Eco*RI-AGC–*Mse*I-CTT) as the likely cause for the assignment of SBTX and SC individuals to the same genetic cluster. This was confirmed by a genetic cluster analysis that excluded loci from the primer pair in question, although the similarity between SC and Gulf samples was still present to a smaller degree when the number of assumed genetic clusters was increased (Figure 2). Because one primer pair appears to be responsible for this grouping, it is possible that the pattern may be an artifact of the AFLP procedure. Errors in peak scoring are a common source of AFLP genotyping error (Bonin et al. 2004), so we used an established method to reduce scoring subjectivity and error in our AFLP data by setting peak height thresholds for inclusion of loci and phenotype calling (Whitlock et al. 2008). This resulted in a conservative data set that performed well in our tests for reproducibility (98.8%).

Another concern in AFLP studies is fragment size homoplasy, which can occur as a result of the anonymous nature of AFLP loci. Equal-length DNA fragments are scored as a presence on the same locus, regardless of whether the fragments are homologous in genomic origin. Amplified fragment length polymorphism homoplasy can result in the underestimation of differentiation between populations and may provide one potential explanation for the unforeseen similarity we observed between SC and SBTX Southern Flounder. Homoplasy bias increases with the number of loci scored per primer pair and with smaller fragment sizes (<125 bp; Vekemans et al. 2002; Caballero et al. 2008). We reduced the numbers of AFLP loci analyzed per primer pair by applying peak height thresholds as previously described (only 38 of 139 loci initially detected were analyzed for *Eco*RI-AGC–*Mse*I-CTT), but small fragment sizes were not excluded in this study. The potential influence of AFLP homoplasy on genetic diversity estimates must be considered, but experimental quantification of the degree of homoplasy is not within the scope of most AFLP studies (Meudt and Clarke 2007).

Alternatively, it may be possible for the geographically disjunct individuals from SC and the Gulf to have similar molecular profiles based on shared adaptive traits. Like other genome-scale molecular data sets, AFLP loci are anonymous and consist mostly of neutral loci, but they often include loci that are linked to genes under selection (Luikart et al. 2003). To investigate any patterns of distribution that may be related to adaptive evolution, future studies would benefit from a genome scan, which detects AFLP loci that show abnormally high among-population differentiation; such outliers are a signature of selection (Beaumont and Balding 2004).

The results of this study confirmed a clear divergence between Gulf and Atlantic Southern Flounder. There was also evidence of genetic structure among geographic populations of Southern Flounder at a finer spatial scale, which was only revealed upon analysis of AFLPs. We found intriguing patterns of differentiation and genetic clustering that did not correspond with obvious geographic patterns. However, there was no differentiation between samples from NC and those from the other Atlantic states when examining hierarchical population structure. Southern Flounder in NC are currently managed as a unit stock. The present molecular analysis suggests that a single-stock designation within NC should be adequate, as there is no clear evidence for additional subdivisions at a finer spatial scale (i.e., between regions within the state). However, the potential for high levels of mixing throughout the Atlantic basin, as evidenced by the presence of only weak genetic structure within the basin, suggests that interstate cooperation will be necessary to achieve a comprehensive population assessment of Southern Flounder at an appropriate spatial scale. To sustain fishery yields and meet conservation objectives for Southern Flounder, the spatial scale of stock assessment will need to be more closely aligned with the spatial scale of important ecological processes that determine stock dynamics. Our present findings suggest that these processes likely operate over broad spatial scales and that future management of Southern Flounder could be improved through a synthesis of biological information encompassing the species' range.

ACKNOWLEDGMENTS

This research was funded by the North Carolina Division of Marine Fisheries (NCDMF) through the Marine Resources Fund. Several individuals provided help with tissue sample collection and genetic data. Southern Flounder from NC were collected by L. Paramore, K. West, J. Rock, M. Seward, and C. Collier as part of NCDMF's fishery-independent sampling program. The SC fish were provided by B. Roumillat and S. Arnott (South Carolina Department of Natural Resources); GA fish were provided by J. Page and E. Robillard (Georgia Department of Natural Resources); and FL fish were provided by R. Brodie (Florida Fish and Wildlife Conservation Commission). S. Midway collected additional fish from Florida and was essential in providing the organizational framework for this study. J. Anderson (Texas Parks and Wildlife Department) graciously provided fish tissue samples and genomic DNA from the Gulf. L. Jarvis helped with initial development of molecular methods and generated a preliminary data set. We also thank three anonymous reviewers for providing comments that helped us to improve the presentation of our findings. The interpretation of the data and the conclusions expressed in this paper are those of the authors and do not necessarily represent the views of the NCDMF.

REFERENCES

Anderson, J. D., and W. J. Karel. 2012. Population genetics of Southern Flounder with implications for management. North American Journal of Fisheries Management 32:656–663.

Anderson, J. D., W. J. Karel, and A. C. S. Mione. 2012. Population structure and evolutionary history of Southern Flounder in the Gulf of Mexico and western Atlantic Ocean. Transactions of the American Fisheries Society 141:46–55.

Avise, J. C. 1992. Molecular population structure and the biogeographic history of a regional fauna: a case history with lessons for conservation biology. Oikos 63:62–76.

Ballard, J. W. O., and M. C. Whitlock. 2004. The incomplete natural history of mitochondria. Molecular Ecology 13:729–744.

Bandelt, H. J., P. Forster, and A. Röhl. 1999. Median-joining networks for inferring intraspecific phylogenies. Molecular Biology and Evolution 16:37–48.

Beaumont, M. A., and D. J. Balding. 2004. Identifying adaptive genetic divergence among populations from genome scans. Molecular Ecology 13:969–980.

Begg, G. A., K. D. Friedland, and J. B. Pearce. 1999. Stock identification and its role in stock assessment and fisheries management: an overview. Fisheries Research 43:1–8.

Bensch, S., and M. Åkesson. 2005. Ten years of AFLP in ecology and evolution: why so few animals? Molecular Ecology 14:2899–2914.

Blandon, I. R., R. Ward, T. L. King, W. J. Karel, and J. P. Monaghan Jr. 2001. Preliminary genetic population structure of Southern Flounder, *Paralichthys lethostigma*, along the Atlantic coast and Gulf of Mexico. U.S. National Marine Fisheries Service Fishery Bulletin 99:671–678.

Bonin, A., E. Bellemain, P. Bronken Eidesen, F. Pompanon, C. Brochmann, and P. Taberlet. 2004. How to track and assess genotyping errors in population genetics studies. Molecular Ecology 13:3261–3273.

Caballero, A., H. Quesada, and E. Rolán-Alvarez. 2008. Impact of amplified fragment length polymorphism size homoplasy on the estimation of population genetic diversity and the detection of selective loci. Genetics 179:539–554.

Campbell, D., P. Duchesne, and L. Bernatchez. 2003. AFLP utility for population assignment studies: analytical investigation and empirical comparison with microsatellites. Molecular Ecology 12:1979–1991.

Conover, D. O., L. M. Clarke, S. B. Munch, and G. N. Wagner. 2006. Spatial and temporal scales of adaptive divergence in marine fishes and the implications for conservation. Journal of Fish Biology 69:21–47.

Craig, J. K., and J. A. Rice. 2008. Estuarine residency, movements, and exploitation of Southern Flounder (*Paralichthys lethostigma*) in North Carolina. North Carolina Sea Grant, Final Report for Fishery Resource Grant 05-FEG-15, Raleigh.

Crandall, K. A., O. R. P. Bininda-Emonds, G. M. Mace, and R. K. Wayne. 2000. Considering evolutionary processes in conservation biology. Trends in Ecology and Evolution 15:290–295.

Earl, D., and B. vonHoldt. 2012. STRUCTURE HARVESTER: a website and program for visualizing STRUCTURE output and implementing the Evanno method. Conservation Genetics Resources 4:359–361.

Evanno, G., S. Regnaut, and J. Goudet. 2005. Detecting the number of clusters of individuals using the software STRUCTURE: a simulation study. Molecular Ecology 14:2611–2620.

Excoffier, L., and H. E. L. Lischer. 2010. Arlequin suite version 3.5: a new series of programs to perform population genetics analyses under Linux and Windows. Molecular Ecology Resources 10:564–567.

Excoffier, L., P. E. Smouse, and J. M. Quattro. 1992. Analysis of molecular variance inferred from metric distances among DNA haplotypes: application to human mitochondrial DNA restriction data. Genetics 131:479–491.

Falush, D., M. Stephens, and J. K. Pritchard. 2007. Inference of population structure using multilocus genotype data: dominant markers and null alleles. Molecular Ecology Notes 7:574–578.

Fu, Y.-X. 1997. Statistical tests of neutrality of mutations against population growth, hitchhiking and background selection. Genetics 147:915–925.

Gilbert, C. R. 1986. Species profiles: life histories and environmental requirements of coastal fishes and invertebrates (south Florida)–Southern, Gulf, and Summer Flounders. U.S. Army Corps of Engineers, Coastal Ecology Group, Technical Report 82 (11.54), Vicksburg, Mississippi.

Ginsburg, I. 1952. Flounders of the genus *Paralichthys* and related genera in American waters. U.S. National Marine Fisheries Service Fishery Bulletin 52:267–351.

Gruenthal, K. M., L. K. Acheson, and R. S. Burton. 2007. Genetic structure of natural populations of California red abalone (*Haliotis rufescens*) using multiple genetic markers. Marine Biology 152:1237–1248.

Hauser, L., and G. R. Carvalho. 2008. Paradigm shifts in marine fisheries genetics: ugly hypotheses slain by beautiful facts. Fish and Fisheries 9:333–362.

Hedrick, P. W. 1999. Perspective: highly variable loci and their interpretation in evolution and conservation. Evolution 53:313–318.

Hutchinson, W. F., C. van Oosterhout, S. I. Rogers, and G. R. Carvalho. 2003. Temporal analysis of archived samples indicates marked genetic changes in declining North Sea Cod (*Gadus morhua*). Proceedings of the Royal Society of London Series B: Biological Sciences 270:2125–2132.

Jakobsson, M., and N. A. Rosenberg. 2007. CLUMPP: a cluster matching and permutation program for dealing with label switching and multimodality in analysis of population structure. Bioinformatics 23:1801–1806.

Knutsen, H., P. E. Jorde, C. André, and N. C. Stenseth. 2003. Fine-scaled geographical population structuring in a highly mobile marine species: the Atlantic Cod. Molecular Ecology 12:385–394.

Knutsen, H., E. M. Olsen, P. E. Jorde, S. H. Espeland, C. André, and N. C. Stenseth. 2011. Are low but statistically significant levels of genetic differentiation in marine fishes "biologically meaningful"? A case study of coastal Atlantic Cod. Molecular Ecology 20:768–783.

Lee, W.-J., J. Conroy, W. H. Howell, and T. Kocher. 1995. Structure and evolution of teleost mitochondrial control regions. Journal of Molecular Evolution 41:54–66.

Lesack, K., and C. Naugler. 2011. An open-source software program for performing Bonferroni and related corrections for multiple comparisons. Journal of Pathology Informatics 2:52–52.

Librado, P., and J. Rozas. 2009. DnaSP v5: a software for comprehensive analysis of DNA polymorphism data. Bioinformatics 25:1451–1452.

Liu, J.-Y., Z.-R. Lun, J.-B. Zhang, and T.-B. Yang. 2009. Population genetic structure of Striped Mullet, *Mugil cephalus*, along the coast of China, inferred by AFLP fingerprinting. Biochemical Systematics and Ecology 37:266–274.

Liu, Y.-G., S.-L. Chen, B.-F. Li, Z.-J. Wang, and Z. Liu. 2005. Analysis of genetic variation in selected stocks of hatchery Flounder, *Paralichthys olivaceus*, using AFLP markers. Biochemical Systematics and Ecology 33:993–1005.

Luikart, G., P. R. England, D. Tallmon, S. Jordan, and P. Taberlet. 2003. The power and promise of population genomics: from genotyping to genome typing. Nature Reviews Genetics 4:981–994.

Maguire, T. L., R. Peakall, and P. Saenger. 2002. Comparative analysis of genetic diversity in the mangrove species *Avicennia marina* (Forsk.) Vierh. (Avicenniaceae) detected by AFLPs and SSRs. Theoretical and Applied Genetics 104:388–398.

McCairns, R. J. S., and L. Bernatchez. 2008. Landscape genetic analyses reveal cryptic population structure and putative selection gradients in a large-scale estuarine environment. Molecular Ecology 17:3901–3916.

Meudt, H. M., and A. C. Clarke. 2007. Almost forgotten or latest practice? AFLP applications, analyses and advances. Trends in Plant Science 12:106–117.

Meyer, A., J. M. Morrissey, and M. Schartl. 1994. Recurrent origin of a sexually selected trait in *Xiphophorus* fishes inferred from a molecular phylogeny. Nature 368:539–542.

Mickett, K., C. Morton, J. Feng, P. Li, M. Simmons, D. Cao, R. A. Dunham, and Z. Liu. 2003. Assessing genetic diversity of domestic populations of Channel Catfish (*Ictalurus punctatus*) in Alabama using AFLP markers. Aquaculture 228:91–105.

Midway, S. R., S. X. Cadrin, and F. S. Scharf. 2014. Southern Flounder (*Paralichthys lethostigma*) stock structure inferred from otolith shape analysis. U.S. National Marine Fisheries Service Fishery Bulletin 112:326–338.

Mock, K. E., T. C. Theimer, O. E. Rhodes, D. L. Greenberg, and P. Keim. 2002. Genetic variation across the historical range of the wild turkey (*Meleagris gallopavo*). Molecular Ecology 11:643–657.

Monaghan, J. P. Jr. 1996. Life history aspects of selected marine recreational fishes in North Carolina, study 2: migration of paralichthid flounders tagged in North Carolina. North Carolina Division of Marine Fisheries, Grant F-43, Segments 1–5, Completion Report, Morehead City.

NCDMF (North Carolina Division of Marine Fisheries). 2005. North Carolina fishery management plan: Southern Flounder (*Paralichthys lethostigma*). North Carolina Department of Environment and Natural Resources, Technical Report, Morehead City.

NCDMF (North Carolina Division of Marine Fisheries). 2013. North Carolina Southern Flounder (*Paralichthys lethostigma*) fishery management plan amendment 1. North Carolina Department of Environment and Natural Resources, Technical Report, Morehead City.

O'Reilly, P. T., M. F. Canino, K. M. Bailey, and P. Bentzen. 2004. Inverse relationship between F_{ST} and microsatellite polymorphism in the marine fish, Walleye Pollock (*Theragra chalcogramma*): implications for resolving weak population structure. Molecular Ecology 13:1799–1814.

Parchman, T. L., C. W. Benkman, and S. C. Britch. 2006. Patterns of genetic variation in the adaptive radiation of New World crossbills (Aves: *Loxia*). Molecular Ecology 15:1873–1887.

Polzin, T., and S. V. Daneshmand. 2003. On Steiner trees and minimum spanning trees in hypergraphs. Operations Research Letters 31:12–20.

Pritchard, J. K., M. Stephens, and P. Donnelly. 2000. Inference of population structure using multilocus genotype data. Genetics 155:945–959.

Rosenberg, N. A. 2004. DISTRUCT: a program for the graphical display of population structure. Molecular Ecology Notes 4:137–138.

Ruzzante, D. E., C. T. Taggart, and D. Cook. 1999. A review of the evidence for genetic structure of cod (*Gadus morhua*) populations in the NW Atlantic and population affinities of larval cod off Newfoundland and the Gulf of St. Lawrence. Fisheries Research 43:79–97.

Ryman, N., S. Palm, C. André, G. R. Carvalho, T. G. Dahlgren, P. E. Jorde, L. Laikre, L. C. Larsson, A. Palmé, and D. E. Ruzzante. 2006. Power for detecting genetic divergence: differences between statistical methods and marker loci. Molecular Ecology 15:2031–2045.

Sønstebø, J. H., R. Borgstrøm, and M. Heun. 2007. A comparison of AFLPs and microsatellites to identify the population structure of Brown Trout (*Salmo trutta* L.) populations from Hardangervidda, Norway. Molecular Ecology 16:1427–1438.

Svensson, E. I., L. Kristoffersen, K. Oskarsson, and S. Bensch. 2004. Molecular population divergence and sexual selection on morphology in the banded demoiselle (*Calopteryx splendens*). Heredity 93:423–433.

Tajima, F. 1989. Statistical method for testing the neutral mutation hypothesis by DNA polymorphism. Genetics 123:585–595.

Takade-Heumacher, H., and C. Batsavage. 2009. Stock status of North Carolina Southern Flounder (*Paralichthys lethostigma*). North Carolina Division of Marine Fisheries, Technical Report, Morehead City.

Tamura, K., G. Stecher, D. Peterson, A. Filipski, and S. Kumar. 2013. MEGA6: molecular evolutionary genetics analysis version 6.0. Molecular Biology and Evolution 30:2725–2729.

Taylor, J. C., J. M. Miller, and D. Hilton. 2008. Inferring Southern Flounder migration from otolith microchemistry. North Carolina Sea Grant, Final Report for Fishery Resource Grant 05-FEG-06, Raleigh.

Timmermans, M. J. T. N., J. Ellers, J. Mariën, S. C. Verhoef, E. B. Ferwerda, and N. M. Van Straalen. 2005. Genetic structure in *Orchesella cincta* (Collembola): strong subdivision of European populations inferred from mtDNA and AFLP markers. Molecular Ecology 14:2017–2024.

Toews, D. P. L., and A. Brelsford. 2012. The biogeography of mitochondrial and nuclear discordance in animals. Molecular Ecology 21:3907–3930.

Toews, D. P. L., and D. E. Irwin. 2008. Cryptic speciation in a Holarctic passerine revealed by genetic and bioacoustic analyses. Molecular Ecology 17:2691–2705.

Vekemans, X., T. Beauwens, M. Lemaire, and I. Roldán-Ruiz. 2002. Data from amplified fragment length polymorphism (AFLP) markers show indication of size homoplasy and of a relationship between degree of homoplasy and fragment size. Molecular Ecology 11:139–151.

Vos, P., R. Hogers, M. Bleeker, M. Reijans, T. van de Lee, M. Hornes, A. Friters, J. Pot, J. Paleman, M. Kuiper, and M. Zabeau. 1995. AFLP: a new technique for DNA fingerprinting. Nucleic Acids Research 23:4407–4414.

Waples, R. S. 1998. Separating the wheat from the chaff: patterns of genetic differentiation in high gene flow species. Journal of Heredity 89:438–450.

Waples, R. S., A. E. Punt, and J. M. Cope. 2008. Integrating genetic data into management of marine resources: how can we do it better? Fish and Fisheries 9:423–449.

Ward, R. D. 2000. Genetics in fisheries management. Hydrobiologia 420:191–201.

Ward, R. D., M. Woodwark, and D. O. F. Skibinski. 1994. A comparison of genetic diversity levels in marine, freshwater, and anadromous fishes. Journal of Fish Biology 44:213–232.

Watterson, J. C., and J. L. Alexander. 2004. Southern Flounder escapement in North Carolina, July 2001–June 2004. North Carolina Division of Marine Fisheries, Final Performance Report F-73, Morehead City.

Weetman, D., A. Ruggiero, S. Mariani, P. W. Shaw, A. R. Lawler, and L. Hauser. 2007. Hierarchical population genetic structure in the commercially exploited shrimp *Crangon crangon* identified by AFLP analysis. Marine Biology 151:565–575.

Wenner, C. A., W. A. Roumillat, J. E. Moran Jr., M. B. Maddox, L. B. I. Daniel, and J. W. Smith. 1990. Investigations on the life history and population dynamics of marine recreational fishes in South Carolina: part 1. South Carolina Wildlife and Marine Resources Department, Marine Resources Research Institute, Final Report F-37, Charleston.

Whitehead, A., S. L. Anderson, K. M. Kuivila, J. L. Roach, and B. May. 2003. Genetic variation among interconnected populations of *Catostomus occidentalis*: implications for distinguishing impacts of contaminants from biogeographical structuring. Molecular Ecology 12:2817–2833.

Whitlock, R., H. Hipperson, M. Mannarelli, R. K. Butlin, and T. Burke. 2008. An objective, rapid and reproducible method for scoring AFLP peak-height data that minimizes genotyping error. Molecular Ecology Resources 8:725–735.

Wright, S. 1978. Evolution and the genetics of populations: variability within and among natural populations, volume 4. University of Chicago Press, Chicago.

Zink, R. M., and G. F. Barrowclough. 2008. Mitochondrial DNA under siege in avian phylogeography. Molecular Ecology 17:2107–2121.

Permissions

List of Contributors

Jody L. Callihan, Julianne E. Harris and Joseph E. Hightower
North Carolina Cooperative Fish and Wildlife Research Unit, Department of Applied Ecology, North Carolina State University, 127 David Clark Labs, Campus Box 7617, Raleigh, North Carolina 27695, USA

Jody L. Callihan
Federal Energy Regulatory Commission, 888 First Street Northeast, Washington, D.C. 20426, USA

Julianne E. Harris
Present address: U.S. Fish and Wildlife Service, Columbia River Fisheries Program Office, 1211 Southeast Cardinal Court, Suite 100, Vancouver, Washington 98683, USA

Erik T. Lang
Riverside Technology (Contracting for the National Marine Fisheries Service), Panama City Laboratory, 3500 Delwood Beach Road, Panama City, Florida 32408, USA Louisiana Department of Wildlife and Fisheries, 2000 Quail Drive, Baton Rouge, Louisiana 70808, USA

Gary R. Fitzhugh
National Marine Fisheries Service, Panama City Laboratory, 3500 Delwood Beach Road, Panama City, Florida 32408, USA

Judson M. Curtis and Matthew W. Johnson
Harte Research Institute for Gulf of Mexico Studies, Texas A&M University–Corpus Christi, 6300 Ocean Drive, Corpus Christi, Texas 78412-5869, USA

Sandra L. Diamond
Department of Biology, Texas Tech University, Lubbock, Texas 79409, USA; and
School of Science and Health, Hawkesbury Campus, University of Western Sydney, Locked Bag 1797, Penrith, NSW 2751, Australia

Gregory W. Stunz
Harte Research Institute for Gulf of Mexico Studies, Texas A&M University–Corpus Christi, 6300 Ocean Drive, Corpus Christi, Texas 78412-5869, USA

Matthew W. Johnson
Bureau of Ocean Energy Management, 1201 Elmwood Park Boulevard, New Orleans, Louisiana 70123, USA

Joseph D. Schmitt
Department of Fish and Wildlife Conservation, Virginia Polytechnic Institute and State University, 100 Cheatham Hall, Blacksburg, Virginia 24060, USA

Todd Gedamke, William D. DuPaul and John A. Musick
Virginia Institute of Marine Science, College of William and Mary, Post Office Box 134, Gloucester Point, Virginia 23062, USA

Theodore S. Switzer
Florida Fish and Wildlife Conservation Commission, Fish and Wildlife Research Institute, 100 8th Avenue Southeast, St. Petersburg, Florida 33701, USA

Derek M. Tremain
Florida Fish and Wildlife Conservation Commission, Fish and Wildlife Research Institute, Indian River Field Laboratory, 1220 Prospect Avenue, Suite 285, Melbourne, Florida 32901, USA

Sean F. Keenan, Christopher J. Stafford, Sheri L. Parks and Robert H. McMichael Jr.
Florida Fish and Wildlife Conservation Commission, Fish and Wildlife Research Institute, 100 8th Avenue Southeast, St. Petersburg, Florida 33701, USA

Nathan M. Bacheler
National Marine Fisheries Service, Southeast Fisheries Science Center, 101 Pivers Island Road, Beaufort, North Carolina 25887, USA

Joseph C. Ballenger
South Carolina Department of Natural Resources, Marine Resources Research Institute, 217 Fort Johnson Road, Post Office Box 12559, Charleston, South Carolina 29412, USA

Mara S. Zimmerman
Washington Department of Fish and Wildlife, 600 Capitol Way North, Olympia, Washington 98501, USA

Meghan O'Neill
407-960 Inverness Road, Victoria, British Columbia V8X 2R9, Canada

James R. Irvine
Fisheries and Oceans Canada, Pacific Biological Station, 3190 Hammond Bay Road, Nanaimo, British Columbia V9T 6N7, Canada

Joseph H. Anderson
Washington Department of Fish and Wildlife, 600 Capitol Way North, Olympia, Washington 98501, USA

Correigh M. Greene
National Oceanic and Atmospheric Administration, National Marine Fisheries Service, Northwest Fisheries Science Center, Fish Ecology Division, 2725 Montlake Boulevard East, Seattle, Washington 98112, USA

Joshua Weinheimer
Washington Department of Fish and Wildlife, 600 Capitol Way North, Olympia, Washington 98501, USA

Marc Trudel
Fisheries and Oceans Canada, Pacific Biological Station, 3190 Hammond Bay Road, Nanaimo, British Columbia V9T 6N7, Canada; and Department of Biology, University of Victoria, Post Office Box 1700, Station CSC, Victoria, British Columbia V8W 3N5, Canada

Kit Rawson
Swan Ridge Consulting, 3601 Carol Place, Mount Vernon, Washington 98273-8583, USA

David D. Chagaris
Florida Fish and Wildlife Conservation Commission, Fish and Wildlife Research Institute, 100 8th Avenue Southeast, St. Petersburg, Florida 33701, USA; and Department of Fisheries and Aquatic Sciences, University of Florida, 7922 Northwest 71st Street, Gainesville, Florida 32653, USA

Behzad Mahmoudi
Florida Fish and Wildlife Conservation Commission, Fish and Wildlife Research Institute, 100 8th Avenue Southeast, St. Petersburg, Florida 33701, USA

Carl J. Walters
Department of Fisheries and Aquatic Sciences, University of Florida, 7922 Northwest 71st Street, Gainesville, Florida 32653, USA; and Fisheries Centre, University of British Columbia, 2202 Main Mall, Vancouver, British Columbia V6T 1Z4, Canada

Micheal S. Allen
Department of Fisheries and Aquatic Sciences, University of Florida, 7922 Northwest 71st Street, Gainesville, Florida 32653, USA

Valerie A. Hall
Maria Mitchell Association, Department of Natural Sciences, 4 Vestal Street, Nantucket, Massachusetts 02554, USA; and Department of Fisheries Oceanography, School for Marine Science and Technology, University of Massachusetts Dartmouth, 200 Mill Street Suite 325, Fairhaven, Massachusetts 02719, USA

Chang Liu and Steven X. Cadrin
Department of Fisheries Oceanography, School for Marine Science and Technology, University of Massachusetts Dartmouth, 200 Mill Street Suite 325, Fairhaven, Massachusetts 02719

Laurie A. Weitkamp
National Oceanic and Atmospheric Administration, National Marine Fisheries Service, Northwest Fisheries Science Center, Conservation Biology Division, Newport Field Station, 2032 Marine Science Drive, Newport, Oregon 97365, USA

David J. Teel
National Oceanic and Atmospheric Administration, National Marine Fisheries Service, Northwest Fisheries Science Center, Conservation Biology Division, Manchester Field Station, Post Office Box 130, Manchester, Washington 98353, USA

Martin Liermann
National Oceanic and Atmospheric Administration, National Marine Fisheries Service, Northwest Fisheries Science Center, Fish Ecology Division, 2725 Montlake Boulevard East, Seattle, Washington 98112, USA

Susan A. Hinton
National Oceanic and Atmospheric Administration, National Marine Fisheries Service, Northwest Fisheries Science Center, Fish Ecology Division, Point Adams Field Station, 520 Heceta Place, Hammond, Oregon 97121, USA

Donald M. Van Doornik
National Oceanic and Atmospheric Administration, National Marine Fisheries Service, Northwest Fisheries Science Center, Conservation Biology Division, Manchester Field Station, Post Office Box 130, Manchester, Washington 98353, USA

Paul J. Bentley
National Oceanic and Atmospheric Administration, National Marine Fisheries Service, Northwest Fisheries Science Center, Fish Ecology Division, Point Adams Field Station, 520 Heceta Place, Hammond, Oregon 97121, USA

Ryan R. Easton
College of Earth, Ocean, and Atmospheric Sciences, Oregon State University, 104 CEOAS Administration Building, 101 Southwest 26th Street, Corvallis, Oregon 97331, USA; and Oregon Department of Fish and Wildlife, Marine Resources Program, 2040 Southeast Marine Science Drive, Newport, Oregon 97365, USA

Selina S. Heppell
Department of Fisheries and Wildlife, Oregon State University, 104 Nash Hall, Corvallis, Oregon 97331, USA

Robert W. Hannah
Oregon Department of Fish and Wildlife, Marine Resources Program, 2040 Southeast Marine Science Drive, Newport, Oregon 97365, USA

Sandra L. Parker-Stetter, John K. Horne and Samuel S. Urmy
School of Aquatic and Fishery Sciences, University of Washington, Box 355020, Seattle, Washington 98195-5020, USA

Ron A. Heintz
National Oceanic and Atmospheric Administration, National Marine Fisheries Service, Alaska Fisheries Science Center, Auke Bay Laboratories, 17109 Point Lena Loop Road, Juneau, Alaska 99801, USA

Lisa B. Eisner
National Oceanic and Atmospheric Administration, National Marine Fisheries Service, Alaska Fisheries Science Center, 7600 Sand Point Way NE, Seattle, Washington 98115, USA

Edward V. Farley
National Oceanic and Atmospheric Administration, National Marine Fisheries Service, Alaska Fisheries Science Center, Auke Bay Laboratories, 17109 Point Lena Loop Road, Juneau, Alaska 99801, USA

Sandra L. Parker-Stetter
National Oceanic and Atmospheric Administration, National Marine Fisheries Service, Northwest Fisheries Science Center, Fisheries Resource Assessment and Monitoring Division, 2725 Montlake Boulevard East, Seattle, Washington 98112, USA

Samuel S. Urmy
School of Marine and Atmospheric Sciences, Stony Brook University, 239 Montauk Highway, Southampton, New York 11968, USA

Anna M. Henry and Teresa R. Johnson
School of Marine Sciences, University of Maine, 200 Libby Hall, Orono, Maine 04469, USA

Rachel Gjelsvik Tiller
SINTEF, Fisheries and Aquaculture, Strindvegen 4, 7034 Trondheim, Norway; and Department of Sociology and Political Science, Norwegian University of Science and Technology, Høgskoleringen 1, 7491 Trondheim, Norway

Jarle Mork
Department of Biology, Norwegian University of Science and Technology, Høgskoleringen 1, 7491 Trondheim, Norway

Yajie Liu
SINTEF, Fisheries and Aquaculture, Strindvegen 4, 7034 Trondheim, Norway

Åshild L Borgersen
Department of Biology, Norwegian University of Science and Technology, Høgskoleringen 1, 7491 Trondheim, Norway

Russell Richards
University of Queensland, St. Lucia QLD 4072, Brisbane, Australia

Ellen M. Yasumiishi, S. Kalei Shotwell, Dana H. Hanselman, Joseph A. Orsi and Emily A. Fergusson
National Marine Fisheries Service, Alaska Fisheries Science Center, 17109 Point Lena Loop Road, Juneau, Alaska 99801, USA

Maria M. Criales
Rosenstiel School of Marine and Atmospheric Science, University of Miami, 4600 Rickenbacker Causeway, Miami, Florida 33149, USA

Laurent M. Cherubin
Harbor Branch Oceanographic Institute, Florida Atlantic University, 5600 U.S. 1 North, Fort Pierce, Florida 34946, USA

Joan A. Browder
National Oceanic and Atmospheric Administration, National Marine Fisheries Service, Southeast Fisheries Science Center, 75 Virginia Beach Drive, Miami, Florida 33149, USA

Verena H. Wang, Michael A. McCartney and Frederick S. Scharf
Department of Biology and Marine Biology, Center for Marine Science, University of North Carolina Wilmington, 601 South College Road, Wilmington, North Carolina 28403, USA

Michael A. McCartney
Aquatic Invasive Species Research Center, Department of Fisheries, Wildlife, and Conservation Biology, University of Minnesota, 2003 Upper Buford Circle, St. Paul, Minnesota 55108, USA